The Theory and Practice of Bulk Materials Handling

Alan W. Roberts

The Theory and Practice of Bulk Materials Handling

Springer

Alan W. Roberts
TUNRA Bulk Solids
University of Newcastle Australia
Shortland, NSW, Australia

ISBN 978-3-032-19700-9 ISBN 978-3-032-19702-3 (eBook)
https://doi.org/10.1007/978-3-032-19702-3

This Springer imprint is published by the registered company Springer Nature Switzerland AG
The registered company address is: Gewerbestrasse 11, 6330 Cham, Switzerland

First and foremost, to my family, particularly my wife Betty, thank you for your patience, encouragement, and unwavering belief in me. Your emotional and physical support sustained me during the most challenging phases of my career and life. I am especially grateful for the way you embraced my passion for the bulk solids handling field, taking an interest in the work and standing alongside me every step of the way.

Foreword

In the early 1980s, I was researching failure and monitoring of mine conveyors at CSIRO Applied Physics. My work was new and well suited for submission to a Ph.D. programme. Colleagues at Sydney University referred me to Prof. Alan W. Roberts at the University of Newcastle. Alan became my Ph.D. supervisor in belt conveying. Over the years, he became a lifelong friend and a kindred spirit in material bulk handling. Since the early 80s, we have journeyed through many industry projects, attending conferences together in many countries, published our research and conducted courses worldwide.

What impressed me the most about Alan is his humble nature and his nurturing approach to all people. Alan mentored and encouraged diversity in engineering, understanding of industry's needs, balanced with practical academic research. He has a unique insight into how materials behave and how to define the equations governing their behaviour, he is not only a brilliant engineer but also a generous mentor with a willingness to share knowledge; that is his nature. Those of us who worked alongside him are inspired by his vision and his commitment to advancing the discipline of bulk solids handling. This book reflects that passion and stands as a seminal and comprehensive understanding of the discipline stemming from his long and remarkable career.

Alexander Harrison, B.E., Ph.D., FRSN
Former Professor of Mechanical Engineering
Berkeley Computational Research
Newcastle, Australia

Preface

This book presents the fundamental concepts required for the design of efficient and reliable bulk material handling systems. A thorough understanding of the way these complex and varied bulk materials behave under storage, flow and handling operations is presented. The handling of bulk materials requires a basic knowledge of the science and technology of particles and their bulk properties and the way these properties are influenced by process and environmental conditions such as temperature, moisture and consolidation. In view of the high component costs of bulk handling in industrial operations, it is essential that the handling of bulk materials be performed as efficiently and reliably as possible. The needs for strict environmental controls over emissions and dust generation must also be considered.

The handling and processing of bulk materials, often referred to as bulk solids, are key operations in a great number of industries worldwide. Such industries are wide ranging, including pharmaceutical, food production, manufacturing, agriculture, power generation and, at the 'heavy' end, mining and mineral production. The motivation for this book is based on the practical experience of the author related to the industries outlined above. This motivation stems from many years of fundamental research and industrial consulting problems. The inclusion of practical, case-study examples is an important feature. Bulk materials predominately comprise particles of granular form ranging from micron size, such as fine powders and industrially created dust, to large rocks, a metre or more in size as occur in Run-of-Mine ore in the mining and resources industries. Other bulk materials are non-granular in form, such as paper pulp, silage, domestic waste, biomass and wood chips.

In the past, the design and operation of handling systems for bulk solids has, all too often, been treated very empirically. This applied particularly to storage bins and surface storage facilities such as stockpiles and gravity fed reclaim systems. There have been many instances of costly flow interruptions such as those caused by the bulk solid forming stable cohesive arches over the openings in the bottoms of storage bins and stockpiles or holding-up against the bin walls without discharge taking place. Other problems that commonly occur in bulk solids storage facilities include solids aggregation, erratic feeding and flooding. In most cases, such flow interruptions are directly attributable, to incorrect design with little or no regard to the flow properties of the material being handled. On other occasions, catastrophic structural failures of bins and silos have occurred due to a lack of appreciation, at the design stage, of the magnitudes of the dynamic wall pressures that occur during discharge. It is important to acknowledge that in recent years the theory and experience has increased the knowledgebase so that such problems are now being minimised.

The aim of the book is to provide readers with the fundamental knowledge required to analyse the design and optimisation of materials handling systems. The book is ideal for engineers, designers and researchers concerned with all aspects of bulk materials handling and processing from fundamental research through to industrial operations.

Shortland, Australia Alan W. Roberts

Acknowledgements

This book marks the culmination of a challenging yet rewarding academic journey in the field of Bulk Materials Handling. It would not have been possible without the support, guidance and encouragement of many individuals to whom I am deeply grateful.

I would like to acknowledge the engineers and pioneers in the field of bulk materials handling, particularly to my friend and colleague Andrew Jenike, whose fundamental research on flow and handling of bulk solids was an inspiration to me.

My appreciation also goes to my colleagues and fellow researchers, notably Emeritus Professor Peter Arnold of Wollongong University who supported me in establishing bulk solids handling, research and industrial development in Australia.

Finally, I am grateful to the staff of TUNRA Bulk Solids, for the many discussions, collaborative spirit and moral support throughout my lifelong journey. The exchange of ideas and shared problem-solving made a significant difference. Your collaboration helped bridge the gap between theoretical analysis and real-world application.

To Dr. Jayne O'Shea, who has played an instrumental role in the editing and production of this book.

Dr. Alex Harison whose long-term contribution to belt conveying and bulk handling have significantly impacted industry worldwide. Thanks for proofreading this book and lifelong friendship.

With heartfelt thanks,

Alan W. Roberts

Contents

About the Author

Emeritus Professor Alan W. Roberts of the University of Newcastle and Founding Director of TUNRA Bulk Solids. He is internationally recognised as a pioneer in bulk solids handling, working on industry projects worldwide, as well as presenting many courses and papers at conferences.

He commenced his engineering career in the mid-1940s earning a B.E. with Honours in 1956 and a Ph.D. at the University of NSW in 1962. His ongoing contributions have also been honoured with two doctorates, Doctor of Science from the University of Wollongong and Doctor of the University from the University of Newcastle.

He has authored 6 design manuals, contributed 13 book chapters and published over 500 research, conference and design papers, establishing modern design methodologies for the field. In 1975, he founded TUNRA Bulk Solids, a world-class research and consulting facility that continues to influence global industry practice. He also helped create the Key Centre for Bulk Solids and Particulate Technologies, fostering collaboration between academia and industry.

Professor Roberts' contributions earned numerous honours, including the A. G. M. Michell Medal, the Peter Nicol Russell Memorial Medal and appointment as a Member of the Order of Australia. His legacy endures through ongoing research, education and industry engagement.

Chapter 1
Introduction—Storage and Flow of Bulk Solids

1.1 Introductory Remarks

The handling and processing of bulk materials, commonly referred to as "bulk solids" are key operations in a great number of industries throughout the world. Such industries are wide ranging, including pharmaceutical, food production, manufacturing, agriculture, power generation and, at the 'heavy' end, mining and mineral production. As illustrated in Fig. 1.1, bulk solids predominantly comprise particles of granular form ranging from micron size, such as fine powders and industrially created dust to large rocks, a metre or more in size as occur in Run-of-Mine (ROM) ore in the mining and resources industries. Other bulk materials are non-granular in form, such as paper pulp, silage, domestic waste, biomass and wood chips.

Fundamental to the design of efficient and reliable handling systems is a thorough understanding of the way these complex and varied bulk materials behave under storage, flow and handling operations. This requires a basic knowledge of the science and technology of particles and their bulk properties and the way these properties are influenced by process and environmental conditions such as temperature, moisture and consolidation. The need for strict environmental controls over emissions and dust generation must also be a major consideration.

The costs of handling operations are very substantial, representing a major component of the overall production costs. It is important, therefore that all bulk handling and storage facilities be designed and operated with a view to obtaining both maximum reliability and efficiency coupled with maximum economy. For this goal to be achieved, it is necessary that attention be given to both the design details and operating procedures of all aspects of the handling plant and associated transportation systems.

In the past the design and operation of handling systems for bulk solids have, all too often, been treated empirically. This applied particularly to storage bins and surface storage facilities such as stockpiles and gravity fed reclaim systems. There have been many instances of costly flow interruptions such as those caused by the

A. W. Roberts, *The Theory and Practice of Bulk Materials Handling*,
https://doi.org/10.1007/978-3-032-19702-3_1

Fig. 1.1 Examples of bulk solid—from powders to particles to rocks

bulk solid forming stable cohesive arches over the openings in the bottoms of storage bins and stockpiles or holding up against the bin walls without discharge taking place. Other problems that can arise include incorrect hopper and feeder interfacing, erratic feeding and flooding and poorly designed transfer chutes. From the point of view of quality control, product degradation due to segregation and attrition need to be avoided. In most cases such flow interruptions are directly attributable to incorrect design with little or no regard to the flow properties of the material being handled.

On other occasions catastrophic structural failures of bins and silos have occurred, as illustrated in Fig. 1.2. In some cases, there has been a lack of appreciation, at the design stage, of the magnitudes of the dynamic wall pressures that occur during discharge. Other instances include retrofit modifications to plant equipment, such as bins and silos, without due consideration of the original bin design specifications. Examples of this include retrofit installations of offloading chutes in the side walls of tall, cylindrical silos designed to operate by axisymmetric discharge through a centrally located outlet in the bin base. The use of an offloading chute results in an eccentric discharge flow pattern which gives rise to severe cracking in the walls in the case of bins or silos of reinforced concrete construction and severe buckling of the walls of bins of steel construction. In either case total bin failure is the potential outcome. A seemingly innocuous example concerns the taking of a sample from a tall, axisymmetric grain silo. From misguided convenience the sample is taken from an inspection port in the outer region of the base adjacent to the silo wall with the silo being substantially full at the time of the sampling. The outflow sample flow, albeit small, induces an eccentric flow pattern with a high probability of total silo collapse

Fig. 1.2 Silo failures

and, sadly, loss of life as has been known to occur. These foregoing examples stress the importance of plant safety in which operation personnel have a basic understanding of the behaviour of bulk solids during storage, flow and handling.

The current level of knowledge and professional development in Bulk Solids Technology and Particle Technology is a result of these two areas being recognised as disciplines in their own right. They have strong overlapping roles, so much so that reference to one often implies the other. Focusing on Bulk Solids Technology, reliable test procedures for determining the strength and flow properties of bulk solids have been developed and analytical methods have been established to aid the design of bulk solids storage, handling and conveying equipment. There has been wide acceptance by industry of these tests and design procedures and, as a result, there are numerous examples throughout the world of modern industrial bulk solids handling installations which reflect the technological advances that have taken place. The advancements that have been made over the period following their inception in the latter nineteenth century have been highlighted in the three review papers of Roberts, Refs. [1–3].

Notwithstanding the current situation, the level of sophistication required by industry demands a better understanding of the behaviour of bulk solids and the associated performance criteria for handling plant design. Experience indicates that the solution of one problem, which leads to an improvement in plant performance, often exposes other problems which need to be solved. Problems in industry frequently multiply at a faster rate than research outcomes. The importance of industrial orientated research cannot be too strongly emphasised.

Current developments in the science and technology of bulk solids handling are now highly focused on computer-based simulation methods—most notably Discrete

Element Modelling (DEM), along with Finite Element Analysis (FEA) and Computational Fluid Dynamics (CFD). These advancements have introduced new challenges in particle characterisation and associated analytical and numerical procedures. Notwithstanding, the importance of the well-established, and proven "classical engineering" approach to problem solving and design based on fundamental mechanics should not be overlooked. It is this latter reason that has provided the motivation for the writing of this book.

1.2 Brief Historical Overview

1.2.1 Early Contributions

The foundations of bulk solids and particle technologies may be traced back many centuries. For instance, the author's early research on screw conveyors for grain handling has a link back to work of Archimedes in the period 287–212 BC. Archimedes was a person of exceptional ability as a mathematician and scientist as well as possessing inventive skills aimed at solutions to practical, real-world problems. Amongst other things, he invented the helical screw pump, a clever, innovative device used for irrigating the deserts. This invention is the forerunner of the screw conveyors that are widely used in modern industrial plants for the feeding and conveying of bulk solids.

One may speculate that Archimedes had an interest in the packing density of particles. In a work called the *Psammites* (Sand-Reckoner) he boasted that he could write down a number greater than the number of grains of sand required to fill the universe. Based on the size of the universe as it was then perceived, now referred to as the Aristarchian Universe, and assuming the grains of sand were not smaller than a poppy seed, his answer was 10^{63}. It is more likely that Archimedes saw this as a challenging problem in mathematics rather than a bulk solids problem. As a matter of interest, it may be noted that number of grains of sand to fill the earth is in the order of 10^{26} give or take a few!

Prior to the latter stage of the nineteenth century where the development of bulk solids and powder technology research began, the fundamentals of mathematics and science had made significant advances. Of particular relevance are the work of Coulomb [4] on friction, Rankine [5] who studied the frictional behaviour of sand and Reynolds [6, 7] who observed the dilatancy effect of sand while undergoing deformation. A classic example of this phenomenon is observed when a person walks along the sand at the water's edge of any beach. The sand is normally critically consolidated with the water exposed as a film on the surface. The compressive pressure due to the person's foot causes the sand around the foot to expand or dilate resulting in an increase in the voidage allowing the water to drain away. Other examples include natural phenomena such as sand and dust storms which have a similarity with industrial applications of pneumatic conveying as well as avalanches which have a similarity with industrial slumping problems such as in large storage bins and stockpiles.

1.2.2 Bulk Solids Handling—Foundations and Developments

Throughout the nineteenth century, the emerging mining, manufacturing and agricultural industries, gave rise to an increased need to store and handle bulk materials in large quantities. While the focus for this activity may have been North America, the United Kingdom and Europe, industrial developments were also being made elsewhere in the world, such as, for example, continents and countries of the southern hemisphere, including South America, Southern Africa and Australia. Since communications had not been widely established at that time, these latter developments were not widely known, if at all. For example, in the case of Australia, coal was discovered in 1796 some eight years after colonisation. The first export of coal occurred from the newly established port in the city of Newcastle in 1801, a load of some 600 tonnes bound for India. Newcastle, situated some 160 km north of Sydney, becoming a major bulk export port. In terms of tonnage capacity, it has been acknowledged as being the largest coal exporting port of the world.

Figure 1.3 summarises the various stages in the development of Bulk Solids Handling spanning some 160 years commencing in the latter part of the 19 century to the present period of the twenty-first century. This diagram is the basis of the discussion that follows.

1.3 Early Silo Load Research—1880s to 1920s

The need to store grain in large quantities provided the impetus for research into silo wall loads with a series of papers on this subject commencing in the 1880's and spanning a period of some thirty years, into the first decade of the twentieth century. This work was significant in that it recognised some fundamental aspects of internal and boundary friction which limit the magnitude of the loads on silo floors and walls. These papers, which emanated mainly from England, Germany, Canada, the USA, and Argentina, laid the foundations of subsequent bin and silo load research that followed. The principal contributors to this research are summarised below.

1.3.1 Isaac Roberts (1882–84)

Isaac Roberts of the UK had been aware of the developments in silo technology in North America and had knowledge of a new structure for Liverpool, U.K. Motivated by these developments, Roberts constructed four, small scale model silo cells of

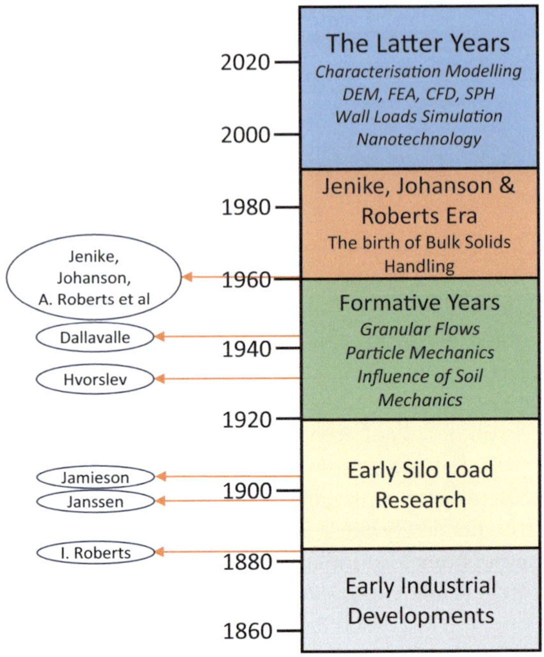

Fig. 1.3 Developments in bulk solids handling

square and hexagonal cross-sections. The test cells were supported on legs, while the cell bottom was supported on a weigh bridge to allow the grain pressure on the base to be measured. Measurements showed that the pressure on the base ceased to increase when the depth of grain exceeded about twice the width of the bin. Roberts had observed the internal and wall friction effect in limiting the pressure on the base of a silo. He published his findings in two papers, the first in "Engineering" of 27 October 1882 [8], and the second in the Proceedings of the Royal Society of London, January 31, 1884 [9].

1.3.2 H. A. Janssen (1895)

The most widely known work emanating from the early period of silo research is that due to the German Engineer, H. A. Janssen. Janssen, who conducted his research in Bremen, published his epic paper "Versuche über Getreidedruck in Silozellen" (On the Measurement of Pressures in Grain Silos), [10]. Like Isaac Roberts before him, Janssen had also observed the growing trend in grain exports from the grain producing countries of the world to the developed countries of Europe. He acknowledged the developments in silo construction in the USA where, because of their large dimensions, silos could store full train loads of grain and transfer these quantities to river or sea going vessels.

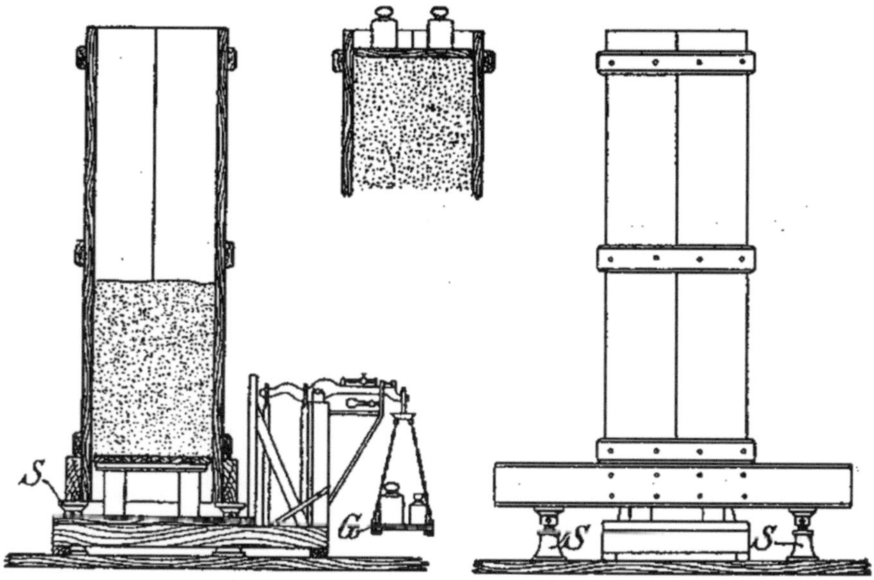

Fig. 1.4 Janssen's model test bin arrangement [11]

Janssen conducted experiments on four cells or bins made of wood, the bins being of square cross-section with sides measuring, respectively, 200, 300, 400, and 600 mm. The test set up, is shown in Fig. 1.4. The floors of the experimental bins were independently mounted on a weigh bridge to permit the grain pressures to be measured. The predominant test grain was wheat. Janssen recorded the pressure on the base of the bins for various fill levels of grain in each test bin.

From his test results and undoubtedly a knowledge of differential calculus, Janssen went on to derive his well-known equation for silo wall pressures:

$$p = \frac{s\gamma}{4k}\left(1 - e^{-4k^{x}/s}\right) \tag{1.1}$$

where

p = vertical pressure of grain
s = length of square side of bin
x = height of grain in bin
γ = specific weight of grain
$k = \frac{P_s\mu}{p}$ = pressure ratio
P_s = side wall pressure
μ = friction coefficient at wall.

It is noted that the parameter k expresses the wall shear stress, $P_s\mu$, as a ratio of the vertical pressure p. The more meaningful pressure ratio is the ratio of side

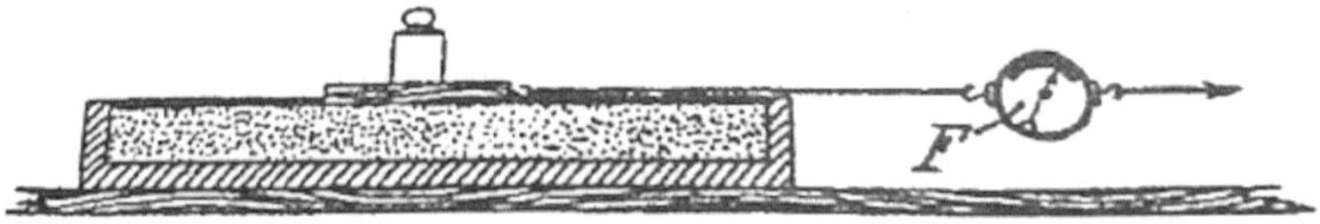

Fig. 1.5 Apparatus used to measure wall friction [11]

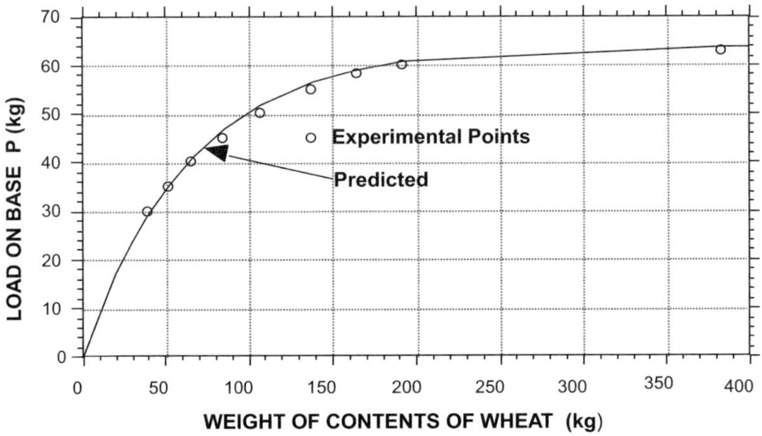

Fig. 1.6 Comparison between measured and calculated results. $\gamma = 0.8$; $k = 0.203$; $\lambda = 0.68$ [1–3]

wall pressure P_s divided by the vertical pressure p for which Janssen described by $\lambda = \frac{k}{f} = \frac{p_s}{p}$.

Janssen used the apparatus shown in Fig. 1.5 to measure the friction coefficient μ. The container was filled with wheat and covered with a loaded wooden lid which was pulled laterally with the shear force F measured by a spring balance. He measured friction coefficients ranging from $\mu = 0.302$ to 0.346.

An example of Janssen's measured and predicted results is shown in Fig. 1.6.

Janssen noted that for a bin of square cross-section the parameter λ is not constant as would be the case of a bin of circular cross-section. Rather, for the square cross-section, he concluded that the lateral pressures are higher at the centre of the sides than at the corners.

Janssen's equation continues to be used to predict the initial filling, static stored loads in the cylindrical sections of silos. It does not apply to the load prediction during discharge.

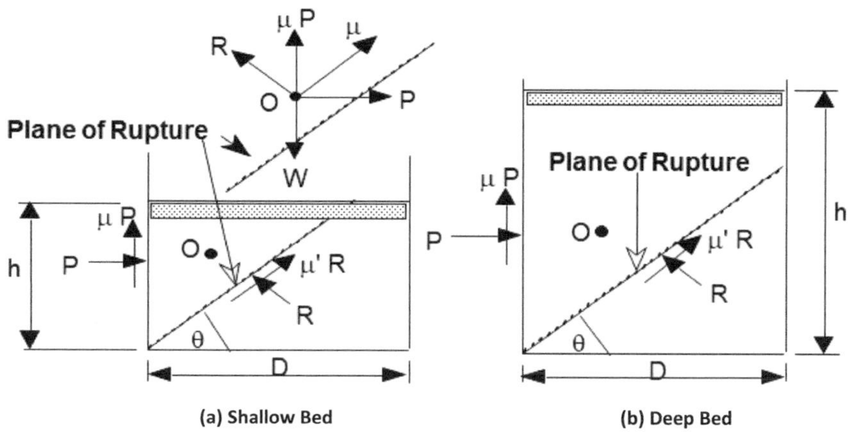

Fig. 1.7 Airy's model for silo pressures [1–3]

1.3.3 Airy (1897)

The work of Wilfred Airy was presented to the Institution of Civil Engineers, London and published in the Proceedings Vol.CXXXI, 1897 [12]. Airy established a theory for explaining the grain pressures in silos and included experimental results for his measurement of the angle of repose of grain and friction coefficient for grain on the bin walls. Airy's model, shown in Fig. 1.7 has some similarity with retaining wall analysis. It assumes a rectangular shaped silo in which a wedged shaped segment of granular material, bounded by a rupture plane at slope θ, causes maximum pressure against the left-hand side wall. The point "O" defines the centre of gravity of the wedged-shaped segment. The segment is of unit thickness. Airy considered two cases, one for shallow bins where the rupture plane does not intersect the bin wall but rather passes through the surface, and the other where the plane intersects the bin wall on the opposite side.

As an example, Fig. 1.8 compares Airy's model with that of Janssen's for a particular silo. Despite the entirely different approaches of the two procedures, Airy's predictions correlated quite well with those of Janssen.

1.3.4 Prante (1896)

Prante, whose research was reported in 1896 [13], conducted experiments on two full size bins, 1.5 m and 3.8 m diameters and 19 m tall. His measurement of lateral wall pressure of wheat at rest gave values slightly less than those given by Janssen's equation. However, he also observed the wall pressures during discharge of wheat at a velocity of 1 mm/s increased quickly, reaching values as much as four times the

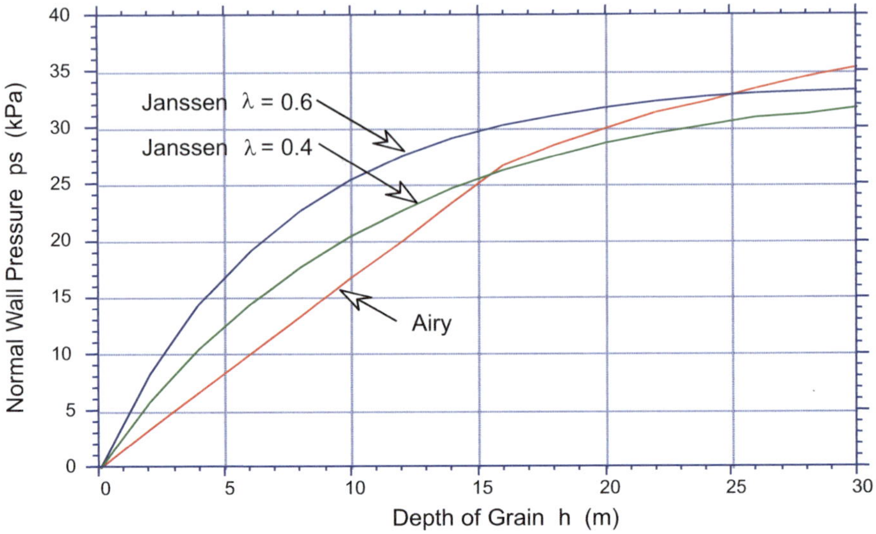

Fig. 1.8 Comparison between solutions of Airy and Janssen silo diameter D = 10 m; $\mu = 0.577$: $\mu' = 0.7$; $\rho = 0.8\ \mathrm{t/m^3}$ [1–3]

static pressures. It is significant that the discharge gate was located near the bottom on one side, probably opposite the pressure measuring device. This would explain the high magnitude of the flow pressures. Prante's results were not considered as reliable in Germany as those predicted by Janssen's method. This was a mistaken conclusion since it overlooked the fact that Janssen's equation did not predict the overpressures during silo discharge. However, his work showed that discharge should be symmetrical, with the gates located as near to the centre of the bin as possible.

1.3.5 Toltz (1903)

Max Toltz, a mechanical engineer in the USA, had followed Prante's work when designing large grain silos in Buffalo, New York in 1897 and in West Superior, Wisconsin. His research, published in 1903 [14], involved a series of experiments on a wooden 4.27 m square bin, 19.8 m high. For pressure measurements, the bin had an opening 0.46 m × 0.91 m in one wall near the bottom. A steel plate was fitted to the opening, being held rigidly at the sides but being free at the top and bottom edges. Using a series of steel plates of varying thicknesses and measuring the deflections of these plates, the pressures during filling and emptying were determined for a range of different heights of grain. The location of the discharge point is not known, but it is presumed that the loading was symmetrical.

1.3.6 Jamieson (1900–1904)

J. A. Jamieson of Montreal, Canada, in 1900 performed a series of experiments on a full-size bin of the Canadian Pacific Railway Elevator in West St John, New Brunswick [15, 16]. The bin was of timber crib construction, 3.66 m by 4.11 m in cross-section and 20.6 m high. The grain used in the tests was wheat from Manitoba. On the full size bin Jamieson used pressure cells consisting of a thin chamber with a rubber diaphragm containing water, the pressure against the diaphragm being measured using a mercury manometer. For the lateral pressure measurement, the cell was placed a short distance above the hopper bottom. For the measurement of vertical pressure on the base, the cell was placed on a small platform attached to the hopper bottom.

The bin was progressively loaded with wheat in stages, with pressure readings taken at each load. After the bin was filled, it was allowed to stand 18 h, during which time the pressure reading was constant. The grain was drawn off through the discharge gate in the bottom of the bin in 13.6 tonne lots at the rate of 3.4 tonnes/hour. Jamieson observed that the pressures during discharge fluctuated considerably, the maximum pressure during discharge being about 4% larger than that during filling. He also observed that closing the valve during emptying gave a slight increase in pressure while opening the valve gave a slight reduction in pressure.

Jamieson was familiar with the work of Janssen and used Janssen's equation to compute the pressures in the bin. Jamieson's measurements showed that the friction coefficient at the wall of the bin was $\mu = 0.441$. He also determined the ratio of lateral to vertical pressure to be $\lambda = 0.596$. With $\lambda = 0.6$, Janssen's equation gave results in very close agreement with the measured values.

Following on from the full-scale bin studies, Jamieson conducted experiments on seven model bins using the test arrangement shown in Fig. 1.9. The bins were 1.98 m tall and 0.3 m and 0.15 m square or round. They were constructed of smooth wooden boards cribbed, flat steel plates and troughed or corrugated steel plates the corrugations running horizontally. Six hydraulic diaphragm type pressure cells were constructed and used to measure the pressures. For the measurement of the vertical pressures on the floor of the bins, the pressure cells formed the whole of the base of each model bin. For lateral pressure measurement, the pressure cells were located just above the base and made part of the bin wall, the face being set flush with the inside surface of the bin. In one set of tests, the grain was in direct contact with the rubber diaphragm, while in another set of tests, the rubber was faced with a wooden board to distribute the pressure over the surface. Jamieson's experiments were performed for the two discharge arrangements:

- symmetrical, central discharge with the outlet in the silo base centrally located and
- asymmetrical, eccentric discharge with the outlet located near one wall, as illustrated in Fig. 1.9 where the outlet is shown on the right hand side of the base.

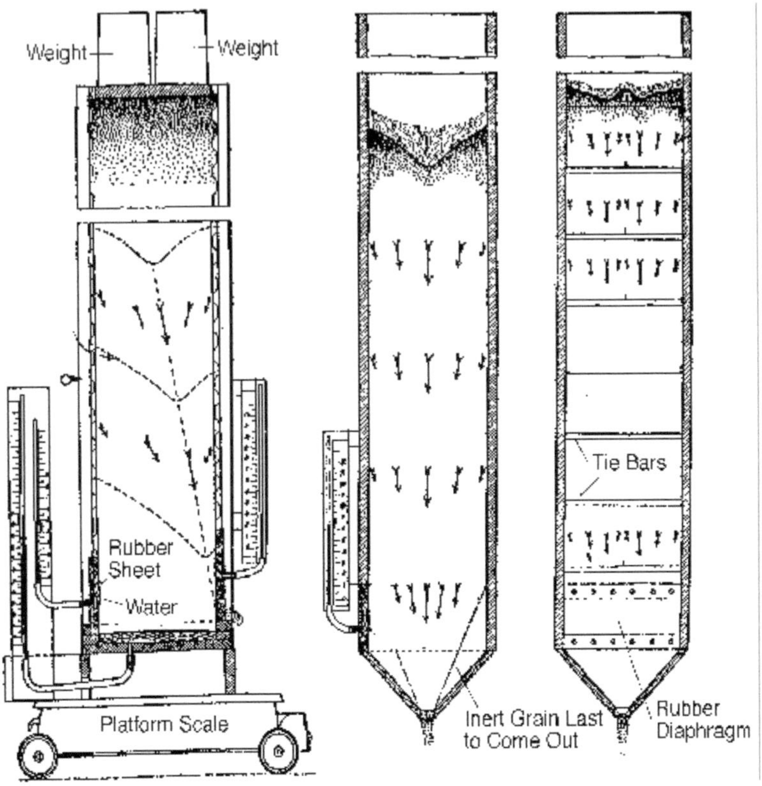

Fig. 1.9 Arrangement of apparatus for model bin tests used by Jamieson [15]

After filling, tapping the test bin with a hammer near the base showed that the pressure on the bottom decreased. No doubt this was due to an arching condition set up near the bottom of the bin. When tapping was continued on all sides from the bottom to the top, Jamieson observed that the grain settled some 50 to 75 mm and the pressure on the base increased slightly. The placement of the 22.7 kg (50 lb) loads on top of the grain in the bin showed that the base pressure increased only slightly with each load added, the pressure becoming normal when the load was removed.

Representative sets of results from Jamieson experiments displaying the filling and flow pressures are shown in Fig. 1.10 for the central discharge and in Fig. 1.11 for eccentric discharge. For both the central and eccentric discharge, the initial filling pressures correlated quite well with Janssen's predictions. As can be observed, for the central discharge case, the discharge pressures were larger than the initial filling pressures for the fill heights of 0.5 to 2.0 m and slightly lower over the fill heights of zero to the 0.5 m fill height. What is particularly significant are the variations of the discharge pressures during eccentric discharge. The pressures on the wall nearest the outlet decreased below the initial pressures while on the wall opposite the outlet increased above the initial filling pressures.

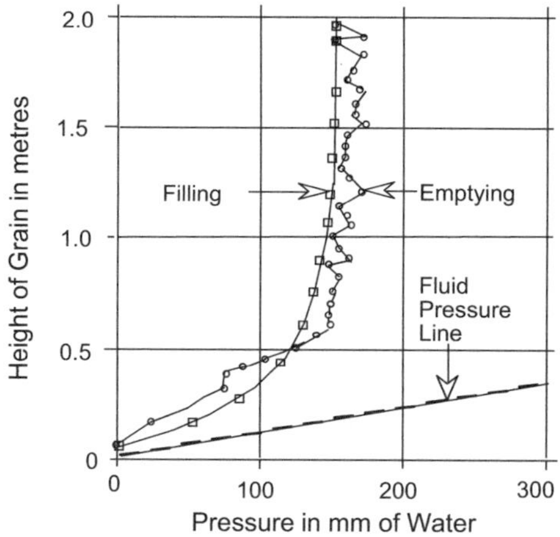

Fig. 1.10 Lateral pressure of wheat due to central filling and emptying [15]

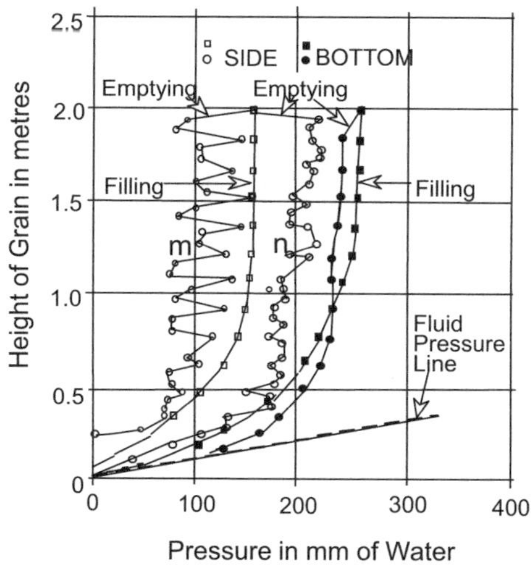

Fig. 1.11 Grain pressures due to filling and emptying—opening at lower side. Curve 'm'—side adjacent to outlet, curve 'n'—side opposite [15]

When viewed overall, Jamieson complemented the work of Janssen thus demonstrating a very strong understanding of the complex nature of silo wall loadings. Jamieson's work, while lesser known than Janssen's, had identified the significance of the over-pressures and eccentric loads caused by non-central discharge that, in subsequent years, have been the cause of several silo failures.

1.3.7 Lufft (1902–1904)

The widening international interest in grain handling is evidenced by the work of Eckhardt Lufft, who conducted experiments on two full size bins in Buenos Aires, Argentina, during the period 1902–04 [17]. The two test bins measured 7.26 m and 3.15 m diameters respectively; both bins were 16.7 m high. Lufft, whose work was performed before Jamieson's research had been published, measured the pressures exerted on the lower walls during both filling and emptying. He also observed the effect of a slow pressure reduction during grain settlement when filling was interrupted. He reported the pressure reduction continuing over periods from several hours to days. He concluded that this was due to an increase in both the angle of repose (presumably internal friction) and the wall friction.

1.3.8 Pleissner (1902–1905)

Pleissner conducted a comprehensive series of experimental work on deep bins to study the pressures generated during filling and during discharge [18]. One of his conclusions was that the lateral to vertical pressure ratio was not constant but varied with location in the bin and increased with depth or consolidation pressure. For wheat, he measured values of λ ranging from 0.3 to 0.5 for a range of bin constructions where different wall friction conditions existed. For rye, he measured values ranging from 0.23 to 0.45.

1.3.9 Bovey (1904)

The work of Jamieson no doubt aroused the interest of Professor Henry T. Bovey of McGill University in Montreal, who carried out a comprehensive series of experiments to check on Jamieson's results. His work, performed in 1904 [19], involved the measurement of vertical and lateral pressures in silos during filling and emptying. He used pressure cells similar to those used by Jamieson.

1.3.10 Ketchum (1902–1920)

Milo S. Ketchum, whose early research into silo wall loads was carried out in 1902–1903 at the University of Illinois, made a major contribution to the field of silo design with the publication, in 1907, of his book "The Design of Bins, Walls and Grain Elevators". Subsequent editions of his book appeared in 1911 and 1919 [20]. In Chapter XVI he includes a review of Janssen's work. While Janssen had considered a horizontal slice in the derivation of his equation, Ketchum rightly pointed out that the vertical and lateral pressures at any location are constant on the surface of a dome rather than across the horizontal slice. However, he accepted Janssen's assumption of constant pressures over the slice and derived Janssen's equation for the normal wall pressure, p, expressing it in the modified form.

$$p = \frac{\gamma R}{\mu\lambda}\left(1 - e^{-\mu\lambda\frac{y}{R}}\right) \tag{1.2}$$

where.

$\gamma =$	bulk specific weight.
$R = \frac{A}{U} =$	'hydraulic' radius
$A =$	area of cross-section
$U =$	perimeter
$\mu =$	coefficient of wall friction
$\lambda-$	pressure ratio
$y =$	depth below loaded surface.

It needs to be noted that the term 'characteristic radius' is now more commonly accepted rather than 'hydraulic radius'.

Ketchum had generalised Janssen's equation to be applicable to silos of any cross section. Experiments were performed by Ketchum on a 30 mm × 30 mm × 2.6 m high model silo constructed of dressed white pine. Using wheat, his results showed λ to be very nearly 0.4.

1.3.11 Summary Remarks

Undoubtedly, the best-known work of this period is that of Janssen [11]. The centenary of Janssen's work was celebrated in the conference "100 Years of Janssen" held in Nuremburg, Germany, in 1995 [3]. While Janssen's contribution dealt with silo loads in flat bottom bins or silos under static storage conditions, the lesser-known contributions of the other researchers included studies of silo loads under dynamic, discharge flow conditions. The research of Jamieson [15], for example, was particularly important since it demonstrated the influence of overpressures on silo walls during symmetrical discharge and the profound influence on silo walls due to eccentric discharge. These dynamic load effects were also reported by others, including Lufft [17]. The influence of load settlement following silo filling was yet another important outcome of the early silo load research.

1.4 The Formative Years—Early to Mid 1900s

Following on from the earlier work on silo loadings, which was mainly focused on the grain industry, the developments in the broader areas of bulk materials was starting to happen. This marked the commencement of what is now known as particle technology.

1.4.1 Developments of Particle Technology

(i) **Phillips (1910)**

As an aid to the understanding of how the internal and boundary friction of free-flowing bulk materials control the pressures in tall, cylindrical containers the contribution of Major C.E.S Phillips is cited. In 1910 Phillips presented a paper entitled "The Remarkable Properties of Sand" to the Royal Institution, the paper being published in the Institution Proceedings [21]. He conducted an effective demonstration to show that the application of a downward force to the top of a column of sand is not transmitted to the bottom if the column. He used a thin cigarette paper to close the bottom of a tube. The tube could then be filled with sand without bursting the paper. Furthermore, he demonstrated that with a piston located on top of the sand and loaded with a large weight, the cigarette paper remains unbroken.

What Phillips had demonstrated was that due to internal friction in the sand and the friction generated between the sand and the tube wall, the major proportion of the load is supported by the tube wall and the load on the paper is no more than that equivalent to a shallow layer of sand. In effect, Phillips had discovered what Janssen, and others had discovered a decade or more earlier. This demonstration has been employed by the author for many years in the teaching of bulk solids handling courses to students and engineers. These subsequent demonstrations involved a 45 mm internal diameter "Perspex" tube enclosed at the bottom with tissue paper held in place by an elastic band, as illustrated in Fig. 1.12.

(ii) **Shaxby and Evans (1922)**

Shaxby, who was the Director of the Viriamu Jones Physical Laboratory, University College, Cardiff, U.K., together with Evans, his student, inspired by the paper of Phillips [21], conducted research, no doubt aimed at establishing a theory to predict and quantify the results shown by Phillips. Their research included "bench top" experiments involving the measurement of the base pressures in vertical tubes filled, mainly, with fine and coarse emery powder. The test tubes had radii of 10.3, 14.6, 15.1 and 19.3 mm. The experimental results correlated quite well with Shaxby's equation. The paper of Shaxby and Evans, "On the Properties of Powders. The Variation of Pressure with Depth in Columns of Powders", received by the Faraday Society on 3 November 1922, was read before the Society on 23 April, 1923 [22]. It is not surprising that Shaxby's equation was virtually identical to Janssen's equation

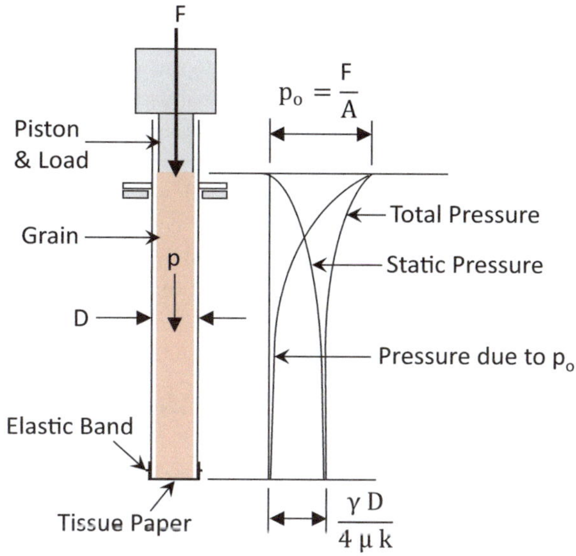

Fig. 1.12 Demonstration based on Philips experiment. F = surcharge load; P_o = surcharge pressure; P = vertical pressure; k = pressure ratio

of 1895, some 27 years earlier. It is most likely that Shaxby had no knowledge of Janssen's work.

(iii) **Dallavalle (1943)**

The contribution of J.M. Dallavalle in his book entitled "Micromeritics", which was first published in 1943 is of particular significance [23]. As Dallavalle wrote in the Preface to the First Edition, the title "Micromeritics" was coined to represent the "science of small particles". The subject matter included in the text is broad ranging including such subjects as: dynamics; shape and size; particle-size measurement; packings; behaviour of particles under pressure; diffusion; electrical, optical, sonic, surface and chemical, properties; thermodynamics of particles; flow of fluids though packings; infiltration and particle-moisture relationships; capillarity; particle surface determination; muds and slurries; transport of particles; dust clouds; atmospheric and industrial dust; collection and separation of particulate matter from air; theory of fine grinding; sampling.

Dallavalle also wrote, "Some users of this text may feel that it touches upon too many apparently unrelated fields. This may be true, but the inclusion of such fields was made purposely to indicate the wide applications of a subject which should receive recognition as deserving a place in the engineering sciences".

Clearly this statement reinforced the interdisciplinary nature of this new named science. While the name "Micromeritics" remains, it has provided the foundation for what we understand to be been embraced by the title 'Particle Technology'. There can be no doubt that Particle Technology is now very firmly established amongst the engineering sciences.

(iv) **Research into Granular Flows**

During the period from the 1920s to the 1960s a number of studies were performed concerning the flow rates of bulk solids through orifices in the bottom of bins and through transfer chutes. This work relied heavily on experimental techniques, with empirical type performance equations being derived from experimental results. Also, the emphasis was mainly on free flowing, non-cohesive granular type materials. While there are several names that could be mentioned, the work of Brown and Hawksley [24], Fowler and Glastonbury [25], Brown and Richards [26, 27], Rose and Tanaka [28] and Brown [29] are of particular note.

1.4.2 The Influence of Soil Mechanics

Soil mechanics as a field of science and technology had already been well developed. Therefore, it is not surprising that this field of study had a significant influence on the research into various aspects of bulk solids handling. Since soil mechanics is mainly concerned with retaining walls and foundation design, naturally, the internal stresses are much higher than those encountered in bulk solids handling. Furthermore, the main concern of soil mechanics is with the conditions existing within the soil prior to failure, whereas the primary interest in bulk solids handling is with the conditions under which failure and flow can occur. Nevertheless, the general similarities between the two fields of study permit some important comparisons to be made.

The work of Hvorslev [30] is of particular importance to the understanding of the mechanism of consolidation and flow of bulk solids. Hvorslev, who studied the stress condition in cohesive soils, showed that the peak shear stress at failure is a function of the effective normal stress on and the voids ratio (or density) in the plane of failure; this condition is independent of the stress history of the sample. The work of Hvorslev was further extended by Roscoe et al. [31], who established the concept of a failure surface in the three dimensional space of shear stress, normal stress and voids ratio. They also showed the existence of a critical void ratio boundary at which unlimited deformation could take place without change in the stress condition and voids ratio. The Hvorslev-Roscoe et al. diagram is shown in Fig. 1.13.

1.5 The 1960s Onward—The Jenike, Johanson and Roberts Era

The flow of cohesive bulk solids from storage bins and associated handling operations are complex problems and it was not until the mid 1950’s before any real progress into the fundamental behaviour of such materials began to take place. The modern developments are very largely due to the foresight and pioneering research of Dr. Andrew W. Jenike, together with Dr. Jerry R. Johanson, who commenced

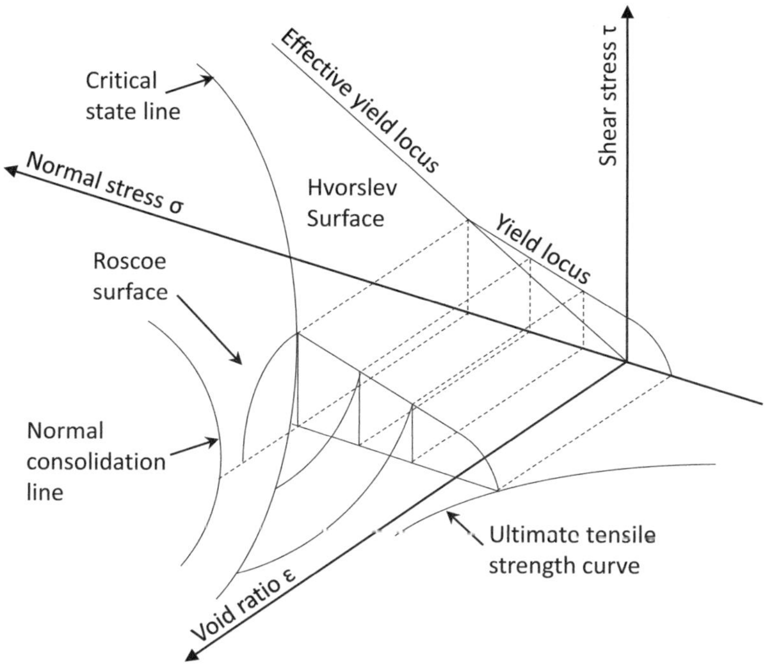

Fig. 1.13 Hvorslev/Roscoe diagram

his research as a Ph.D. student of Jenike. A comprehensive historical review of the Jenike/Johanson story has been written by Johanson [32].

The original research was conducted at the University of Utah, with the three University of Utah publications, Bulletins 108, 116 and 123, [33–35], laying the foundations of the modern theory of bulk solids storage and flow.

Jenike also saw the need to learn from the research conducted in the field of soil mechanics and relied heavily on the work of the Russian author Sokolovski who had written a book entitled "The Statics of Soil Media" [36]. As Johanson reports [32], the particular chapter of Sokolovski's book concerned with the critical equilibrium of a "wedge" in which the soil stress variation is directly proportional to the distance from a point is considered, helped to explain some experimental results Jenike had obtained earlier. It provided the key to the radial stress theory [34, 35].

Jenike also recognised the need to explore plasticity theory in order to explain the flow or yielding conditions in deforming bulk solids. Both he and Johanson studied Prager's book "The Theory of Plasticity" [37] as well as Hill's "The Mathematical Theory of Plasticity" [38]. Jenike collaborated with Professor R.T. Shield of Brown University in the U.S.A. to establish the basis for the Effective Yield Locus as well as exploring the mechanisms for solids flow [39].

The significant results and outcome of this work included the following:

- Establishment of the two principal flow modes, Mass-Flow and Funnel-Flow
- Radial stress theory describing the converging flow in mass-flow hoppers and limits for mass-flow
- Flow/No-Flow criteria
- Direct shear apparatus for the determination of the flow properties of bulk solids.

This marked the beginning of Bulk Solids Handling being recognised as a research, technological and professional discipline in its own right. There is no doubt that the pioneering work of Jenike and Johanson and their follow up work along with their co-workers on silo wall loads and other bulk handling topics provided the foundation for the great deal of ongoing worldwide research and industrial awareness in the various fields of bulk solids handling. It also created the stimulus for the discipline of bulk solids handling being integrated into university engineering courses and research programmes. Associated industry consultancies were also established—for example, during the 1960s and 1970s at the Universities of Wollongong and Newcastle in Australia.

A parallel stream of Australian research emerged during the same period, investigating the mechanics of screw conveyors transporting grain. Early studies into the performance of grain augers examined the relationships between auger geometry, material flow patterns, and conveying capacity, leading to a 1962 publication by Roberts and Willis on grain auger performance [40]. Subsequent work by Roberts in 1963 [41] analysed vortex motion within vertical grain augers, identifying the characteristic flow structures that develop during vertical conveying and their influence on throughput and power demand.

This research trajectory was later extended into the study of granular flow behaviour within transfer chutes, leading to two notable publications by Roberts in 1967 [42] and 1969 [43], which examined the dynamics of non-cohesive granular materials flowing through curved and straight chutes. These studies provided early quantitative descriptions of the forces, flow regimes, and particle trajectories governing chute performance.

Together, these contributions advanced the emerging body of knowledge on bulk solids motion in mechanical handling equipment and helped establish a more rigorous engineering framework for the design and analysis of screw conveying systems and gravity-flow transfer chutes.

1.6 From the 20th to 21st Century

Developments have occurred on several fronts including research into the properties and behaviour of particulate solids during storage and flow, further work on bin loads and applied research aimed at improving the efficiency of industrial operations. Important 'breakthroughs' have been made possible through more sophisticated scientific equipment for experimental work and modern computer technology to assist finding the solution of complex problems. From the initial focus on silo design, the subject of bulk solids storage, flow and conveying has become all-embracing.

By way of example, a 'snap-shot' of some of these many interesting developments with selected references is presented:

- Re-examination of the mass-flow and funnel-flow limits taking account of the surcharge head at the hopper/cylinder transition and the establishment of the conditions for "intermediate-flow" [44].
- Development of test equipment for characterising bulk solids and powders in terms of their stress/strain relationships and flow properties including boundary or wall friction [45–61].
- Analysis of vibration of bulk solids in relation to flow promotion [62–64].
- Studies of friction, adhesion and wear in bulk solids handling operations [1, 65–68].
- Flow property measurement and reactor vessel design for handling and processing of stringy bulk solids, such as domestic waste [69], and wet bulk solids.
- Dust control and measurement of 'dustiness' [70].
- Flow rate predictions for fine powders discharging from mass-flow hoppers [71–73].
- Flow characteristics in hoppers and discharge equipment in relation to anti-segregation and mixing including the application of inserts [74–77].
- A new look at the prediction of rat-hole geometry in funnel-flow bins [78].
- Wall load predictions for symmetrical mass and funnel-flow bins [79–95] as well as for eccentric discharge.
- Analysis of 'silo quaking' in bins of various geometrical configurations [96–102].
- Investigations of pressures acting at the base of stockpiles and procedures for predicting the live capacity during gravity reclaim [103–113].
- Hopper/feeder interfacing for optimum draw-down and feeder load prediction and drive torque control [114–117].
- Performance characteristics of various feeder types including belt, apron, vibratory, screw, rotating table, tube, oscillating plate and rotary valves [117–129].
- Discrete and continuum approaches to the modelling of bulk solids flow [130–133].
- Pilot scale testing of bulk solids equipment using dynamic simulation [134, 135].
- Studies of the flow of bulk solids through transfer chutes and development of models for chute design, including analytical and numerical methods for optimising chute profiles for minimum wear [43, 136–144].
- Mechanical conveying—belt, special belt, screw, bucket, chain [145–150].
- Pneumatic conveying—lean phase, dense phase, slug and plug flow [151–153].

The twenty-first century is witnessing a major focus on simulation methods, notably DEM along with CFD, SPH and FEA as applied to bulk solids handling research and industrial plant design and problem solving. DEM is now well established but, as a precautionary note, it is important that the "classical" fundamental mechanics approach is not overlooked.

Along with the foregoing comments, research into the widely varying areas of Particle Technology has been proceeding at a very impressive rate. Of particular note are the quite exciting developments in nanotechnology, such as in applications to medical science, particle flow analysis and computer technology. The spin-off from this research to the broader areas of particle and bulk solids technology will continue to be of great value.

1.6.1 Inter Disciplinary Roles

The interrelation between Particle Technology and Bulk Solids Technology requires clarification. While clearly there is a great deal of overlap, perhaps the main distinction lies in the range of particle sizes involved and the associated fields of application. Particle Technology is associated more with finer particles commonly less than a few millimetres down to micron size and currently, down to nanometres. On the other hand, Bulk Solids Technology, while concerned with fine particles in the micron size range when dealing with powders and dust, the size range often extends to much larger 'particles' which may exceed one metre in size. Such is the case when dealing with ROM ores in mining operations.

In considering the interdisciplinary nature of the subjects of Particle and Bulk Solids Technologies, it is important not to overlook a third research discipline, namely, Geomechanics, which embraces Soil Mechanics. All three share a common core area which embraces particle characterisation and modelling. The interactions between these three principal research disciplines are illustrated in Fig. 1.14.

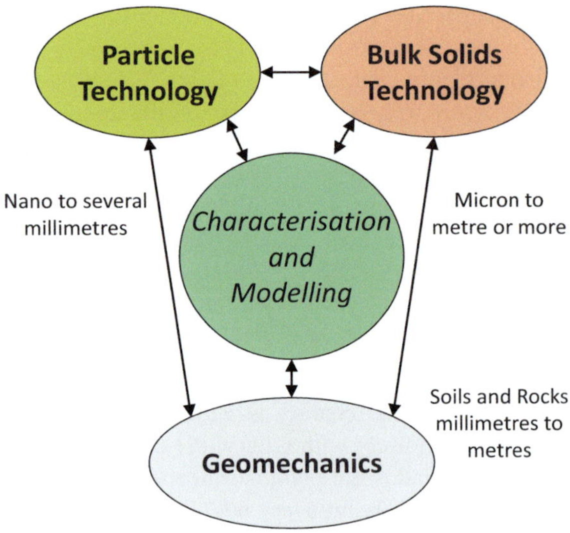

Fig. 1.14 Interdisciplinary roles of particle technology

1.7 Flowability of Bulk Solids

1.7.1 Some General Characteristics

When considering the flow of bulk solids, it is important to realise that the behaviour of a solid during flow is not like that of a liquid. In fact, the properties of solids and liquids differ so much that the mechanisms for flow in the two cases are quite different. The principal differences are:

i. Bulk solids can transfer shearing stresses under static conditions whereas liquids do not.
ii. Many solids when consolidated possess cohesive strength and retain their shape under pressure.
iii. The shearing stresses that occur in a slowly deforming or flowing bulk solid can usually be considered independent of the rate of shear and dependent on the mean pressure acting within the solid. In a liquid the situation is reversed, the stresses are dependent on the rate of shear and independent of the mean pressure.

The above differences suggest that a bulk solid must be regarded as a plastic and not a visco-elastic continuum.

There are a great many terms used from time to time to describe the properties of bulk materials. By way of illustration, [85] identified a number of material properties, many of which can only be used in a qualitative, descriptive and empirical way and are difficult to define precisely and even more difficult to measure. Often, too, some properties are used incorrectly, particularly in relation to the storage and flow of bulk solids. The angle of repose has frequently been misused as a measure of angle of internal friction and flowability of a bulk solid.

The angle of repose is the angle formed by the sloping surface of the pile formed when a loose, unconsolidated bulk solid is deposited on a horizontal surface, as illustrated by the angle θ_R in Fig. 1.15. This angle can vary with the moisture content, internal friction, shape, cohesion, size distribution and compaction of the material. However, variation is generally small despite the wide variation of the internal friction angles and degree of cohesion. For example, the angle of reposed for a dry, free flowing bulk solid such as wheat is approximately 33°, whereas the angle of repose for a cohesive bulk solid such as coal or iron ore is approximately 38°. On the other hand, the internal friction angles will be very different, around 33° for the wheat and 50° or more for coal and iron ore.

The fact that materials can be piled in this manner is due mainly to shear Coulomb frictional properties and cohesion and demonstrates their major difference with liquids. However, the angle of repose does not provide a measure of flowability; its use should be restricted to the determination of contours of piles. It is worth noting that in stockpiling, segregation usually occurs; the large particles are freer flowing than the fines and tend to separate out as indicated in Fig. 1.15.

Fig. 1.15 Angle of repose

1.7.2 Pressure Distributions Under Storage and Flow

The difference in behaviour of bulk solids and liquids can be further illustrated by considering the pressure distribution in a column of bulk material at rest in a cylindrical container as illustrated in Fig. 1.16a. This is an example of the impact of Janssen's research. Owing to the frictional properties the lateral pressure P_n at the section considered is less than the vertical pressure P_z. It is assumed that:

$$\frac{P_n}{P_z} = k = constant \tag{1.3}$$

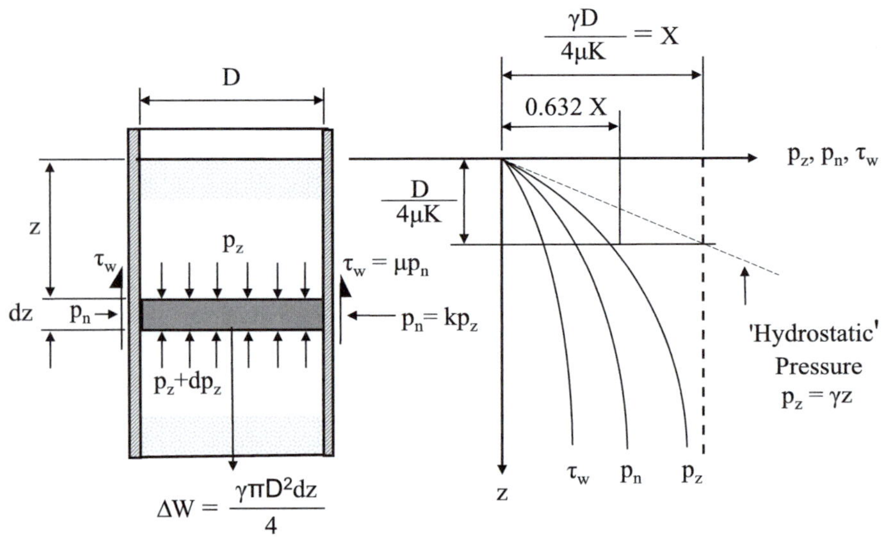

Fig. 1.16 Force analysis of column of bulk material

The factor k depends on the internal frictional properties of the bulk solid. A lower bound value is $k = 0.35$. In the absence of other information, it may be assumed that $k = 0.4$.

To simplify the analysis, some assumptions may be made:

- The pressure P_z at depth z is uniform over the cross section
- The coefficient of friction μ is a constant along the walls
- The specific weight γ is constant throughout the bulk solid
- The factor k relating P_n to P_z is constant along the walls.

None of the above conditions is strictly true. For instance, the coefficient of friction μ varies with the consolidation of the bulk solid which, in turn, varies with the depth. However, for most cases, the variation of μ is small particularly as the consolidation pressure increases. The specific weight also varies with the consolidation but approaches asymptotically, a constant value as the consolidation pressure increases. While k depends on the effective angle of internal friction of the bulk solid which, in turn, depends on the degree of consolidation, for many cases k is substantially constant.

From an equilibrium analysis of the elemental slice (Fig. 1.16) it can be shown that the following differential equation arises:

$$\frac{dP_z}{dz} + \frac{4\mu k}{D} P_z = \gamma \quad (1.4)$$

The solution of (1.4) yields the well-known Janssen equation:

$$P_z = \frac{\gamma D}{4\mu k}\left[1 - e^{\frac{-4\mu k z}{D}}\right] \quad (1.5)$$

where

$\gamma =$ Specific weight of the bulk material
$D =$ Cylinder diameter
$\mu =$ Coefficient of friction at the walls.

The slope of the Janssen curve is obtained by differentiating (1.5) to obtain

$$\frac{dP_z}{dz} = \gamma e^{\frac{-4\mu k z}{D}} \quad (1.6)$$

When $z = 0$, the slope of the P_z curve is γ.

Thus, the pressure P_z has a decaying exponential rise approaching, asymptotically, the value of $\frac{\gamma D}{4\mu k j}$ as shown in Fig. 1.16b. Also shown in Fig. 1.16b are the normal pressure P_n and shear stress τ_w. By comparison, in the case of a liquid, there is no vertical support provided by the walls and the pressure is the hydrostatic pressure γz which increases linearly with depth. Furthermore, at any depth z the pressure in a fluid is the same in all directions. The hydrostatic pressure variation is shown in Fig. 1.16b.

The foregoing analysis which explains how the support is provided by the walls when a bulk solid is confined by a tube is the basis of the classical demonstration conducted by Phillips as described in Sect. 1.4.1. The characteristics of bulk solid behaviour described here indicate that a column of material can be supported in a tube which has its lower end immersed in an open reservoir of the material. This would not be possible in the case of a liquid.

1.7.3 Forced Flow in a Vertical Cylindrical Tube

While the frictional properties of bulk solids offer advantages in the storage of materials, particularly in view of the support given by the walls of the bin or silo, these properties have an adverse effect when bulk materials are to be elevated. To illustrate this point, consider the problem of forcing a column of bulk material up a circular tube, as illustrated in Fig. 1.17.

In this case the shear stress at the wall acts downward in a direction opposing the motion and consequently provides a positive feedback effect. That is the greater the conveying pressure P_z, the greater the lateral pressure P_n acting against the walls and the greater the frictional drag becomes. This in turn means that the greater P_z must become.

Fig. 1.17 Analysis of forced flow conditions

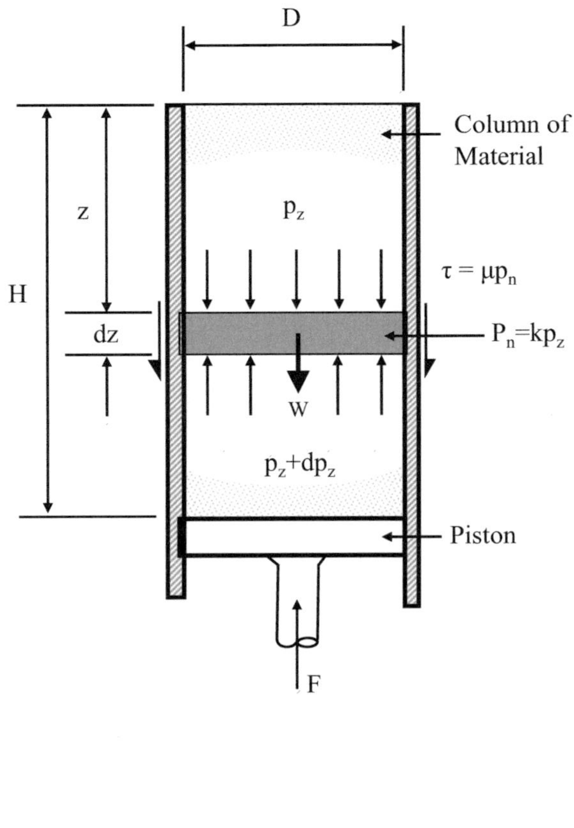

From the equilibrium of the element of the bulk solid shown in Fig. 1.17; the following differential equation is derived:

$$\frac{dP_z}{dz} - 4\mu\frac{k}{D}P_z = \gamma \tag{1.7}$$

Solving (1.7) yields:

$$P_z = \frac{\gamma D}{4\mu k}\left[e^{\frac{4\mu kz}{D}} - 1\right] \tag{1.8}$$

The conveying force F_H for a column of height H is:

$$F_H = \frac{\gamma \pi D^3}{16\mu k}\left[e^{\frac{4\mu kH}{D}} - 1\right] \tag{1.9}$$

Thus, the force F_H increases exponentially with column height H.

By way of illustration, Fig. 1.18 shows the force required to initiate the upward motion of a column of millet seed through a 45 mm internal diameter Perspex tube. The full line shows the theoretical curve computed using Eq. 1.9 while the plotted points indicate the experimental results. It should be noted that for the longer column length, once flow commences, the force for a particular column length increases beyond the value indicated in Fig. 1.18. This behaviour is due to such factors as irregularities in the internal surface of the tube and to the radial deformation of the tube due to the lateral pressure.

For comparison purposes the hydrostatic force is also shown in Fig. 1.18. The results for the bulk solid clearly indicate that bulk solids cannot be 'pumped' in a single-phase mode in the same sense as a fluid. In the case of fluids, the pressure to be overcome is that due to the hydrostatic head plus the frictional losses due to the pipe walls and bends, the latter losses being dependent on the fluid velocity. However, these losses are insignificant when compared with those arising in the case of bulk material under forced flow.

The exponentially increasing force with column length indicates that for the vertical conveying of bulk solids, the column length must be limited requiring the use of continuous conveyors such as, for example, the cable and disk conveyor or the enclosed screw or auger conveyor shown, respectively, in Fig. 1.19a and b. The cable and disk conveyor consists of a series of circular disks attached at equal spacings on an endless cable, conveying bulk material with linear velocity through the tube. The screw conveyor elevates bulk material via the rotational motion of the helical screw flight. In each case the bulk material column height is governed by the fill height "h" and pitch "L" of the disks and screw, respectively. Hence the conveying force is a linearly increasing function of the conveying force per pitch, as illustrated in Fig. 1.20.

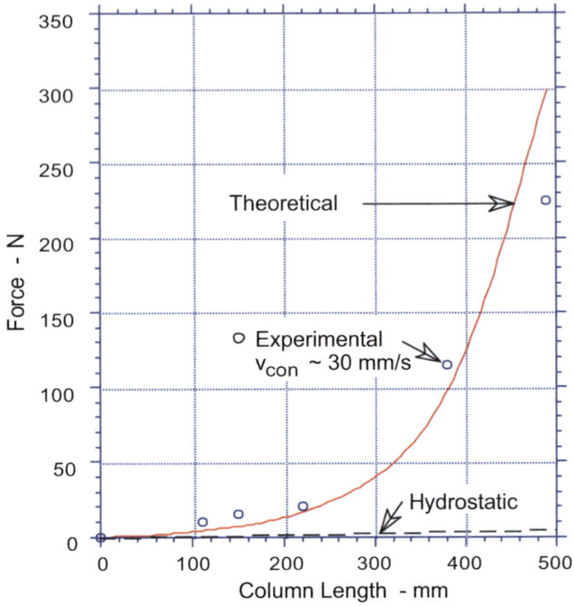

Fig. 1.18 Force to convey millet seed vertically through a 45 mm acrylic tube

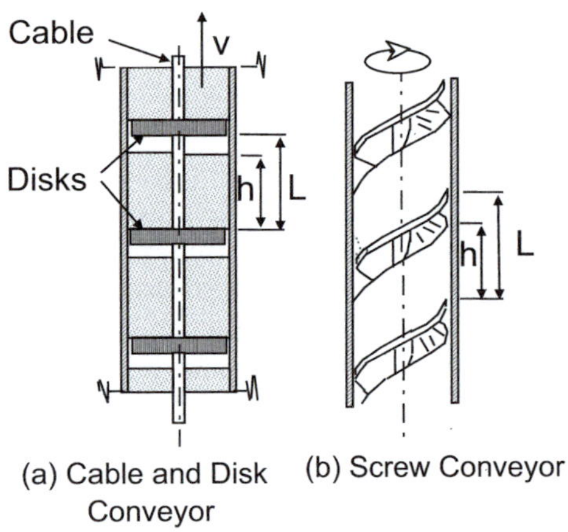

Fig. 1.19 Cable and disk and screw conveyors

The force to move the material on one disk is, from Eq. (1.9):

$$\Delta F = \frac{\gamma \pi D^3}{16\mu k}\left[e^{\frac{4\mu k h}{D}} - 1\right] \tag{1.10}$$

where

$h =$ height of material on one disk.

The total conveying force is given by Eq. 1.11.

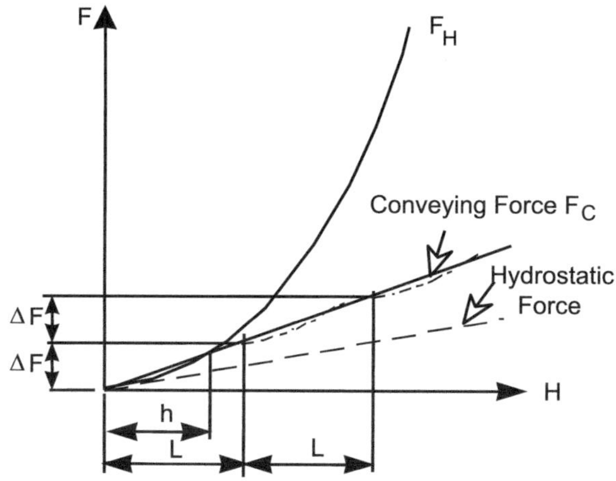

Fig. 1.20 Comparison of forces to convey bulk material through vertical tube

$$FC = \Delta F \frac{H}{L} \tag{1.11}$$

where

$H =$ height of lift
$L =$ pitch.

The actual conveying force will, of course, be larger than that given by Eq. (1.11) to overcome such losses such as churning in the conveyor intake.

The forced flow characteristics of bulk solids can be greatly improved by reducing both the internal friction of the material and wall friction. This is the basis of hydraulic and pneumatic conveying where the bulk solid is conveyed in the two-phase mode. For the single-phase mode, it is necessary to convey the bulk solid by means of a carrying vehicle such as a belt conveyor or bucket elevator which has low mechanical friction.

1.7.4 Block Diagram Representation

The analyses of the pressures generated in a cylinder under static storage and forced flow conditions may be represented by block diagrams as in Fig. 1.21. The driving or forcing function in each case is the specific weight γ of the bulk material and the responding variable is the pressure P_z. For the static storage case, which gives rise to the Janssen pressure distribution, the feedback arising from wall friction is negative, this being consistent with a stable system in a control sense, as per Fig. 1.21a. For the forced flow case, the feedback is positive which is consistent with an unstable system in a control sense, as per Fig. 1.21b; in this case, the pressure continues to

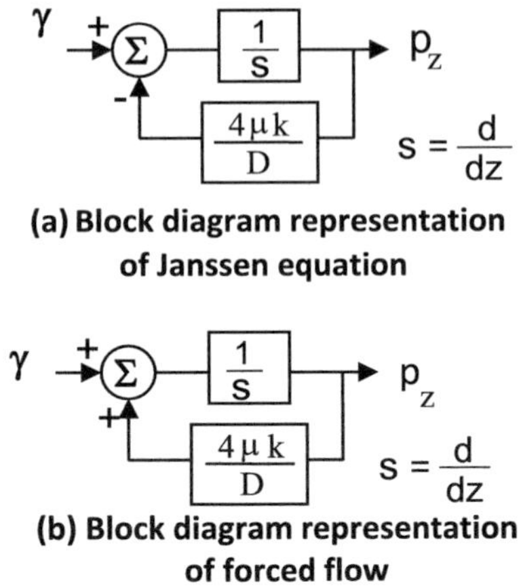

Fig. 1.21 Block diagram representation of cylinder pressures

increase with an increase in depth and does not reach a finite limiting value as occurs in the Janssen analysis.

Based on linear systems theory, the transfer functions for the above cases are given by

$$\frac{P_z(s)}{\gamma(s)} = \frac{D}{4\mu k}\left[\frac{1}{Zs \pm 1}\right] \tag{1.12}$$

where

$Z = \frac{D}{4\mu k}$ = Depth constant
$s =$ Laplace Transform Variable
$-$ sign corresponds to negative feedback
$+$ sign corresponds to positive feedback.

1.7.5 Two-Phase Forced Flow in Vertical, Cylindrical Tube

Following on the discussion in Sect. 1.7.3, the research conducted by Roberts in 1966 on force flow of grain in a vertical tube with varying amounts of air permeation is reviewed [154]. The test rig developed for that study is illustrated in Fig. 1.22. It comprised a 45 mm bore by 3.66 m long 'Perspex' tube which was fitted with a perforated piston and piston rod, the piston serving to force columns of grain up the tube. The top end of the piston rod was guided centrally in the tube. Vent holes in this top guide allowed air to be expelled from the tube when air permeation was used. The tube was supported in a tubular steel framework which may be lowered to the ground when it is necessary to empty grain from the cylinder. The loading and piston raising device comprised a small winch, which was driven through a reduction worm gear box and friction clutch by a variable speed motor. The winch cable passed from the winding drum up and over the pulley at the top of the tubular framework and was connected to the force measuring load cell mounted on the piston.

A regulated air supply was employed for tests with fluidisation of the grain column. A calibrated orifice meter permitted measurement of the air flow. The air enters the Perspex tube via the lower flange member and passes through the holes in the piston to the grain column. The exhaust air is discharged via the holes drilled in the top flange member. The pressure drop across the grain column was measured by means of a manometer as well as by a pressure transducer connected to a recorder. The testing work was carried out using millet as the granular material.

To examine the feasibility of forced flow with fluidisation, tests were performed on a range of grain columns for a number of piston speeds utilising varying amounts of air up to and slightly exceeding the amount required for fluidisation. Allowance

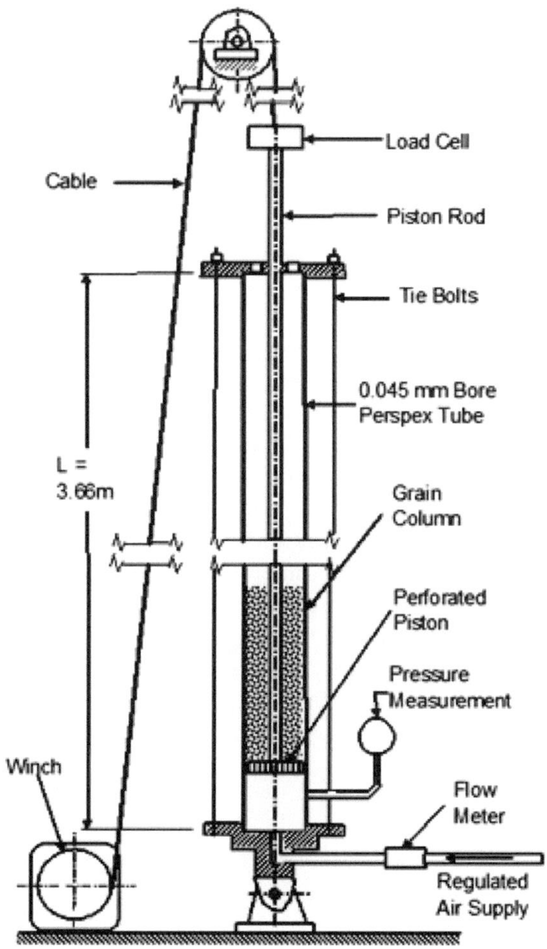

Fig. 1.22 Forced flow test rig [154]

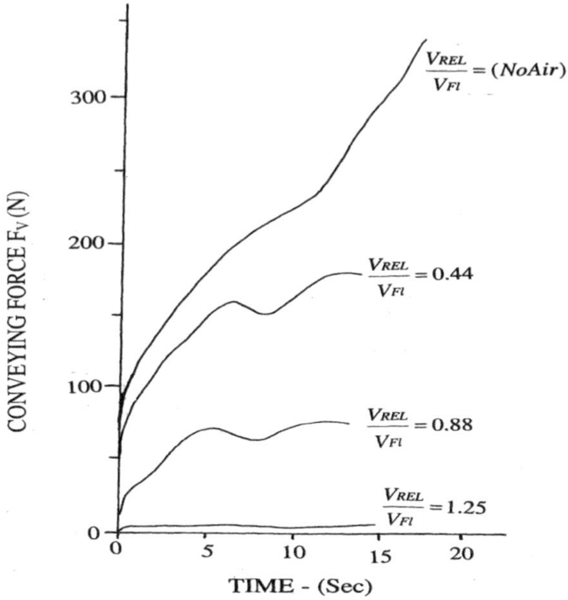

Fig. 1.23 Conveying force for vertical conveying with air permeation H = 0.394 m—Millet [154]

was made for the piston speed by specifying the air flow in terms of its velocity, V_{Rel}, relative to the piston.

At zero and low permeation levels, the conveying force increases with time as the upward velocity of the piston results in a significant increase in compaction of the grain in forced motion. The reduction in conveying force as air permeation is increased is quite substantial as indicated by a typical set of force records for a column height of H = 0.394 m, for which L/D = 8.76, as shown in Fig. 1.23.

Below fluidisation, the conveying force is both column length and time dependent, as before. As the air flow approaches that for fluidisation (that is $V_{Rel}/V_{FL} = 1.0$), the force becomes negligibly small. Once fluidisation has been obtained, it is apparent from the tests that the force to maintain motion is independent of column length. An indication of this is given in Fig. 1.24, which compares the force records of 1.22 m and 1.83 m columns. Again, extremely small conveying forces are involved. The fluctuations or cyclical nature of the force measurements are a result of the slugging action occurring once fluidisation has been exceeded.

Follow up research by Roberts and Jones based on the original paper was published in 2003 [155].

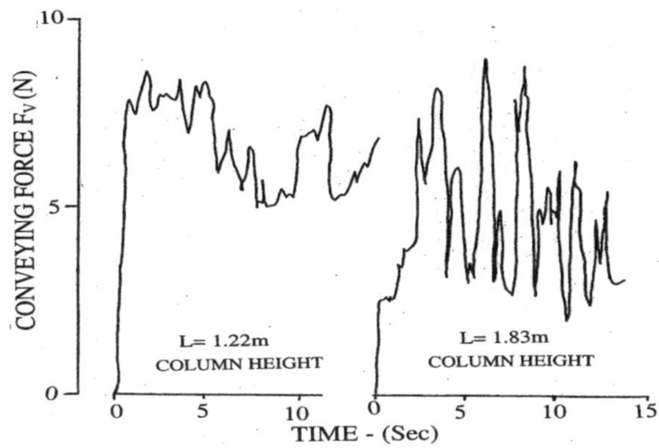

Fig. 1.24 Force flow with air permeation $\frac{V_{REL}}{V_{FL}} = 1.25$—Millet [154]

1.8 Handling Plant Design and Performance

1.8.1 Flow Properties

The foregoing material highlights some basic characteristics which distinguish the behaviour of bulk solids. There are a number of parameters, embraced by the term's *strength* and *flow properties* which are necessary to fully describe the general class of cohesive bulk solids. Cohesive bulk solids are those which gain bulk strength under the action of consolidating pressures and retain this strength when the consolidating pressures cease to act. Free flow, non-cohesive bulk solids, such as dry sand or grain are a special case; these materials gain in shear strength with an increase in consolidating pressures but lose this strength once the pressures are released. The various flow properties of relevance to handling plant design are described in Chaps. 2 and 3.

In the context of this book, handling plant comprises such equipment as bins, hoppers, silos, stockpiles, feeders, discharge equipment, transfer chutes, standpipes and flow promotion devices. For the design of such equipment, it is important that the relevant strength and flow properties of the bulk solid be determined for the conditions pertaining to the handling plant. Normally, this requires establishment of the worst likely condition to be encountered. In the case of bulk strength, this requires the 'most difficult' sample rather than the average sample to be selected. The test procedures involve the establishment of the worst likely moisture content for which the bulk strength is maximum.

While these procedures are to guarantee flow for all eventualities, it is important to also recognise that plant design will require judicious selection of parameters relevant to the particular design calculations. For instance, the highest wall friction for a hopper lining material will be necessary to determine the required opening dimension for flow; on the other hand, for the determination of wall loads, the normal pressures at the wall must be calculated using the lower bound of wall friction. This, of necessity, means that it is the kinetic rather than static coefficient that is of interest, allowing for the fact that some polishing of the hopper surface may take place during operation. In contrast, the in-plane loads in the bin wall are computed using the upper bound wall friction.

Again, wall friction, in relation to hopper half-angle for flow, becomes important when designing bins which ensure that flow instabilities, such as the silo-quaking problem, may be avoided. On other occasions, such as in the handling of weathered or stock-piled coal, which is in the unwashed state, consideration must be given to the changing properties due to the weathering process. In many cases, the change in relevant flow properties may be quite small, and even barely perceptible, but nevertheless can have a profound influence on whether a handling plant performs satisfactorily or not.

Furthermore, Bulk Solids Handling is often thought of as being 'end on' to the process when the operations of storage and transport are only considered. This is erroneous. While not diminishing the importance of storage and transportation, bulk solids technology is very much an integral part of most, if not all, industrial process operations.

1.8.2 Feedback and Quality Control—A Design Necessity

The design and commissioning of a new bulk solids handling plant or the upgrading and improvement of an existing plant is basically a four-step process as illustrated in Fig. 1.25.

Often the process embodied in the four phases is treated as an open loop system. When this occurs, there is the danger of 'drift' or errors occurring due to designs

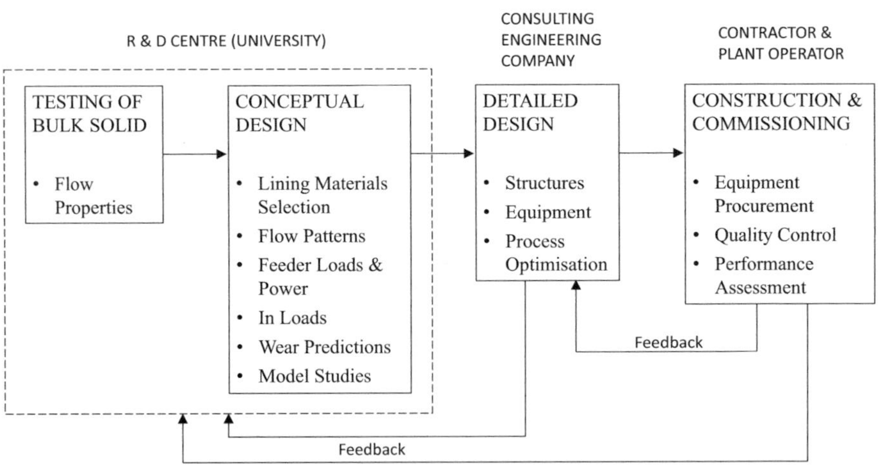

Fig. 1.25 Plant design and commissioning—need for feedback

and plant details departing from those recommended from the conceptual design and testing phases. This may seriously impair the plant performance. Sometimes small variations in such items as hopper linings where the surface roughness does not comply with that of the sample tested or where a lining is poorly fitted with projecting ledges can cause flow blockages. Departures in hopper geometries and/or in feeder loads from the conceptual design may lead to unsatisfactory performance which ultimately becomes expensive to rectify.

Inadequate or total lack of flow property data may jeopardise the plant design and performance. For these reasons, the importance of feedback in the total process is of the utmost importance for quality control and to ensure that the plant will operate to the required performance specifications. Feedback is also most important as a part of the 'learning process' for all those involved in the various phases of the project. The costs of the testing and conceptual design phases and of maintaining the necessary feedback are quite small in relation to the overall capital and running costs of the plant.

1.8.3 Problem Recognition

In solving problems in an existing bulk solids handling plant, the need to recognise the primary cause of the problem is essential. This often requires a 'trained eye' and it does not necessarily follow that those involved in the plant operation are in the best position to effect a solution. Often those involved may be too close to the problem or may not have the necessary background in solids handling theory and practice to perform a correct diagnosis. All too often, the solution to bulk solids handling problems are taken for granted and are mistakenly "180 degrees out of phase with the real, correct solution". Plant safety is of the utmost importance.

References

1. Roberts, A.W., *Bulk solids handling; Recent developments and future directions.* Bulk Solids Handling;(Germany, FR), 1991. **11**(1).
2. Roberts, A.W., *An historical overview and current developments.* Bulk Solids Handling, 2006. **26**(6).
3. Roberts, A.W. *100 Years of Janssen.* 1995.
4. Coulomb, C.A., *Eassi Sur une Application des Regles des Maximis et Minimis a Quelques Problems de Statique Relatifs de l' Architecture".* Mem. Math. Phys., 1773: p. 343–381.
5. Rankine, W., *On the Stability of Loose Earth.* Phil. Trans., 1857.
6. Reynolds, O., *On the Dilatancy of Media Composed of Rigid Particles in Contact.* Phil. Mag., 1855. **20**: p. 469–481.
7. Reynolds, O., *Experiments Showing Dilatancy, a Property of Granular Material, Possibly Connected with Gravitation.* Proc. Roy. Inst. Gr. Brit., 1887. **11**: p. 354–363.
8. Roberts, I., *The pressure of wheat stored grain.* Engineering, 1882. **34**: p. 399.
9. Roberts, I., *Determination of the vertical and lateral pressures of granular substances. Part I - Wheat and peas.* Royal Society of London. Proceedings., 1884. **36**: p. 225–240.

10. Janssen, H.A., *The Measurement of Pressures in Grain Silos.* Z. Ver. Dtsch. Ing. Beih, 1895. **35**: p. 1045–1049.
11. Janssen, H.A., *Tests on Grain Pressure Silos.* Z. Ver. Dtsch. Ing. Beih, 1895. **35**: p. 1045–1049.
12. Airy, W., *The Pressure of grain.* Proc. Inst. Civ. Eng. Paper, 1897. **3049**.
13. Prante, N., *Messungen des getreidedruckes gegen silowandungen.* Zeitschrift des Verein Deutscher Ingenieure (Germany), 1896. **40**(39): p. 1122–1125.
14. Toltz, M., *Discussion on grain pressures in deep bins.* Trans. Can. Soc. Civ. Eng, 1903. **17**: p. 641.
15. Jamieson, J.A., *Grain Pressures in Deep Bins.* Trans. Can. Soc. Civ. Eng., 1903. **17**(554–654).
16. Jamieson, J.A., *Grain Pressures in Deep Bins.* Engineering News, 1904. **LI, No.10**: p. 236–243.
17. Lufft, E., *Tests on Grain Pressure in Deep Bins at Buenos Aires, Argentina.* Eng. News, 1904. **52**: p. 531.
18. Pleissner, *Versuche zur ermittlung der boden-und seitenwanddrucke in getreide silos.* Zeitschrift des vereins deutscher ingenieure, 1906. **50**(25): p. 976–986.
19. Bovey, H.T., *Experiments on grain pressure in deep bins and strength of wooden bins.* eNG. nEWS, 1904. **52**(2): p. 32–34.
20. Ketchum, M.S., *The Design of Walls, Bins and Grain Elavator, 3d ed.* 1919: McGraw-Hill, New York.
21. Phillips, C.E.S., *The Flow of Sand.* Nature, 1910. **82**(2104): p. 487–487.
22. Shaxby, J.H., J.C. Evans, and V. Jones, *On the properties of powders. The variation of pressure with depth in columns of powders.* Transactions of the Faraday Society, 1923. **19**(July): p. 60–72.
23. Dallavalle, J.M., *Micromeritics: The Technology of Fine Particles.* 1943: Pitman Publishing Co.
24. Brown, R.L. and P.G.W. Hawksley, *Effect of Container Walls on Packing Density of Particles.* Nature, 1946. **157**(3992): p. 585–585.
25. Fowler, R.T. and J.R. Glastonbury, *The flow of granular solids through orifices.* Chemical Engineering Science, 1959. **10**(3): p. 150–156.
26. Brown, R.L. and J.C. Richards, *Exploratory Study of the Flow of Granules through Apertures.* Trans. Inst. of Chem. Engnrs, 1959. **37**(2).
27. Brown, R.L. and J.C. Richards, *Profile of Flow of Granules through Apertures.* Trans. Inst. of Chem. Engnrs, 1960. **38**(5).
28. Rose, H.F. and T.Tanaka, *Rate of Discharge of Granular Materials from Bins and Hoppers.* The Engineer (London), 1959. **208**(Oct. 23).
29. Brown, R.L., *Minimum Energy Theorem for Flow of Dry Granules Through Apertures.* Nature, 1961. **191, July 29**.
30. Hvorslev, M.J., *On the Physical Properties of Distributed Cohesive Soils.* Ingeniorvidensk. Skr, 1937. **45**.
31. Roscoe, K.H., A.N. Schofield, and C.P. Wroth, *On the Yielding of Solids.* Geotechnique, 1958. **8**: p. 22–53.
32. Johanson, J.R., *Theory of Bulk Solids Flow - A Historical Perspective.* Int. J. Bulk Solids Storage in Solis, 1987. **3**(1): p. 1–15.
33. Jenike, A.W., *Gravity Flow of Bulk Solids. Bull 108.* The Univ of Utah, Engng Exp Station, USA, 1961.
34. Jenike, A.W., *Storage and Flow of Solids. Bull. 123.* The Univ of Utah, Engng Exp Station, USA, 1964.
35. Johanson, J.R., *Stress and Velocity Fields in Gravity Flow of Bulk Solids. Bull. 116.* The Univ of Utah Engng Exp Station, USA, 1962.
36. Sokolovski, V.V., *Statics of soil media.* 1960: Butterworths Scientific Publications.
37. Prager, W. and G. Hodge, *Theory of Perfectly Plastic Solids.* New York, Dover Publications, 1968. **(Original work, 1951)**.
38. Hill, R., *The Mathematical Theory of Plasticity.* Oxford: Clarendon Press, 1983.

39. Jenike, A.W. and R.T. Shield, *On the Plastic Flow of Coulomb Solid Beyond Original Failure.* A.S.M.E., Jnl. of App. Mech, 1959.
40. Roberts, A.W. and A.H. Willis, *Performance of Grain Augers.* Proceedings, The Institution of Mechanical Engineers (UK), 1962. **197**(8).
41. Roberts, A.W., *An Investigation of Grain Vortex Motion with Relation to the Performance within Vertical Grain Augers.* Proceedings of the Institution of Mechanical Engineers, 1963. **178**(1): p. 293–307.
42. Roberts, A.W., *The Dynamics of Granular Material Flow through Curved Chutes.* Mech. and Chem. Eng. Transactions, The Instn. of Engrs. Aust., 1967. **MC3**(2).
43. Roberts, A., *An Investigation of the Gravity Flow of Noncohesive Granular Materials Through Discharge Chutes.* Transactions of A.S.M.E, Journal of Engineering for Industry, 1969. **91**: p. 373.
44. Benink E.J., *Flow and stress analysis of cohesionless bulk materials in silos related to codes.* 1989, Ph.D. thesis, Univ. of Twente, Enschede, The Netherlands.
45. Bagster, D.F., *A Study of the Caking Behaviour of Raw Sugar under Storage*, in *Proc. Int. Conf. on Bulk Materials Storage, Handling and Transportation.* 1983: Newcastle, Australia.
46. Schwedes, J., *Testers for measuring flow properties of particulate solids.* Powder Handling and Processing, 2000. **12**: p. 337–354.
47. Solids, E.W.P.o.t.M.o.P., *Standard shear testing technique for particulate solids using the Jenike shear cell 46 pp.* 1989, IChemE. p. 46.
48. Schulze, D. *A new ring shear tester for flowability and time consolidation measurements.* in *Proc. 1st International Particle Technology Forum.* 1994.
49. 3880–1991., A.S.A., *Bin Flow Properties of Coal.* Standards Association of Australia.
50. D6128–16, A., *Standard Test Method for Shear Testing of Bulk Solids Using the Jenike Shear Tester.* ASTM International, 2016.
51. 16, A.D.-. *Standard Test Method for Bulk Solids Using Schulze Ring Shear Tester.* ASTM International, 2016.
52. Johanson, J., *The johanson indicizer [tm] system vs. the jenike shear tester.* Bulk Solids Handling, 1992. **12**: p. 237–237.
53. Bell, T., B. Ennis, and R. Grygo, *Practical Evaluation of the Johanson Hang-Up Indicizer [R].* Bulk Solids Handling, 1994. **14**: p. 117–117.
54. Harder, J. and J. Schwedes, *The development of a true biaxial shear tester.* Particle & Particle Systems Characterization, 1985. **2**(1–4): p. 149–153.
55. Schwedes, J. and J. Harder. *Influence of stress history on the yield limit of bulk solids.* in *Second International Conference on Bulk Materials Storage, Handling and Transportation: 1986; Preprints of Papers.* 1986. Institution of Engineers, Australia Barton, ACT.
56. Maltby, L.P., *Investigation of the behaviour of powders under and after consolidation, Dr. Ing.* 1993, thesis Høgskolen i Telemark.
57. Maltby, L. and G. Enstad, *Uniaxial tester for quality control and flow property characterization of powders.* Bulk Solids Handling, 1993. **13**: p. 135–135.
58. Janssen, R., M. van der Kraan, and B. Scarlett, *Evaluation of the flexible wall biaxial tester for the measurement of BCR-limestone.* Preprints of the 1st European symposium process technology in pharmaceutical and nitritional sciences, 1998: p. 209–215.
59. Kwade, A., D. Schulze, and J. Schwedes, *Determination of the stress ratio in uniaxial compression tests-Part 2.* powder handling and processing, 1994. **6**(2): p. 199–204.
60. Lahlouh, E., et al. *Assessment of coal handlability.* in *World Congress on Particle Technology 3.* 1998.
61. Wiche, S.J., A.W. Roberts, and W.A. McBride, *Flowability Tester for Characterising Bulk Solids*, in *Proc. 8th International Conference on Bulk Materials Storage, Handling and Transportation, The Instn. of Engrs.* 2004: Australia, Wollongong, NSW, Australia. p. 79–83.
62. Roberts, A.W., M. Ooms, and O.J. Scott, *Influence of vibrations on the strength and boundary friction characteristics of bulk solids and the effect on bin design and performance* 1986. **6**: p. 161–169.

63. Roberts A.W., *Vibrations of powders and bulk solids.* Chapter 6 in *Handbook on Powder Science & Technology*, 1984: Van Nostrand.
64. Kollmann, T. and J. Tomas, *The Influence of Vibrations on Flow Properties of Cohesive Powders*, in *Proc. 7th International Conference on Bulk Materials Storage, Handling and Transportation.* 2001, The Instn. of Engrs.: Australia, Wollongong, NSW, Australia. p. 417–427.
65. Roberts, A.W., M. Ooms, and S.J. Wiche, *Concepts of Boundary Friction, Adhesion and Wear in Bulk Solids Handling Operations.* Intl. Jnl. of Bulk Solids Handling, 1988. **10**(2).
66. Roberts, A., L. Sollie, and S. De Silva, *The interaction of bulk solid characteristics and surface parameters in surface or boundary friction measurements.* Tribology international, 1993. **26**(5): p. 335–343.
67. Johanson, J.R. and T. Royal, *Measuring and use of wear properties for predicting life of bulk materials handling equipment.* Bulk Solids Handling, 1982. **2**(3): p. 517–523.
68. Wiche, S., S. Keys, and A. Roberts, *Abrasion wear tester for bulk solids handling applications.* Wear, 2005. **258**: p. 251–257.
69. Roberts, A.W., et al., *Controlling Consolidation Pressures, Bulk Density and Permeability in Storage Vessels for Compressible Bulk Material*, in *Proc 8th Int. Conf. on Bulk Materials, Storage, Handling and Transportation.* 2004: Australia, Wollongong. p. 424–428.
70. 4156.6–2000, A.S.A., *Coal preparation, Part 6: Determination of dust/moisture relationship for coal.* Standards Association of Australia, 2000.
71. McLean, A.G., *Flow Rates Simple Bulk Solids from Mass-Flow Bins.* 1979, The University of Wollongong, Australia.
72. Arnold, P.C. and Z.H. Gu, *The Effect of Permeability on the Flowrate of Bulk Solids from Mass-Flow Bins.* The Intl. Jnl. of Powder Handling and Processing, 1990. **2**(3): p. 229–238.
73. Gu, Z., P. Arnold, and A. McLean, *A simplified model for predicting the particle flowrate from mass flow bins.* Powder technology, 1993. **74**(2): p. 153–158.
74. Johanson, J., *In-bin blending.* Chemical Engineering Progress, 1970. **66**(6): p. 50–&.
75. Johanson, J., *Controlling flow patterns in bins by use of an insert.* Bulk Solids Handling, 1982. **2**: p. 495–498.
76. Craig, D., et al., *Utilising the impulse response concept to predict blending capabilities of silos.* Powder Handling and Processing, 1994. **6**(4): p. 373–378.
77. Kirkwood, C., A. Roberts, and D. Craig, *Silo Technology-The Anti-Segregation and Blending Characteristics of a Mass-Flow Hopper and Rotary Valve.* Powder Handling and Processing, 1999. **11**(1): p. 71–74.
78. Johanson, J.R., *Predicting rathole stability in funnel flow bins*, in *Proceedings of 3rd European Symposium: Storage and Flow of Particulate Solids, PARTEC.* 1995: Nurnberg. p. 21–23.
79. Jenike, A.W. and J.R. Johanson, *On the Theory of Bin Loads.* Jnl. of Engng for Industry, Trans. A.S.M.E., Series B, 1969. **90**(2): p. 339.
80. Jenike, A.W., J.R. Johanson, and J.W. Carson, *Bin Loads - Part 2: Concepts.* Jnl. of Engng for Industry, Trans. A.S.M.E., Series B, 1973. **95**(1).
81. Jenike, A.W., J.R. Johanson, and J.W. Carson, *Bin Loads - Part 3: Mass-Flow Bins.* Jnl. of Engng for Industry, Trans. A.S.M.E., Series B, 1973. **95**(1).
82. Jenike, A.W., J.R. Johanson, and J.W. Carson, *Bin Load - Part 4: Funnel-Flow Bins.* Jnl. of Engng for Industry, Trans. A.S.M.E., Series B, 1973. **95**(1).
83. Walker, D.M., *An Approximate Theory for Pressure and Arching in Hoppers.* Chem. Engng. Sci., 1966. **21**: p. 975.
84. Walters, J.K., *Theoretical Analysis of Stresses in Silos With Vertical Walls.* Chem. Eng. Science, 1973. **28**: p. 779.
85. Arnold, P.C., A.G. McLean, and A.W. Roberts, *Bulk Solids: Storage, Flow and Handling.* 2nd ed. 1979: The University of Newcastle Research Associates (TUNRA).
86. Theimer, O.F., *Failures of Reinforced Concrete Grain Silos.* Journal of Engng. for Industry, trans. ASME, Series B, 1960. **91**(2): p. 460.
87. Jenike, A.W., *Load assumptions and distributions in silo design.* 1977: Jenike & Johanson Incorporated.

88. Ooms, M. and A. Roberts, *The reduction and control of flow pressures in cracked grain silos.* Bulk Solids Handling, 1985. **5**(5): p. 1009–1016.
89. Roberts, A.W. *Some aspects of grain silo wall pressure research: influence of moisture content on loads generated and control of pressures in tall multi-outlet silos.* in *Proceedings of the Powder and Bulk Solids Conference, Rosemont, Illinois.* 1988.
90. AS3774–1990, A.S., *Loads on Bulk Solids Containers.* Standards Association of Australia, 1990.
91. EN 1091–4, *Eurocode 1 - Action on Structures. Part 4: Silos and tanks.* European Committee for Standardization, Central Secretariat, Brussels, Belgium, 2005.
92. Rombach, G.A. and J. Eibl. *Numerical simulation of filling and discharging processes in silos.* in *3rd International Conference on Bulk Materials.* 1989.
93. Ooi, J. and J. Rotter, *Elastic and plastic predictions of the storing pressures in conical hoppers.* Transactions of the Institution of Engineers, Australia. Mechanical engineering, 1989. **14**(3): p. 165–169.
94. Wu, Y.H., *Static and dynamic analysis of the flow of bulk materials through silos.* 1990.
95. Ruckenbrod, C. and J. Eibl. *Silo loading.* in *Proc Fourth World Congress of Chemical Engineering, Karlsruhe, Germany.* 1991.
96. Roberts, A.W., *Shock loads in silos--the" silo quaking" problem.* Bulk Solids Handling(Switzerland), 1996. **16**(1): p. 59–70.
97. Roberts, A.W. and C. Wensrich, *Flow dynamics or 'quaking' in gravity discharge from silos.* Chemical Engineering Science, 2002. **57**: p. 295–305.
98. Wensrich, C.M., *Analytical and numerical modelling of quaking in tall silos.* 2002, University of Newcastle.
99. Wensrich, C., *Experimental behaviour of quaking in tall silos.* Powder Technology, 2002. **127**(1): p. 87–94.
100. Roberts, A. *Review of the "silo quaking" problems in bins of various geometrical shapes and flow patterns.* 2003.
101. Tejchman, J. and G. Gudehus, *Silo-music and silo-quake experiments and a numerical Cosserat approach.* Powder Technology, 1993. **76**(2): p. 201–212.
102. Gudehus, G. and J. Tejchman, *Silo Music and Silo Quake.* Forschung und Praxis Tagung, Karlsruhe University, 1992: p. 103–110.
103. Smid, J. and J. Novosad. *Pressure distribution under heaped bulk solids.* in *Powtech Conference.* 1981.
104. Roberts, A.W. and L.H. Teo, *Design Considerations for Maximum Reclaim Capacity of Conical Stockpiles.* Int. Journal of Bulk Solids Handling, 1990. **10**(1).
105. Li, E. and D. Bagster, *An idealised three-dimensional model of heaped granular materials.* Powder technology, 1993. **74**(3): p. 271–278.
106. Bagster, D. and E. Li. *Further developments in the 3-D microscopic modelling of a heap of granular material.* in *NATIONAL CONFERENCE PUBLICATION-INSTITUTION OF ENGINEERS AUSTRALIA NCP.* 1995. Institution of Engineers, Australia.
107. Baxter, J., U. Tüzün, and D.M. Heyes, *Modelling of granular interactions in the formation of stockpiles*, in *Proc Chemeca 97.* 1997: Rotorua, New Zealand. p. 245–256.
108. Wittmer, J., et al., *An explanation for the central stress minimum in sand piles.* Nature, 1996. **382**(6589): p. 336–338.
109. Wittmer, J., M. Cates, and P. Claudin, *Stress propagation and arching in static sandpiles.* Journal de Physique I, 1997. **7**(1): p. 39–80.
110. Watson, A., *Searching for the sand-pile pressure dip.* Science, 1996. **273**(5275): p. 579–580.
111. McBride, W.A., *Mechanics of Bulk Solids Stockpiles.* 2001, The University of Newcastle, Australia.
112. Jeong, H.Y., *Numerical Simulations of Stresses Under Stockpiled Mass Over Ground With or Without Loadout Tunnel.* 2005, The University of Western Ontario, Canada.
113. Roberts, A.W., *Characterisation for Hopper and Stockpile Design*, in *Characterisation of Bulk Solids.* 2005. p. 85–131.

114. Roberts, A., M. Ooms, and K. Manjunath, *Feeder load and power requirements in the controlled gravity flow of bulk solids from mass-flow bins.* Trans. IE Aust., Mechanical Engineering, Vol. ME9, 1984(1): p. 49–61.
115. Manjunath, K.S. and A.W. Roberts, *Wall Pressure-Feeder Load Interactions in Mass-Flow Hopper/Feeder Combinations -Part 1.* Bulk Solids Handling, 1986. **6**(4): p. 769–775.
116. Schulze, D. and J. Schwedes, *Bulk solids flow in the hopper/feeder interface.* Powder Handling and Processing, 1993. **5**: p. 341–341.
117. Roberts, A.W., *Overview of feeder design focusing on belt and apron feeders.* 2001. **21**: p. 13–24.
118. Roberts, A., *Predicting the Performance of Screw Conveyors*, in *In: International Conference on Bulk Materials Storage, Handling and Transportation (5th : 1995 : University of Newcastle, N.S.W.). 5th International Conference on Bulk Materials Storage, Handling and Transportation: Proceedings. Barton, ACT: Institution of Engineers, Australia, 1995: 507–515.* 1995.
119. Roberts, A.W., *Predicting the Volumetric and Torque Characteristics of Screw Feeders.* Bulk Solids Handling, 1996. **16**(2): p. 233–244.
120. Wiche, S. and B. McBride, *Investigations into Smooth Discharge Screw Feeders*, in *In: National Conference on Bulk Materials Handling (1996 : Melbourne, Vic.). 1996 National Conference on Bulk Materials Handling: Preprints of Papers. Barton, ACT: Institution of Engineers, Australia, 1996: 47–51.* 1996.
121. Roberts, A.W., M. Ooms, and O.J. Scott, *Performance Characteristics of a Ross Type Chain Feeder*, in *10th Anniversary Powder and Bulk Solids Conference.* 1985: Chicago, USA.
122. Roberts, A.W., M. Ooms, and K.S. Manjunath, *Performance of Dump Hopper and Apron Feeder using Inserts to Control Feeder Loads.* Journal of Bulk Solids Storage in Silos, 1990.
123. Roberts, A.W., *Recent developments in feeder design and performance.* Handbook of Powder Technology, 2001. **10**: p. 211–223.
124. Roberts, A. and M. Ooms, *Performance Characteristics of Reciprocating Plate Feeders*, in *In: International Conference on Bulk Materials Storage, Handling and Transportation (3rd : 1989 : Newcastle, N.S.W.). Third International Conference on Bulk Materials, Storage, Handling and Transportation: Preprints of Papers. Barton, ACT: Institution of Engineers, Australia, 1989: 369–377.* 1989.
125. Roberts, A.W., A. Harrison, and S. Wiche, *Performance Characteristics of Plough Feeders*, in *15th Powder and Bulk Solids Conference.* 1990: Chicago. U.S.A.
126. Roberts, A.W., *Mechanics of Shear-Gate Type Plate Feeder*, in *17th Powder and Bulk Solids Conference.* 1992: Chicago, U.S.A. p. 287–302.
127. Roberts, A.W. and B. McBride, *Some Aspects of Tube Feeder Performance*, in *19th Powder and Bulk Solids, Conference.* 1994: Chicago, U.S.A. p. 157–170.
128. Manjunath, K.S. and A.W. Roberts, *Investigation into the Performance of Tube Type Feeders*, in *Proc. 4th International Conference for Conveying and Handling of Bulk Solids, (CHOPS).* 2003: Budapest, Hungary. p. 14.25–14.30.
129. Roberts, A.W., *Performance of orbiting screw reclaimers.* Bulk Solids Handling, 2006. **26**: p. 24–31.
130. Deresiewicz, H., *Mechanics of Granular Matter**This study was supported by the Office of Naval Research under contract Nonr-266(09) with Columbia University*, in *Advances in Applied Mechanics*, H.L. Dryden and T. von Kármán, Editors. 1958, Elsevier. p. 233–306.
131. Trollope, D.H., *Mechanics of Discontinua of Clastic Mechanics.* Department of Engineering, Research Bulletin No.2, University College of Townsville (Now James Cook University), 1967.
132. Cleary, P.W., *Ball Mill Simulation Using DEM: Charge Motion, Power Draw, Segregation and Wear*, in *Proc. 8th International Conference on Bulk Materials Storage, Handling and Transportation*, T.I.o.E. Australia, Editor. 1998: Wollongong, NSW, Australia. p. 381–387.
133. Tüzün, U., *Computer Simulations of Bulk Solids; a Critical Review of Discrete Element Techniques*, in *Proc. 8th International Conference on Bulk Materials Storage, Handling and Transportation,*, T.I.o.E. Australia, Editor. 1998: Wollongong, NSW, Australia. p. 35–39.

134. Roberts, A.W., *Dynamic Modelling of Bulk Solids Storage, Discharge and Handling Systems.* Bulk Solids Handling, 2000. **20**(2): p. 183–189.
135. Roberts, A.W., *Predicting the Performance of Open Front Inclined Apron Feeders*, in *Proc. From Powder to Bulk Conference*, T.I.o.M. Engineers, Editor. 2000: London. p. 443–452.
136. Chiarella, C., W. Charlton, and A.W. Roberts, *Optimum chute profiles in gravity flow of granular materials: a discrete segment solution method.* 1975.
137. Chiarella, C. and W. Charlton, *Chute profile for minimum transit time in gravity flow of granular material.* Journal of Agricultural Engineering Research, 1972. **17**(2): p. 149–153.
138. Savage, S.B., *Gravity flow of cohesionless granular materials in chutes and channels.* Journal of Fluid Mechanics, 1979. **92**(1): p. 53–96.
139. Roberts, A. and O. Scott, *Flow of bulk solids through transfer chutes of variable geometry and profile.* Bulk Solids Handling, 1981. **1**(4): p. 715–727.
140. Roberts, A.W., *Chute Design Considerations for Feeding and Transfer*, in *Proc. Beltcon 11, International Materials Handling Conference.* 2001: Johannesburg, South Africa.
141. Hill, J.M. and X.M. Zheng, *Dilatant Double Shearing Theory Applied to Granular Chute Flow;.* Acta Mechanica 118, 1996. **1-4**: p. 97–108.
142. Roberts, A.W., *Chute Performance and Design for Rapid Flow Conditions.* Chemical Engineering & Technology, 2003. **26**(2): p. 163–170.
143. Roberts, A.W., C.M. Wensrich, and R.E. Stratton, *Influence of Flow Properties and Geometry on Chute Design and Performance*, in *Proc. Particle Systems Analysis Conference.* 2005: Stratford-Upon Avon, England.
144. Wensrich, C.W., *Evolutionary Optimisation in Chute Design.* Powder Technology, 2003. **138**: p. 118–123.
145. Roberts, A.W., et al., *Dust Reduction in Delivery of Particulate Commodities*, U.P.O.a.C.P.O. Provisional Patent filed with Australian Patent Office, Editor. 2006.
146. Roberts, A.W., *Mechanical Conveying*, in *Proc. From Powder to Bulk Conference*, T.I.o.M. Engineers, Editor. 2000: London. p. 377–410.
147. Roberts, A.W., *Bulk Solid and Conveyor Belt Interactions for Efficient Transportation Without Spillage.* Bulk Solids Handling, 1998. **18**(1): p. 49–57.
148. Wheeler, C.A., *Analysis of Main Resistances of Belt Conveyors; PhD Thesis.* 2003, The University of Newcastle, Australia.
149. Wheeler, C.A., A.W. Roberts, and M.G. Jones, *Calculating the Flexure Resistance of Bulk Solids Transported on Belt Conveyors.* Particle & Particle Systems Characterization, 2004. **21**(4): p. 340–347.
150. Roberts, A.W., *Design and Performance Criteria for Screw Conveyors in Bulk Solids Operations.* Bulk Solids Handling, 2002. **21**(6): p. 436–444.
151. Klinzing, G., *Pneumatic conveying: transport solutions, pitfalls, and measurements.* Handbook of Powder Technology, 2001. **10**: p. 291–301.
152. Wypych, P., *Dilute-phase pneumatic conveying problems and solutions.* Handbook of Powder Technology, 2001. **10**: p. 303–318.
153. Jones, M., *Characterisation for Pneumatic Conveyor Design*, in *Characterisation of Bulk Solids.* 2005. p. 151–180.
154. Roberts, A.W., *Developments in Materials Handling Research at Wollongong University College*, in *2nd Annual Conference, The Illawarra Group, The Institution of Engineers, Australia.* 1966: Wollongong, Australia.
155. Roberts, A.W. and M. Jones, *Analysis of Forced Flow of Granular Materials in Vertical Pipes without and with Air Permeation.* Particulate Science and Technology—PARTICULATE SCI TECHNOLOGY, 2003. **21**: p. 25–44.

Chapter 2
Storage and Handling Cohesive Bulk Solids—Basics of Bin Design

2.1 Introductory Remarks

Cohesive bulk solids may be stored in open or enclosed storage bins, open or enclosed stockpiles or in bulk containers. The type of storage system depends on the characteristics of the bulk solid, capacity and the available space. Bulk solids, such as food products and others that degrade with time, need special storage systems. Most cohesive bulk materials increase in bulk strength during long periods of undisturbed storage. The storage properties of these materials need to be assessed and taken into account in the overall design. It is widely accepted that the modern approach to storage bin and related equipment design has emanated from the pioneering work of Dr. Andrew W. Jenike and his co-worker Dr. J. R. Johanson in the 1960's at the University of Utah. Of particular note are the three University publications, Bulletins 108 and 123 of Jenike [1, 2] and Bulletin 116 of Johanson [3].

The classical, fundamental approach to bin design is focused on axi-symmetric and plane symmetric bins which are widely used in industrial applications. This explains why the design of bins of symmetrical geometrical shapes can be undertaken with the focus on the 2-dimensional planar stress field analysis of the consolidation stresses during storage and flow. For plane-flow bins with wedge-shaped hoppers, transverse planes perpendicular to the centreline plane of symmetry are the basis for the analysis. For axi-symmetric bins with conical hoppers, the transverse, diametrical or meridian planes passing through the symmetrical centreline are the specific planes for the analysis. In this latter case, the stresses in the third or circumferential direction also require consideration. All other cases, particularly those involving eccentric loadings, require particular analytical procedures as are illustrated through case study design examples presented in later chapters of this book.

It is noted that the application of Mohr Circle stress field analysis is widely adopted throughout this book.

A. W. Roberts, *The Theory and Practice of Bulk Materials Handling*,
https://doi.org/10.1007/978-3-032-19702-3_2

2.2 Storage and Handling System Design Philosophy

A philosophy for storage bin or stockpile and associated handling plant design is basically a four-step process:

(i) Determine the strength and flow properties of the bulk solids for the worst likely conditions expected to occur in practice.
(ii) Determine the bin or stockpile configuration to give the required storage capacity, flow patterns and discharge rates with acceptable characteristics and to ensure that discharge is reliable and predictable.
(iii) Consider a design to suit available space and estimate loadings on hopper and bin walls and on feeders under operating conditions.
(iv) Design and detailing of the bin or stockpile and associated handling equipment.

It is important that all storage and handling plant design problems follow the above procedures. When investigating the required geometry of storage bins and stockpiles it should be assumed that gravity will provide reliable flow from storage. Not until it has been demonstrated that the gravity forces available are insufficient to provide reliable flow should more sophisticated reclaim methods be investigated. Unless otherwise indicated throughout this chapter, it is assumed that the bins and hoppers are symmetrical in shape.

2.3 Symmetrical Storage Bin Geometries and Flow Patterns

2.3.1 Principal Modes of Flow and Bin Geometrical Shapes

Following the definitions of Jenike, there are two basic modes of flow, mass-flow and funnel-flow. These are illustrated in Fig. 2.1. In mass-flow the bulk material is in motion at substantially every point in the bin whenever material is drawn from the outlet. The material flows along the walls with the bin and hopper (that is, the tapered section of the bin) forming the flow channel. Mass-flow is the ideal flow pattern and occurs when the hopper walls are sufficiently steep and smooth and there are no abrupt transitions or inflowing valleys. Mass-flow guarantees complete discharge of the bin contents at predictable flow rates. It is a 'first-in, first-out' flow pattern, that is, the bulk solids that are first loaded into the empty bin is always the first to discharge. When properly designed, a mass-flow bin can re-mix the bulk solid during discharge should the solid become segregated upon filling of the bin.

Mass-flow is particularly suitable for fine powders that are prone to aeration. When such powders are loaded into a bin, sufficient time should be allowed to elapse to permit deaeration to take place before the powder is discharged.

Funnel-flow occurs when the hopper is not sufficiently steeply sloped, and the walls of the hopper are not smooth enough for mass-flow to occur. Funnel-flow bins are characterised either by their squat hopper proportions or their flat bottoms. In

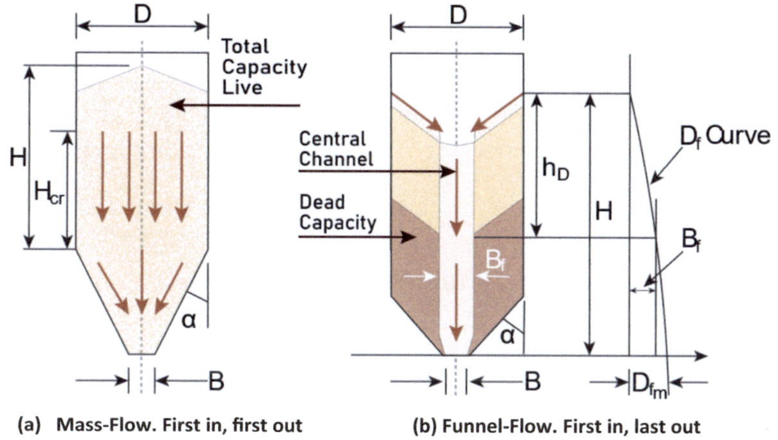

Fig. 2.1 Modes of flow

this case, the bulk solid sloughs off the top surface and falls through the vertical flow channel that forms above the opening. Flow is generally erratic and gives rise to segregation problems. Flow will continue until the level of the bulk solid in the bin drops an amount h_D equal to the draw-down. At this level, the bulk strength of the contained material is sufficient to sustain a stable hollow pipe or 'rathole' of nominal diameter B_f as illustrated in Fig. 2.1b. Once the level defined by h_D is reached, there is no further flow and the material below this level represents 'dead' storage. This is a major disadvantage of funnel-flow. For complete discharge, the bin opening needs to be at least equal to the critical rathole dimension determined at the bottom of the bin corresponding to the bulk strength at this level. However, for many cohesive bulk solids and for the normal consolidation heads occurring in practice, particularly in large scale operations such as in mining and mineral production, ratholes measuring several metres in cross-section are known to occur, as well as being predictable at the design stage.

Funnel-flow has the advantage of providing wear protection of the bin walls, since the material flows against stationary material. However, it is a 'first-in, last-out' flow pattern which is unsatisfactory for bulk solids that degrade with time. It is also unsatisfactory for fine bulk solids of low permeability. Such materials may aerate during discharge through the flow channel and this can give rise to flooding problems or uncontrolled discharge.

Explanatory Comments regarding Funnel-Flow:

- **Terminology**: Jenike adopted the name "pipe" and the associated flow as "piping" to describe the hollow channel discharge flow. However, the term "rathole" is now widely accepted, as is the case in this present book. The term arose from the excellent training film entitled "The Million Dollar Rathole" produced in the early 1970's by the US Steel Research Centre, in Monroeville, Pennsylvania. The purpose of the film was to highlight the high cost due to production losses when bulk solids failed to flow as a result of blockage problems.

- While the rathole in Fig. 2.1b is shown with, nominally, vertical walls, the walls are generally steeply sloped as is explained in later chapters of this book. In effect, the rathole behaves as a pseudo mass-flow hopper formed within the store bulk-solid.

2.3.2 Symmetrical Mass-Flow Bins and Definitions

Symmetrical Mass-Flow bins are of two principal geometric shapes, axi-symmetric with conical hoppers, or plane-flow with wedge-shaped hoppers, as illustrated in Fig. 2.2. The 'bin' refers to the overall container, while the hopper is the lower tapered section of the bin. The upper or parallel section of the bin, whether circular, rectangular or square in cross-section, is referred to as the 'cylinder'. The slope of the hopper is defined by the 'hopper half-angle, α', which is the angle of the hopper wall with respect to the vertical. The opening dimension is denoted by B, the diameter or width by D and, in the case of plane-flow, the length by L. The subscripts 'c' and 'p' refer to 'conical' and 'plane-flow' respectively. For plane-flow, L should be greater than $3 \times B_p$. The height of the cylinder is denoted by 'H', and the aspect ratio is expressed by H/D.

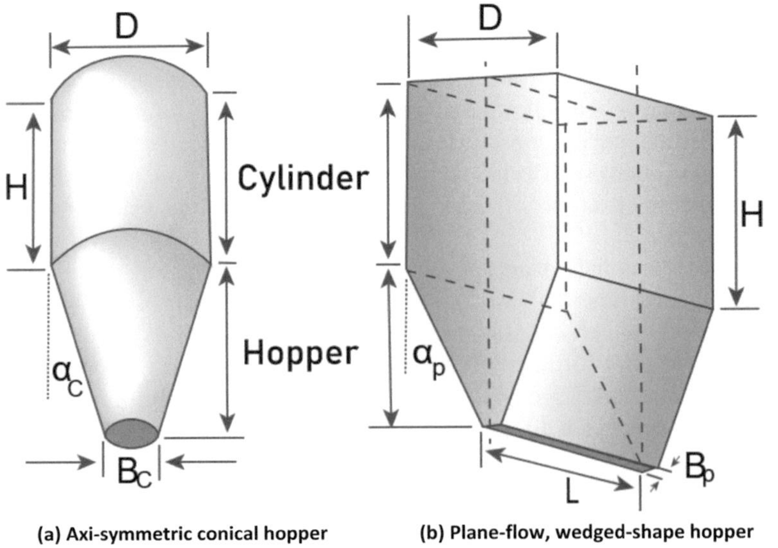

Fig. 2.2 Symmetrical mass-flow bin geometrical shapes

For a given bulk solid, the discharge through the axi-symmetric, conical hopper involves two-dimensional flow convergence, whereas for the plane-flow wedged-shaped hopper the flow convergence is one-dimensional. The net result is that, typically, the hopper half-angle α_c for the conical hopper is generally 8°–10° smaller than the corresponding half-angle α_p for the plane-flow hopper, while the required outlet diameter B_c of the conical hopper is approximately twice the opening slot width B_p of the plane-flow hopper. Hence the plane-flow, wedged-shaped hopper is more efficient for storage and discharge flow than the corresponding conical hopper. However, there is need to employ a feeding device, such as a belt or apron feeder, to control and feed the outflow along the slot to the discharge end of the hopper. This requires particular attention to be given to the hopper and feeder interface to achieve the desired uniform draw-down of bulk solids along the hopper.

The terms '*silo*' and '*bunker*' are sometimes used as an alternative to '*bin*'. Generally, a *silo* is a tall bin having an aspect ratio greater than that commonly associated with grain storage, a *bunker* typically refers to long, plane flow bins with low aspect ratios, often consisting solely of a hopper.

2.3.3 Variations of Mass-Flow Bins

Typical variations of mass-flow bin designs are illustrated in Fig. 2.3. The transition hopper of Fig. 2.3a combines the advantages of plane-flow and conical hopper geometries to achieve a more acceptable opening hopper slot length L that would otherwise be necessary under plane-flow alone. Figure 2.3b depicts a chisel-shaped hopper to achieve plane-flow in a cylindrical bin. Figure 2.3c illustrates the pyramidal-shaped hopper which approximates axi-symmetric flow from a bin with a square cross-section. While simple to manufacture, such hoppers have the potential for build-up of bulk solids to occur in the corners of the in-flowing valleys, particularly when the bulk solids possess cohesive properties. The build-up problem may be overcome by radiusing the corners of the hopper or by fitting converging shaped gusset plates in the valleys as illustrated. The design of pyramidal hoppers is based on axi-symmetric flow with the hopper half-angle being applied to the valleys. Hence, the slopes of the side walls will be steeper than that required for the equivalent conical hoppers.

2.3.4 Funnel-Flow Bins

Typical Funnel-Flow bins are shown in Fig. 2.4. Such bins may have flat or tapered bottoms, but in the case of the latter, the hopper half-angles are larger than for mass-flow bins. Since funnel flow bins are prone to 'ratholing', they often require the use of flow promotion aids to initiate and maintain flow. In the case of the funnel-flow bin with the long-slotted opening of Fig. 2.4b, the width of slot B should be large enough to prevent a stable arch from forming across the opening.

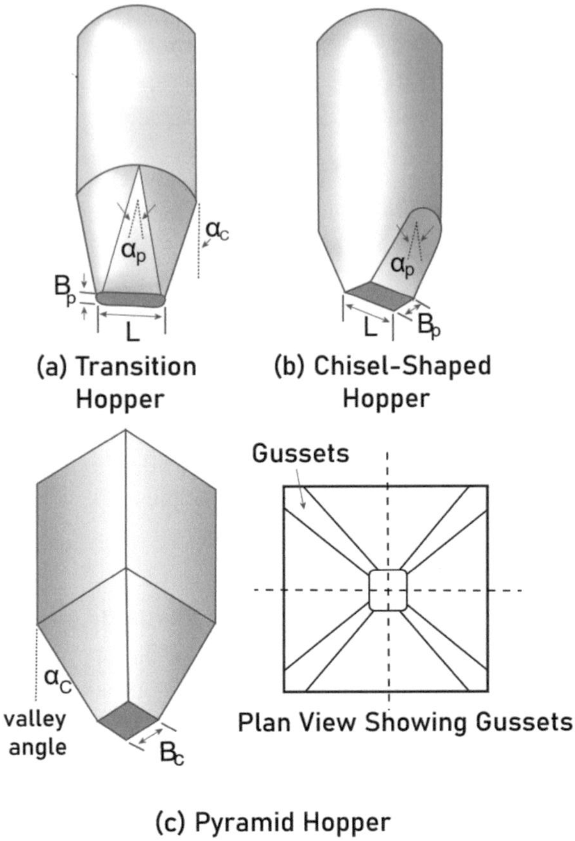

Fig. 2.3 Typical variations in mass-flow bins

2.3.5 Expanded-Flow Bins

Where large quantities of a bulk solid are to be stored, the expanded-flow bin, as illustrated in Fig. 2.5 offers considerable advantages. The bin combines the storage efficiency of the funnel-flow bin with the reliable discharge characteristics of the mass-flow hopper. It is necessary for the mass-flow hopper to have a diameter at least equal to the critical pipe or rathole dimensions D_f at the transition with the funnel flow section of the bin; this ensures that the flow of bulk solid from the funnel-flow upper section of the bin can be fully expanded into the mass-flow hopper.

For a given storage volume, the expanded-flow bin permits head heights, to be limited, as well as providing wear protection for the main upper section of the bin. Wear is confined mainly to the mass-flow hopper outlet when flow along the walls is necessary to obtain reliable discharge. The expanded-flow bin concept may also be used to advantage in the case of bins or bunkers with multiple outlets.

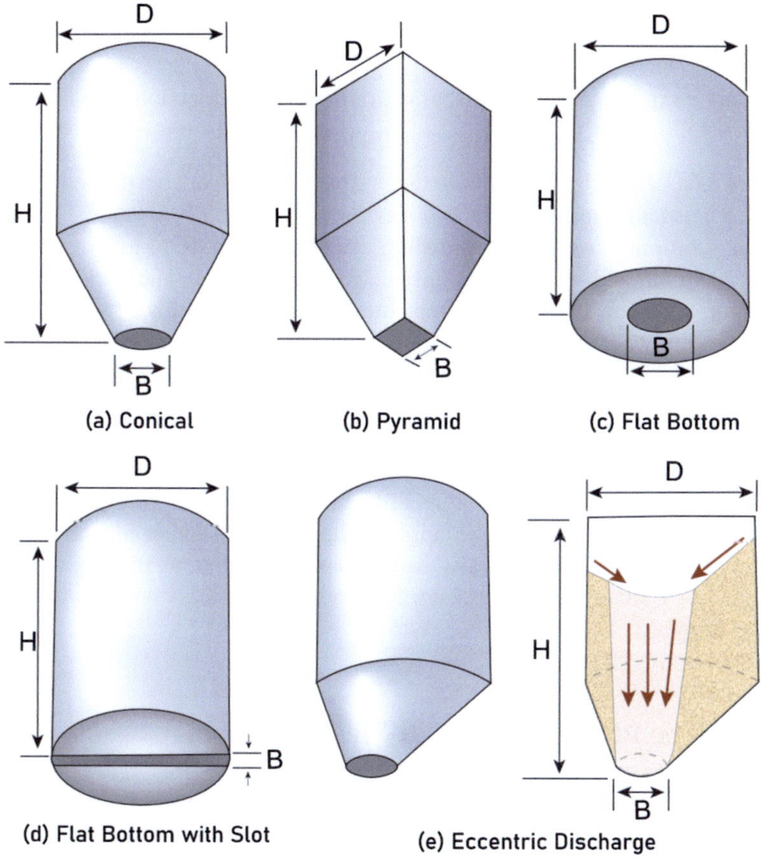

Fig. 2.4 Typical funnel-flow bins

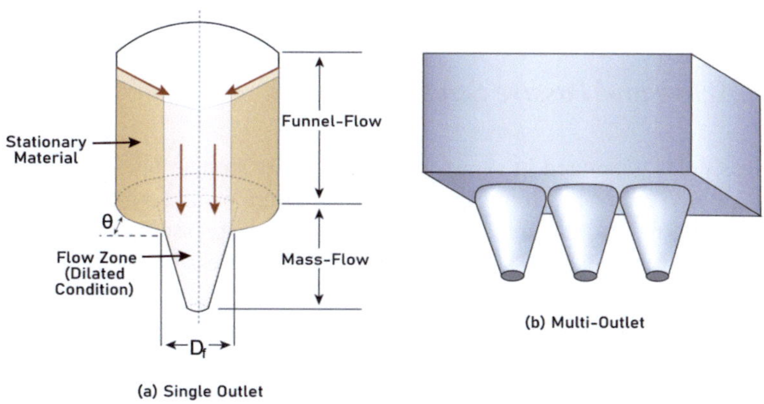

Fig. 2.5 Expanded-flow bins

2.3.6 Storage Bin Examples

The variations in bin geometry discussed in the previous sections are illustrated in Fig. 2.6 as follows:

- Figure 2.6a Large Mass-Flow lime storage bin with hopper constructed of stainless steel. The smaller mass-flow day-bin is also shown.
- Figure 2.6b Funnel-Flow train loading coal bin with a flood-loading, telescopic chute. The bin and hopper are constructed of mild steel. The hopper half-angle α_c is too large for mass-flow to be achieved. Since the bin is employed for train loading with wagons in continuous motion, the bin can operate more as a chute with the coal in continuous outflow.
- Figure 2.6c Well designed mass-flow train loading bin with a variable geometry, mass-flow hopper. The lower hopper sections are constructed of stainless steel and the upper hopper section is a mild steel. In this way the required storage capacity of the bin is achieved by judicious application of the coal flow properties. The overall advantage of limiting the height of the bin can be achieved with this design.
- Figure 2.6d Expanded-flow train loading bin for continuous loading of iron ore.
- Figure 2.6e Expanded-flow coal storage bin with twin mass flow hoppers loading belt conveyors.

2.4 Bulk Solid Stress Fields in Mass-Flow Bins

For bin design, knowledge of the bulk strength, internal friction and hopper wall friction characteristics is of importance. These bulk solid parameters must be related to the desired flow modes developed in the particular flow channels formed within hoppers and associated handling plant such as feeders and chutes.

2.4.1 Active and Passive Stress Fields

The stress fields generated in symmetrical mass-flow storage bins are indicated in in Fig. 2.7. The major and minor principal stresses are σ_1 and σ_2 respectively. The stress σ_1 is referred to as the major consolidating stress or pressure which governs the bulk strength of the bulk solid. There are two primary stress fields namely "active" or peaked, and "passive" or arched as illustrated in Fig. 2.7. When the bin is filled from the empty condition, an active stress field is developed as shown in Fig. 2.7a. In this case the major stress acts primarily in the vertical direction curving away slightly when the stress direction lines intersect the bin walls. For this stress field to be fully formed, it is assumed that there is no downward settlement of the bulk solid in the hopper. That is the bulk solid is highly incompressible and the gate preventing outflow is very stiff.

(a) Mass-flow lime storage bin and day bin

(b) Funnel-flow coal bin

(c) Variable geometry mass-flow coal bin

(d) Expanded-flow iron ore bin

(e) Expanded-flow coal bin

Fig. 2.6 Storage bin examples

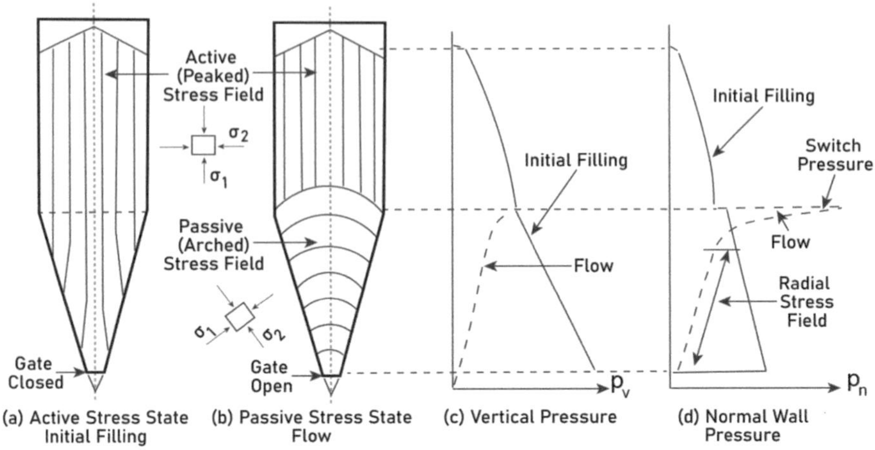

Fig. 2.7 Stress fields for initial filling and flow in mass-flow bins

Once flow has been initiated, the vertical support provided by the gate is removed, with the downward, converging flow of the bulk solid in the hopper resulting in a substantial component of the vertical load being transferred to the hopper walls. As a result, the stress field in the hopper changes from active to passive (arched) as illustrated in Fig. 2.7b. During the transient phase of the discharge flow, the switching from active to passive commences at the hopper outlet and travels up the hopper becoming locked in at the cylinder/hopper transition with the discharge continuing at a constant rate. The wall loads for the active and passive stress fields are illustrated in Fig. 2.7c, d, respectively. It needs to be noted that the passive stress field is quite stable and is retained even if the discharge is stopped before the bin is empty. The only way the active stress field can be re- established is for the bin to be completely emptied and again re-filled.

2.4.2 Combined Stress Fields

Following on from the foregoing discussion concerning the two principal stress fields active and passive, it is important that the influence of the combined stress fields in the hopper, as depicted in Fig. 2.8, be considered.

Combined stress fields arise under the following conditions:

- The upward motion of the switch stress during the transient phase of discharge flow initiation as described in (a) above.
- The bin discharges leaving a level of bulk solids in the hopper which retains the passive, arched stress field which acts as a cushion for the active stress field that is established on re-filling.

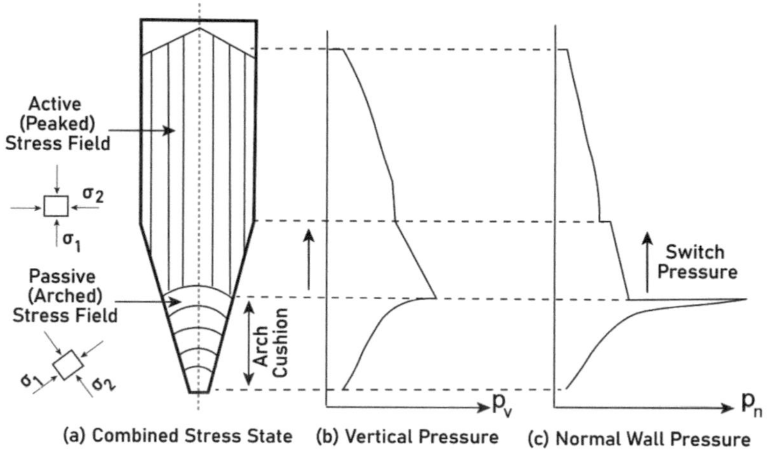

Fig. 2.8 Combined stress fields during filling, load settlement and flow in mass-flow bins

- When the bin is filled from the empty condition the settlement of the bulk solids causes the passive stress field to be partially or wholly established as illustrated in Fig. 2.8.

The stress fields have a major influence on both the magnitude and distribution of the pressures acting on the bin walls and the magnitude of the loads acting on discharge gates and feeders.

2.5 Consolidation Stress and Yield Functions

2.5.1 Yield Loci Defining Flow of Bulk Solid

When a bulk solid flows, it does so by internal shear or yielding. Within the mass of the solid itself, there is internal shear, whereas at the boundary wall of a hopper or chute, there is boundary shear. Consider an element of bulk solid flowing by gravity in a symmetrical hopper as depicted Fig. 2.9. Since an arched stress field exists, the major and minor principal stresses σ_1 and σ_2, respectively, are as shown.

The relationship between the normal and shear stresses during flow is defined by the Yield Locus **YL**, or more specifically the Instantaneous Yield Locus **IYL**, which is tangential to the two Mohr semi-circles as indicated. The major Mohr semi-circle is defined by the principal stresses or pressures σ_1 and σ_2, while the Mohr semi-circle through the origin tangential to the Yield Locus defines the unconfined yield strength σ_c. The extrapolation of the Yield Locus to the shear stress axis allows the cohesive stress τ_o to be determined; τ_o is the shear stress at zero normal stress.

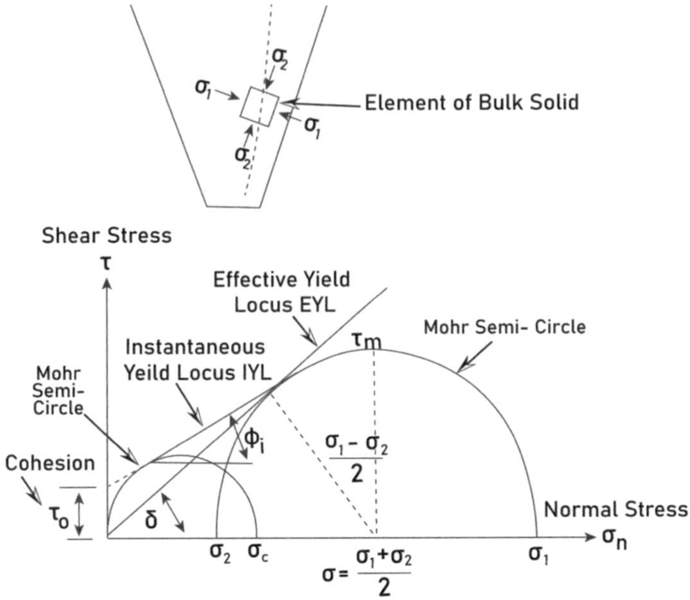

Fig. 2.9 Yield loci for consolidated bulk solid

The straight line through the origin and tangential to the major Mohr semi-circle is the Effective Yield Locus **EYL**. The angle δ defining the slope of the EYL is called the Effective Angle of Internal Friction. For a free flowing non cohesive bulk solid such as dry sand or grain, both τ_o and σ_c are zero. Hence the internal friction angles δ and φ_i are the same. For a cohesive bulk solid, the EYL may be thought of as the Yield Locus for an equivalent free flowing bulk solid having similar properties to that of the actual bulk solid. In bulk solids analysis, the effective angle of internal friction δ is a more convenient parameter to use than is the cohesive stress τ_o. The angle δ defines the major and minor principal stresses σ_1 and σ_2 as follows:

$$\sin\delta = \frac{\sigma_1 - \sigma_2}{\sigma_1 + \sigma_2}, \text{ for which } \frac{\sigma_2}{\sigma_1} = \frac{1 - \sin\delta}{1 + \sin\delta} \tag{2.1}$$

The mean stress σ is given by Eq. (2.2).

$$\sigma = \frac{\sigma_1 + \sigma_2}{2} = \frac{1}{1 + \sin\delta} \tag{2.2}$$

The maximum shear stress τ_m, which corresponds to the mean stress, is given by

$$\tau_m = \frac{\sigma_1 - \sigma_2}{2} \tag{2.3}$$

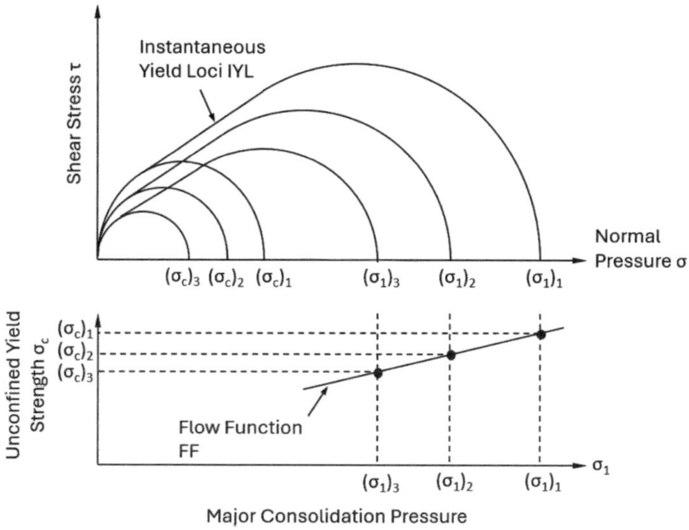

Fig. 2.10 Instantaneous yield loci and flow functions

2.5.2 *Instantaneous Yield Loci and Flow Function*

Figure 2.10 shows the instantaneous yield loci **IYL** for a bulk solid for three consolidation conditions defined by the major consolidation stresses $(\sigma_1)_1$, $(\sigma_1)_2$, and $(\sigma_1)_3$. The instantaneous condition is that corresponding to the instant when the consolidation stress acts. In the case of a storage bin, it is the instant following the initial filling of the bin.

Referring to Fig. 2.10, the Mohr stress semi-circles through $(\sigma_1)_1$, $(\sigma_1)_2$, and $(\sigma_1)_3$ are tangential to the respective yield loci as shown. The Mohr semi-circles through the origin and tangential to the yield loci define the unconfined yield strength or stress corresponding to each consolidation condition. The unconfined yield strength is denoted by σ_c. The three strength values $(\sigma_c)_1$, $(\sigma_c)_2$, and $(\sigma_c)_3$ correspond, respectively, to the three consolidation pressures $(\sigma_1)_1$, $(\sigma_1)_2$, and $(\sigma_1)_3$. The bulk strength, represented by the Flow Function FF, is the unconfined yield strength σ_c expressed as a function of the major consolidation stress σ_1. It is obtained from the IYL as shown in Fig. 2.10.

2.5.3 *Time Yield Loci and Flow Function*

Many cohesive bulk solids will gain strength during undisturbed storage. Referring to Fig. 2.11, the application of the major consolidation stress σ_1 over an extended period of time will cause the unconfined yield strength to increase from σ_c to σ_{ct}.

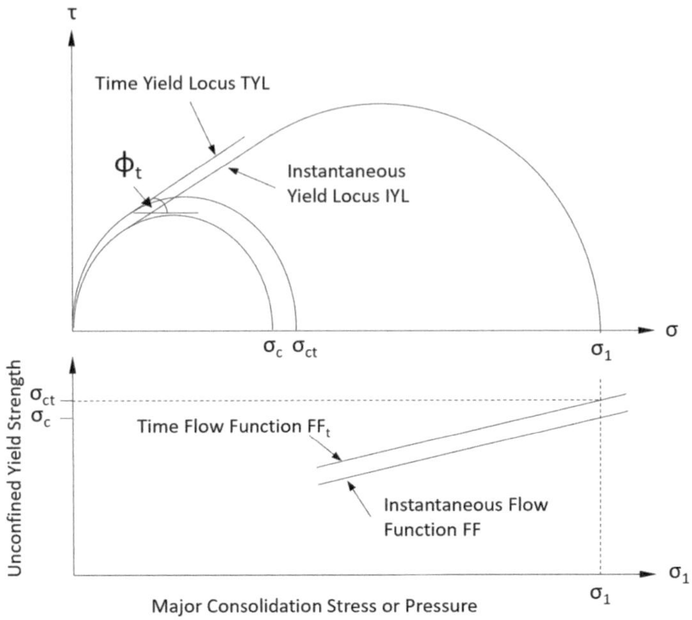

Fig. 2.11 Instantaneous and time yield loci and flow functions

The Time Yield Locus **TYL** defines the yielding condition following undisturbed time storage. The angle φ_t is called the static angle of internal friction. It is the angle defined by the slope of the TYL at the point of tangency with the Mohr semi-circle through σ_{ct}.

2.5.4 Classification of Flow Functions for Bulk Solids

The various forms of Flow Functions for bulk solids are illustrated in Fig. 2.12. Most cohesive bulk solids tend towards a limiting strength as the consolidation stress increases. Their Flow Functions are convex upward in shape as indicated by curve (a) of Fig. 2.12. Simple bulk solids have a linear Flow Function over the range of consolidating pressures considered as in curve (b). Free flowing bulk solids have zero strength and their Flow Function coincides with the horizontal axis as depicted by curve (c). Occasionally bulk solids are encountered in which the bulk strength increases non-linearly with an increase in consolidating stress as in curve (d). Such bulk solids are free flowing under low consolidation stresses but tend to cake or bond together as the consolidating stresses increase. A typical material exhibiting these characteristics is ammonium nitrate prill [4]. Bulk solids with Flow Functions of the form given by curve (d) are very difficult to handle.

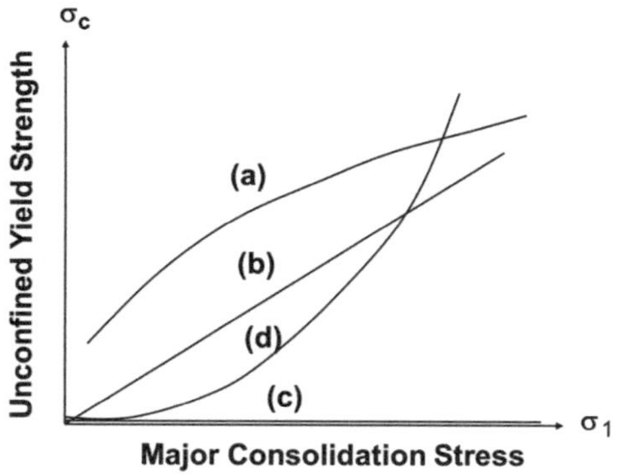

Fig. 2.12 Flow functions for bulk solids

2.5.5 *Wall Yield Loci and Wall or Boundary Friction*

At a boundary surface, such as that of a hopper or chute, the flow is characterised by the Wall Yield Locus **WYL** which relates the shear stress at the wall to the corresponding normal stress. These two stresses define the wall friction angle. The Wall Yield Locus and stress conditions at the boundaries of a hopper and chute are illustrated in Fig. 2.13. Figure 2.13a, b show the flow channels in hoppers and chutes respectively. Figure 2.13c shows the bulk solid yield functions and Mohr semi-circle which relates the normal and shear stresses at a hopper wall to the principal stresses or pressures of the bulk solid at that location. Typically, the **WYL** for most bulk solids, is linear or slightly convex as shown in Fig. 2.13c. In some cases, the **WYL** does not pass through the origin, indicating adhesion forces or forces of attraction between the bulk solid and the wall surface. This is highlighted in Fig. 2.14 which shows, diagrammatically, the low pressure or stress end of the **WYL** extended into the tensile normal stress region. The tensile stress for zero shear stress is given by σ_0 and is usually referred to as the adhesion. The cohesive stress τ_0 is the shear stress corresponding to zero normal stress.

The wall friction angle φ is defined by

$$\varphi = \tan^{-1}\left[\frac{\tau_w}{\sigma_w}\right] \tag{2.4}$$

where

$\tau_w =$ shear stress at the wall
$\sigma_w =$ pressure acting normal to the wall.

It is evident from Figs. 2.13 and 2.14 that, as the normal pressure σ_w increases, the wall friction angle φ_w decreases as illustrated in Fig. 2.15. At low values of normal pressure, the wall friction angle φ_w can be very high, significantly influencing the geometry of mass-flow hoppers in the region of the outlet as well as the performance of gravity flow chutes. In reality, wall or boundary friction can only exist for values

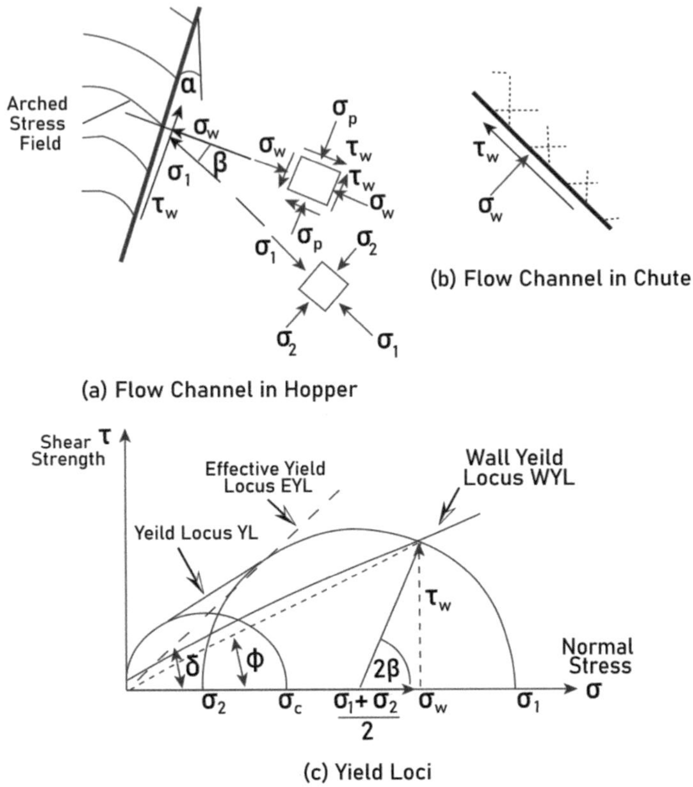

Fig. 2.13 Bulk solid yield and flow channels

Fig. 2.14 Wall yield locus, cohesion and adhesion

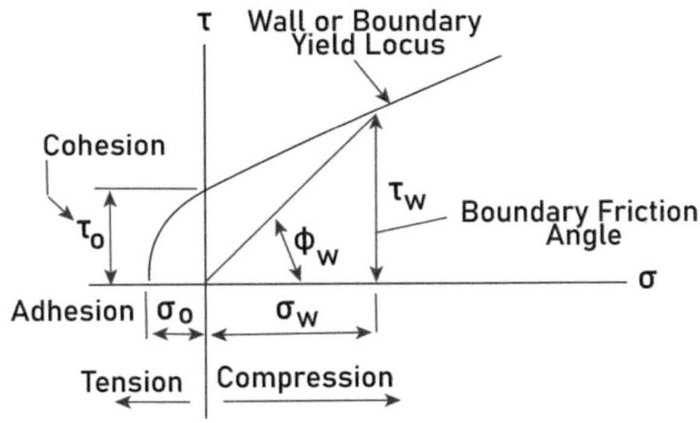

of φ_w equal to or less than the angle of internal friction as defined by δ as illustrated in Fig. 2.15.

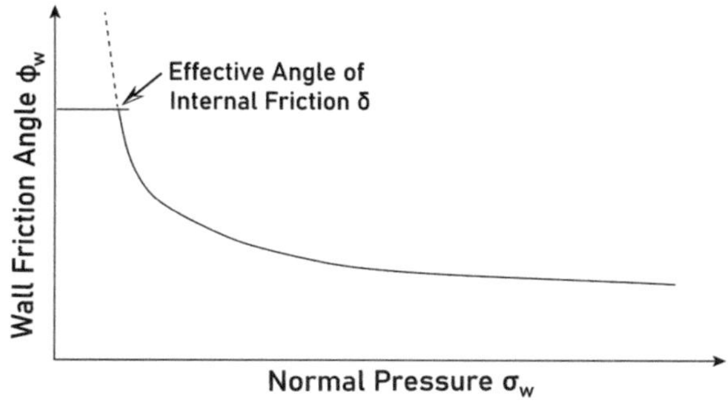

Fig. 2.15 Boundary or wall friction variation with normal friction

2.6 Overview of Jenike Method for Mass-Flow Bin Design

The design of hoppers for mass-flow requires determination of the two basic parameters.

- Hopper half-angle α for the chosen hopper lining material. This is obtained from the mass-flow, funnel-flow limits using the measured wall friction angle obtained from the flow property tests.
- The critical or minimum opening dimension B_{CR} to prevent a cohesive arch from forming and hence to prevent discharge flow.

2.6.1 *Mass-Flow and Funnel-Flow Limits*

The limits for mass-flow and funnel-flow have been established by Jenike [1, 2] on the assumption that a radial stress field exists in the hopper during gravity flow. The limits depend on the hopper half angle α, the wall friction angle φ and the effective angle of internal friction δ. Once the wall friction angle and effective angle of internal friction have been obtained, the hopper half-angle may be determined. In functional form:

$$\alpha = f_1(\varphi, \delta) \tag{2.5}$$

In the derivation of the theory of flow in converging channels, Jenike [5] showed that the bounds between mass-flow and funnel-flow in conical or axi-symmetric channels are given by:

$$\alpha = \frac{\pi}{2} - \frac{1}{2}\cos^{-1}\frac{(1 - \sin\delta)}{2\sin\delta} - \beta \tag{2.6}$$

where

$$\beta = \frac{1}{2}\left\{\varphi + \sin^{-1}\frac{(\sin\varphi)}{\sin\delta}\right\} \tag{2.7}$$

Since the flow in the hopper is based on radial stress field theory, the influence of the head in the cylinder on the flow patterns, particularly in the transition region, is not taken into account. For plane-flow hoppers, the bounds between mass-flow and funnel-flow are much less severe than for conical hoppers. However, in plane-flow bins, much larger hopper half-angles are possible, which means that the discharging bulk solid will undergo a significant change in direction as it moves from the cylinder to the hopper. In view of this Jenike [1, 2] has adopted recommended design limits in order to select the hopper half-angle. These design limits may be represented by the empirical equation:

$$\alpha = \frac{1}{0.725(\tan\delta)^{0.2}}\left\{exp\left[3.75(1.01)^{(0.1\delta-3)}\right] - \varphi\right\} \tag{2.8}$$

where δ and φ are in degrees.

The bounds for conical and plane-flow hoppers for five values of δ are plotted in Fig. 2.16. In the case of conical hoppers, it is recommended, as a safety precaution, that the hopper half angle be 3 degrees less than the limiting value. For plane-flow, the design limit may be selected; if the transition of the cylinder and hopper is sufficiently radiused, then it is even possible to select a hopper half angle 3 to 4 degrees larger than the design limit.

To illustrate the influence of wall friction on mass-flow hopper design, consider, for example, a conical hopper handling coal with $\delta = 50°$. As shown in Fig. 2.17, if the hopper is fabricated from mild steel with mill scale finish, the friction angle is likely to be approximately 30°. On this basis, $\alpha = 14°{-}3° = 11°$, which makes

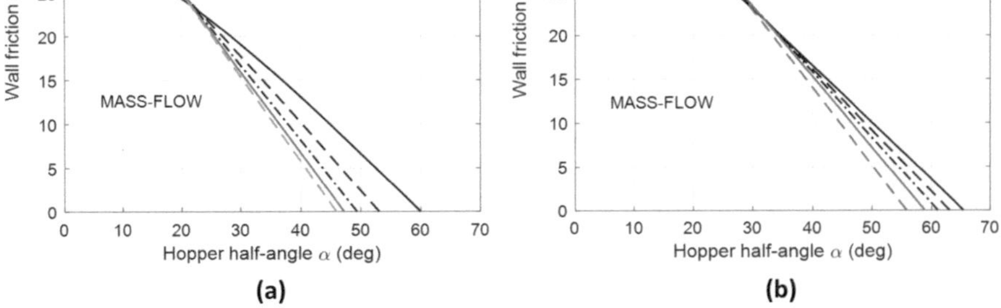

Fig. 2.16 Limits for **a** mass-flow conical and **b** plane-flow channels

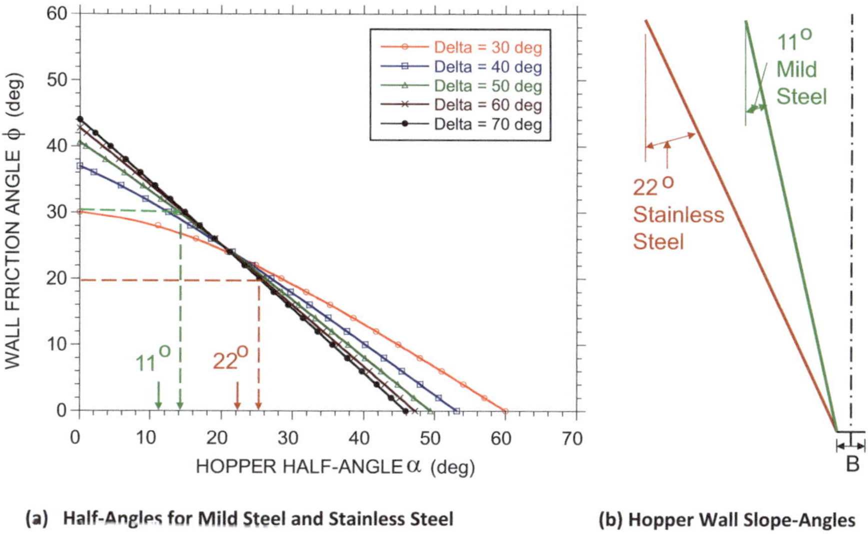

Fig. 2.17 Conical hopper design, half-angle selection

for a very steep hopper as shown in Fig. 2.17b. On the other hand, If the hopper is constructed of stainless steel type 304 with 2B finish, for which the friction angle φ is 20° or even less, the hopper half-angle $\alpha = 25° - 3° = 22°$. Reducing the friction angle from 30° to 20° leads to an increase in the limiting hopper half-angle, resulting in a significantly shallower hopper design."

The corresponding angles for plane-flow are $\alpha = 22°$ for $\varphi = 30°$ and $\alpha = 35°$ for $\varphi = 20°$. The importance of the wall friction angle in determining the mode of flow is clearly demonstrated. For a given hopper geometry, a small increase in φ can change a conical hopper operating under mass-flow into a hopper operating under funnel-flow.

2.6.2 *Critical Arching Dimension for Mass-Flow*

The opening dimension for mass-flow must be large enough to prevent a cohesive arch from forming. A cohesive arch is one which forms as a result of the bulk strength due to the consolidation of the bulk solid. It is to be distinguished from a mechanical arch which occurs as a result of the interlocking of solid particles. Mechanical and cohesive arches are illustrated in Fig. 2.18. In order to prevent a mechanical arch from forming, the opening dimension B should be at least four times the maximum particle or lump size.

The aim in mass-flow design is to determine the hopper half-angle α for flow along the walls and the minimum hopper opening dimension B to prevent a cohesive

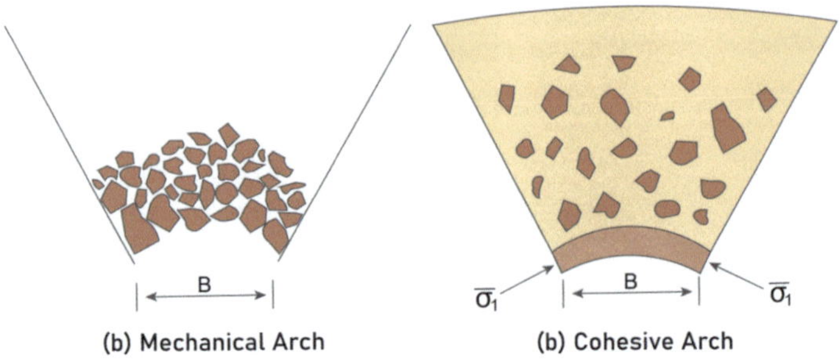

Fig. 2.18 Mechanical and cohesive arches

arch from forming. Normally the actual opening dimension will be selected larger than the minimum value to ensure that the required flow rate can be obtained.

Consider a bulk solid flowing in the mass-flow bin of Fig. 2.19. The solid was unconsolidated when it was first deposited in the bin, but as it flows downward, it becomes consolidated under the action of the major consolidating pressure σ_1. The variation of σ_1 is as indicated in Fig. 2.19. As a result of this consolidation, the strength σ_c of the bulk solid which is related to σ_1 by the Flow Function *FF* also varies as shown. The stress $\overline{\sigma}_1$ set up in a stable arch is assumed to be proportional to the width *B* of the arch. The variation of $\overline{\sigma}_1$ is also shown in Fig. 2.19. Whenever $\overline{\sigma}_1 > \sigma_c$, the bulk solid has insufficient strength to support a cohesive arch. When $\overline{\sigma}_1 < \sigma_c$, the stress in the arch is too small to cause the arch to collapse. Hence the critical or minimum opening dimension is determined for the critical condition $\overline{\sigma}_1 = \sigma_c$.

Fig. 2.19 Critical arch determination for mass-flow

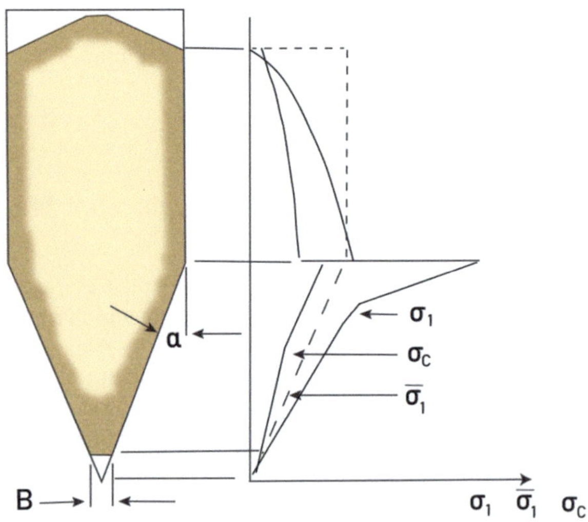

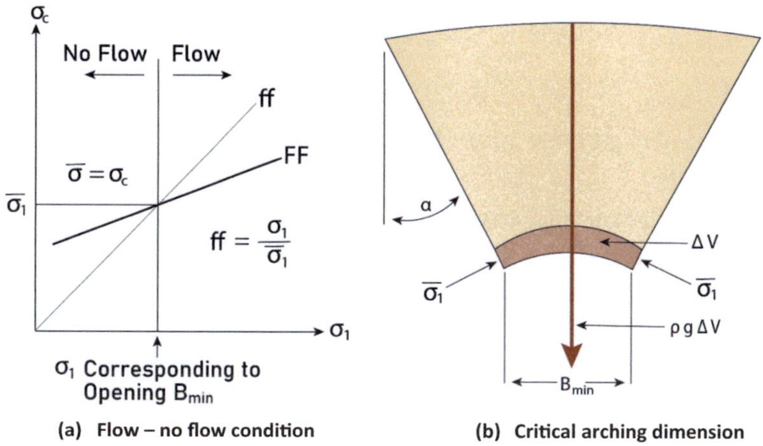

Fig. 2.20 Flow/no flow condition for mass-flow design

For design purposes, the foregoing is translated into the diagram in Fig. 2.20. In this figure the relationship between the unconfined yield strength σ_c and the major consolidating pressure σ_1 is given by the Flow Function FF, which is a bulk solid parameter. The stress $\overline{\sigma}_1$ acting in an arch is related to the major consolidating pressure σ_1 by the 'Flow Factor' **ff**; the Flow Factor is a flow channel parameter and describes the stress condition in the hopper during flow. The Flow Factor is given by

$$\mathrm{ff} = \frac{\sigma_1}{\overline{\sigma}_1} \tag{2.9}$$

In functional form, the initial opening dimension B is given by

$$B = f_2(\delta, \alpha, \varphi, \sigma_1, \rho) \tag{2.10}$$

where

$\rho =$ Bulk density

The other parameters are as previously defined.

Referring to Fig. 2.20, the critical opening dimension B is obtained for the condition $\sigma_c = \overline{\sigma}_1$ defined by the intersection point of the Flow Function and Flow Factor. The expression for B is

$$B = \frac{\overline{\sigma}_1 H(\alpha)}{\rho g} \tag{2.11}$$

where

g= acceleration due to gravity

The function H(α) is plotted in Refs. [2, 5, 6]. It is reproduced in Fig. 2.23.

2.7 Review of Jenike Flow Factor Determination

2.7.1 Introductory Remarks

The design of hoppers for mass-flow requires the determination of the following two basic parameters.

- Hopper half-angle a for the chosen hopper lining material. This is obtained from the mass-flow, funnel-flow limits using the measured wall friction angle from flow property tests.
- The critical or minimum opening dimension B_{CR} which is the smallest opening size that prevents the formation of cohesive arches and ensures reliable discharge flow.

Once these parameters are established, the hopper opening dimension can be increased and the hopper half-angle can be adjusted, (usually increased) within the bounds of mass-flow to achieve the plant operational requirements such as the required bulk solids discharge rate. In the absence of any specific requirements, a rule-of-thumb suggestion is to select the actual hopper opening dimension nominally 10% to 20% larger than the critical opening dimension, that is, $B \approx k_B\ B_{CR}$, where k_B ranges from 1.1 to 1.2.

The key to the determination of the critical arching, or more specifically, the doming opening diameter of an axi-symmetric or conical hopper and the corresponding critical arching width of a plane-flow or wedged shaped hopper are the relevant flow factors *ff* for the modes of flow. The determination of flow factors for mass-flow is one of the major outcomes of the work of Jenike whose studies of hopper flow involved an in-depth "classical" equilibrium analysis of the various stress states governing the storage and discharge of bulk solids [1, 2, 6]. The assumptions adopted by Jenike are.

- Quasi steady state. That is, static storage and discharge at a constant velocity without acceleration or inertia effects
- The stress field in a converging channel or hopper approaches a radial stress field in the vicinity of the channel vertex or hopper opening
- The bulk solid is incompressible, or more specifically the bulk density remains constant during loading and during discharge

It has been established from flow property tests that the bulk density is adequately represented by the equation

$$\rho = \rho_o\ {\sigma_1}^b \tag{2.12}$$

where

ρ_o = constant
σ_1 = major consolidation stress.
b = Compressibility constant.

2.7.2 *Overview of Flow Factor Theory*

As an aide to an understanding of Jenike's contribution to hopper design, salient aspects of his quite complex theory are summarised in "conceptual" form in the following discussion. With the focus on mass-flow in axi-symmetric or conical and plane symmetric or wedged-shaped hoppers, Jenike introduced the coefficient "m" as a convenient way to represent the equilibrium stress field conditions by two equations.

$$m = 0 \text{ for plane flow}$$

$$m = 1 \text{ for axial symmetry}$$

It is noted that Jenike's analysis includes both asymmetric and symmetric flow channels. The present discussion is confined to the symmetric case and to the mean stress "σ", which is the average of the major and minor principal stresses σ_1 and σ_2 respectively. That is

$$\sigma = \frac{\sigma_1 + \sigma_2}{2} = \frac{\sigma_x + \sigma_y}{2} = \frac{\sigma_r + \sigma_\theta}{2} \tag{2.13}$$

As is usual when describing bulk solids systems, compressive stresses are assumed to be positive and tensile stresses negative. The following two systems of co-ordinates are adopted:

(i) Polar spherical system as shown in Fig. 2.21 using r, θ, ε co-ordinates
(ii) Plane-cartesian polar cylindrical system using x,y,z co-ordinates

The application of the first of these co-ordinate systems is illustrated in Fig. 2.21 which examines the stresses in the transverse plane of a plane-flow hopper and the meridian plane of an axi-symmetric hopper. The principal co-ordinates are "r" and "θ". The co-ordinate "ε", which is not shown, is associated with the circumferential co-ordinate of an axi-symmetric hopper, for which, in accordance with the Haar and von Karman hypothesis [7], σ_ε is equal to the major consolidation stress of the meridian plane r, θ. That is, $\sigma_\varepsilon = \sigma_1$. For a plane-symmetric hopper, the transverse plane depicted in Fig. 2.21 is such that $\varepsilon = 0$.

Jenike analysed the equilibrium of the system depicted in Fig. 2.21 and derived the following two, hyperbolic partial differential equations:

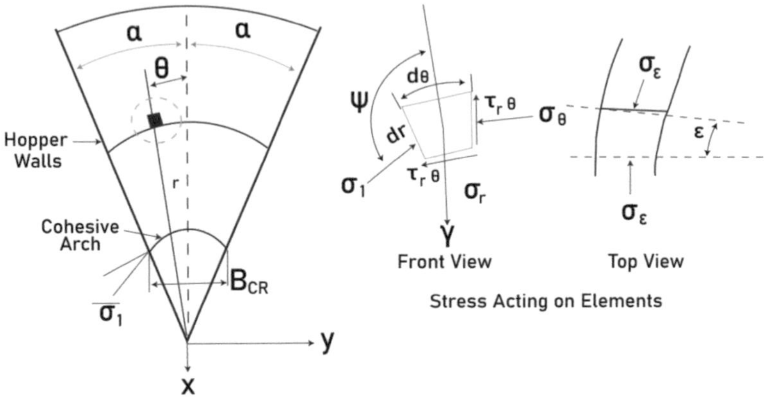

Fig. 2.21 Co-ordinate systems in plane flow and in the meridian plane of axial symmetry

$$\frac{\partial \sigma_r}{\partial r} + \frac{1}{r}\frac{\partial \tau_{r\theta}}{\partial \theta} + \frac{1}{r}[(\sigma_r - \sigma_\theta) + m(\sigma_r - \sigma_\varepsilon) + m\tau_{r\theta}\cot\theta] + \gamma\cos\theta = 0 \quad (2.14)$$

$$\frac{\partial \tau_{r\theta}}{\partial r} + \frac{1}{r}\frac{\partial \sigma_\theta}{\partial \theta} + \frac{1}{r}[m(\sigma_\theta - \sigma_\varepsilon)\cot\theta + (2+m)\tau_{r\theta}] - \gamma\sin\theta = 0 \quad (2.15)$$

The Mohr Circle analysis diagram linked to Fig. 2.21 is shown in Fig. 2.22. The component stresses are defined by the following equations:

$$\sigma_r = \sigma(1 + \sin\delta\, cos\, 2\,\psi) \quad (2.16)$$

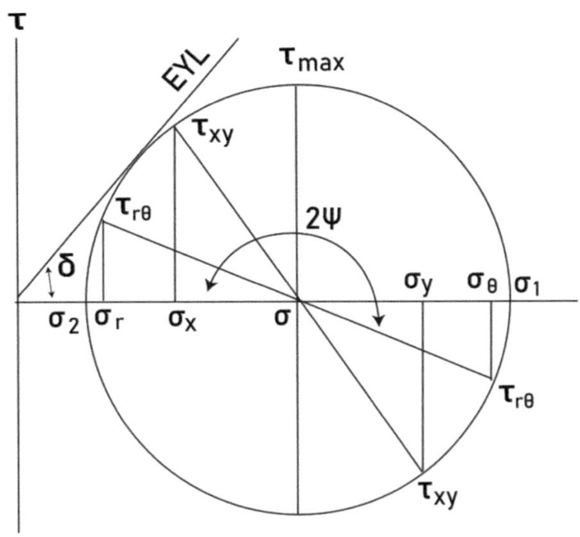

Fig. 2.22 Mohr circle diagram for component stresses during flow

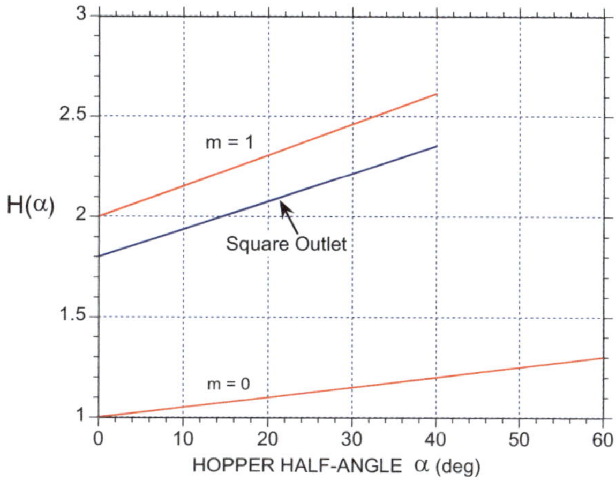

Fig. 2.23 Arch thickness parameter for mass-flow

$$\sigma_\theta = \sigma(1 - \sin\delta\cos 2\psi) \tag{2.17}$$

$$\tau_{r\theta} = \sigma\sin\delta\cos 2\psi \tag{2.18}$$

where ψ is the angle between the directions of σ_1 and of the ray r.

The principal stresses are:

$$\sigma_1 = \sigma(1 + \sin\delta) \text{ and } \sigma_2 = \sigma(1 - \sin\delta) \tag{2.19}$$

2.7.3 *Variation of the Mean Stress*

The overall aim is to obtain a solution of the Eqs. (2.14) and (2.15) leading to the determination of the variation of the mean stress "σ" as a function of both "r" and "θ". That is, adopting the nomenclature of Jenike

$$s(\theta) = s(m, \delta, r, \theta, \varphi, \gamma) \tag{2.20}$$

where

$s =$ stress function
m = symmetry factor
m = 0 for Plane-flow
m = 1 for axi-symmetric flow
$\delta =$ effective angle of internal friction
$\varphi =$ hopper wall friction angle

$\gamma = \rho g$ = bulk specific weight
ρ = bulk density,
g = gravitational acceleration
r, θ are co-ordinates

For hopper design, the boundary conditions defined at the hopper walls for which $\theta = \alpha$ are of particular interest. It is noted that the analysis assumes that the variables δ, φ and γ are each constant.

The variables defined in Eqs. (2.16)–(2.18), together with their derivatives, $\frac{\partial\sigma}{\partial r}, \frac{\partial\sigma}{\partial\theta}, \frac{\partial\sigma_r}{\partial r}, \frac{\partial\sigma_\theta}{\partial\theta}, \frac{\partial\tau_{r\theta}}{\partial r}, \frac{\partial\tau_{r\theta}}{\partial\theta}$, are derived and are substituted in the equilibrium Eqs. (2.14) and (2.15). In view of the large number of terms in the expanded equilibrium equations, they are presented in functional form in Eq. (2.21) and (2.22). Full details can be found in McLean [8] and Arnold, McLean & Roberts [9].

$$\frac{\partial s}{\partial\theta} + s f(r, \theta) + g(r, \theta) = 0 \tag{2.21}$$

$$r\frac{\partial s}{\partial r} + s\, h(r, \theta) + j(r, \theta) = 0 \tag{2.22}$$

2.7.4 *Convergence to Radial Stress Field*

Towards the hopper outlet, the stress field approaches a radial stress field where the stresses along a given ray defined by "θ" are proportional to the radial distance "r" from the vertex. That is, the angle "ψ " defining the major consolidation stress "σ_1" is given by $\psi = \psi(\theta)$. Also $\frac{\partial\psi}{\partial\theta} = \frac{d\psi}{d\theta}$ and $\frac{\partial\psi}{\partial r} = 0$.

Following further mathematical manipulation, the two equilibrium Eqs. (2.21) and (2.22) are transformed into two simultaneous, ordinary differential equations which are expressed in functional form as Eqs. (2.23) and (2.24).

$$\frac{d\psi}{d\theta} = F_\theta(\theta, \psi, s) \tag{2.23}$$

$$\frac{ds}{d\theta} = F_s(\theta, \psi, s) \tag{2.24}$$

These equations are solved numerically for a given set of boundary conditions to obtain the solutions $\psi(\theta)$ and $s(\theta)$. For symmetrical mass-flow hopper analysis, the boundary conditions of particular interest are

$$\psi = \frac{\pi}{2} \text{ for } \theta = 0^\circ \tag{2.25}$$

and

$$\psi = \frac{1}{2}\left[\pi + \sin^{-1}\left(\frac{\sin\varphi}{\sin\delta}\right) + \varphi\right] \text{ for } \theta = \alpha \tag{2.26}$$

2.7.5 Flow Factors to Prevent Arching and Doming

The transitional flow of bulk solids from the radial stress field within the hopper to the discharge zone immediately beyond the hopper outlet can influence the profile of doming for a conical or axi-symmetric hopper and arching for a plane-flow, wedged-shaped hopper. In the original work of Jenike and Johanson, a simple arch or dome of uniform thickness was assumed to determine the critical opening B_{CR} as follows:

$$B_{CR} = \frac{2\overline{\sigma}_1(m+1)}{\gamma} \tag{2.27}$$

where

$m = 0$ for plane-flow and $m = 1$ for axi-symmetric flow
$\gamma = \rho g$ bulk specific weight
$\rho =$ bulk density

Equation (2.27) assumes that $\overline{\sigma}_1$ acts at 45° to the vertical in order to maximise the value of B_{CR}.

Subsequently, Jenike and Leser [10] carried out a more fundamental analysis of the arch formation taking into account the variation of the arch thickness. Their solution is not known explicitly but may be represented by the equation:

$$H(\alpha) = \left[\frac{130+\alpha}{65}\right]^m \left[\frac{200+\alpha}{200}\right]^{1-m} \tag{2.28}$$

It is presented in graphical form as shown in Fig. 2.23. The modified equation for the minimum, critical arching dimension B_{CR} is

$$B_{CR} = \frac{\overline{\sigma}_1 H(\alpha)}{\gamma} \tag{2.29}$$

For convenience, this equation is often written in terms of the major consolidation stress and the flow factor as follows

$$B_{CR} = \frac{\sigma_1 H(\alpha)}{ff\gamma} \tag{2.30}$$

It is noted that for $\alpha = 0$, $H(\alpha) = 1$ for plane-flow and $H(\alpha) = 2$ for axi-symmetric flow. These values correspond to all values of α presented in the original method given by Eq. (2.28). The $H(\alpha)$ line for a square opening is also included in Fig. 2.23.

The equation for the flow factor *ff* in functional form, as derived by Jenike is

$$\mathrm{ff} = \frac{(1 + \sin\delta)\, s(m, \delta, \alpha, \varnothing)}{2 \sin\alpha} \, H(\alpha) = \frac{\sigma_1}{\bar{\sigma}_1} \tag{2.31}$$

2.7.6 Design Charts

The derivation of the flow factors, which involves the solution of the differential Eqs. (2.14) and (2.15), involves some tedious complex algebra. To assist with the application to hopper design, Jenike produced two sets of flow factor charts, one set for axisymmetric or conical hoppers and the other set for plane-flow hoppers. These charts were plotted for effective angles of internal friction δ ranging from 30° to 70° in steps of 10°. Arnold et al. [9] have, independently, also produced the two sets of charts over the range of δ equal to 30°–70°, but in this case in 5° steps for δ. Examples of the flow factor charts for axi-symmetric and plane flow for $\delta = 50°$ are presented in Figs. 2.24 and 2.25.

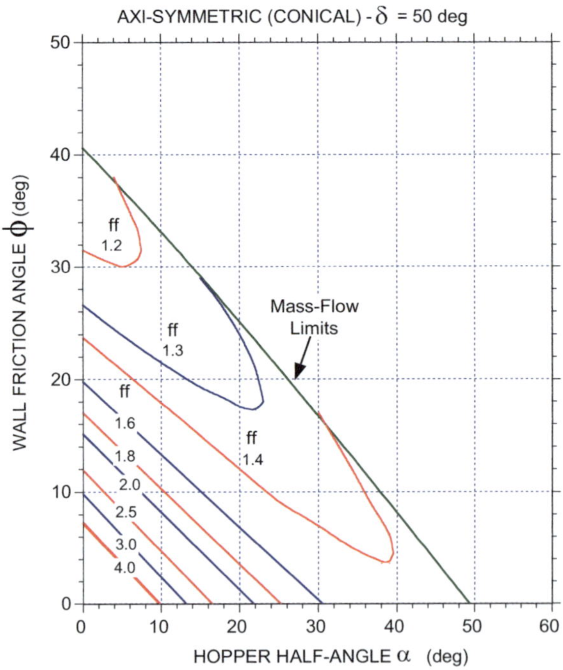

Fig. 2.24 Flow factor chart for axi-symmetric (conical) hopper, $\delta = 50°$

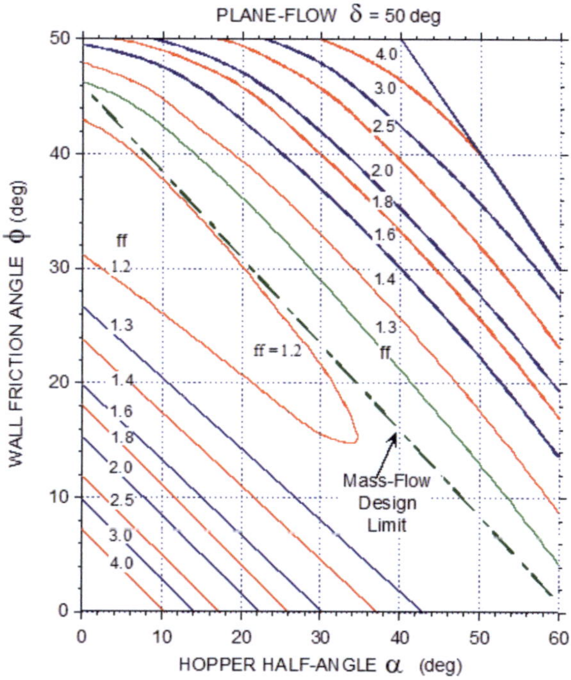

Fig. 2.25 Flow factor chart for plane-flow hopper, $\delta = 50°$

In the case of axi-symmetric or conical hoppers, the mass-flow and funnel-flow limits are analytically determined. It is usual to include a 3° safety margin when selecting the half-angle α. In the case of plane-flow, the design limits are based on the need to avoid build-up of bulk material at the transition due to the normally large transition angle at the intersection level of the upper cylindrical section and the lower hopper. It is possible to increase the hopper half-angle α by 3°–4° by radiusing the transition or by employing gusset plates, as illustrated in Fig. 2.26.

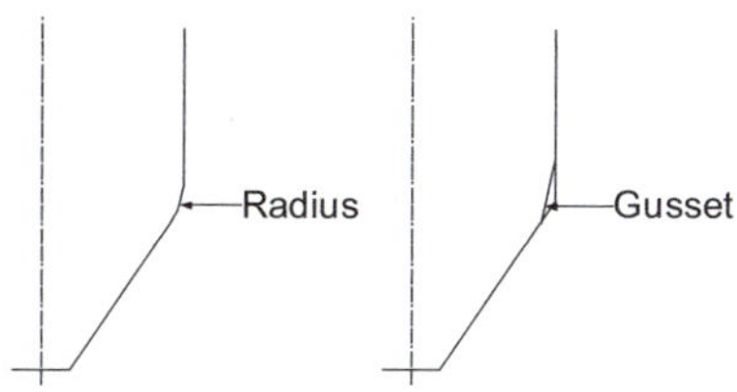

Fig. 2.26 Reducing the build-up of bulk solids

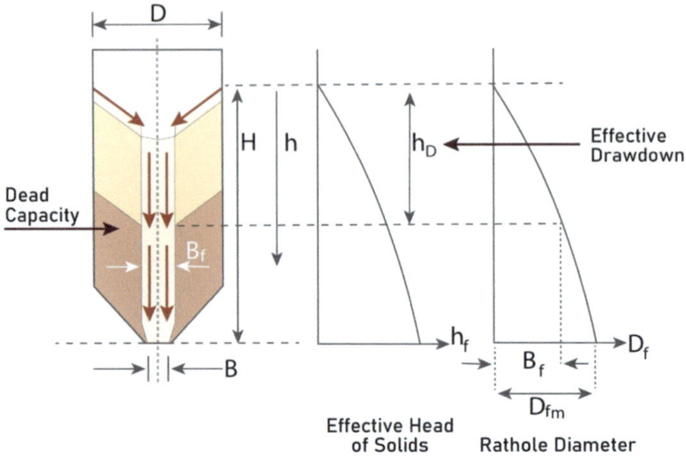

Fig. 2.27 Funnel-flow bin analysis showing effective draw-down

2.8 Funnel-Flow Bin Geometry—An Introductory Overview

As previously indicated, an attractive feature of funnel flow in symmetrical bins is the protection it offers to abrasive wear of bin and hopper walls. However, this mode of flow generally significantly limits the live capacity of the bin. The performance of the bin in terms of draw-down and live capacity depends on the size of the bin opening, or combined size in the case of multiple outlets, in relation to the critical pipe or rathole diameter D_f [6, 11]. Referring to Fig. 2.27, the variation of critical rathole dimension with head h is determined as follows:

In functional form, D_f is expressed as:

$$D_f = f_3(\sigma_1, \varphi_t, \varphi, \rho, h/D) \tag{2.32}$$

where

$\varphi_t =$ static angle of internal friction
$h =$ head of solids
$D =$ diameter or width of the bin

The actual expression used to determine D_f is

$$D_f = \frac{\overline{\sigma}_1 G(\varphi_t)}{\rho g} \tag{2.33}$$

The function $G(\varphi_t)$ is plotted in Refs. [5, 6]. The stress acting in the rathole $\overline{\sigma}_1$ equals the unconfined yield strength σ_c corresponding to the major consolidation pressure σ_1 at the location considered. Knowing h, the effective head of solids h_f is

Fig. 2.28 Time flow function. Funnel-flow design

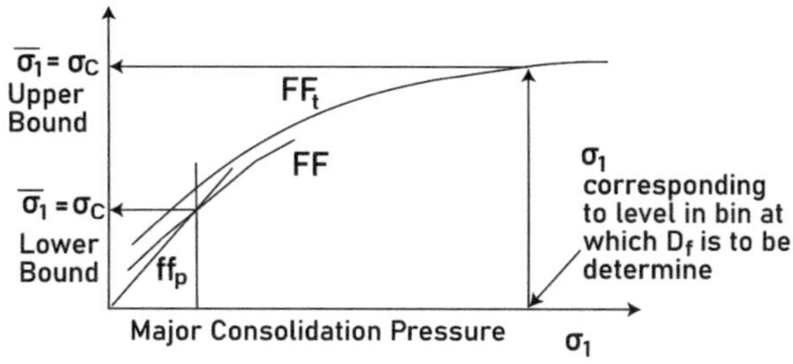

$$h_f = \frac{R}{K_j \tan\varphi}\left(1 - e^{-K_j \tan\varphi^{h/R}}\right) \tag{2.34}$$

where $R = \frac{\text{area of bin cross-section}}{\text{perimeter of bin cross-section}}$

$\varphi =$ wall friction angle
$k_j =$ ratio of lateral to vertical stress, assumed $= 0.4$

σ_1 is determined from

$$\sigma_1 = \rho g h_f \tag{2.35}$$

Then, $\overline{\sigma}_1$ is obtained from the bulk strength or Flow Function (usually the Time Flow Function FF_t) in accordance with Fig. 2.28. FF_t is determined in laboratory tests.

If the effective head of solids h_f varies with the head h, as indicated, the critical rathole D_f diameter also varies with the head. As bulk material discharges it does so by sloughing off the top surface and flowing down through a flow channel which forms above the bin outlet. The diameter of the flow channel B_f is slightly larger than the bin opening dimension B, depending on the properties of the particular bulk solid being handled. Flow will continue until the head drops an amount h_D referred to as the draw-down. At this level, the rathole dimension B_f equals the critical rathole dimension D_f. Flow then ceases, the material remaining in the bin being *dead* capacity.

In order to completely empty a funnel-flow bin, it is necessary for the bin opening dimension B to be such that $B_f = D_{fm}$, which is the critical pipe or rathole dimension corresponding to the maximum consolidation pressure which occurs at the bottom of the cylindrical section of the bin. As previously stated, for many cohesive bulk solids under the normal range of consolidation pressures experienced during storage, critical rathole dimensions several metres in magnitude are determined. In such cases funnel-flow is not a practical solution since the draw-down is limited and the live capacity correspondingly reduced.

2.9 Concluding Remarks

The material presented in this chapter has highlighted the basic concepts concerning the design of storage bins for bulk solids. The stress fields and associated stress functions relevant to bin design have been defined. They form the basis of the flow property characterisation tests which is the subject of Chap. 3 that follows.

With respect to the review of mass-flow bin design presented herein (in this chapter), the following comments are offered:

(i) The surcharge load acting at the cylinder and hopper transition is neglected, assuming that the radial stress field is fully developed
(ii) The determination of the critical opening dimension is based on the weight of material in the cohesive arch that forms over the opening and neglecting the material in the hopper above the arch.

In view of the complexity of the flow factor design equations requiring *tedious* numerical solutions, it is readily apparent that the design of mass-flow bins relies heavily on the use of the flow factor charts such as those illustrated in Figs. 2.24 and 2.25. Hence, the motivation to develop alternative analytical solutions which are more readily applicable to mass-flow bin design. This becomes the principal focus of Chap. 4 which integrates the flow property results from Chap. 3 into the adopted modified design methodology. The influence of the matters raised in (i) and (ii) above are taken into consideration.

The theory of funnel-flow bins and stockpiles is reviewed and updated in the light of relevant research and industrial design experience. This is basis of the material presented in Chap. 5.

The remaining chapters covering the wider areas of bulk solids handling follow from the foregoing chapters.

References

1. Jenike, A.W., *Gravity Flow of Bulk Solids. Bull 108.* The Univ of Utah, Engng Exp Station, USA, 1961.
2. Jenike, A.W., *Storage and Flow of Solids. Bull. 123.* The Univ of Utah, Engng Exp Station, USA, 1964.
3. Johanson, J.R., *Stress and Velocity Fields in Gravity Flow of Bulk Solids. Bull. 116.* The Univ of Utah Engng Exp Station, USA, 1962.
4. Roberts, A.W., O.J. Scott, and W. Tiedemann, *Bulk Handling of Ammonium Nitrate: Flow Properties and Storage Bin Design Criteria*, in *International Powder and Bulk Solids Conference.* 1979: Philadelphia, U.S.A.
5. Jenike, A.W., *Storage and Flow of Solids.* Utah Agric. Exp. St. Bull, 1964. **123**.
6. Jenike, A.W., *Quantitative Design of Mass Flow Bins.* Powder Tech., 1967. **1**: p. 237.
7. Haar, A. and T. Kármán, *Zur Theorie der Spannungszustände in plastischen und sandartigen Medien.* Nachrichten von der Gesellschaft der Wissenschaften zu Göttingen, Mathematisch-Physikalische Klasse, 1909. **1909**: p. 204–218.

8. McLean, A.G., *Flow Rates Simple Bulk Solids from Mass-Flow Bins*. 1979, The University of Wollongong, Australia.
9. Arnold, P.C., A.G. McLean, and A.W. Roberts, *Bulk Solids: Storage, Flow and Handling*. 2nd ed. 1979: The University of Newcastle Research Associates (TUNRA).
10. Jenike, A.W. and T. Leser, *A Flow - No Flow Criterion in the Gravity Flow of Powders in Converging Channels*, in *Proc. 4th Intl. Congress on Rheology*. 1963.
11. Jenike, A.W. and J.R. Johanson, *Bin Loads*. Journal of Structural Div., ASCE, 1968. **95**(ST4): p. 1011.

Chapter 3
Basic Material Properties

3.1 Introductory Remarks

In Chaps. 1 and 2, the basic characteristics of bulk solids were discussed and the relevant properties that define their behaviour during storage and flow were indicated. There are now well-defined testing methods that reproduce the various conditions pertaining to bulk solids and which allow the design of handling plants to be performed on a correct analytical basis. The principal tests used to determine the flow properties of bulk materials are outlined in this chapter.

3.2 Bulk Solids Flow Properties

In order to design storage bins and associated handling systems it is essential that the flow properties be determined by the testing of a representative sample. The testing, which duplicates field conditions in the laboratory, provides the designer with such parameters as:

(i) Flow functions *FF* for instantaneous and time storage conditions. These represent the strength versus consolidation occurring under storage and flow.
(ii) Effective angle of internal friction δ.
(iii) Wall friction angles ϕ for different bin wall materials and finishes.
(iv) Bulk density ρ as a function of consolidation.
(v) Critical pipe or "rathole" diameter D_f as a function of effective head of solids.
(vi) Abrasive wear rates for different wear liners.
(vii) Impact wear characteristics for different wear liners as a function of impact angle.

For fine powders, flow rate predictions may be critical. For the determination of flow rates, it is necessary to measure

A. W. Roberts, *The Theory and Practice of Bulk Materials Handling*,
https://doi.org/10.1007/978-3-032-19702-3_3

(viii) Solids density.
(ix) Permeability of the solids as a function of consolidation.

3.3 Shear Test Apparatus

The strength and flow characteristics of bulk solids are based on the various yield loci obtained from laboratory tests. Such tests involve a consolidation phase, in which the sample is brought to the critical state condition in terms of voidage, followed by a shear phase. In general, the shear versus consolidation characteristics may be obtained by direct shear using either the linear or rotational type shear testers, or indirectly using the biaxial or triaxial cells [1–4]. The shear test methods are indicated schematically in Fig. 3.1.

These tests are applicable to bulk solids over the range of moisture conditions up to, but not beyond, the saturated moisture content.

(a) **Jenike Type Direct Shear Tester**

Figure 3.2a shows a Jenike shear tester built in 1964 where measurements have pronounced slip stick due to the use of a "soft" pneumatic load cell. A modern Jenike type shear tester is pictured in Fig. 3.2b which uses a loadcell with a stiffness of 1 kN/mm.

Of the above methods, the linear direct shear tester of the type developed by Jenike [5, 6], is probably the most widely used; it allows simple measurement of the relevant bulk solid parameters, including wall or boundary friction. The tester is equipped with a shear cell of circular shape in plane, as illustrated in Fig. 3.3.

The shear cell is composed of a base which is located on the machine, a shearing ring which rests on top of the base, and a cover which has a shear force loading bracket attached to it. The shear cell is mounted in the test machine and shear strain is applied by means of the loading stem. The sample of material contained in the cell is subjected to normal pressure through the applied force V; the shear force (and hence stress) is measured by means of a load transducer and is recorded continuously with time. Generally, the cell used has an inside diameter of 95.5 mm giving the cell a cross-sectional area of 1/140 m^2. For high consolidation conditions, it is usual to use a cell of 63.5 mm diameter having a cross-sectional area of 1/315 m^2. The applied strain rate is approximately 2 mm/min.

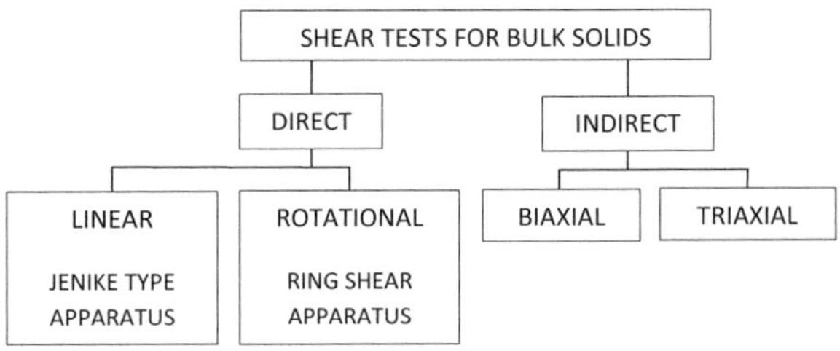

Fig. 3.1 Shear tests for bulk solids

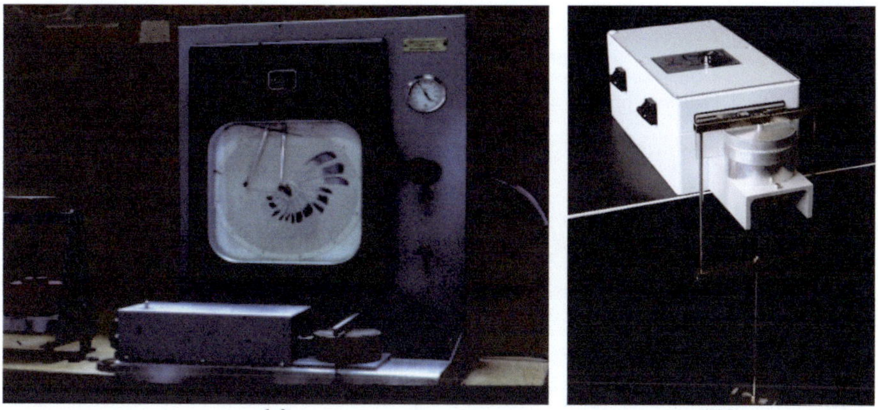

(a) (b)

Fig. 3.2 Jenike shear testers **a** Original version with pneumatic load cell and **b** Current version with electronic load sensor (TUNRA Bulk Solids, The University of Newcastle)

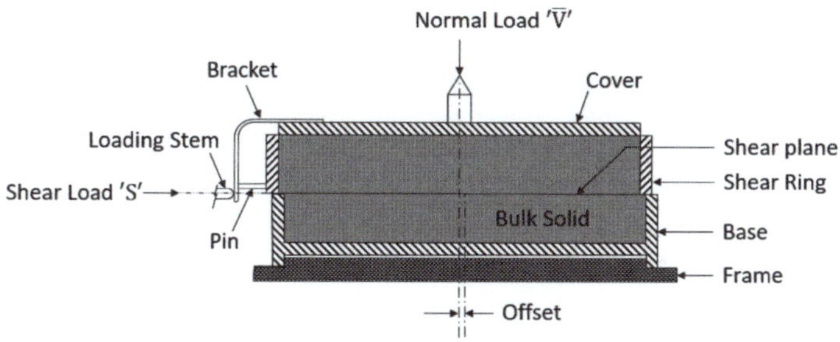

Fig. 3.3 Jenike type shear cell

As indicated in Fig. 3.3, the shear ring is first offset, the objective being that during the shear consolidation phase of the test, consolidation to the critical state condition should be completed by the time the shearing has moved to a position of concentricity with the base. In view of the limited travel for this condition to be reached, it is necessary to commence each test with a pre-consolidation phase using Fig. 3.4. The pre-consolidation phase involves the application of the prescribed consolidation force V_t and, at the same time, the twisting top is given a pre-determined number of oscillatory twists through an amplitude of approximately, $\pm 10°$. On completion of pre-consolidation, the mould ring is removed and the sample screeded level with the top of the shear ring to remove excess material. The cell is then set-up as indicated in Fig. 3.3 for the shear consolidation followed by shear test.

Where time consolidation tests are required, samples contained in their shear cells are first consolidated in the normal way using the direct shear tester and are then placed in consolidating benches of the type illustrated in Fig. 3.5. The samples are stored for the requisite period of time under pressure equal to the major consolidating

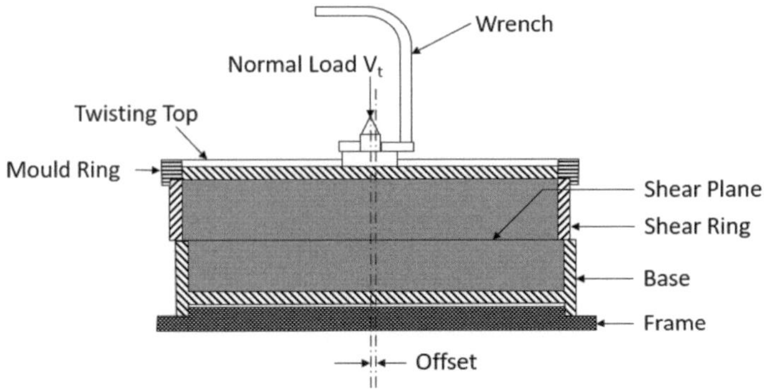

Fig. 3.4 Pre-consolidation of sample

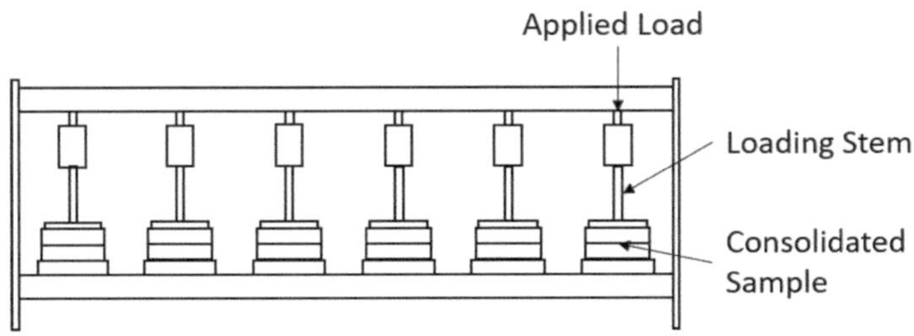

Fig. 3.5 Consolidating bench

pressure. After time storage, the shear cells are placed back in the direct shear tester to complete the shear tests.

In performing shear tests on bulk solids, it is usual to test the −4 mm size fraction. Where bulk solids composed of particles of a large size range from coarse to fines, it is the fine particles that contribute to the solid's cohesive strength; the coarse particles are generally free flowing.

Furthermore, when a storage bin is being filled, the coarse particles tend to roll to the outside while the fines congregate in the centre, particularly in the region of the outlet, where a cohesive arch may form. This supports the rationale in only testing the −4 mm size fraction.

For projects requiring multiple Time Yield Loci tests to be performed such as industrial bin design applications, the multiple series of consolidation benches have been developed, as shown in Fig. 3.6. The load applied by each cell is set at the required value and, by means of feedback control, maintained at that value for the requisite undisturbed period of time.

(b) **Torsional or Ring Shear Apparatus**

A disadvantage of the Jenike type direct shear tester is the limited travel it offers during the consolidation phase. As is evident from Fig. 3.3, travel is limited, effectively, by the thickness of the shear ring. The torsional or ring shear testers such as

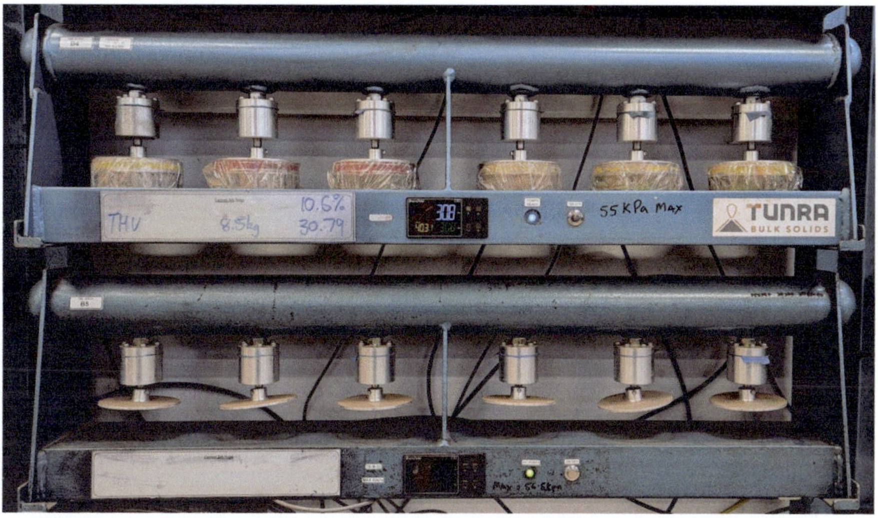

Fig. 3.6 Series of consolidation benches (TUNRA Bulk Solids, The University of Newcastle)

the Walker cell, virtually have unlimited travel. However, ring shear testers have the following disadvantages:

- Erroneous measurements may occur due to particles becoming wedged in the space between the top and bottom halves of the cell.
- Due to the rotational motion, shear strain is a function of the radius and hence is not uniform for all regions of the bulk solid sample.
- Time consolidation tests are difficult to perform.

An alternative torsional shear apparatus is the Schultze tester [1], which offers improved control over normal stress and shear strain during testing. It is particularly useful for characterising powders and bulk solids under low consolidation stresses.

(c) **Triaxial Test Apparatus**

This method has certain advantages, particularly for the testing of samples containing larger size particles. The test is, perhaps, a little less simple to perform than the direct shear test of Jenike and, for this reason, is not widely used for bulk handling design applications. In addition, the test is not so readily applicable for performing time consolidation tests. On the other hand, the triaxial test, by its very nature, has better control over the accuracy of determination of the major consolidating pressure, since this is measured directly rather than induced as in the direct shear test.

(d) **Biaxial Testing Machine**

Research in powder characterisation in the 1980's in Germany generated a growing interest in the biaxial testing machine. The German research and development, for example, was initiated by Professor J. Schwedes at Braunschweig University [2,

7]. The basis of the test is that controlled strain may be applied independently in either of the two orthogonal directions in the horizontal plane, strain in the vertical direction being zero. The potential as a research tool was realised by DeSilva and Enstad of POSTEC Research, Porsgrunn, Norway [8]. Following research focused on the research led by Scarlet at Delft University in the Netherlands aimed at further extending the control of strain testing [9, 10]. In view of the flexibility of independent strain control, the biaxial cell offers considerable scope as a sophisticated research tool rather than a machine used to evaluate flow properties on a routine basis for bulk solids handling equipment design.

3.4 Application of Direct Shear Test

(a) Hvorslev/Roscoe Diagram

The Hvorslev/Roscoe Diagram shown in Fig. 3.7 conveys important information concerning the Jenike test procedures. Hvorslev [11], who studied the stress condition in cohesive soils, showed that the peak shear stress at failure is a function of the effective normal stress and the voids ratio (or density) in the plane of failure. This condition is independent of the stress history of the sample. The work of Hvorslev was further extended by Roscoe et al. [12], who established the concept of a failure surface in the three dimensional space of shear stress, normal stress and voids ratio. They also showed the existence of a critical void ratio boundary at which unlimited deformation could take place without change in the stress condition and voids ratio.

The first phase of the Jenike test is to bring the under consolidated bulk solids sample in the shear cell to the critical state consolidated condition by the process of shear consolidation. Because of the limited travel of the shear cell of one cell wall thickness, it is often necessary, particularly for cohesive bulk solids, to first pre-consolidate the sample by means of an oscillatory twisting process under a pre-determined normal load and then complete the process by shear consolidation to the critical state condition. This is deemed to occur when the consolidation line approaches, *asymptotically* the Critical State Line as shown in Fig. 3.7. If this is not the case and the sample is over consolidated as shown by the red line in Fig. 3.7, the test is invalid and must be repeated. The same applies if the sample is under consolidated. It is most important that the operator be thoroughly trained in the test procedures.

(b) Instantaneous Yield Loci and Flow Function

The procedure for using the Jenike direct shear apparatus is detailed by Jenike in Bulletin 123 [6] and Arnold et al. [13]. A brief review of the test procedure is now given. As indicated in Sect. 3.3(a), the sample of bulk solid is first pre-consolidated followed by consolidation under shear in which the material is made to flow under the consolidation stresses until a steady state is reached or approached. Figure 3.8

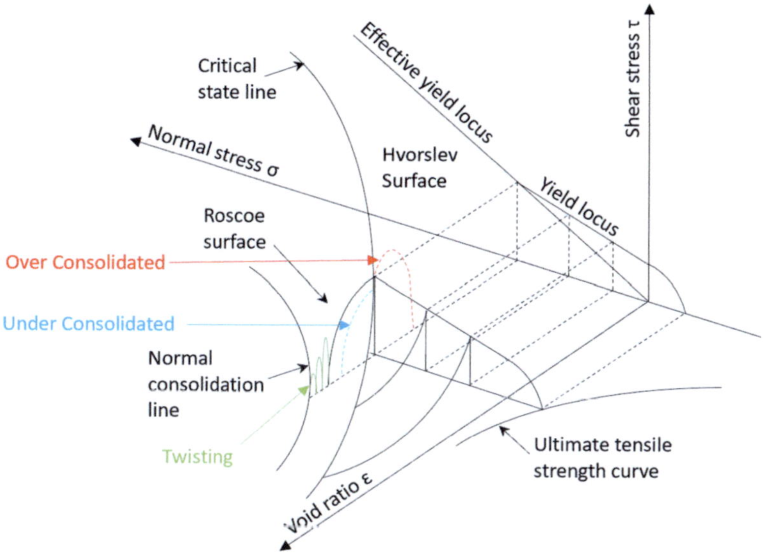

Fig. 3.7 Hvorslev/Roscoe diagram

indicates the consolidation point as (σ, τ). Shearing of the sample is now performed at a normal stress σ_n, smaller than σ.

At least three, but preferably six samples (allowing for repeats), are prepared and sheared in this way at three values of σ_n to produce the Yield Locus of Fig. 3.8 for the one consolidation stress σ. By drawing a Mohr semi-circle through the origin and tangential to the Yield Locus (YL) the point σ_c is found and defined as the unconfined yield strength. A Mohr semi-circle through point (σ,τ) and tangential to the Yield Locus (YL) defines σ_1 the major consolidation principal stress. The values σ_1 and σ_c are both important for design purposes and represent one point on the Flow Function relationship describing the strength of the bulk solid being tested.

The slope of the YL at any point defines the angle of kinematic friction φ_k as shown in Fig. 3.8. By repeating this procedure for other consolidating loads, a series

Fig. 3.8 Determination of yield locus using direct shear tester

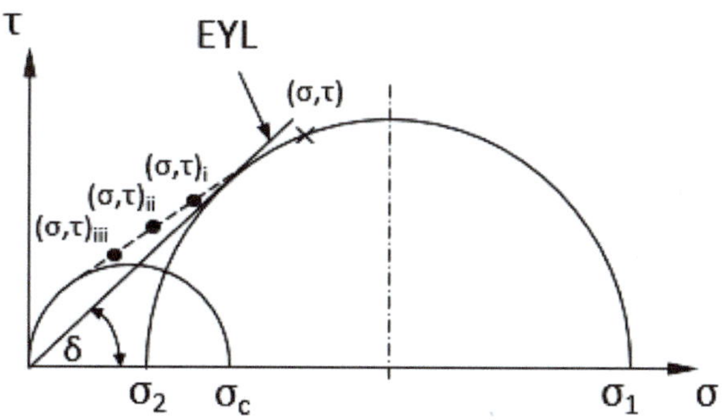

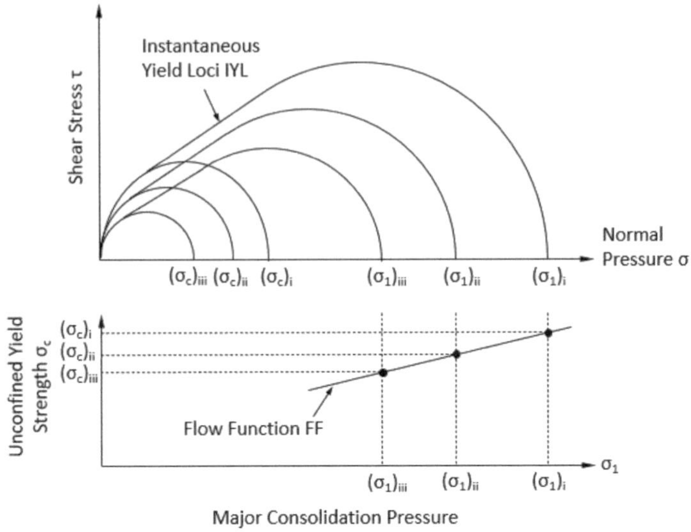

Fig. 3.9 Family of yield loci for a cohesive bulk solid

of YL curves may be obtained to find additional values of σ_c and σ_1 for the Flow Function (FF) relationship. Following the definitions given in Sect. 3.2, the straight line drawn through the origin and tangential to the Mohr circle defining the major and minor principal forces or pressures is called the Effective Yield Locus EYL and the angle δ is called the Effective Angle of Internal Friction. A family of YL are shown in Fig. 3.9. These graphs are plotted in stress units; where force units are used, as is sometimes the case, conversion to stress or pressure units is achieved by dividing by the area of the shear cell, which is normally 1/140 m^2.

(c) **Time Yield Loci**

The Time Yield Loci (TYL) are obtained by storing the consolidated samples of bulk solids in the consolidation bench for the requisite period of storage time, the samples being stored under the major consolidating pressure σ_1. After time consolidation, the samples are sheared in the shear test apparatus. A typical TYL is shown in Fig. 3.10. The slope of the TYL at the point of tangency with the Mohr semi-circle through the origin defines the Static Angle of Internal Friction φ_t.

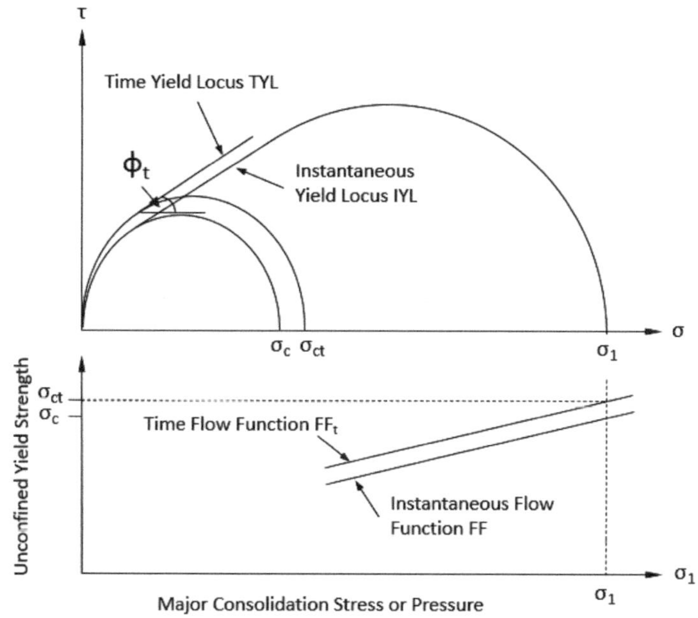

Fig. 3.10 Time yield locus

3.5 Flow Functions for Bulk Solids

The various types of flow for bulk solids are illustrated in Fig. 3.11. For most cases, flow functions for cohesive bulk solids will be concave down in shape as in curve (a) or straight lines through the origin as in curve (b). Bulk solids depicting the latter characteristic are referred to as "simple bulk solids". Free flowing bulk solids have no cohesion and hence no strength (that is, $\sigma_c = 0$); their flow function coincides with the horizontal axis as in the case of curve (c). The strength of some materials increases more rapidly as consolidation pressure increases and in this case the flow function will depict a concave upward shape as in curve (d). A typical material exhibiting these characteristics is ammonium nitrate prill [14], a material used as an explosive in mining operations.

Several factors influence the strength and hence flow function of bulk materials. These include the moisture content, temperature, storage time, particle size distribution and external factors such as mechanical vibrations.

The influence of moisture content is illustrated in Fig. 3.12, which shows the instantaneous flow properties of a typical run-of-mine coal. The plotted results depict the bulk density ρ, effective angle of internal friction δ and flow function FF for the three moisture contents 5, 10 and 15% (wet basis). Referring specifically to the flow functions, the strength increases significantly with moisture content, the increase being most pronounced over the range 5–10%. Beyond 15%, the strength could start to level out and even decrease somewhat.

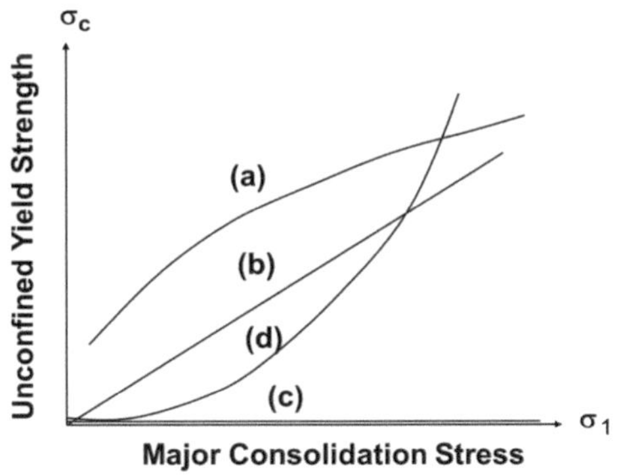

Fig. 3.11 Flow functions for bulk solids

Most cohesive bulk solids gain strength with consolidation during prolonged storage. This is illustrated in Fig. 3.13 for a run-of-mine coal at 10% moisture content. An appreciable but not unacceptable increase in strength is obtained after two days storage, that is storage over a weekend period. However, after 5 days the strength gained is very considerable indeed. While it would be possible to design a satisfactory mass-flow bin to handle the coal for two days storage, it would be impossible to design a satisfactory storage bin to discharge the coal by gravity after 5 days storage or longer. Some form of flow promotion would be necessary. Alternatively periodic withdrawal of some coal (say 2 to 3 min every 24 h) would help to reconstitute the coal and reduce the influence of inter-particle bonding; in this way, the influence of prolonged storage on the gain in bulk strength is minimised.

With respect to particle size distribution, as the percentage of fines increases so too will the strength increase. This is illustrated in Fig. 3.14 for pyrophyllite, a material used in the manufacture of refractories. The graph indicates an increase in shear strength for a given consolidation pressure with decrease in size. It is interesting to note that the strength of the composite sample is similar to that due to the fine fraction.

The strength of some bulk solids is influenced by temperature. This is particularly the case where moist materials are stored at sub-zero temperatures giving rise to freezing. Elevated temperatures can also have an influence such as ammonium nitrate prill which has a critical phase change temperature around 38°C [14].

3.6 Shear Tests for Wet Solids

Wet solids handling is an area of increasing interest. At present, it seems to fall 'in no man's land' between rheology and bulk solids. Applications commonly concern the design of vessels for the storage and gravity flow of super-saturated bulk solids. Where the solids may settle out of suspension during storage, it is necessary to ensure that gravity discharge may occur without blockages due to arching or 'ratholing'.

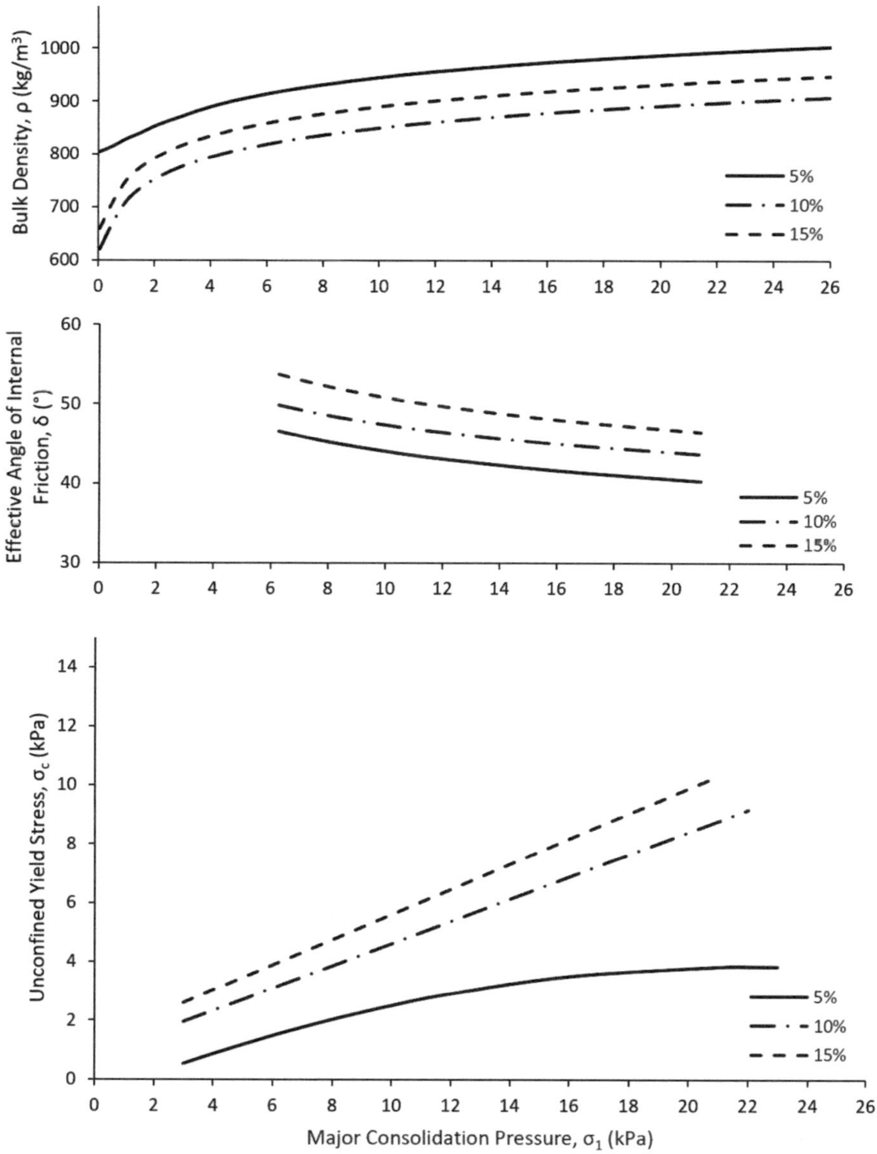

Fig. 3.12 Flow properties showing the influence of moisture

For this reason, the storage vessel should be designed for mass-flow. The required hopper geometrical parameters may be determined for saturated bulk solid samples using submerged shear tests as depicted in Fig. 3.15.

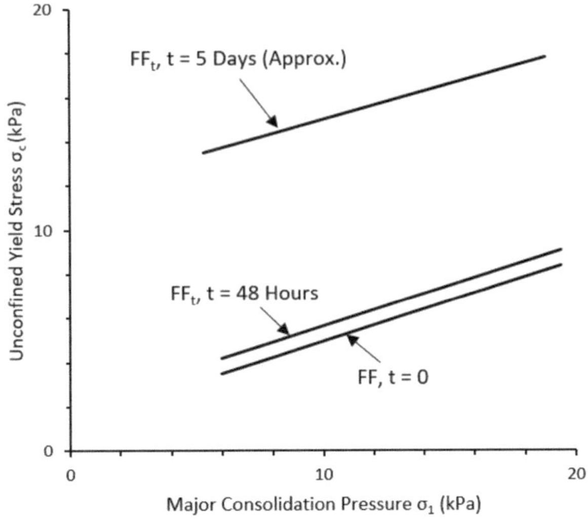

Fig. 3.13 Effect of storage time on the strength of an unwashed R.O.M. coal 10% M.C.

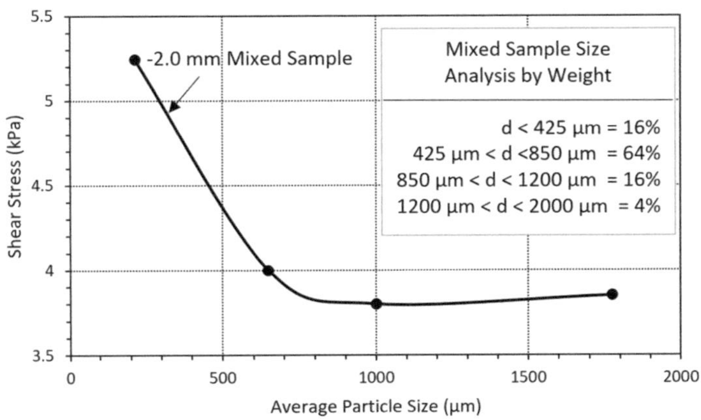

Fig. 3.14 Effect of particle size on shear strength of pyrophyllite at 5% M.C. (d.b.) $\sigma = 4.76$ kPa during consolidation

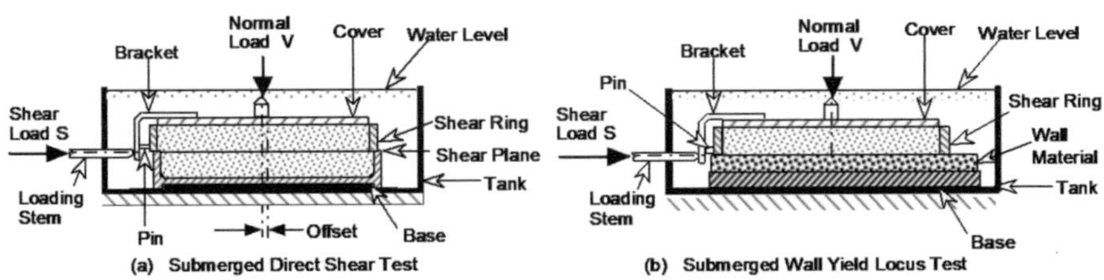

Fig. 3.15 Submerged shear tests

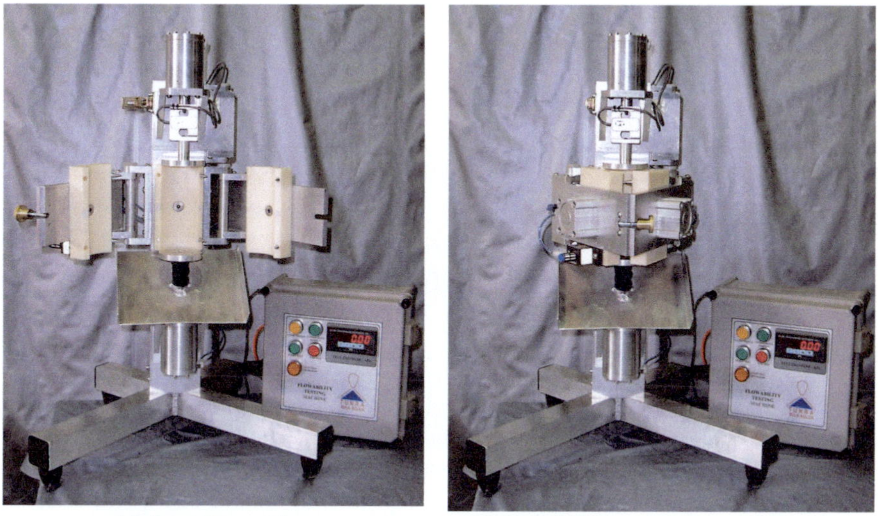

(a) Arms Swung Clear (b) Mould Segments Clamped Together

Fig. 3.16 Flowability tester

3.7 Flowability Tester

A flowability tester, developed by University of Newcastle, is depicted in Fig. 3.16 [15]. In effect, this is an unconfined compression test in which the lateral pressure is controllable during the consolidation phase by means of pneumatic actuators attached to the three segments of the mould cylinder.

Figure 3.16a shows the segments of the mould cylinder retracted and their support arms swung clear. Figure 3.16b shows the arms lock in place and the segments clamped together to contain the bulk solid or powder sample during the consolidation phase of the test. For the unconfined phase of the test, the segments are retracted to leave the consolidated cylindrical sample exposed so that the axial load may be applied to effect failure. When compared with the Jenike direct shear test which is somewhat time consuming, the flowability tester provides a much quicker analysis of the flow properties of a bulk solid, the device is particularly suitable for quality control tests.

3.8 Bulk Density

The bulk density of bulk solids needs to be determined as a function of major consolidation pressure. This is achieved using a compressibility tester. The relationship for the bulk density is:

$$\rho = C(\sigma_1)^b \quad (3.1)$$

where

C= constant
b = compressibility constant

3.9 Wall Yield Loci and Friction

(a) **Introductory Remarks**

Of the various parameters influencing the gravity flow performance of bulk solids in hoppers and chutes, probably the most significant is the friction characteristic at the contact zone between the bulk solid and wall or surface. As discussed in Sect. 2.4 of Chap. 2 when reference was made to Fig. 2.13, the importance of low wall friction in permitting an acceptable hopper half angle for mass-flow in hopper design is emphasised.

Primarily, wall friction depends on the interaction between three groups of variables, those relating to the bulk solid, those relating to the wall or boundary surface and those which arise from the loading and environmental conditions. This interaction is illustrated, diagrammatically, in Fig. 3.17.

The relevant properties in each of the three groups are summarised as follows:

(i) **Bulk Solid Characteristics:**

These include:

- Particle size and size distribution
- Particle shape
- Particle hardness
- Moisture content
- Particle density
- Bulk density
- Surface chemistry characteristics
- Temperature
- Undisturbed storage time

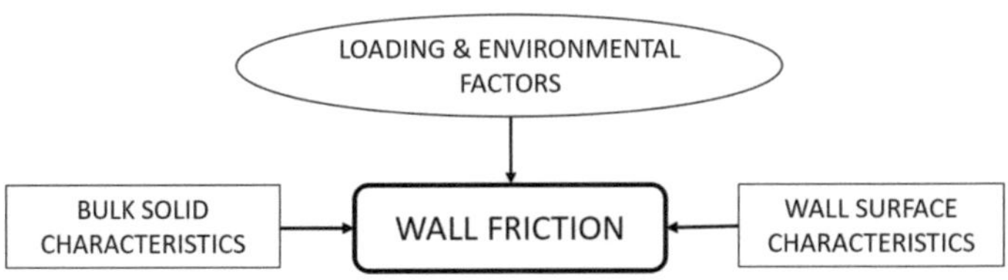

Fig. 3.17 Wall friction interactions

(ii) **Wall Surface Characteristics**

These include:

- Surface roughness
- Hardness
- Chemical composition

(iii) **Loading and Environmental Factors:**

These include:

- Normal pressure between bulk solid and wall surface
- Relative rubbing or sliding velocity
- Temperature and humidity or moisture conditions
- Wall vibrations

(b) **Surface or Wall Friction**

The wall friction characteristics are displayed by the wall yield locus W.Y.L., the wall friction angle φ is defined as follows:

$$\varphi = \tan^{-1}\left[\frac{\tau_w}{\sigma_w}\right] \tag{3.2}$$

where

τ_w = shear stress at the wall
σ_w = corresponding normal stress at the wall

Wall or boundary friction can be measured using the Jenike direct shear apparatus. Figure 3.18a shows the type of arrangement used. The inverted shear tester of Fig. 3.18b developed at the University of Newcastle permits low values of consolidating pressures to be measured. With this device, it is possible to measure the WYL into the tensile stress zone. The normal load V is varied to obtain a locus of points for the values of shear force S for each normal load V. The curve obtained is called the wall yield locus (WYL). Some wall yield loci are shown in Fig. 3.19 which illustrates the variability obtained from different types of wall materials.

Figure 3.19a shows the wall loci for black coal at 10% moisture content on three surfaces, namely, stainless steel type 304 with 2B finish, mild steel polished and rusted mild steel, all for the instantaneous or zero storage time condition. In the case of the polished mild steel surface, the wall yield locus measured after 72 h undisturbed contact or storage time with coal is also shown; the increase in friction in this case is quite considerable with corrosion and adhesion or bonding of coal particles to the steel surface occurring.

Figure 3.19b compares the wall yield loci for black coal at 18.7% moisture content and brown coal at 65% moisture content on two surfaces, stainless steel type 304

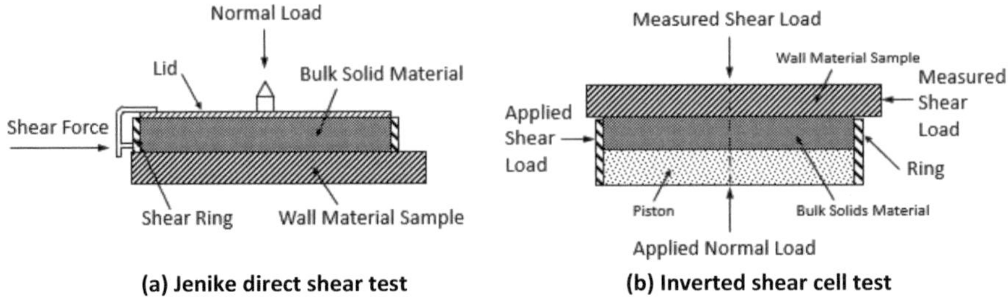

Fig. 3.18 Test arrangement for determination of wall yield loci

with 2B finish and TIVAR 88, an ultra-high molecular weight glass filled polyethylene material. While in absolute terms the moisture contents of the two coals are significantly different, in relative terms with respect to their composition and saturated moisture conditions, they are comparable. For the black coal, the wall yield loci for the stainless steel and TIVAR surfaces are similar, both exhibiting low friction, this also being the case for the brown coal on the TIVAR surface. However, as is clear, the friction characteristics of brown coal on the stainless-steel surface leads to abnormally high friction angles rendering this surface entirely unsuitable as a hopper or chute lining material for brown coal. Even though the TIVAR wall material performed well for the black coal in this example as it does indeed for a great many black coals, occasions can occur where ultra-high polyethylene wall materials may prove to be unsuitable.

Figure 3.19c illustrates the influence of particle size on the wall yield loci. The graphs show the wall yield loci for two particle size ranges of the same black coal on two different wall materials. As indicated, a significant difference occurs for the black mild steel surface, whereas for the bright stainless-steel surface, the change in particle size range has produced practically no change in the respective wall yield loci. The behaviour depicted in Fig. 3.19c may be explained in terms of the interrelation between the particle size distribution and wall roughness.

The adhesive/cohesive characteristics of the WYL discussed in Sect. 2.5.5 (Chap. 2) is exhibited in the various WYL shown in Fig. 3.19. It is to be noted that the Jenike direct shear type apparatus does not permit adhesion to be measured directly. This is because there is always positive pressure between the bulk solid and the test surface due to the weight of the shear ring, lid and contained bulk solid sample. The inverted shear cell of Fig. 3.18b allows for accurate direct measurement of both adhesion (defined as the normal stress for zero shear) and cohesion (defined as the shear stress for zero normal stress or pressure). The device allows the shear to be conducted at zero and negative (that is tensile) pressures.

Fig. 3.19 Typical wall yield loci

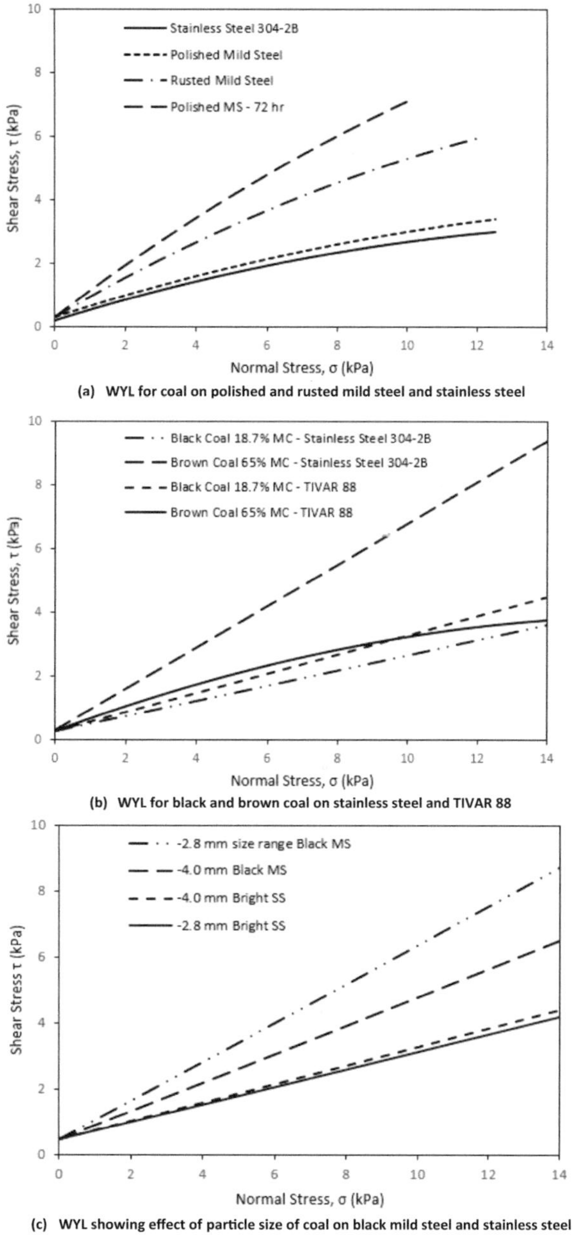

3.10 Surface Roughness

Surface roughness is an important factor in terms of its influence on wall friction. Yet the specification of surface roughness in terms of appropriate parameters which adequately describe the surface is a complex matter requiring careful and detailed consideration. The measurement and analysis of surfaces has been discussed in some detail by Thomas [16] and Nowicki [17]. Nowicki discusses the evolution of surface roughness measurement and lists some 32 parameters that may be used. These may be grouped into six categories, namely:

(a) Height parameters
(b) Horizontal parameters
(c) Shape parameters
(d) Shape spatial extent and amplitude
(e) Statistical parameters
(f) Three dimensional Characteristics of Surface Roughness

It is not feasible in this report to discuss these parameters in detail; rather, a brief overview is given highlighting those parameters that are more commonly used.

(a) **Height Parameters**

These are perhaps the more commonly used parameters; they are simple to describe but are more qualitative than quantitative in the information they provide. The height parameters include:

- Arithmetic mean deviation R_a or centreline average roughness or CLA
- Ten-point height of irregularities or R_z Number
- Maximum height of profile or R_{max}
- Root Mean Square (RMS) roughness R_q
- Mean height of peaks R_p and mean depth of valleys R_v
- Peak to valley height R
- Mean of maximum height of profile R_{tm}
- Profile solidity factor $k = R_v/R_{max}$

Of the above, the R_a and R_q numbers are perhaps the more commonly used. Referring to Fig. 3.20, these two parameters are described as follows:

$$R_a = \frac{1}{L}\int_0^L |y(x)|dx \tag{3.3}$$

where

y(x) is the coordinate height measured from the mean centre line shown.

The RMS roughness has the advantage in that it weights the deviation from the centre line more heavily than that in the R_a determination:

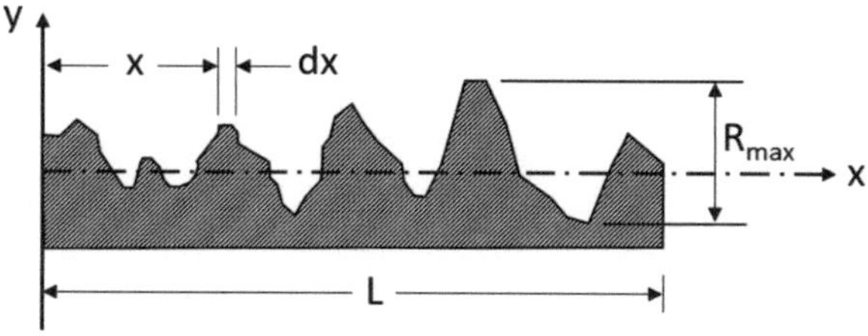

Fig. 3.20 Surface profile

$$R_q = \sqrt{\frac{1}{L}\int_0^L [y(x)]^2 dx} \tag{3.4}$$

In addition to the R_a and R_q numbers, the R_{max} and R_z are also commonly used in practice.

(b) **Horizontal Parameters:**

These include:

- Mean spacing of profile irregularities S_m
- Number of intersections of profile with mean line n(0)
- Number of peaks in profile m

(c) **Shape Parameters Associated with Microroughness**

These include:

- Profile slope at centre line
- RMS profile slope Δ_q
- Radius of asperity r
- Curvature radius of profile o

(d) **Parameters Associated with Shape, Spatial Extent and Amplitude**

These include:

- Relative length of profile l_o
- Mean slope of profile Δ_a
- High spot count HSC

(e) **Statistical Parameters**

The tools of statistical analysis provide an important means for analysing and specifying surface profiles. Parameters commonly used include:

- Auto-correlation function $R_y(\Delta_x)$
- Spectral density G(fo)
- Distribution of ordinates of profile $f_h(y)$
- Distribution of peaks $f_w(y)$

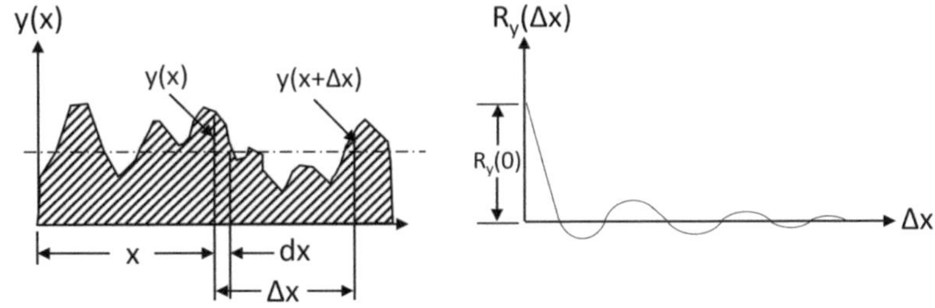

Fig. 3.21 Auto-correlation function

The use of statistical parameters such as the Auto-correlation Function and Spectral Density provide information on the amplitude versus 'waviness' of surfaces. These parameters are defined as follows:

(i) **Auto-Correlation Function:**

The Auto-correlation Function, which is shown in Fig. 3.21, is given by

$$R_y(\Delta x) = \frac{1}{L}\int_0^L y(x)y(x + \Delta x)dx \tag{3.5}$$

It is to be noted that, when $\Delta x = 0$, the auto-correlation function equals the mean square value. Hence the RMS value is given by:

$$\text{RMS value } R_q = \sqrt{R_y(0)} \tag{3.6}$$

(ii) **Spectral Density:**

One parameter which describes the vertical and horizontal roughness is the Amplitude-Frequency spectrum obtained from the Fourier Transform of the surface roughness profile

$$Y(f) = \int_{-\infty}^{\infty} y(x)e^{-j2\pi fx}dx \tag{3.7}$$

where

f= frequency = $1/\lambda$
λ = wave length

$j = \sqrt{-1}$

Alternatively, the Spectral Density, which is obtained from the Fourier Transform of the Auto-correlation Function, may be used. The Spectral Density is defined as,

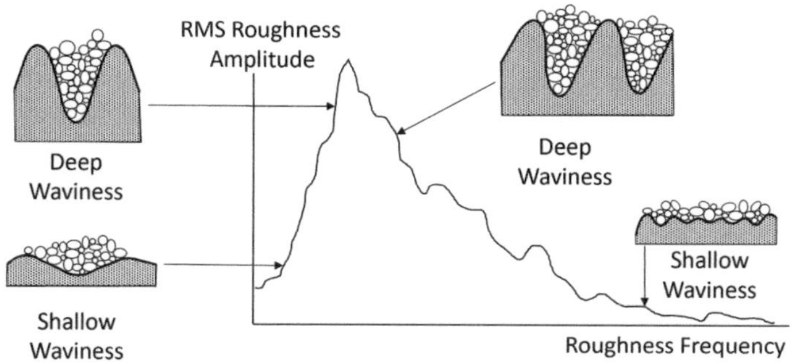

Fig. 3.22 Typical amplitude frequency spectrum of a random surface

$$G(f) = \int_{-\infty}^{\infty} R_y(\Delta x) e^{-j2\pi fx} dx \tag{3.8}$$

The Frequency spectrum has been used by Ooms and Roberts [18] to describe the surfaces of wall materials used in hoppers and chutes. The information obtained from Spectral Density diagrams is illustrated in Fig. 3.22. Using this diagram, ready correlation with the data on bulk solid particle size distribution can be obtained. In this way, it is possible to assess the interaction between the bulk solid and wall surface with respect to relative sliding motion.

By examining the amplitude versus wavelength (i.e., inverse of frequency) for different zones of the Spectral Density diagram and relating this information to the particle size distribution, it is possible to assess the extent to which solids to surface sliding and solids to solids sliding take place. The latter occurs when the particles are smaller than the roughness grooves formed in the boundary surface; this is likely to result from the deep waviness zones at the low frequency end of the Spectral Density diagram as illustrated in Fig. 3.22.

By way of illustration, Fig. 3.23 compares the wall yield loci, surface profiles and surface roughness amplitude frequency spectra for black coal on two samples of stainless-steel type 304 with 2B finish.

One stainless steel sample is from a 3 mm plate and the other from a 5 mm plate; while both have the same centreline or R_a index, the plates were produced by different rolling processes. It is quite evident that the frequency spectra are quite different, and this is reflected in the wall friction characteristics of the 5 mm plate, the rougher surface of the two, being the higher.

The interrelation of particle size distribution and surface roughness has an important bearing on the wall or boundary friction. In order to illustrate this matter, Fig. 3.24, which is based on the work of Ooms and Roberts [18], shows the wall friction angle as a function of mean particle size for black coal on stainless steel 3CR12. The results correspond to one normal consolidation pressure; the coal sample was sieved to form discrete size ranges with the mean values as plotted. The increase in wall friction with decrease in particle size is clearly shown.

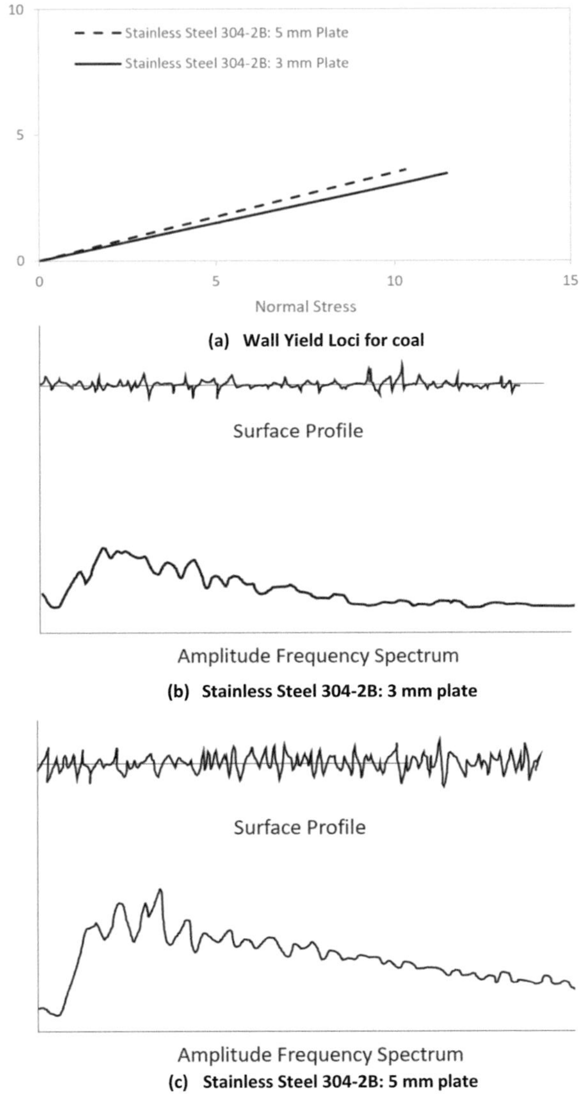

Fig. 3.23 Wall yield loci, surface profile and roughness frequency spectra for coal on stainless steel. (Redrawn from [18])

The effect of increased fines on hopper and chute performance may be directly observed. For hopper design and performance, wall friction has by far the major influence, significantly affecting the magnitude of the hopper half angle. For comparison purposes, the mass-flow half-angles for conical and plane-flow hoppers, corresponding to the friction angles measured, are also plotted in Fig. 3.24. In the case of the conical hoppers, the half-angles incorporate a 3° design margin. As may be seen,

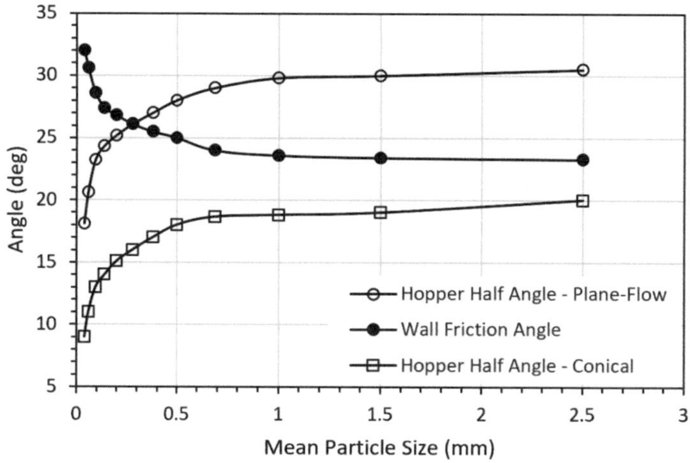

Fig. 3.24 Variation of wall friction angle and mass-flow hopper half-angle with mean particle size. Black coal $\delta = 50°$; Lining material = Stainless steel 3CR12

a decrease in particle size will require a significant increase in hopper half angle if mass-flow is to be maintained.

(f) **Three Dimensional Characteristics of Surface Roughness**

Although it is obvious, it is, nonetheless, necessary to draw attention to the fact that the surface roughness is three dimensional, with two horizontal or lateral coordinates in addition to the height coordinate. Thus, for a given bulk solid, the wall friction characteristics may vary according to the direction of sliding motion. Thomas [16], for example, has examined the three dimensional roughness characteristics of surfaces.

3.11 Roughness Classification for Surfaces Used in Bulk Solids Handling

The publication AS3774-1996 Loads on Bulk Solids Containers [19] proposed four surface roughness characteristics, designated D1 to D4, ranging from polished to very rough or corrugated surfaces. In order that these surface classifications may be quantified in some way, Ooms of The University of Newcastle (Australia) [20] proposed that the four categories of surfaces be grouped in terms of roughness bands based on the Mean Centreline Roughness or R_a number. Surfaces commonly used in Bulk Solids Handling have been classified according to this procedure and presented in Fig. 3.25. The four main groups are:

D1 Polished e.g. Polished stainless steel, Extruded H.D.P.E., Galvanised carbon steel.

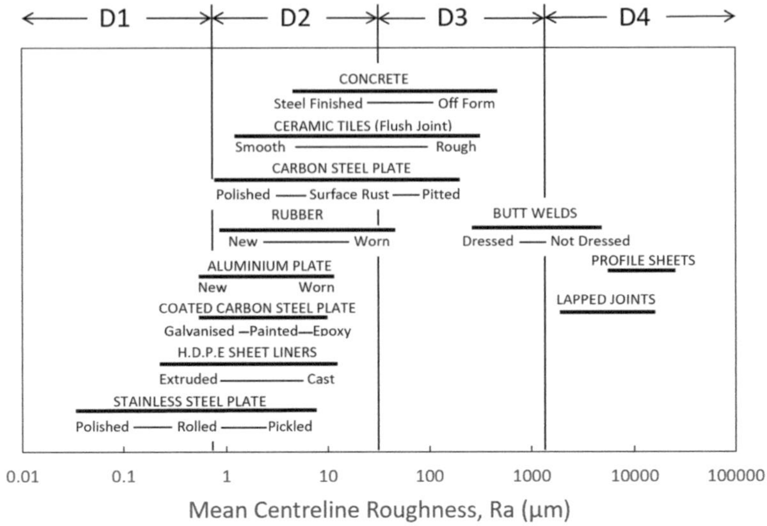

Fig. 3.25 Roughness classification proposed by Ooms (Re-drawn from [20])

D2 Smooth	e.g. Pickled stainless steel, cast H.D.P.E., painted carbon steel, carbon steel with light surface rust, smooth ceramic tiles, steel finished concrete.
D3 Rough	e.g. Off form concrete, pitted carbon steel, coarse ceramic tiles.
D4 Corrugated	e g. Profiled sheeting with horizontal lag.

From the discussion in Sect. 3.10, it is apparent that the proposed form of classification is somewhat restrictive in terms of the limited information conveyed by the R_a number. The limitations have been pointed out by Ooms [20], who indicates that the wall roughness may not necessarily remain constant and should be considered as a variable. Furthermore, the interaction effect referred to in Sect. 3.9 has also been noted by Ooms. For instance, a polished or lightly rusted carbon steel surface may become deeply pitted and change from group D2 to group D3. An aluminium surface is easily scored and may change from group D1 to group D2. On the other hand, some stainless-steel surfaces will polish during service and may change from group D2 to group D1. It must be noted that the lateral characteristics of surface roughness are two dimensional and, during service, wear due to motion of a bulk solid may be more pronounced in one lateral direction than the other.

The changes referred to in the previous paragraph will depend on the relevant properties of the wall material and the bulk solid. However, as Ooms points out, they are somewhat predictable within the broad roughness groups D1–D3. The influence of discontinuities such as ledges or protrusions offered by butt welds and lap joints should be considered.

The interaction between particle size and wall roughness, as discussed in Sect. 3.3, requires consideration. For instance, in the case of larger particles, group D2 and D3 surfaces may appear as polished or smooth surfaces, respectively. On the other hand, for extremely fine powders, most surfaces may appear as Group D3 or rough surfaces.

The influence of adhesion and moisture effects requires comment. The adhesion component of total wall friction is generally higher for polished or group D1 surfaces than it is for group D2 and D3 surfaces at low normal pressures. The effect of moisture will accelerate the deterioration of carbon steel surfaces.

3.12 Direct Shear Test for Elastic and Rheological Properties

3.12.1 Coulomb Elastic Model

As discussed in Sect. 3.3, the Jenike shear cell test focuses on the measurement of the shear strength as a function of time. Additional information of importance to the dynamic behaviour of bulk solids is readily available from the shear cell tests when the shear strength is measured as a function of shear deformation. The shear test model is illustrated in Fig. 3.26.

As shown in Fig. 3.26, the shear force S is composed of two zones:

- Linear zone comprising a "step" response F_f due to the cohesion of the bulk solid followed by a "ramp kx" response due to the elasticity of the consolidated solid
- A non-linear response as the sample approaches and reaches shear failure.

The results for this shear test example are shown in Fig. 3.27. The sample was consolidated under the applied normal load of V = 59.5 N (σ = 8.33 kPa) and then sheared under the applied normal load V = 20.3 N (σ= 2.85 kPa). The initial cohesive component of the shear phase of the test was F_f = 13 N (τ = 1.8 kPa) and the shear load at failure S = 29 N (τ = 4.1 kPa). For the linear elastic zone during shear the bulk stiffness was k = 60N/mm (k = 8.4kPa/mm).

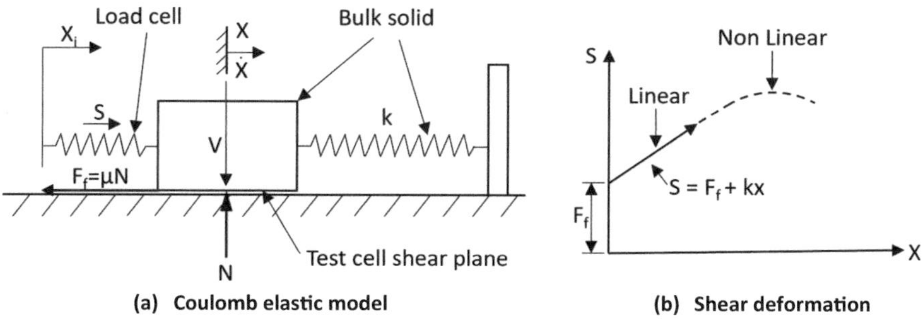

Fig. 3.26 Shear test model

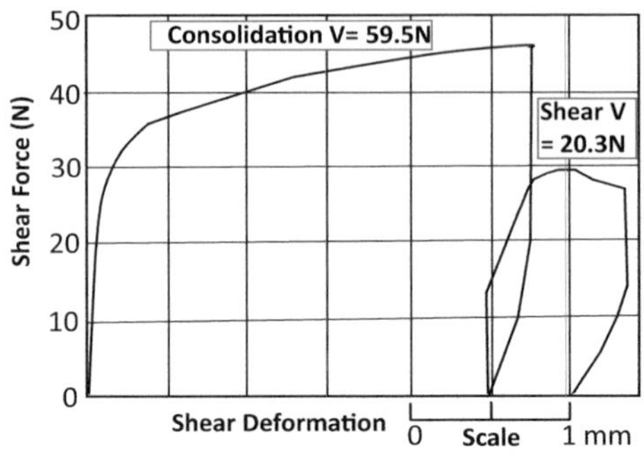

Fig. 3.27 Sample test results showing the consolidation and shear failure. Pyrophyllite (a refractory clay); 5% moisture content (db)

It is interesting to note that after shear failure and the shear force is relaxed (i.e. S = 0), shear force relaxes with a step plus ramp response, the reverse image of the loading response.

3.13 Influence of Vibrations

(a) Introductory Remarks

Roberts et al. [21–24] have studied the influence of mechanical vibrations on the strength and flow properties of bulk solids. This work has led to the following conclusions:

- Bulk strength can be appreciably reduced by the application of mechanical vibrations the best results being achieved by the application of higher frequencies (>100 Hertz) and small amplitudes (approximately 0.1 mm)
- A theory of failure has been established which shows that the reduction in strength decreases exponentially with the vibration velocity, the limiting condition being governed by the critical state condition.
- The damping is a combination of Coulomb and viscous type damping
- The application of vibrations to a wall or surface can significantly reduce wall friction and therefore promote flow.

A vibrating Shear Cell apparatus has been developed at the University of Newcastle, Australia. This apparatus is an extension of the standard Jenike type shear tester with provision for vibrations to be applied to the shear cell, either without shear or simultaneously with an applied shear force. In the former, it is possible to examine vibration consolidation effects on bulk strength; in the latter, it is possible to measure the reduction in bulk strength due to applied vibrations, the data obtained being of relevance to the design and selection of vibratory flow promotion devices.

The test arrangements using the dynamic shear cell allows for vibrations to be applied to the whole cell or top half only as indicated in Fig. 3.28.

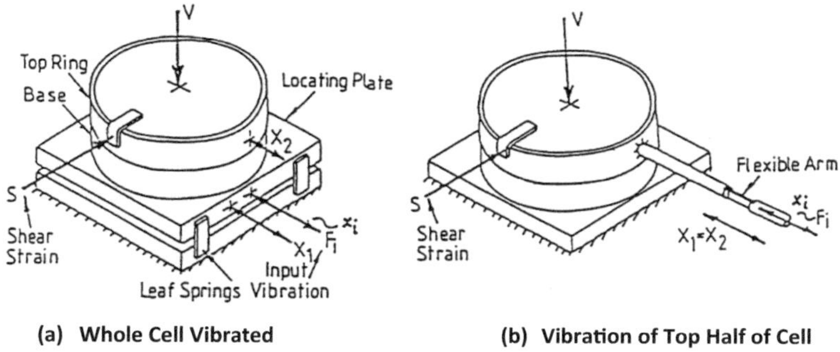

Fig. 3.28 Test arrangements for the vibrating shear cell [24]

(b) Influence of Vibrations on Bulk Strength

The research has shown that the reduction in shear stress of bulk solids subjected to vibrations decreases exponentially with increase in vibration velocity as shown in Fig. 3.29. The vibration velocity on the shear plane correlates with the voidage.

The theory of failure is given by:

$$\tau_f = \tau_\infty + \left(\tau_{fo} - \tau_\infty\right)e^{-2\pi X_r \frac{f}{U}} \tag{3.9}$$

where

$\tau_f =$ Shear stress measured at frequency f.

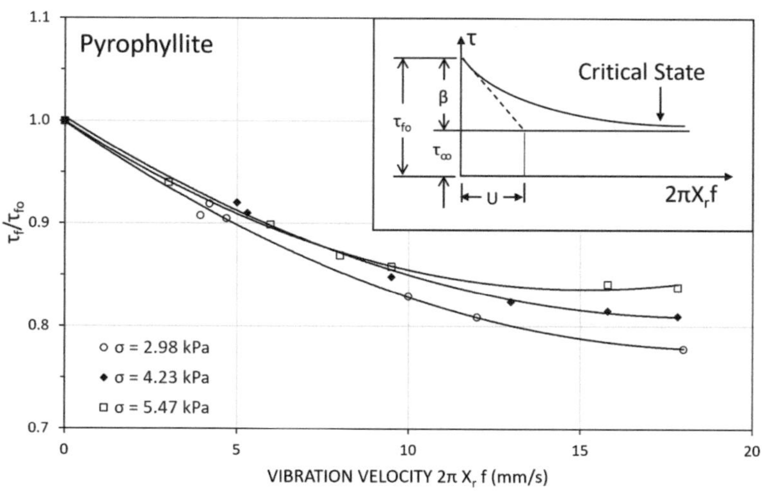

Fig. 3.29 Shear stress attenuation as a function of relative velocity on the shear plane –1 mm pyrophyllite (a refractory clay); 5% moisture Content (db); Normal pressure during consolidation = 7.9 kPa; Amplitude of Applied vibration = 0.1 mm

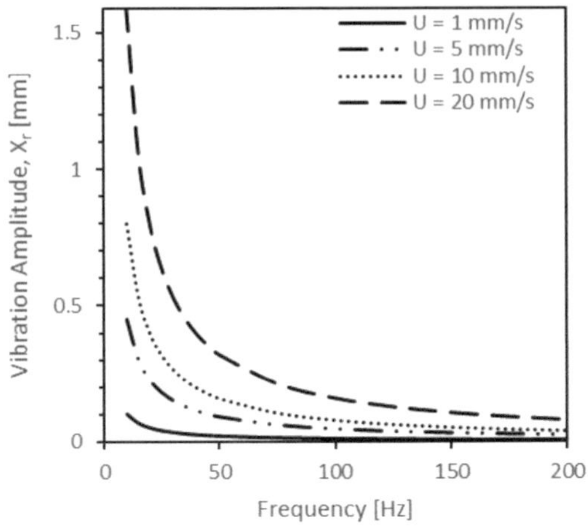

Fig. 3.30 Relationship between $X_r f$ and U for maximum shear strength reduction

$\tau_\infty =$ Limiting shear strength as indicated in Fig. 3.30.
$\beta =$ Shear stress as indicated in Fig. 3.29.
$f =$ Frequency (Hz).
$X_r =$ Vibration amplitude (mm).
$U =$ Vibration velocity constant (mm/s).

Experimental evidence suggests that U is independent of the consolidation pressure and applied normal pressure during vibration. U appears to be constant for a particular bulk solid; for example, for pyrophyllite (a refractory clay) $U = 7$ mm/s and for iron ore $U = 10$ mm/s. Knowing the value of U for a particular bulk solid, the value of the relative amplitude X_r on the shear plane and frequency f for maximum shear strength reduction may be estimated from

$$X_r f = 0.8U \tag{3.10}$$

The form of Eq. (3.10), for a range of values of vibration velocity constant U is given in Fig. 3.30. Experience has shown that the best results for the maximum reduction in shear strength for flow promotion purposes is achieved by using higher frequencies, at least 100 Hz, and lower amplitudes.

For bin design, the influence of vibrations on the flow function is of particular interest. By way of illustration, Fig. 3.31 shows the unvibrated and vibrated flow functions for pyrophyllite; for the latter, a vibration velocity of U ≈ 13mm/s was used, this being close to the value required for maximum shear strength reduction. The reduction in strength shown by the vibrated flow function is quite appreciable. The influence on bin design is discussed in Chap. 6.

(c) **Influence of Vibrations on Wall Friction**

The apparatus of Fig. 3.28 also permits measurement of the vibrated wall yield locus. In this case, a sample of the wall material is used in place of the lower half of the

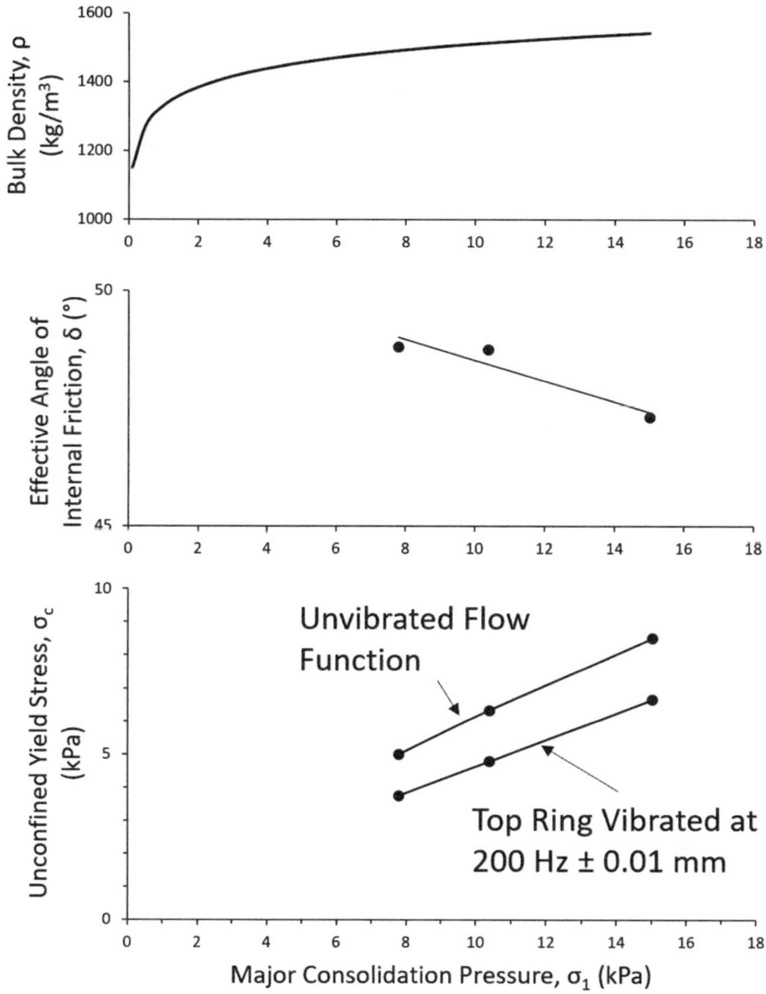

Fig. 3.31 Effect of vibration on flow function for −1 mm pyrophyllite at 5% M.C. (d.b.)

shear cell. The influence of mechanical vibrations on wall friction, is illustrated in Fig. 3.32, which presents the wall yield loci for pyrophyllite on mild steel plate.

As shown in Fig. 3.32, the application of vibrations can significantly reduce wall shear stresses and hence wall friction, particularly as the vibration frequency increases.

Measurements have been made of the dynamic or apparent roughness which indicates the relative surfaces condition when surfaces are vibrated. The apparent roughness is shown to decrease as the vibration frequency increases, this being illustrated for rusty mild steel in Fig. 3.33. Also shown in Fig. 3.33 is the variation of the wall friction angle for coal on the same surface. The reduction in wall friction

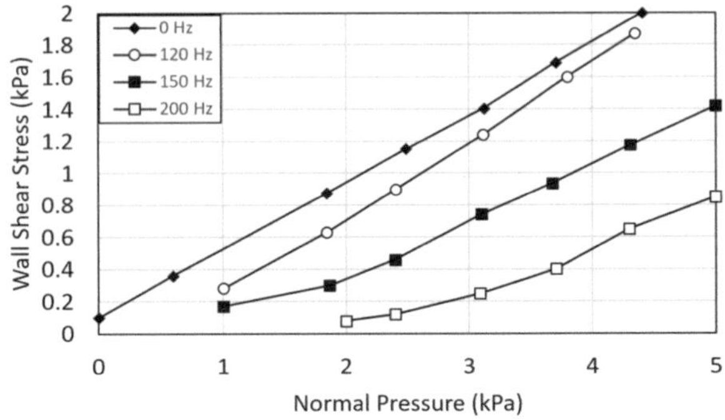

Fig. 3.32 Vibrated wall yield loci for pyrophyllite on mild steel plate. Vibration amplitude $X_r = 0.1$ mm. Moisture content = 5% (w.b.)

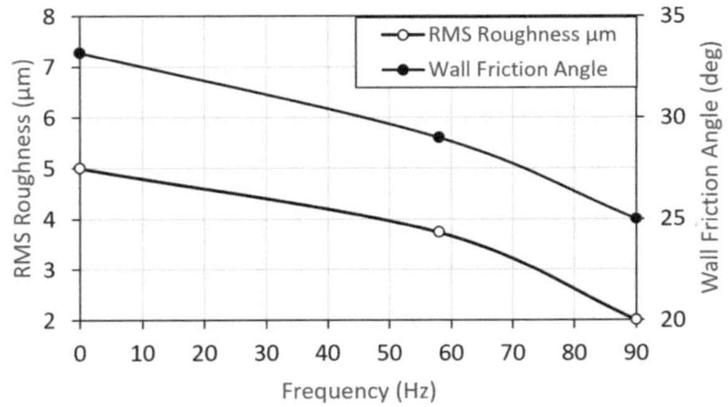

Fig. 3.33 Variation of dynamic or apparent RMS roughness and wall friction angle with vibration frequency for coal on rusty mild steel plate

angle with increase in frequency has the same characteristic shape as the apparent roughness; this indicates a correlation between wall friction and apparent roughness, but further research is necessary to substantiate this conclusion.

3.14 Ageing Effects

Ageing effects, due mostly to extended periods of undisturbed storage, may have a detrimental effect on the operation of a storage system by:

- causing an increase in strength of the bulk solid material in the storage bin or stockpile
- causing hopper wall surface corrosion and deterioration due to wet solids contact.

The influence of undisturbed storage over extended time periods on bulk strength was discussed in Sect. 3.8 and illustrated in Fig. 3.13. It is quite apparent that the gain in strength is due to two principal causes:

(i) 'Mechanical' settling under the action of the stress field established in the bulk solid as the solid approaches, asymptotically, its equilibrium condition.
(ii) Inter-particle bonding due to surface chemistry effects.

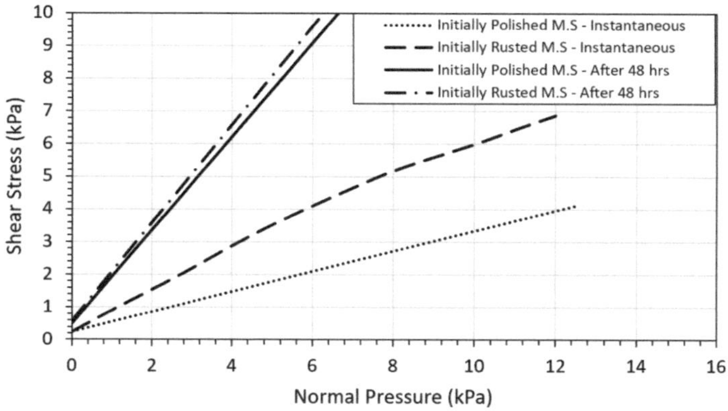

Fig. 3.34 Surface friction of coal on carbon steel surfaces

Impurities in bulk solids, such as clay in coal, can lead to problems in storage and handling. Problems of this type are known to occur when coal is left undisturbed in stockpiles for extended periods of time. Oxidation and dispersion of clays arising from the weathering process are likely to create difficulties due to adhesion and blockages in subsequent handling and storage operations.

For bulk solids in contact with boundary surfaces such as in chutes and hoppers, the influence of prolonged undisturbed contact can be quite dramatic. It has been found that this effect is more severe on carbon steel surfaces in contact with wet solids, particularly black coals which may contain quantities of corrosive agents. The effect of prolonged storage on carbon steel was illustrated in Fig. 3.19a. A further example of this phenomenon is shown in Fig. 3.34. This figure shows comparative results of wall friction tests, using the Jenike direct shear apparatus, of wet black coal on mild steel surfaces for instantaneous conditions and for conditions simulating 48 h undisturbed storage at ambient temperatures.

In applications where temperature gradients or elevated temperatures exist special attention needs to be given to hopper surfaces in relation to corrosion resistance to maintain smooth surfaces and reliable performance. High density polyethylene lining and some stainless steels have been found to perform well where carbon steels have not performed well. In two 3500 tonne flat bottom coal silos, with 7 outlet hoppers constructed of carbon steel, arching and piping occurred when an attempt was made to discharge coal after four days of undisturbed storage during a conveyor breakdown. Inspection of the hoppers showed extreme degradation of the hopper surfaces caused by corrosion. Testing indicated that stainless steel lining (or some other suitable lining unaffected by the corrosive atmosphere and having an appropriate wall friction angle) placed over the existing Mild Steel would prevent a re-occurrence of flow interruptions.

3.15 Phase Separation

During periods of undisturbed storage of wet bulk solids in bins, some phase separation may take place. Materials such as coals and ores with high moisture content may go through some dramatic physical changes. A good design takes this into account by determining the flow properties for a wide range of moisture conditions. In open stockpiles in areas of high rainfall this can be a substantial problem. Arnold et al. [25] reports on the problem of water washing the fines through a coal stockpile and causing 'slumping' into the reclaim tunnel. Correct design taking account of these possibilities will give reliable operation under all conditions.

For extended periods of storage, after the excess of free moisture has either gravitated through the material or been allowed to evaporate, a different type of phase separation problem may arise, particularly in materials which are soluble or have soluble components. Liquid bridges present in wet bulks solid may crystallise forming solid bridges. Some interesting work has been done in an attempt to measure these effects. Bagster [26] has investigated caking of raw sugar while Tomas and Schubert [27] have conducted experiments with fine ores.

3.16 Abrasive Wear

The TUNRA Bulk Solids circular bed wear tester can be used to experimentally measure the wear rate of samples of wall material abraded against the actual wear media seen on site such as iron ore, bauxite, gravel and sand. A description of the development of the test facility is given in [28] with further details described in [29]. The wear media is continuously mixed by a passive plough and grading system and the moisture of the sample is controlled. The circular bed wear tester is shown in Fig. 3.35, where the mass loss of wear samples due to abrasive interaction with the bulk material is monitored over time to determine a steady state wear rate for each sample.

3.17 Impact Wear

The impact wear characteristics of wear lining materials from an abrading bulk solid material can be assessed by conducting impact wear testing in a purpose designed rotary impact wear tester located at TUNRA Bulk Solids, pictured in Fig. 3.36, schematic shown in Fig. 3.37. The test facility consists of a large rotating wheel with buckets attached internally that continuously lift the abrading bulk material up and feed it into a fixed hopper. The material flows through the hopper and down a straight vertical chute, impacting the test specimens upon exit of the hopper. Mass loss of the

Fig. 3.35 Abrasion wear testing system [29]

test specimens due to the impacting stream of abrading bulk material is monitored over time to determine the steady state wear rate.

The specimen mounting plate is shown in Fig. 3.38. It permits one 300 × 300 mm or two 300 × 150 mm wear liners to be mounted to it at angles ranging between 0° and 180°, in 5° increments.

3.18 Dustiness Tests

For obvious environmental reasons, the control of dust in bulk solids handling and processing plants has a high priority. Through proper design, passive (non-energy) dust control can be achieved in process plants such as in conveyor feeding and transfer operation. In open transport operations such as rail wagons and large storage systems notably stockpiles, the control of dust generation due to windage needs to be achieved. This is particularly important in the case of the storage and transport of mineral ores such as coal where the propensity for dust generation will vary with moisture content and coal type.

Australian Standard AS 4156.6–2000 [30] provides the specification of the test equipment and measurement procedures for the determination of the dust versus moisture relationship for coal. The test is also of value for the assessment of surface

Fig. 3.36 Rotary impact wear tester (TUNRA Bulk Solids, The University of Newcastle)

Fig. 3.37 Render of TUNRA Bulk Solids' impact wear tester

Fig. 3.38 Specimen mounting plate

sealing surfactants for controlling dust losses due to windage from open stockpiles. While the Standard refers specifically to coal, it is equally useful as a test for virtually all bulk solids.

The test equipment is shown in Fig. 3.39. It consists of a rotating drum fitted with eight longitudinal vanes or lifters. The test sample is placed in this drum and air is drawn through the sample as it rotates carrying dust particles to the stationary filter collection bag held within the sealed stationary compartment. The specific details of the test procedure are given in the Standard. Besides this test, wind tunnel tests are also used for dustiness tests.

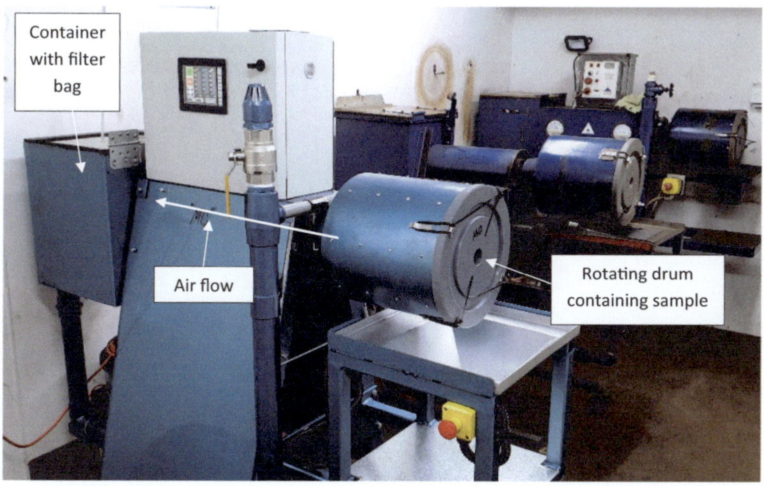

Fig. 3.39 Dustiness tester

3.19 Permeability and Particle Solids Density

In the case of fine powders, such as pulverised coal, permeability and particle solids density are properties that need to be measured.

(a) **Permeability**

The settling and consolidation characteristics of fine powder type bulk solids, when placed in storage, and the gravity discharge characteristics during flow from storage are governed, to a large extent, by the ability of air to permeate through the stored mass. Permeability is determined as a function of consolidation pressure by measuring the pressure drop across a sample as a function of sample height. This is illustrated in Fig. 3.40.

The relationship is given by Darcy's equation:

$$v = c^{dp/dx} \approx c^{\Delta p/h} \tag{3.11}$$

where

v = air velocity.
c = permeability factor.
Δp = pressure drop across sample.
h = sample height.

The permeability factor is given by:

Fig. 3.40 Permeability test

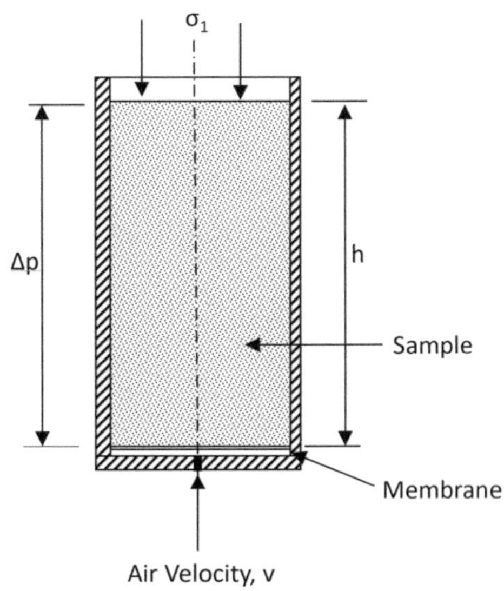

$$c = \frac{c_0}{\sigma_1{}^a} \tag{3.12}$$

where

c_0 = constant
σ_1 = major consolidation pressure
a = permeability constant for the bulk solid

(b) **Solids Density**

In addition to measurements of bulk density, information on the particle solids density is sometimes required. In particular, the estimation of flow rates of fine powders requires knowledge of the particle density.

A simple method for measuring the solids density utilises an air comparison pycnometer. This instrument is illustrated in Fig. 3.41. This instrument, which measures the relative density of the solid particles in air, determines the true particle volume of a powder sample of known mass. The solids density is then

$$\rho_s = \frac{\text{mass}}{\text{particle volume}} \tag{3.13}$$

The air-comparison pycnometer consists basically of two chambers and two pistons, a valve connecting the two chambers, a differential pressure indicator, and a digital counter calibrated for readings in cubic centimetres.

With the connecting valve closed any change of the position of one piston must be duplicated by an identical stroke in the other to maintain the same pressure on each side of the differential pressure indicator. If the connecting valve is closed and both pistons are advanced the same amount to a given position, inserting a sample in the measuring chamber would cause a pressure differential which is brought to

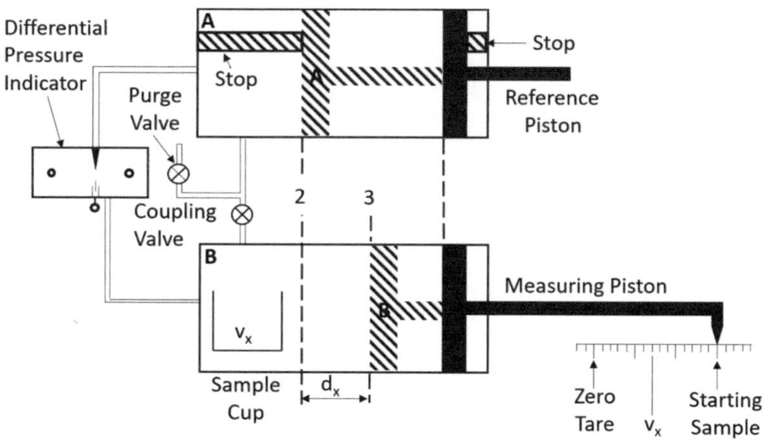

Fig. 3.41 Air comparison pycnometer

zero by withdrawing the piston in this chamber. Using this condition, the distance that the measuring piston differs from its position before inserting the sample will be proportional to the volume being measured. In this way the true volume of a sample is measured.

3.20 Concluding Remarks

The material presented in this Chapter has outlined the principal test procedures that are used to determine the flow properties of a bulk solid. With the aid of these properties, it is possible to design bins and handling plant for reliable performance.

References

1. Schulze, D. *A new ring shear tester for flowability and time consolidation measurements.* in *Proc. 1st International Particle Technology Forum.* 1994.
2. Harder, J. and J. Schwedes, *The development of a true biaxial shear tester.* Particle & Particle Systems Characterization, 1985. **2**(1-4): p. 149–153
3. Johanson, J., *The johanson indicizer [tm] system vs. the jenike shear tester.* Bulk Solids Handling, 1992. **12**: p. 237–237.
4. Solids, E.W.P.o.t.M.o.P., *Standard shear testing technique for particulate solids using the Jenike shear cell 46 pp.* 1989, IChemE. p. 46.
5. Jenike, A.W., *Gravity Flow of Bulk Solids.* Utah Agric. Exp. St. Bull, 1961. **108**.
6. Jenike, A.W., *Storage and Flow of Solids.* Utah Agric. Exp. St. Bull, 1964. **123**.
7. Schwedes, J. and J. Harder. *Influence of stress history on the yield limit of bulk solids.* in *Second International Conference on Bulk Materials Storage, Handling and Transportation: 1986; Preprints of Papers.* 1986. Institution of Engineers, Australia Barton, ACT.
8. Maltby, L. and G. Enstad, *Uniaxial tester for quality control and flow property characterization of powders.* Bulk Solids Handling, 1993. **13**: p. 135–135.
9. Janssen, R., M. van der Kraan, and B. Scarlett, *Evaluation of the flexible wall biaxial tester for the measurement of BCR-limestone.* Preprints of the 1st European symposium process technology in pharmaceutical and nitritional sciences, 1998: p. 209–215.
10. Janssen, R.J.M. and H. Zetzener, *Measurements on Cohesive Powder with two Biaxial Shear Testers.* Chemical Engineering & Technology, 2003. **26**(2): p. 147–151.
11. Hvorslev, M.J., *Ueber die Festigkeitseigenschaften gestörter bindiger Boden.* Ingenioervidensk. Ingenieurvidenskabelige Skrifter, Copenhagen., 1937.
12. Roscoe, K.H., A.N. Schofield, and C.P. Wroth, *On the Yielding of Solids.* Geotechnique, 1958. **8**: p. 22–53.
13. Arnold, P.C., A.G. McLean, and A.W. Roberts, *Bulk Solids: Storage, Flow and Handling.* 2nd ed. 1979: The University of Newcastle Research Associates (TUNRA).
14. Roberts, A.W., O.J. Scott, and W. Tiedemann, *Bulk Handling of Ammonium Nitrate: Flow Properties and Storage Bin Design Criteria,* in *International Powder and Bulk Solids Conference.* 1979: Philadelphia, U.S.A.
15. Wiche, S.J., A.W. Roberts, and W.A. McBride, *Flowability Tester for Characterising Bulk Solids,* in *Proc. 8th International Conference on Bulk Materials Storage, Handling and Transportation, The Instn. of Engrs.* 2004: Australia, Wollongong, NSW, Australia. p. 79–83.
16. Thomas, T.R., *Recent Advances in the Measurement and Analysis of Surface Microgeometry.* Wear, 1975. **33**: p. 205–233.

17. Nowicki, B., *Multiparameter Representation of Surface Roughness.* Wear, 1985. **102**: p. 161–176.
18. Ooms, M. and A.W. Roberts, *The Effect of Surface Roughness on the Design and Performance of Gravity Flow Systems* in *Proc. 10th Anniversary Powder & Bulk Solids* 1985: Chicago, U.S.A.
19. Australia, S.A.o., *Australian Standard AS3774*, in *Loads on Bulk Solids Containers.* 1996.
20. Ooms, M., *A Classification of Roughness of Internal Walls.* 1985, The University of Newcastle, Australia
21. Roberts, A.W. and O.J. Scott, An Investigation into the Effects of Sinusoidal and Random Vibrations on the Strength and Flow Properties of Bulk Solids. Powder Tech., 1978. **21**: p. 45–53.
22. Roberts, A.W., O.J. Scott, and K.W. KLi, *The Influence of Mechanical Vibrations on the Flow of Bulk Solids*, in *Proc. International Conference on Powders and Bulk Solids.* 1979: Philadelphia, U.W.A.
23. Roberts, A.W., *The Effects of Vibration on the Strength and Flow Properties of Bulk Solids in Storage Bin Operation*, in *Proc. International Conference on Design of Silos for Strength and Flow.* 1980: University of Lancaster, England.
24. Roberts, A.W., *Vibration of Fine Powders and Its Application.* In: Fayed, M.E., Otten, L. (eds) *Handbook of Powder Science & Technology.* 1997: Springer, Boston, MA. https://doi.org/10.1007/978-1-4615-6373-0_5.
25. Arnold, P.C., et al., *Improving the Performance of Gravity - Reclaim Stockpile System*, in *Proc. Int. Conf. on Bulk Materials Storage, Handling and Transportation.* 1983: Newcastle, Australia.
26. Bagster, D.F., *A Study of the Caking Behaviour of Raw Sugar under Storage*, in *Proc. Int. Conf. on Bulk Materials Storage, Handling and Transportation.* 1983: Newcastle, Australia.
27. Tomas, J. and H. Schubert, *Flow Behaviour of Moist Bulk Solids*, in *Proc. 8th Int. Congress, CHISA.* 1984: Prague, Czechoslovakia.
28. Wiche, S.J., S. Keys, and A.W. Roberts, *Abrasion wear tester for bulk solids handling applications.* Wear, 2005. **258**(1): p. 251–257.
29. Chen, W., et al., *Abrasion wear resistance of wall lining materials in bins and chutes during iron ore mining.* International Journal of Mineral Processing, 2017. **167**: p. 42–48.
30. 4156.6–2000, A.S.A., *Coal preparation, Part 6: Determination of dust/moisture relationship for coal.* Standards Association of Australia, 2000.

Chapter 4
Design of Storage and Handling Systems for Cohesive Bulk Solids

4.1 Introductory Remarks

The design of storage bins, stockpiles and associated handling plant for the handling and processing of bulk solids has, paradoxically, two seemingly conflicting objectives, the design for "failure" and the design for strength and "non failure". By way of explanation, the storage bin holding the bulk solid, obviously, must be structurally strong enough to safely store the product without failure, but at the same time, have the appropriate geometry to ensure that the internal stresses generated in the bulk solid are sufficient to exceed its yield strength to initiate discharge flow. The governing strength condition is the major consolidation stress σ_1 which influences the bin wall loads while, at the same time governs the bulk solid yield strength which is exceeded when flow is required.

By way of introduction, it is useful to reiterate the four basic steps involved in the design of storage bins and associated handling plant as discussed in Chap. 2:

(i) Determination of the strength and flow properties of the bulk solids for the worst likely conditions expected to occur in practice.
(ii) Determination of the bin or stockpile configuration to give the required storage capacity, flow patterns and discharge rates with acceptable characteristics and to ensure that discharge is reliable and predictable.
(iii) Consider a design to suit available space and estimate loadings on hopper and bin walls and on feeders under operating conditions.
(iv) Design and detailing of the bin and handling system.

The relevant storage and flow properties of bulk solids, presented in Chap. 3, are based on meaningful test procedures designed to simulate actual loading and environmental conditions expected to occur in practice. These include undisturbed storage times, temperature and moisture along with particle size distribution including dust

A. W. Roberts, *The Theory and Practice of Bulk Materials Handling*,
https://doi.org/10.1007/978-3-032-19702-3_4

and, in the case of mineral ore, impurities such as clay. They also include the interaction properties of friction and wear for bulk solids in contact with hopper and bin wall materials and their interaction with boundary surfaces.

The aim of this chapter is twofold:

(i) To develop a design methodology involving appropriate analytical models and associated graphical procedures in combination with relevant research knowledge and industry-based recommendations for the design of mass-flow storage bins and associated handling plant.
(ii) To demonstrate the application of the design procedures to the solution of industry-based problems via practical case study examples.

4.2 Mass-Flow Hopper Design Parameters

As previously stated, the two basic parameters for mass-flow hopper design are the hopper half- angle α and the hopper opening dimension B_{CR} for discharge flow to occur, that is, Jenike's *flow, no-flow criterion.* In view of the relevance of these two parameters to the hopper design theory presented in this Chapter, the following introductory remarks are presented by way of review:

(a) Hopper Half Angle

The hopper half-angle α is obtained by considering the bounds for axi-symmetric or conical and plane-flow hoppers. These bounds were shown in Chap. 2 and, for convenience, are reproduced here. The governing equations, (2.6), (2.7) and (2.8) upon which these bounds are based, also for convenience, are reproduced here as Eqs. (4.1), (4.2) and (4.3) respectively.

In the derivation of the theory of flow in converging channels, Jenike [1] showed that the bounds between mass-flow and funnel-flow in conical or axi-symmetric channels are given by:

$$\alpha = \frac{\pi}{2} - \frac{1}{2}\cos^{-1}\frac{(1 - \sin\delta)}{2\sin\delta} - \beta \tag{4.1}$$

where

$$\beta = \frac{1}{2}\left\{\varphi + \sin^{-1}\frac{(\sin\varphi)}{\sin\delta}\right\} \tag{4.2}$$

Since the flow in the hopper is based on radial stress field theory, no account is taken of the influence of the head in the cylinder on the flow patterns developed, particularly in the region of the transition. For plane-flow hoppers, the bounds between mass-flow and funnel-flow are much less severe than for conical hoppers. In view of this, Jenike [1] has adopted recommended design limits in order to select the hopper half-angle. These design limits may be represented by the empirical equation:

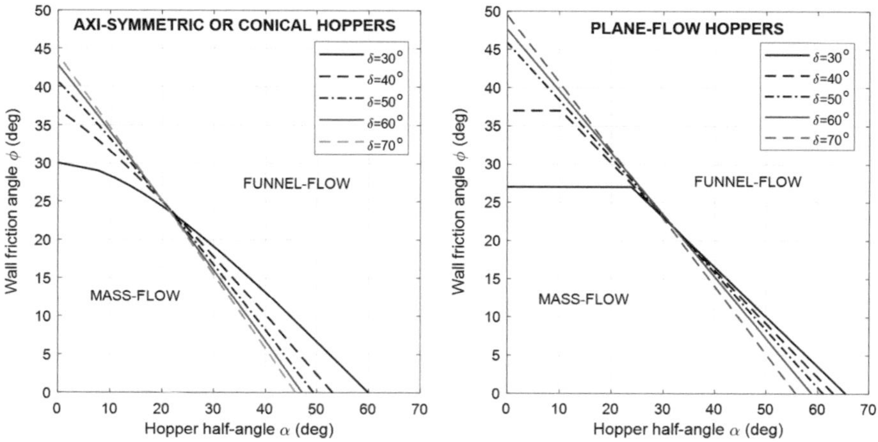

Fig. 4.1 Limits for mass-flow for conical and plane-flow channels

$$\alpha = \frac{1}{0.725(\tan\delta)^{0.2}}\left\{exp\left[3.75(1.01)^{(0.1\delta-3)}\right] - \varphi\right\} \quad (4.3)$$

where δ and φ are in degrees.

The bounds for conical and plane-flow hoppers for five values of δ are plotted in Fig. 4.1. In the case of conical hoppers, it is recommended, as a safety precaution, that the hopper half angle be 3 degrees less than the limiting value. For plane-flow, the design limit may be selected. As discussed in Chap. 2, Sect. 2.7.6 and illustrated in Fig. 2.26, If the transition of the cylinder and hopper is sufficiently radiused or fitted with appropriate gusset plates, then it is even possible to select a hopper half angle 3 to 4 degrees larger than the design limit.

(b) Passive Stress fields and Stable Arch Formation

Having determined the hopper half-angle α based on the hopper wall friction characteristics, the next step in the hopper design is the determination of the hopper opening dimension B which is to be greater than the minimum or critical opening dimension B_{CR} in order to ensure that flow blockages do not occur and the required discharge flow rate is achieved in accordance with the plant operational requirements.

The basic design approach, summarised in Fig. 4.2, involves the interaction between two parameters:

- "Flow Function", *FF*, which is a bulk material parameter representing the functional relationship between the unconfined yield strength σ_c and the major consolidation stress σ_1 That is $FF = f(\sigma_c, \sigma_1)$
- "Flow Factor", *ff*, which is a hopper parameter defined by $ff = \frac{\sigma_1}{\overline{\sigma}_1}$, where $\overline{\sigma}_1$ is the stress acting in the arch as depicted in Fig. 4.2b.

Referring to Fig. 4.2a, the minimum hopper opening dimension B_{CR} for flow to be on the verge of occurring is when the stress in $\overline{\sigma}_1$ acting in the arch equals the

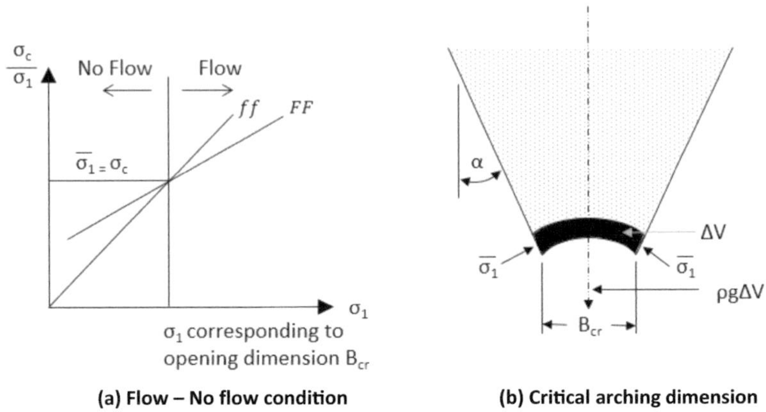

Fig. 4.2 Bulk solid flow function and hopper flow factor

major consolidation stress σ_1. This condition is defined by

$$B_{CR} = \frac{\overline{\sigma}_1 H(\alpha)}{\rho g} \tag{4.4}$$

where

$\rho =$ bulk density
$g =$ gravitational acceleration
$H(\alpha) =$ Arch thickness parameter.

As discussed in Chap. 2, the arch thickness parameter $H(\alpha)$, shown in Fig. 4.3 was derived by Jenike and Leser [2]. As pointed out by Arnold et al. [3], $H(\alpha)$ is not known explicitly, but may be represented by the following relationship

$$H(\alpha) = \left[\frac{130 + \alpha}{65}\right]^m \left[\frac{200 + \alpha}{200}\right]^{1-m} \tag{4.5}$$

where

$m =$ 1 for axi-symmetric or conical hoppers
$m =$ 0 for plane-flow, wedged-shaped hoppers.

For a square opening, $H(\alpha) = 0.9 \times H(\alpha)_{conical}$.

Examples of stable cohesive arches that form when the hopper opening $B < B_{CR}$ are shown in Figs. 4.4 and 4.5. Figure 4.4a shows a stable cohesive arch formed in a plane-flow, wedged-shaped hopper formed in iron ore discharging via a belt feeder. The feeder has been lowered and removed from beneath the hopper to permit a laser scan of the arched profile to be taken. The results of the scan are presented in Fig. 4.4b. This work was part of a research study performed by Guo [4] and subsequently published in Guo et al. [5].

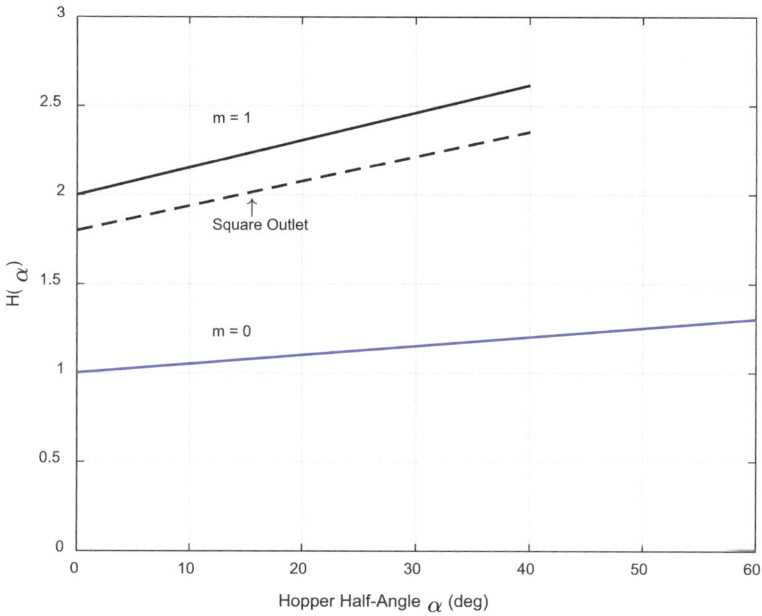

Fig. 4.3 Cohesive arch thickness parameter H(α) [2]

Fig. 4.4 Stable cohesive arch formed in iron ore of hopper and belt feeder [4]

Fig. 4.5 Stable cohesive arch formed in nickel ore of industrial hopper and belt feeder

Figure 4.5 presents an example of a cohesive arch formed in an actual industrial wedged-shaped hopper and belt feeder combination. The bulk material in this case is nickel ore. The similarity of the arched profiles of these examples is quite strong, along with the demonstration of the arched characteristic of the passive stress field.

4.3 A Theoretical Model of a Hopper Passive Stress Field

As indicated in the review of the Jenike bin design theory presented in Chap. 2, there was a perceived need to develop an analytical model to allow "closed form" solutions to facilitate the design of mass-flow hoppers without the need to rely on flow factor charts. In effect the Jenike flow-factor charts could then be used, as required, to "spot-check" the calculated results. This need aroused the attention of Arnold and McLean [6] whose investigations identified the research of Enstad [7] who studied the stress fields in plane-flow, wedged-shaped hoppers. It was found that Enstad's analytical predictions of the Jenike stress fields in plane flow hoppers, particularly the radial stress fields, were most encouraging and acceptable. This led to the extension of Enstad's theory to axisymmetric or conical hoppers and to the establishment of the mass-flow hopper design methodology which is now described. The analytical method has been extended to cover both axi-symmetric and plane-symmetric mass-flow hoppers and presented in detail by Arnold, Mclean and Roberts [3].

4.3.1 Enstad's Theory—Plane Flow, Wedge-Shaped Hoppers

As illustrated in Fig. 4.6a and b, Enstad assumed a passive state of stress formed by circular arch layers. Each arched layer centred on the centreline of the hopper as shown. The major consolidation stress σ_1 is proportional to the radial distance r. At each specific coordinate r, the major consolidation stress is constant in magnitude but variable in direction being tangential to the circular arc defined by the radius r'. The minor principal stress σ_2 acts in the direction of the radius r'. The angle β is the angle between the normal stress σ_n at the wall and the major principal stress σ_1 as depicted in the Mohr diagram, Fig. 4.7 and given by Eq. (4.6).

$$\beta = \frac{1}{2}\left[\varphi + \sin^{-1}\left(\frac{\sin\varphi}{\sin\delta}\right)\right] \tag{4.6}$$

where

$\varphi =$ wall friction angle
$\delta =$ effective angle of internal friction.

The basis of the stress field analysis is the equilibrium of the arch element shown in Fig. 4.6a and b for which the following variables are described:

(a) ΔA is the cross-sectional area of the element taking into account the variation in arch thickness between successive arch profiles separated by Δr. The area

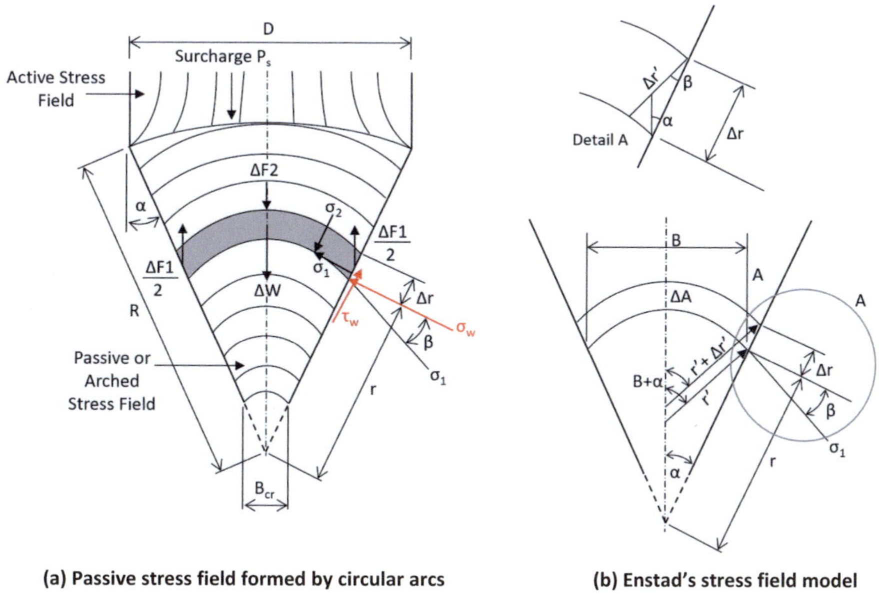

Fig. 4.6 Passive circular arc stress field—redrawn from Enstad [7]

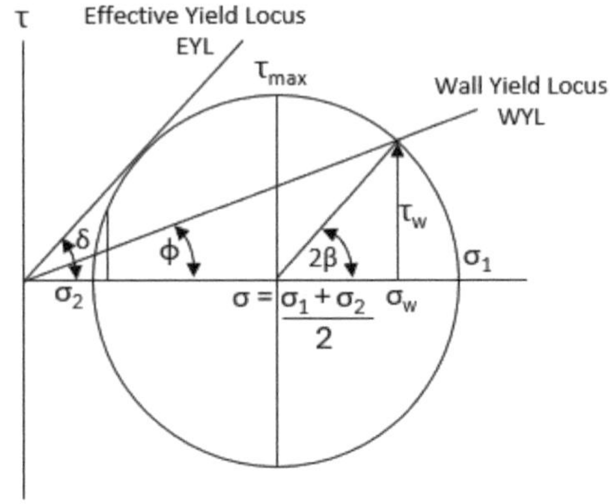

Fig. 4.7 Mohr circle diagram for determination of angle β

ΔA is given by

$$\Delta A = 2r\Delta r\left[\frac{(\sin\alpha)(\beta+\alpha)}{\sin^2(\beta+\alpha)} + \frac{\sin\beta}{\sin(\beta+\alpha)}\right]\sin\alpha \tag{4.7}$$

The weight of the bulk solid element per unit length is

$$\Delta W = 2\gamma r\Delta r\left[\frac{(\sin\alpha)(\beta+\alpha)}{\sin^2(\beta+\alpha)} + \frac{\sin\beta}{\sin(\beta+\alpha)}\right]\sin\alpha \tag{4.8}$$

where

$\gamma = \rho g =$ bulk specific weight
$\rho =$ bulk density
$g =$ gravitational acceleration

(b) For the force balance equilibrium analysis

$$\Delta W = \Delta F_1 - \Delta F_2 \tag{4.9}$$

(c) ΔF_1 is the vertical component of the supporting pressure force provided by the hopper walls.
(d) Through a process of mathematical manipulation, the following differential equation is derived:

$$\frac{d\sigma}{dr} - \frac{X}{r}\sigma = -\gamma Y \tag{4.10}$$

where

$\gamma = \rho g =$ bulk specific weight
$\sigma =$ mean stress $= \sigma_1(1 - \sin\delta)$.

The parameters X and Y are given by

$$X = \frac{\sin\delta}{1-\sin\delta}\left[\frac{\sin(2\beta+\alpha)}{\sin\alpha}+1\right] \tag{4.11}$$

$$Y = \frac{(\beta+\alpha)\sin\alpha + \sin\beta\sin(\beta+\alpha)}{(1-\sin\delta)\sin^2(\beta+\alpha)} \tag{4.12}$$

The solution of Eq. (4.10) is

$$\sigma = \frac{\gamma Yr}{X-1} + \left[\sigma(R) - \frac{\gamma YR}{X-1}\right]\left[\frac{r}{R}\right]^X \tag{4.13}$$

where $\sigma(R)$ = surcharge pressure.

4.3.2 Axi Symmetric or Conical Hoppers

The procedure for the analysis of axi-symmetric or conical hoppers is summarised as follows:

(a) Weight of conical element located at radius r

$$\Delta W = \gamma\pi r^2\Delta r\left[\frac{2\sin\alpha(1-\cos(\beta+\alpha))}{\sin^3(\beta+\alpha)} + \frac{\sin\beta}{\sin(\beta+\alpha)}\right]\sin^2\alpha \tag{4.14}$$

(b) Force balance $\Delta W = \Delta F_1 - \Delta F_2$ as in (4.9)
(c) Derivation of equilibrium equation

$$\frac{d\sigma}{dr} - \frac{X}{r}\sigma = -\gamma Y \tag{4.15}$$

This equation is identical to Eq. (4.10) for the plane flow case except that the parameters X and Y are

$$X = \frac{2\sin\delta}{1-\sin\delta}\left[\frac{\sin(2\beta+\alpha)}{\sin\alpha}+1\right] \tag{4.16}$$

$$Y = \frac{2\sin\alpha(1-\cos(\beta+\alpha)) + \sin\beta\sin^2(\beta+\alpha)}{(1-\sin\delta)\sin^3(\beta+\alpha)} \tag{4.17}$$

4.3.3 Generalising Expressions for X, Y and σ

Employing the mass-flow hopper symmetry factor m, where

m = 0 for plane-flow, wedged-shaped hoppers
m = 1 for axisymmetric, conical hoppers.

The expressions for X and Y become

$$X = \frac{2^m \sin\delta}{1 - \sin\delta}\left[\frac{\sin(2\beta + \alpha)}{\sin\alpha} + 1\right] \tag{4.18}$$

$$Y = \frac{[2(1 - \cos(\beta + \alpha))]^m (\beta + \alpha)^{1-m} \sin\alpha + \sin\beta \sin^{1+m}(\beta + \alpha)}{(1 - \sin\delta)\sin^{2+m}(\beta + \alpha)} \tag{4.19}$$

Note: For the term $(\beta + \alpha)$ in the numerator of Eq. (4.19), the angles α and β are expressed in radians.

The solution of the equilibrium Eq. (4.15) for the mean stress is as follows

$$\sigma = \frac{\gamma Y r}{X - 1} + \left[\sigma(R) - \frac{\gamma Y R}{X - 1}\right]\left[\frac{r}{R}\right]^X \tag{4.20}$$

where $\sigma(R) = \mathrm{p_s}$ = surcharge pressure, $\sigma = \sigma_1(1 - \sin\delta)$ = mean stress.

It is important to note that the first term on the right-hand side of Eq. (4.20) is *linear*, and the second term is *non-linear*. The linear term represents Jenike's *radial stress field* and is independent of the surcharge pressure, while the non-linear term arises as a result of the surcharge load. To illustrate the combined load contributions of these two terms, consider the example of mean stress distributions in a conical mass-flow hopper, as shown in Fig. 4.8.

As illustrated, the non-linear term reduces quite quickly as the surcharge load is transferred to the upper region of the hopper walls. The load transfer then follows the linear reduction pattern as the passive stress field is established in the converging tapered hopper. Thus, for mass-flow hopper design, particularly the determination of the hopper opening dimension to prevent the formation of stable cohesive arches, it is the linear term for the mean stress of Eq. (4.20) that is of particular interest. That is,

$$\sigma = \frac{\gamma Y r}{X - 1} \tag{4.21}$$

Since the major consolidation stress σ_1 is the chosen dominant stress in bin and hopper design, Eq. (4.21) then becomes,

$$\sigma_1 = \frac{\gamma Y r}{X - 1}(1 + \sin\delta) \tag{4.22}$$

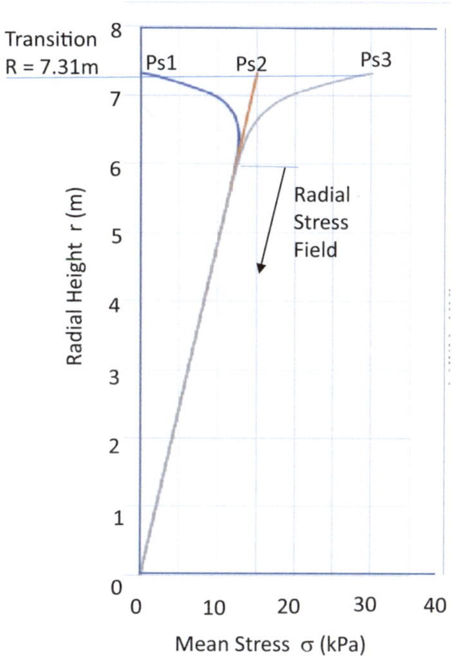

Fig. 4.8 Example solution of Eq. (4.20) for conical hopper. α = 20°, D = 5 m, δ = 50°, ϕ = 22.5°, m = 1, ρ = 1.0 t/m^3, R = 7.31 m Ps1 = 0 kPa, Ps2 = 15.1 kPa, Ps3 = 30 kPa

Noting that the radial distance $r = \frac{B}{2sin\alpha}$, where B is the corresponding hopper opening dimension. The non-dimensional stress function is given by

$$\frac{\sigma_1}{\gamma B} = \frac{Y}{2(X-1)\sin\alpha}(1+\sin\delta) \quad (4.23)$$

4.3.4 *Flow Factor Determination*

As discussed in Sect. 4.2 the flow factor ff is a flow channel parameter which relates the stress $\overline{\sigma}_1$ acting in an arch to the major consolidation stress σ_1 as follows

$$ff = \frac{\sigma_1}{\overline{\sigma}_1} \quad (4.24)$$

Noting that the hopper opening width B corresponding to the cohesive arch is given by

$$B = \frac{\sigma_1 H(\alpha)}{ff\gamma} \quad (4.25)$$

It follows from Eqs. (4.23) and (4.25) that the ff is given by

$$ff = \frac{Y}{2(X-1)\sin\alpha}(1+\sin\delta)H(\alpha) \tag{4.26}$$

where $H(\alpha)$ = arch thickness parameter.

4.3.5 Arch Thickness Parameter

As discussed in Chap. 2 and in Sect. 4.2 of this Chapter, the arch thickness parameter $H(\alpha)$ based on the work of Jenike and Leser [2] has been widely adopted over a number of years for a great many industrial design applications. Despite the complexity of the arch analysis, the simple form of the $H(\alpha)$ Eq. (4.5) and graphs of Fig. 4.3 are amenable to the design process.

For consistency and confirmation purposes, it is considered important to analyse the arch thickness parameter based on the circular arc, passive stress state theory presented in the foregoing sections. The analysis is based on the critical arch model of Fig. 4.9 in which

σ_1 = major consolidation stress
σ_c = unconfined yield stress
$\overline{\sigma}_1$ = stress acting in arch.

Similar to Jenike and Leser's arch analysis, this model considers a single cohesive arch structure. Also, it is assumed that the stress $\overline{\sigma}_1$ acting in the arch at the hopper

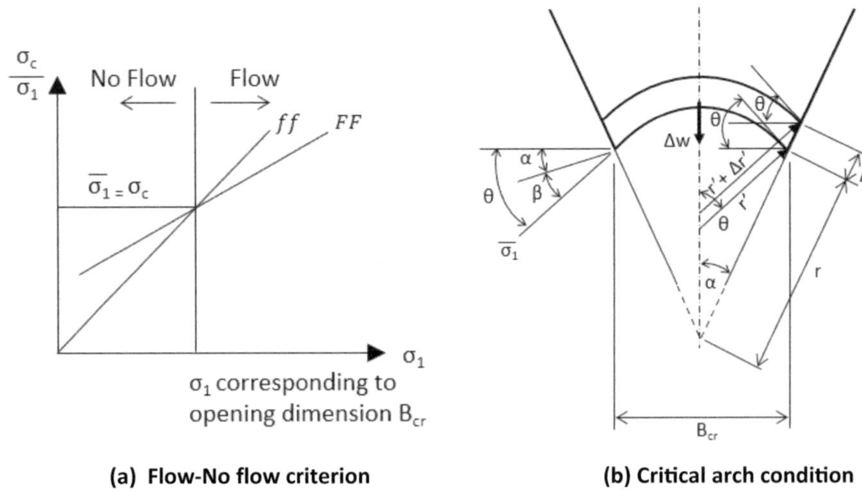

Fig. 4.9 Critical arch determination for mass-flow

wall boundaries is inclined at the angle θ to the horizontal as shown. The angle $\theta = 45°$ is chosen for optimisation purposes. The two symmetry cases are considered:

m = 1 for axisymmetric or conical hoppers
m = 0 for plane flow, wedged shaped hoppers.

The weight of an arch element is given by Eq. (4.27), where $G(\alpha)$ is calculated using Eq. (4.28).

$$\Delta W = 2^{(1-m)}\gamma\pi^{m}r^{(m+1)}\Delta r\sin^{(m+1)}\alpha G(\alpha) \tag{4.27}$$

$$G(\alpha) = \frac{2^{m}\sin\alpha(1-\cos\theta)^{m}\theta^{(1-m)}}{\sin^{(m+2)}\theta} + \frac{\sin(\theta-\alpha)}{\sin\theta} \tag{4.28}$$

The Arch Thickness Parameter is given by Eq. (4.29).

$$H'(\alpha) = \left[\frac{4\cos(\theta-\alpha)\sin\theta}{G(\alpha)}\right]^{m}\left[\frac{2\cos(\theta-\alpha)\sin\theta}{G(\alpha)}\right]^{1-m} \tag{4.29}$$

Figure 4.10 compares the arch thickness parameters obtain from Eq. (4.29) with those of Jenike and Leser. For the plane-flow m = 0 case the arch thickness parameters $H(\alpha)$ and $H'(\alpha)$ are practically identical. For the axi-symmetric case $H'(\alpha)$ values are a little lower, particularly for the hopper half-angle range $0° \leq \alpha \leq 20°$ and a little less conservative than those for $H(\alpha)$. While either arch thickness parameter may be applied, in general, the Jenike and Leser $H(\alpha)$, Eq. (4.5), is recommended.

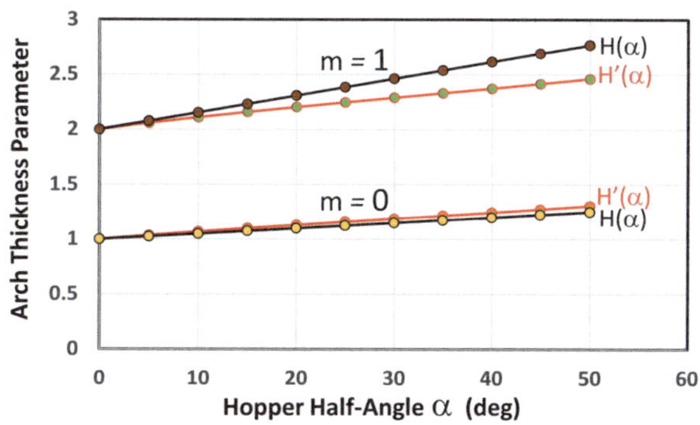

Fig. 4.10 Arch thickness parameters. H'(α) Eq. (4.29) H(α) Eq. (4.5) Jenike and Leser [2]

4.4 Mass-Flow Design Methodology

4.4.1 Design Charts

While accepting the advantages of the analytical approach to hopper design as presented herein, it is important that the link back to the research of Jenike [1] be maintained. To facilitate the application of his research to mass-flow bin design, Jenike presented his results in graphical form for both axi-symmetric, or conical, and plane-flow hoppers for values of δ ranging from 30 to 70° in 10° steps. Arnold et al. also solved the flow factor equations and presented the charts for the same range of δ values but in 5° steps [3].

With a view to simplifying the mass-flow design procedures, Moore and Arnold [8] developed two design charts, one for axi-symmetric hoppers and the other for plane-flow hoppers. The charts covered the values of $\delta = 30°$ to 70°and presented the essential information for hopper design. Employing different axis configurations, Roberts [3] applied Jenike's graphical results to develop two, equivalent design charts for axi-symmetric and plane-flow hoppers. The current charts have been re-plotted from the solutions to the analytical design equations and are presented in Figs. 4.11 and 4.12 for axi-symmetric hoppers and in Figs. 4.13 and 4.14 for plane-flow hoppers. The charts display the critical design data at, and in the region of, the limits for mass-flow and funnel-flow, thereby considerably simplifying the selection of the hopper half-angles and corresponding flow factors.

By way of explanation, the following comments are presented:

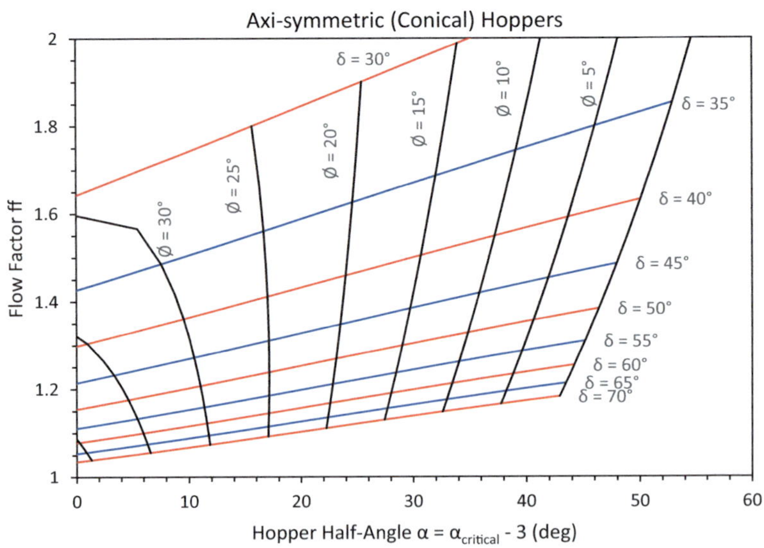

Fig. 4.11 Design chart for axi-symmetric (conical) hoppers—extended parameter range

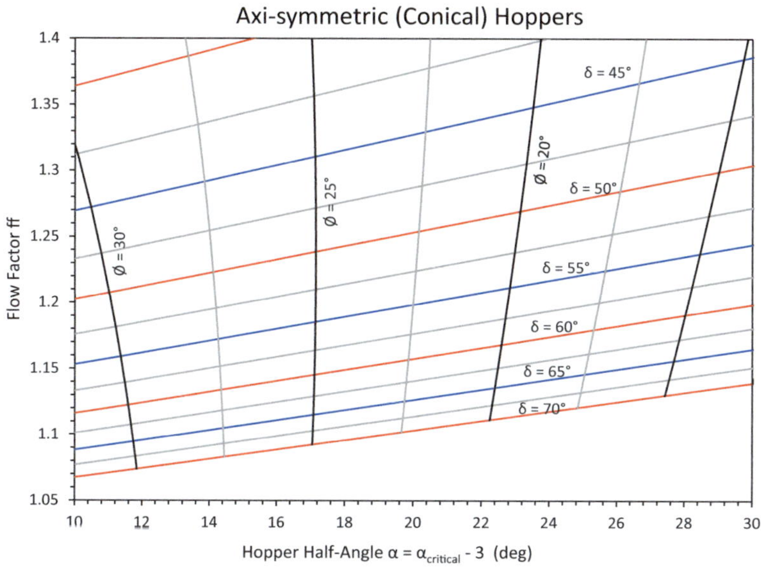

Fig. 4.12 Design chart for axi-symmetric (conical) hoppers—focused section

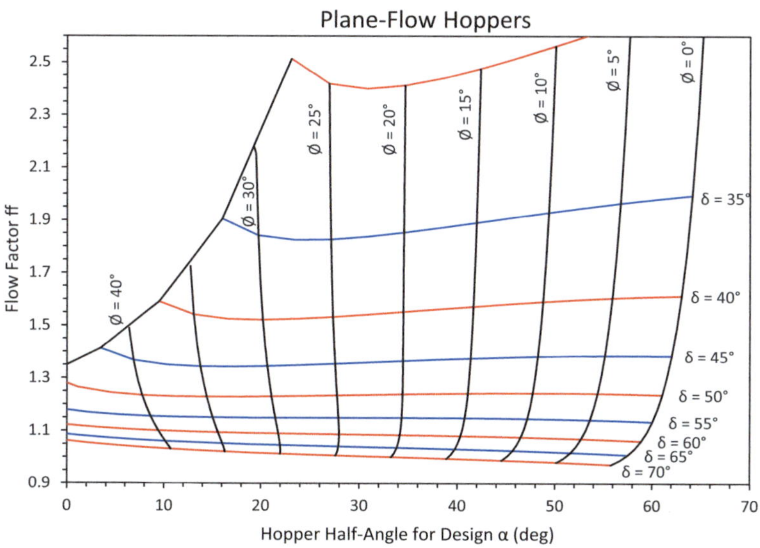

Fig. 4.13 Design chart for plane-flow hoppers—extended parameter range

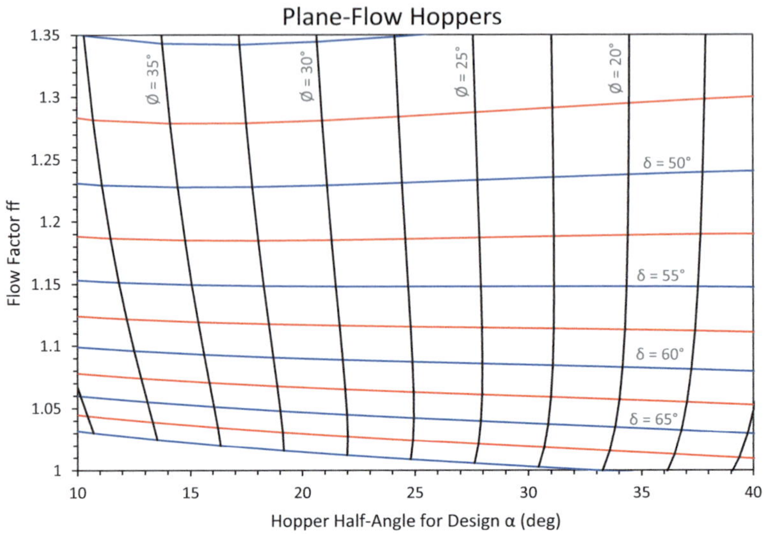

Fig. 4.14 Design chart for plane-flow hoppers—focused section

- The graphs Figs. 4.11 and 4.13 are the *extended* versions covering the full range of delta values and wall friction angles. In the case of cohesive bulk solids, it is more common for delta values to be 45° and higher and wall friction angles seldom less than 20°. Hence, for convenience, *focused sections* of the data are presented in Figs. 4.12 and 4.14 for axi-symmetric and plane-flow hoppers, respectively.
- For axi-symmetric hoppers (Figs. 4.11 and 4.12), the hopper half-angles α have been reduced by 3° from the limiting values as a design safety margin.

4.4.2 Hopper Design Example

To illustrate the procedures for mass-flow hopper design, the case of a mass-flow bin with a conical hopper to store Steaming Coal at 22.6% moisture content is considered. The relevant Flow Properties are presented in Figs. 4.15 and 4.16. Referring to the Flow Function (FF) graphs of Fig. 4.15, the Flow functions are plotted for the instantaneous bulk storage conditions corresponding to the initial filling of the bin and for 3 days undisturbed time storage. As shown, the increase in bulk strength over the 3 day time storage is quite appreciable.

The wall friction angles and Wall Yield Loci (WYL) for Mild Steel with Mill Scale finish and Stainless Steel type 304 with 2B finish are plotted in Fig. 4.16a and b respectively. Since these results are plotted against the normal stress (or pressure) σ_n, they need to be converted to the corresponding major consolidation stress σ_1. Based on the Mohr Circle analysis depicted in Fig. 4.17, the following equation is derived:

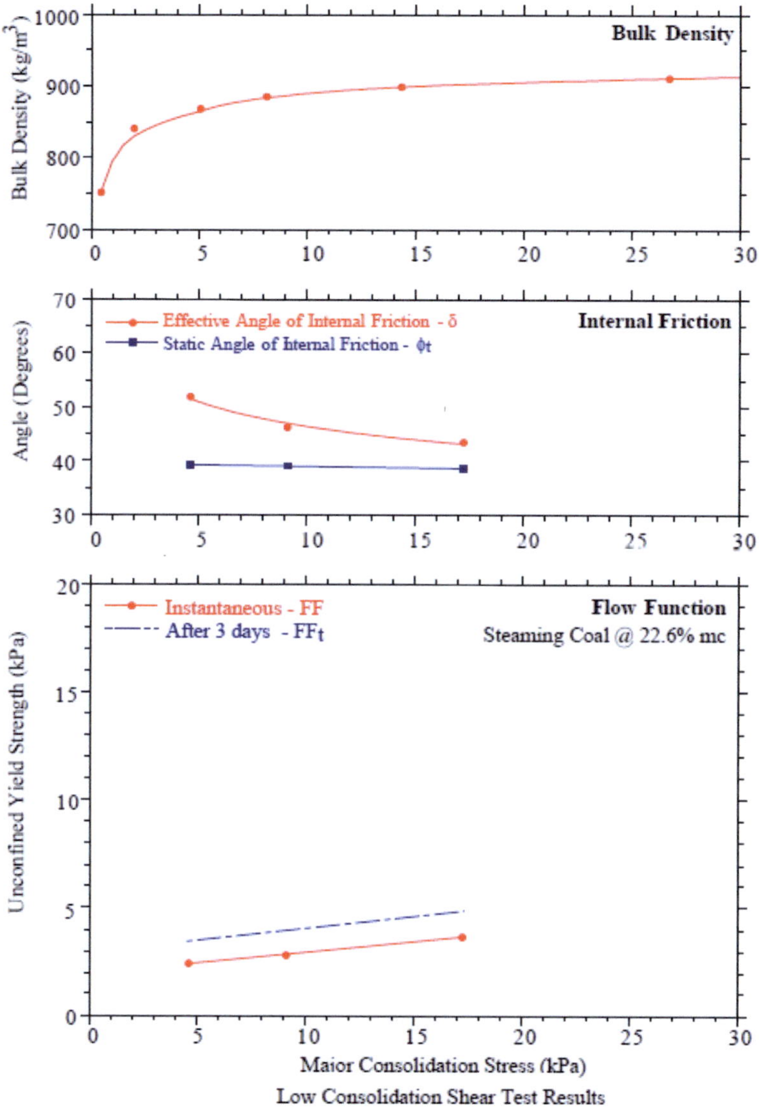

Fig. 4.15 Flow properties for steaming coal. Flow functions FF and FF_t; Effective internal δ and static friction Φ_t angles; Bulk density ρ

$$\sigma_n = \frac{\sigma_1}{1 + \sin\delta}(1 + \sin\delta\cos 2\beta) \tag{4.30}$$

The hopper design procedures are as follows:

(a) **Step 1** Estimate δ and σ_1 at the hopper outlet

Assume $\sigma_1 = 2.5$ kPa for which $\delta = 55^{\circ}$.

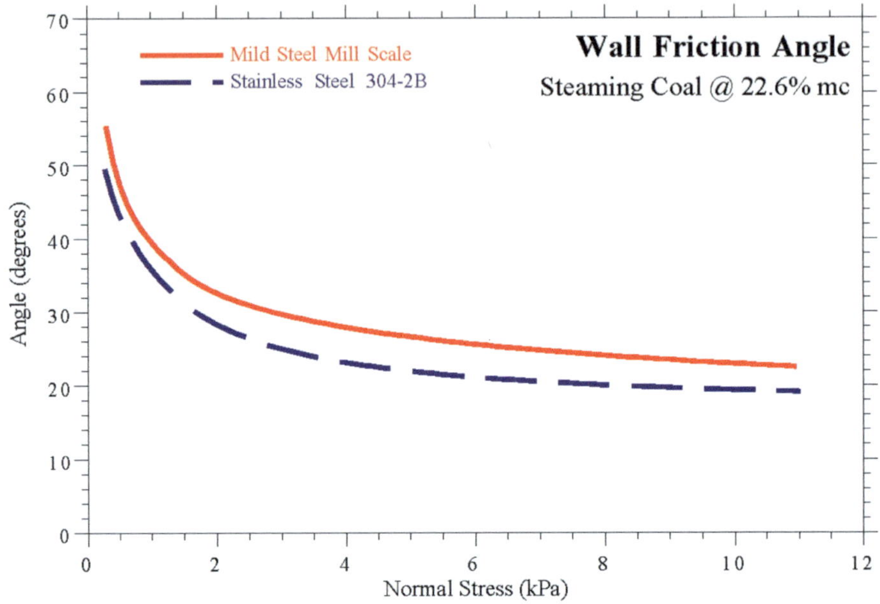

(a)

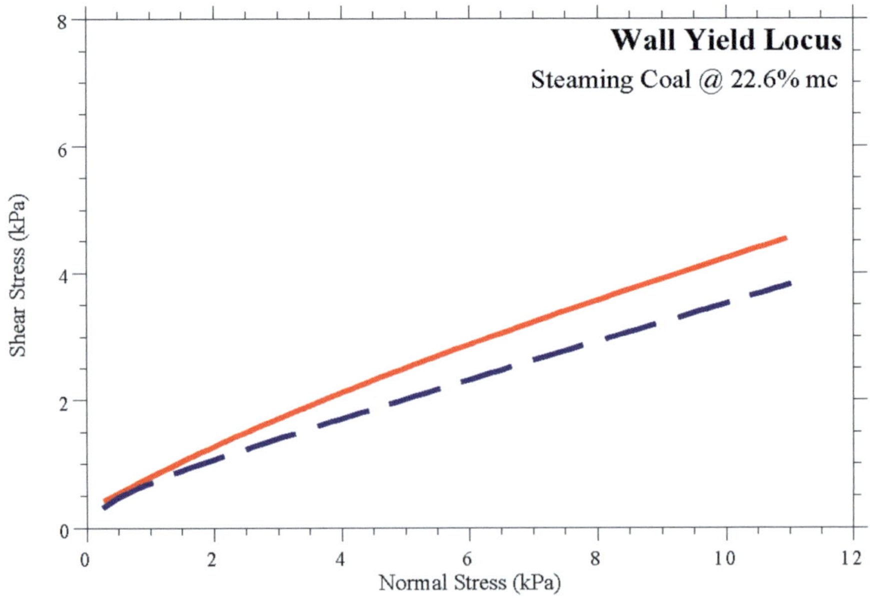

(b)

Fig. 4.16 Flow properties for steaming coal. **a** Wall friction angles φ; **b** Wall Yield Loci (WYL)

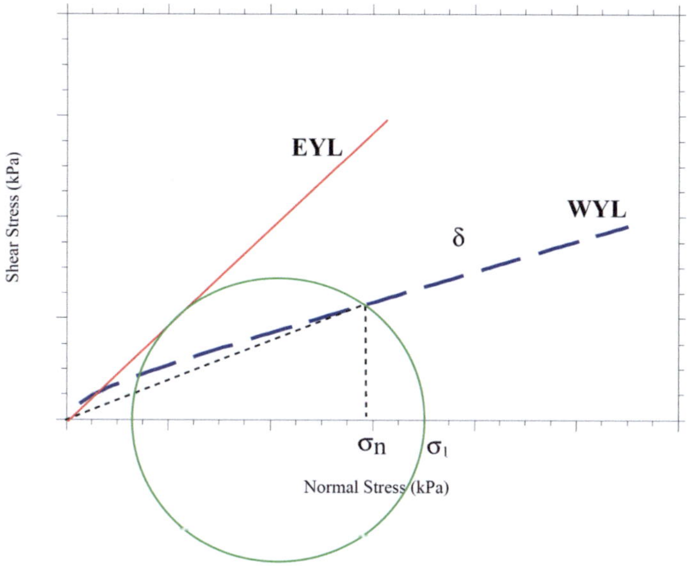

Fig. 4.17 Wall yield locus (WYL) σ_n versus σ_1 relationship

(b) **Step 2** Estimate the wall friction angle φ corresponding to $\sigma_1 = 2.5$ kPa.

It is proposed to manufacture the hopper from Stainless Steel. It is assumed that the wall friction angle $\varphi = 28^\circ$ corresponding to $\sigma_1 = 2.5$ kPa and the effective angle of internal friction $\delta = 55^\circ$. From Eq. (4.6) the angle $\beta = 31.5^\circ$ is determined. The application of Eq. (4.30) yields $\sigma_n = 1.89$ kPa which corresponds to the assumed value of 28° from Fig. 4.16a. The initial assumptions have been confirmed and no iterations are required. Referring to the design graph of Figs. 4.11 through 4.14, the following critical Flow Properties are:

- Major consolidation Stress $\sigma_1 = 2.5$ kPa
- Effective angle of Internal Friction $\delta = 55^\circ$
- Bulk density $\rho = 840\ \text{kg/m}^3 = 0.84\ \text{t/m}^3$
- Wall friction angle $\varphi = 28^\circ$

(c) **Step 3** Determine the Flow Factor ff and Hopper Half-Angle α.

For the instantaneous condition, the analytic solution for the flow factor, Eq. (4.26), gives ff = 1.17. Incorporating a 3 deg safety margin, the hopper half angle $\alpha = 13.7^\circ$, Eq. (4.1). For comparison, the graph of Fig. 4.11 is cited. For the values of $\delta = 55^\circ$ and $\varphi = 28^\circ$, the values of $ff = 1.17$ and $\alpha = 13.5^\circ$. These are in good agreement with the values determined analytically. The plotted values of Fig. 4.11 are illustrated in the diagram of Fig. 4.18.

(d) **Step 4** Draw the $ff = 1.17$ line on the FF graph of Fig. 4.13 and determine B_{CR}.

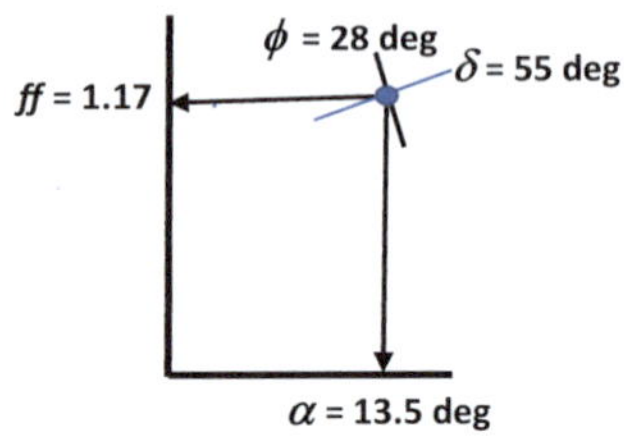

Fig. 4.18 Sub-section of Fig. 4.11 showing design values

The *ff* line is defined by the equation $\sigma_c = \frac{\sigma_1}{ff}$ and is shown in Fig. 4.18. The intersection point occurs at $\sigma_1 = 2.5$ kPa and unconfined yield strength $\sigma_c = 2.14$ kPa. The minimum or critical hopper opening dimension B_{CR} corresponds the stress $\overline{\sigma}_1 = \sigma_c$. From Eq. (4.5), H(α) = 2.21. Substituting the values into Eq. (4.4) $B_{CR} = \frac{\overline{\sigma}_1 H(\alpha)}{\rho g} = \frac{2.14*2.21}{0.84*9.81} = 0.573$ m.

(e) **Step 5** Influence of Time Consolidation

The consolidation condition following 3 days of undisturbed time storage is depicted by the Flow Function FF_t. The following parameters are determined (Fig. 4.19):

- Flow Factor $ff_t = 1.23 \approx 1.2$
- Major Consolidation Stress $\sigma_1 = 3.9$ kPa
- Effective Internal Friction Angle $\delta = 52$ deg
- Hopper Half-Angle $\alpha = 17.5$ deg
- Hopper Opening Dimension $B_{CRt} = 0.86$ m

4.4.3 α Versus B and Flow Rate Graphs

Once the critical opening dimension has been determined, the actual opening dimension may then be selected to meet the plant throughput requirements. In the absence of other information, a "rule of thumb" suggestion is for the opening dimension B to be as follows:

(i) $B \geq 1.2B_{CR}$ based on instantaneous conditions
(ii) $B \geq 1.1B_{CR}$ based on time storage conditions

The larger of (i) and (ii) is selected.

When the time consolidation dominates, once discharge flow commences, dilation and expansion of the bulk solid mass occur resulting in the flowing solids to revert towards the instantaneous condition (Fig. 4.20).

Since the wall friction angle decreases with normal pressure as indicated in Fig. 4.16a, the relationship between the hopper half-angle α and the opening dimension B provides important information for hopper design. The application of Eq. (4.25) in conjunction with the flow property graphs provides the means for determining the α versus B graph which is shown in Fig. 4.21. The values of α and B for plotting this graph are illustrated in the following example.

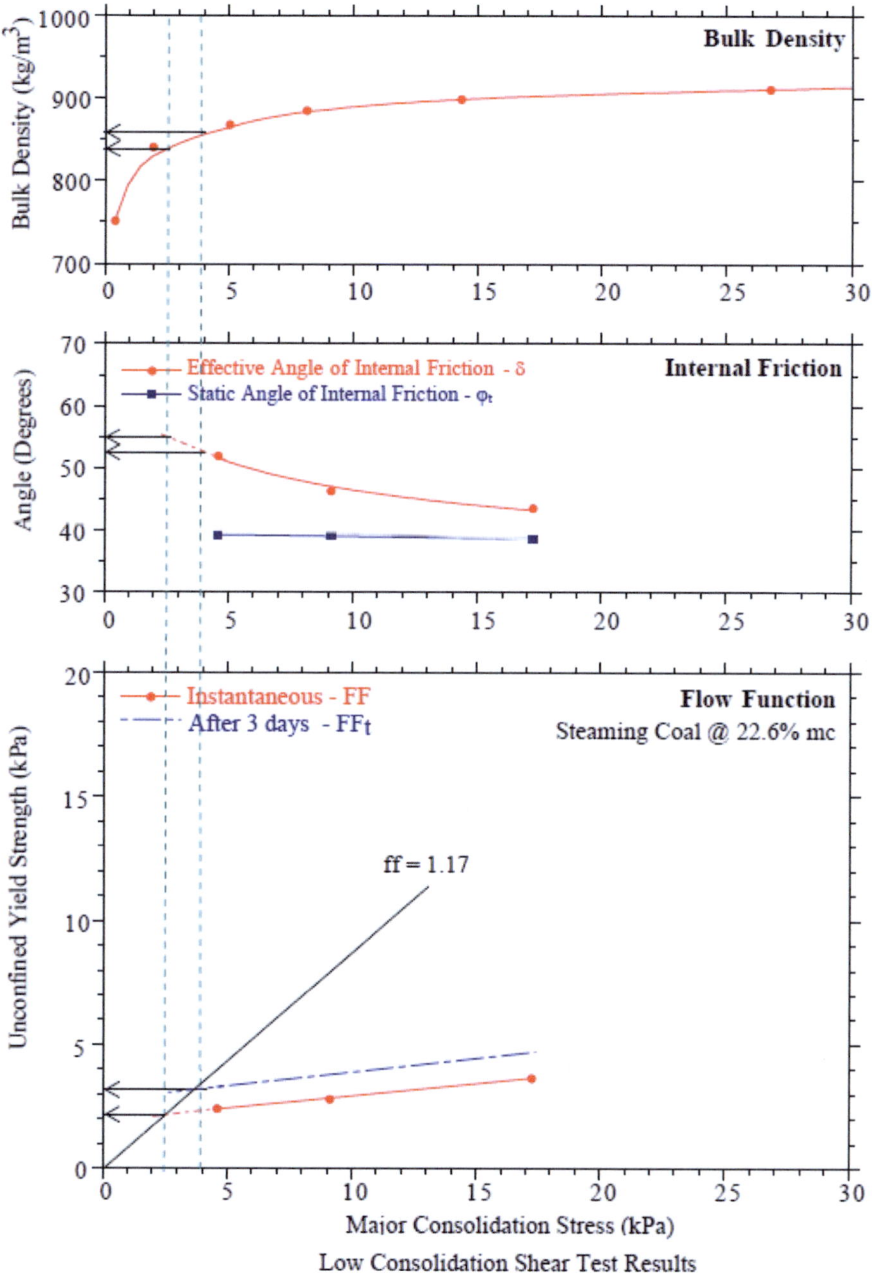

Fig. 4.19 Hopper design steps 1 to 5

Choosing, for example, the major consolidation stress $\sigma_1 = 10$ kPa for which the corresponding values of $\delta = 47°$, bulk density $\rho = 0.89\ \text{t/m}^3$, normal wall pressure $\sigma_n = 8.8$ kPa and wall friction angle $\varphi = 19°$. The values are obtained from the flow property and design graphs of Figs. 4.10 through 4.20. The flow factor $ff = 1.32$ is then determined and plotted on the Flow Function graph of Fig. 4.20, as shown. The value the unconfined yield stress $\sigma_c = \frac{\sigma_1}{ff} = 7.6$ kPa. For $\sigma_1 = 10$ kPa and $\rho = 0.89\ \text{t/m}^3$. The hopper half-angle $\alpha = 24.6$ deg and opening dimension $B = 2.06$ m.

The foregoing procedure is performed for a range of consolidation stresses; the results being summarised in Table 4.1. This table also includes the corresponding, estimated flow rates Q t/s which are based on flow analysis presented in Sect. 4.5 for an assumed coarse bulk solid.

The parameter ff_a is defined as the "actual flow factor" which is the line from the origin intersecting the Flow Function FF at the $\sigma_1 = 10$ kPa point as shown. The value of $ff_a = 3.30$ is obtained. The ratio of the flow factors ff and ff_a provides a useful indication of the degree of discharge flow of the bulk solids from a hopper. When ff_a equals the critical flow factor ff there is no flow. As the ff_a increases relative to ff, increased flow occurs. This is demonstrated in the following expression for the degree of flowability Fl:

$$FI = \left(1 - \frac{ff}{ff_a}\right) \tag{4.31}$$

For the critical arching case defined by the "flow/no-flow" condition $\frac{ff}{ff_a} = 1$ and Fl $= 0$.

For flow to occur, $\frac{ff}{ff_a} < 1.0$.

In the majority of cases, the flow rate Q is for unimpeded discharge in excess of the plant requirements. For this reason, feeders are employed to control the discharge flow to the required value. As previously indicated, the increase in hopper half-angle a follows from a decrease in wall friction that is linked to an increase in wall pressure. The increase in wall pressure follows directly form an increase in the hopper opening dimension B.

4.4.4 Plane-Flow Hopper

Following the same general procedure, the critical hopper geometry for a plane-flow hopper yields

$$\alpha_{CR} = 24° \quad B_{CR} = 0.296\ \text{m}$$

Based on the flow properties, there are several possible combinations of hopper geometries that may be chosen. By way of example, one possible solution is as

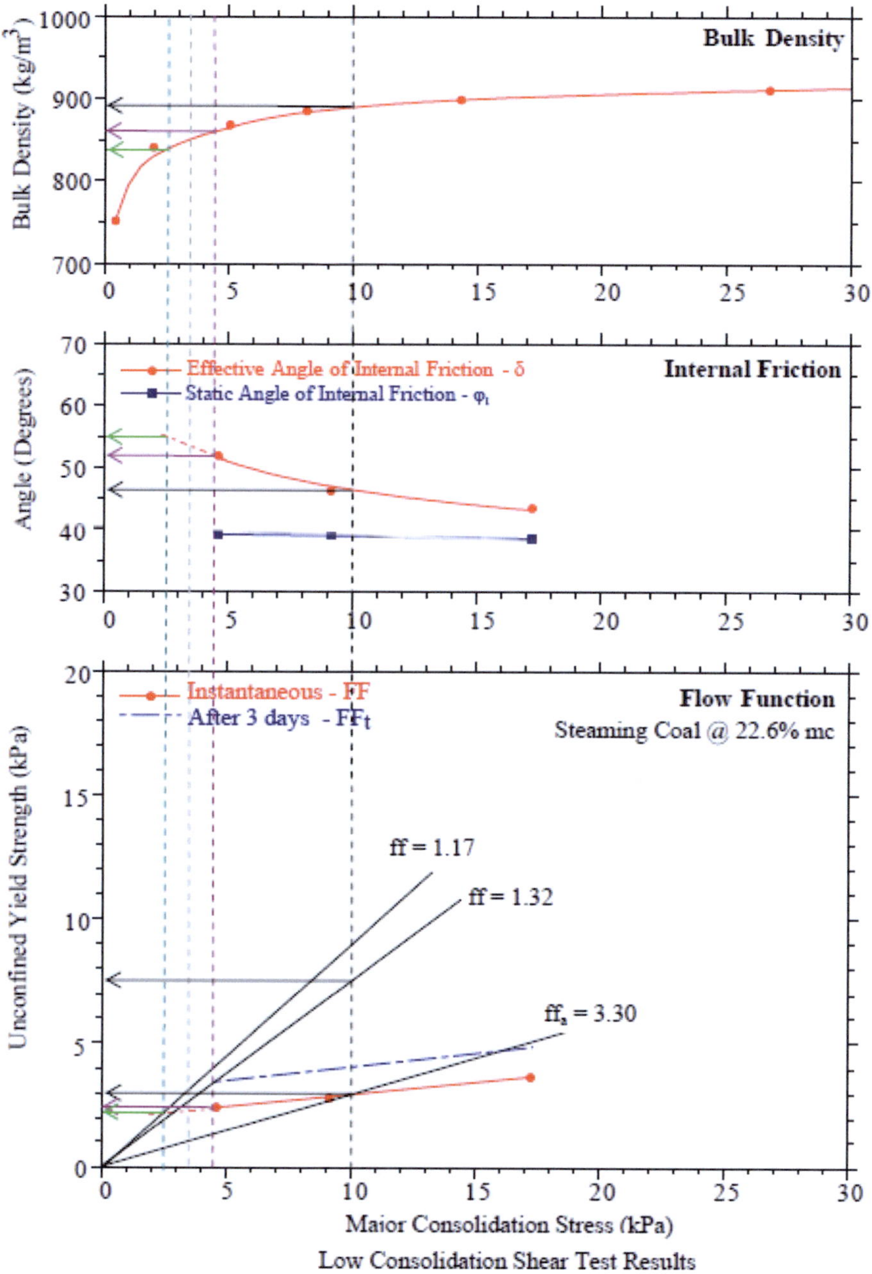

Fig. 4.20 Hopper design

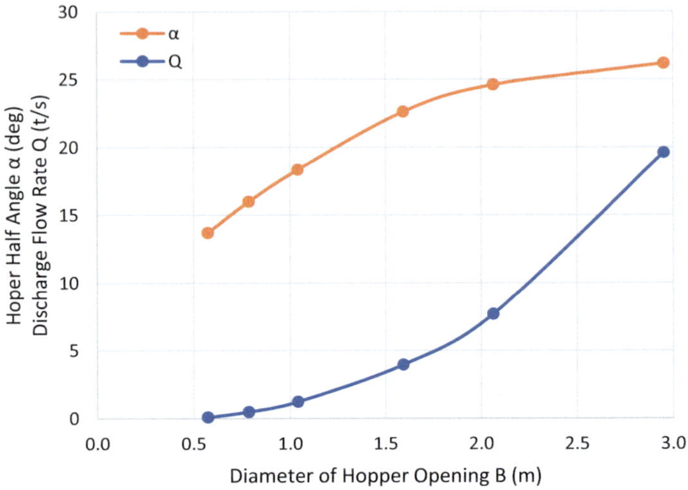

Fig. 4.21 α versus B and flow rate

Table 4.1 Mass flow design example

σ_1 (kPa)	δ (deg)	σ_n (kPa)	φ (deg)	ff	ff_a	ρ(t/m^3)	α (deg)	$H(\alpha)$	B(m)	Q (t/s)
2.5	55	1.9	28.0	1.17	1.20	0.84	13.7	2.21	0.57	0.000
3.5	53	2.7	26.0	1.20	1.50	0.85	16.0	2.25	0.79	0.480
4.7	52	3.8	24.0	1.22	1.92	0.86	18.3	2.28	1.04	1.226
7.5	49	6.4	20.5	1.28	2.77	0.88	22.6	2.35	1.59	3.945
10.0	47	8.7	19.0	1.32	3.30	0.89	24.6	2.38	2.06	7.665
15.0	44	13.2	18.0	1.38	4.38	0.90	26.1	2.40	2.95	19.568

follows:

$$\alpha = 29^\circ \qquad B = 0.5$$
$$L \geq 3 \times B \text{ i.e. } L \geq 1.5\,\text{m}$$

4.5 Hopper Opening and Flowrate

4.5.1 *Coarse Bulk Solids*

Coarse bulk solids are those in which the particle size range is such that the air permeability is sufficiently high to allow air to percolate through the stored solids with a minimum of resistance. Because of its relevance to this particular subject, the 1965 paper published by Johanson is cited [9].

Consider the accelerated flow of a bulk solid in the region of the outlet of the hopper shown in Fig. 4.22. In this case the air pressure gradient $\Delta p_r = 0$. From a dynamic equilibrium analysis of the forces acting on the arch at the hopper outlet, it may be shown that

$$\overline{\sigma}_1 = \frac{\rho g B}{H(\alpha)}\left[1 - \frac{a}{g}\right] \tag{4.32}$$

where

$\rho =$ bulk density
$\overline{\sigma}_1 =$ stress acting in arch at angle 45°
$a =$ acceleration of discharging bulk solid
$B =$ hopper opening
$H(\alpha) =$ arch thickness parameter as previously defined.

The minimum hopper opening to just prevent a cohesive arch from forming occurs when static equilibrium prevails, that is, the acceleration $a = 0$. Equation (3.32) then

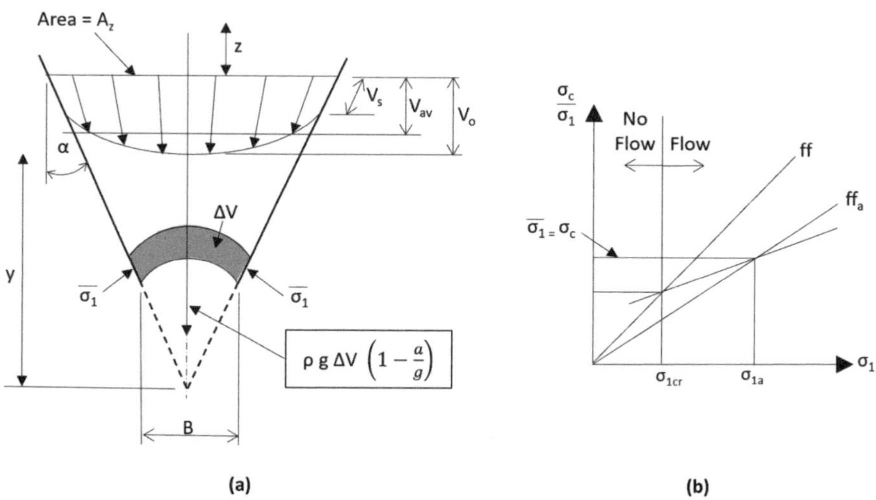

Fig. 4.22 Flow in a hopper

reverts to Eq. (4.4). In this case $\overline{\sigma}_1 = \sigma_c$ (Fig. 4.22b) and $B = B_{CR}$. For accelerated flow to occur, the actual flow factor $ff_a > ff$ and $B > B_{CR}$. It may be shown that the acceleration of the discharging bulk solid is given by Eq. (4.33).

$$a = g\left(1 - \frac{ff}{ff_a}\right) \tag{4.33}$$

where

ff = Critical Flow Factor based on the minimum arching dimension
ff_a = Actual Flow Factor based on the actual opening dimension

$$\text{ff}_a = \frac{\sigma_1}{\sigma_c} \tag{4.34}$$

where σ_1 = major consolidation stress or pressure at the outlet corresponding to dimension B.

While Eq. (4.33) expresses the total acceleration of the bulk solids at the exit level of the hopper, the acceleration needs to be examined in more detail. The total acceleration is composed of the following two components:

- component a_c due to the convergence of the flow in the tapered flow channel formed by the hopper
- component a_v due to the increase in velocity at the commencement of free fall as the bulk solids exits from the hopper.

That is

$$a = a_c + a_v \tag{4.35}$$

(i) Convergent Component of Acceleration a_c

It is assumed that the density of the bulk solids discharging in the converging flow channel is constant. Referring to Fig. 4.22, it may be shown that variation of the cross-sectional area A with the coordinate y is given by

$$\frac{1}{A}\frac{dA}{dy} = \frac{m+1}{y} \tag{4.36}$$

Also, the acceleration $a_c = v\frac{dv}{dy} = v\frac{dv}{dA}\frac{dA}{dy}$, which becomes

$$a_c = \frac{v^2}{A}\frac{dA}{dy} \tag{4.37}$$

At the hopper outlet, $B = 2y\tan\alpha$. It may be shown that the expression for a_c becomes

$$a_c = \frac{2v^2(m+1)}{B}\tan\alpha \quad (4.38)$$

(i) Exit Acceleration Component a_v

This component is given by Eq. (4.39).

$$a_v = g\left[1 - \frac{ff}{ff_a}\right] - \frac{2v^2(m+1)}{B}\tan\alpha \quad (4.39)$$

This shows that as the discharge velocity increases, $a_v \to 0$. Thus, an average terminal discharge velocity v_a is reached. With $a_v = 0$,

$$v_a = \sqrt{\frac{Bg}{2(m+1)\tan\alpha}\left[1 - \frac{ff}{ff_a}\right]} \quad (4.40)$$

and the flowrate is

$$Q_0 = \rho B^{(1+m)} L^{(1-m)} \left[\frac{\pi}{4}\right]^m v_a \quad (4.41)$$

or

$$Q_0 = \rho B^{(1+m)} L^{(1-m)} \left[\frac{\pi}{4}\right]^m \sqrt{\frac{Bg}{2(m+1)\tan\alpha}\left[1 - \frac{ff}{ff_a}\right]} \quad (4.42)$$

where

m = 1 for axi-symmetric hopper
m = 0 for plane-flow hopper
L = Length of slot in case of plane-flow hopper.

The other parameters are as previously defined.

4.5.2 *Velocity Distribution in Hopper*

The velocity distribution in a converging channel has been studied in some detail by Johanson [10]. A practical application of this earlier work of Johanson was presented by Johanson and Royal [11], who were interested in calculating the sliding velocity at a hopper wall in relation to the wear of hopper linings. This concept has also been followed up by Roberts et al. [12–14] who have also examined hopper wall wear, as well as wear in chutes.

The average velocity along the converging hopper (Fig. 4.22) is computed from the flow rate as follows:

$$V_{av} = \frac{Q_m}{\rho A_z} \quad (4.43)$$

where

$Q_m =$ Mass flow rate at section
$A_z =$ Cross-sectional area.

The velocity profile in the hopper is as depicted in Fig. 4.22. The velocity of sliding at the wall is

$$v_s = K_v v_{av} \quad (4.44)$$

The parameter K_v for conical and plane-flow hoppers may be obtained from Refs. [3, 15].

By way of example, for a plane-flow bin, with a wall friction angle of 20° and a corresponding mass-flow angle of 35° (Fig. 4.13) $v_o/v_s = 9.4$. The corresponding value of $K_v = 0.54$. The wall friction angle of 20° would apply, for example, for coal on stainless steel type 304 with 2B finish. If structural steel plate is used, the wall friction angle is more likely to be 30°; on this basis, the hopper half-angle for mass-flow in a plane-flow hopper is, from Fig. 4.13, $\alpha = 22°$. The corresponding value of $K_v = 0.54$ which is also the same as the previous case for the stainless-steel lining.

It is to be noted that velocity ratio v_o/v_s and K_v have a direct bearing on the blending or mixing characteristics of hoppers [1, 6].

4.5.3 Hopper Opening and Flow Rate—Fine Bulk Solids

For fine powders of low permeability, flow rates are very much lower than those computed using the method outlined in Sect. 4.5.1 above will occur. The analysis is much more complex and involves two-phase flow theory. The flow of fine powders has been studied by McLean [3, 16], and more recently by Arnold and Gu [17, 18].

Referring to Fig. 4.23, the net effect is for an air or gas pressure gradient Δp to exist as depicted. The pore pressure in the hopper will then be slightly below atmospheric as shown. Air flow from the atmosphere below the arch will be impeded due to the resistance imposed by the bulk solid or powder. The atmospheric air will hold back the bulk solid inhibiting flow. Some particles may fall like rain drops from the surface; if the pore pressure should equalise with the outside pressure at any stage, then there may be a sudden surge of material or flooding take place. Flow is characterised by an unstable starved/flood type behaviour which is most unsatisfactory.

Where a high proportion of low permeability fines is present, a larger hopper opening is required than for equivalent coarse bulk solids to achieve the required flow rate. In some cases, small quantities of air may be injected into the hopper to bias the pore pressure to at least atmospheric pressure. For the reasons stated, fine

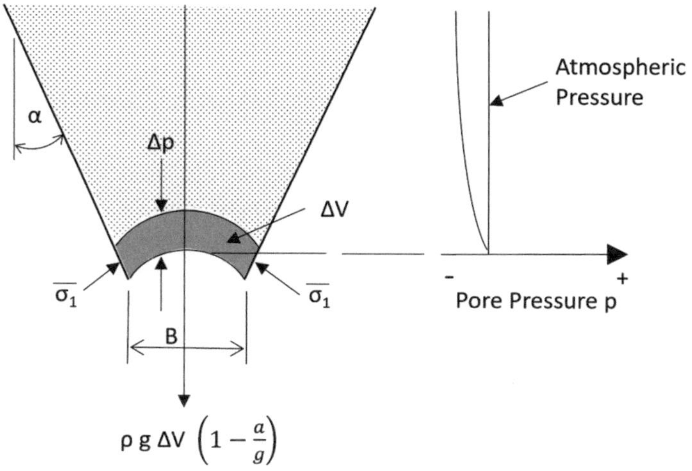

Fig. 4.23 Conditions in a mass-flow hopper governing discharge

powders are prone to flooding and uncontrollable discharge if allowed to aerate so that extreme care must be exercised in designing and installing any air permeation system.

Usually, a feeder is used to control the flow from a hopper. For this reason, particular care must be exercised in ensuring that the interfaces between the hopper and feeder are correctly designed to prevent problems due to flooding. First and foremost, it is essential that a mass-flow hopper be used and that the hopper and feeder be designed as an integral unit [19–21]. Referring to Fig. 4.24, if the speed of the feeder is too fast, then cavitation will occur at the outlet as illustrated in Fig. 4.24a; an uncontrollable 'starve/flood' flow will result. By ensuring that the feeder speed is controlled so that the outlet is always choked as illustrated in Fig. 4.24b, much better control over the feed will then be possible.

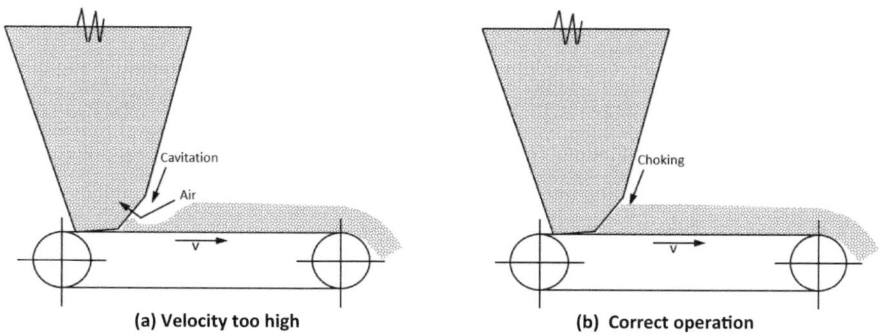

Fig. 4.24 Feeding of fine powders

4.6 Factors Affecting Mass-Flow Geometry

4.6.1 Critical Opening Size B Versus Moisture Content

Changes in moisture content can significantly influence the strength of bulk solids. By way of illustration, the critical hopper opening dimension B for three Hunter Valley (Australia) coals plotted as a function of moisture content are shown in Fig. 4.25 [1, 22]. The critical opening dimension is an indicator of bulk strength.

Sample (1) is a raw open cut coal, Sample (2) a washed version of (1) and Sample (3), a blend of (2) and another washed coal. All samples have low strength for moisture contents up to about 12%, but the strength increases significantly as shown over the range 12% to 17% moisture content. The high strength of the raw unwashed coal is clear. As shown, the bulk strength decreases as moisture content approaches the saturation limit.

It is important to note that washed coals, such those illustrated by samples (2) and (3) in Fig. 4.25, generally exhibit very little, if any, increase in bulk strength due to time storage. However, raw, run-of-mine (R.O.M.) coals will normally gain substantial increases in strength with undisturbed storage owing to inter-particle bonding resulting from surface chemistry effects. Such results are typical of coals containing components of clay. The effect of the increase in strength is to further increase the size of the hopper opening.

The variability of bulk solids such as coal is illustrated in Fig. 4.26. This shows the critical opening dimension versus moisture content for another Hunter Valley coal. In this case the peak strength is obtained at a lower moisture content; further, the reduction in strength beyond the worst moisture condition is quite substantial, which is in contrast with the samples of Fig. 4.25. The examples illustrated in Figs. 4.25

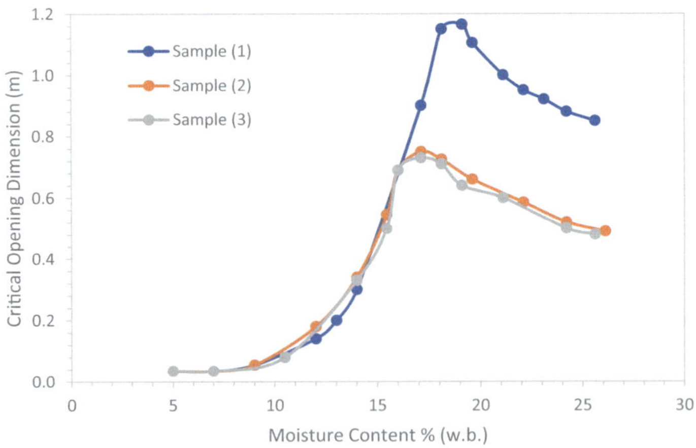

Fig. 4.25 Critical opening dimension B_{CR} as a function of moisture content for three coal samples—stainless steel 304-2B lining

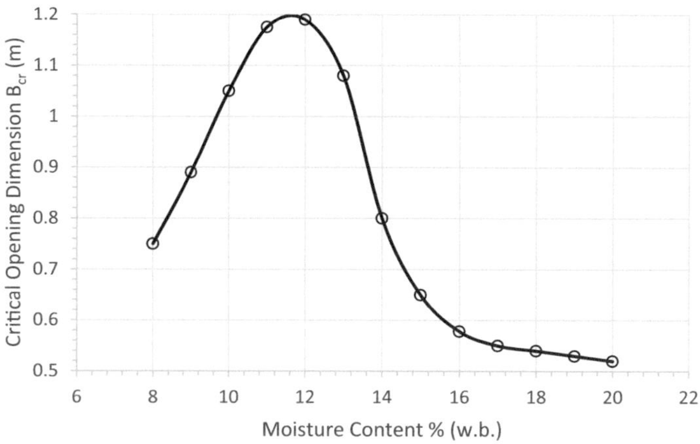

Fig. 4.26 Critical opening dimension B_{CR} as a function of moisture content for a Hunter Valley coal based on conical hopper lined with stainless steel

and 4.26 highlight the variability of bulk solids and emphasise the need for testing each sample.

Since feeders are normally used to control the flow from bins and the cost of feeders increases as the opening increases, there is every incentive to limit the hopper opening size. It is desirable, therefore, that the moisture content of the bulk solid be controlled within acceptable bounds. However, in practice it is usually not possible to achieve this objective.

At the current stage of development, it is not possible to predict the bulk strength versus moisture content characteristics of bulk solids. This information can only be acquired through testing. Of all the bulk solids that have been examined, it is most likely that coal represents the bulk solid that has been examined the most. Despite this, however, it is still necessary to test each coal to determine the bulk strength versus moisture characteristics. As a general guide, it has been found that coals will reach a peak strength somewhere between 70 to 90% of the saturation moisture levels.

4.6.2 *Variation in Wall Friction*

As indicated in Chaps. 2 and 3, it is common for wall yield loci to be slightly convex upward in shape. This effect is often more pronounced in the case of moist bulk solids in contact with polished surfaces where adhesion/cohesion effects are exhibited. The convexity of the wall yield loci means that the wall friction angle decreases with increase in normal contact pressure, as illustrated in Fig. 4.27.

Since an increase in both the normal wall pressure and consolidation pressure accompany an increase in hopper span, the corresponding decrease in wall friction angle will permit the hopper half angles to be increased (in accordance with Fig. 4.11).

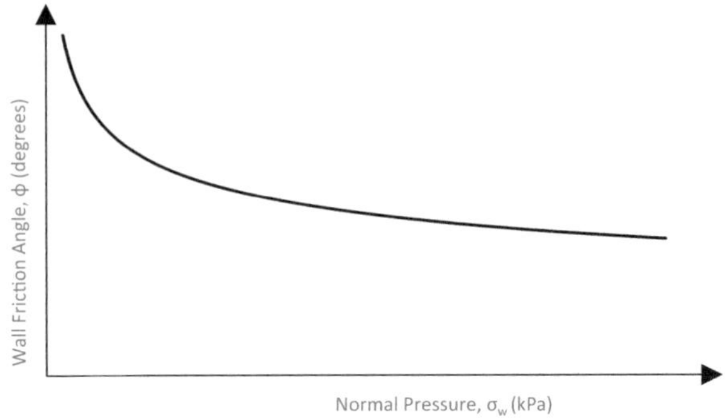

Fig. 4.27 Characteristic surface or wall friction variation with normal pressure

Hence it is possible to calculate a hopper half angle α as a function of hopper span or opening dimension as indicated previously in Fig. 4.21. The foregoing may be used to change the hopper slope as in Fig. 4.28.

The following examples illustrate the application of the measured flow properties to the analysis and design of storage bins:

(a) ***Train Loading Coal Bin with Variable Geometry Hopper***

Figure 4.29 shows a train loading coal bin employed for flood loading coal into rail wagons while the train is in continuous motion. The bin incorporates a variable geometry, mass-flow hopper, composed of three hopper sections as shown. The lower two hopper sections are constructed of stainless steel and the upper hopper section is of mild steel. In this way the required storage capacity of the bin is achieved by judicious application of the coal flow properties. The overall advantage of limiting the height of the bin can be achieved with this design. Had a single hopper been chosen, as depicted in Fig. 4.29, the hopper half-angle would be governed by the smallest value corresponding to the bin outlet.

The choice of mild steel for the upper hopper section requires some explanation. Noting that the bin is being loaded from a belt conveyor, the coarse or large coal particles will tend to roll to the outer region of the bin, becoming into contact with the mild steel hopper walls. During discharge, the larger particles are more likely to roll or slide downward without blocking. The fine coal particles will congregate in the central region, particularly near, and at the outlet. It is normally the finer particles that create the higher friction which governs the hopper half-angle at the bin outlet.

(b) ***Power Station Variable Hopper in Coal Bin***

Figure 4.30 shows a boiler coal feed bin with "infinitely" variable hopper. The discharge from the bin is via a standpipe which delivers the coal to a gravimetric feeder, and then to a pulverizing mill and finally to the boiler. It is rather an expensive solution; normally two or three hopper sections will be adequate.

Fig. 4.28 Increasing hopper half angle in varying hopper slope

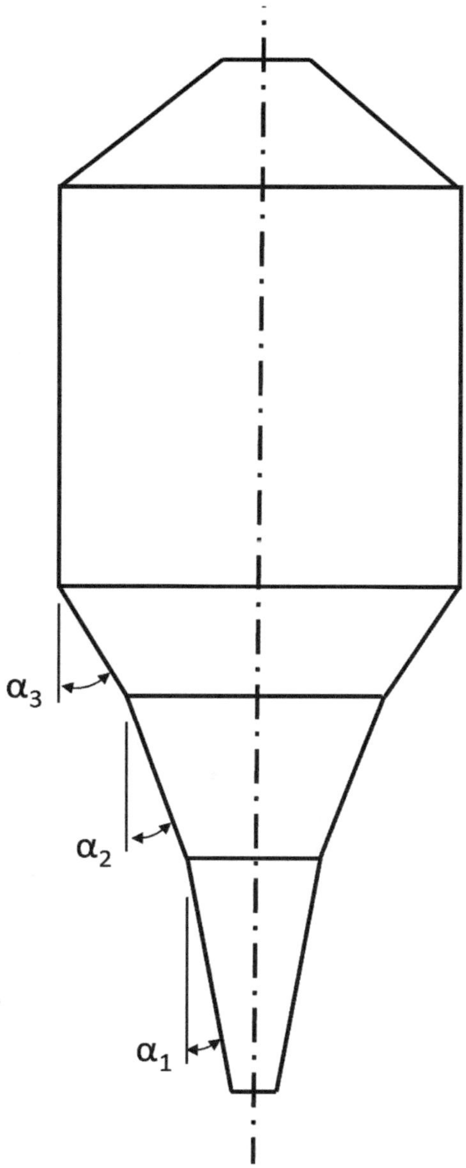

4.6.3 Influence of Bulk Density

While the Flow Function *FF* which governs the bulk strength at the outlet has a major influence on the magnitude of *B*, the bulk density ρ (or bulk specific weight $\gamma = \rho g$) corresponding to the major consolidation pressure at the outlet also has a

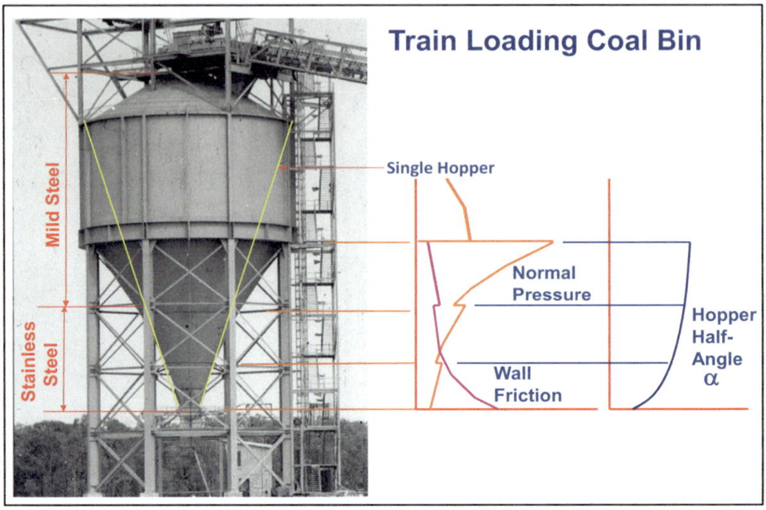

Fig. 4.29 Train loading coal with variable geometry hopper

Fig. 4.30 Variable geometry coal, boiler feed hopper

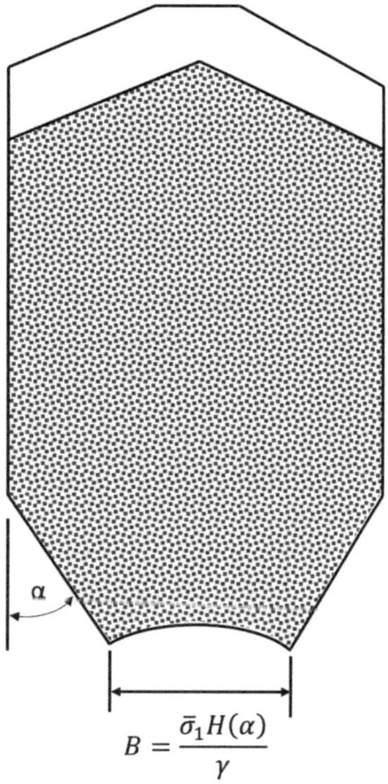

Fig. 4.31 Influence of bulk density on bin opening

significant influence. As indicated by Eq. (4.4), the opening dimension is inversely proportional to the bulk density. This is illustrated in Fig. 4.31.

For instance, a bulk solid such as iron ore has a high bulk density in the order of 2.5 t/m^3. This has the advantage of assisting in controlling the size of the opening dimension and offsetting, to some extent, the influence of the cohesive nature of the ore the bulk strength. On the other hand, many materials used in the process and food industries have very low bulk densities, less than 0.5 t/m^3. This makes the design of mass-flow bins difficult, since the opening dimension often becomes disproportionably large in relation to the bin width. For this reason, bins to handle bulk materials of low density require the use of an appropriate discharge aid such as a live bottom bin discharger.

Similar remarks apply to funnel flow, since the critical rathole dimension D_f is inversely proportional to the bulk density.

4.6.4 Bulk Solids Exhibiting Caking Characteristics

As is already apparent, the strength of most bulk solids is influenced by environmental factors such as moisture and temperature. The influence of moisture in relation to coal and similar mined bulk solids has been discussed. Temperature can have a significant influence, particularly in the case where moist materials are stored at sub-zero temperatures where freezing can occur. Some bulk solids, such as raw sugar and manufactured prill type chemicals, are prone to caking under pressure when exposed to the atmosphere, particularly if the solids are hydroscopic and the humidity high. Elevated temperatures can also have an influence, such as ammonium nitrate prill which has a critical phase change temperature around 38 °C [23]. Bulk solids which exhibit caking characteristics usually depict a concave Flow Function as illustrated in Fig. 4.32.

The concave Flow Function raises some questions about the conditions for mass-flow. The Flow/No-flow condition is depicted in Fig. 4.32 and, as can be observed, the conditions for flow and no-flow are the reverse of the normal situation indicated in Fig. 4.2. The bulk material is quite free flowing at low pressures, but flow is impaired higher up in the hopper due to the significant increase in bulk strength that occurs. The effect of this may be seen more readily in Fig. 4.33 which shows the distribution of the consolidation pressure σ_1, the unconfined yield strength σ_c and the stress acting in an arch $\overline{\sigma}_1$.

The consolidation pressure σ_1 during flow is linear over the lower region of the hopper for which a radial stress field applies. However, towards the transition of the hopper and the cylinder, σ_1 increases non-linearly as shown and σ_c also shows a somewhat similar characteristic. When $\overline{\sigma}_1 > \sigma_c$, as in the lower region of the hopper, a cohesive arch will not form and flow will ensue. An arch will form higher up in the hopper where $\sigma_c > \overline{\sigma}_1$. This problem may be solved in one of two ways. Either eliminate the cylindrical section of the bin and use a hopper only (Fig. 4.34). Alternatively control the quality of the bulk solid such that $\sigma_c < \overline{\sigma}_1$. In the first case σ_1 will reduce to zero at the transition as in Fig. 4.33; σ_c also reduces. In the second case, a limiting or upper bound flow function defined by $\sigma_c = \overline{\sigma}_1$ may be obtained for the given bin geometry. The actual flow function must always be below the upper bound curve for the range of consolidation pressure expected to occur in storage.

Where it is proposed to store the material in a mass-flow hopper, to achieve the stress condition of Fig. 4.33, it should be noted that a plane-flow hopper offers some advantages with respect to storage efficiency, particularly since the hopper half angle is significantly larger (approximately 8 degrees to 10 degrees larger) than the equivalent conical hopper. A disadvantage of the plane-hopper is the long slotted

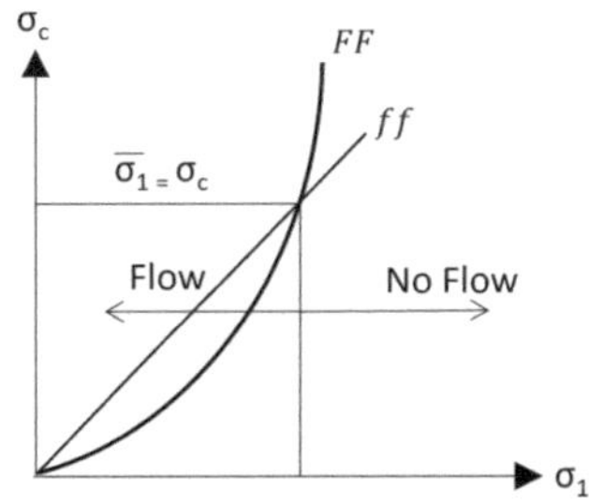

Fig. 4.32 Flow/no-flow condition for bulk solid with concave flow function

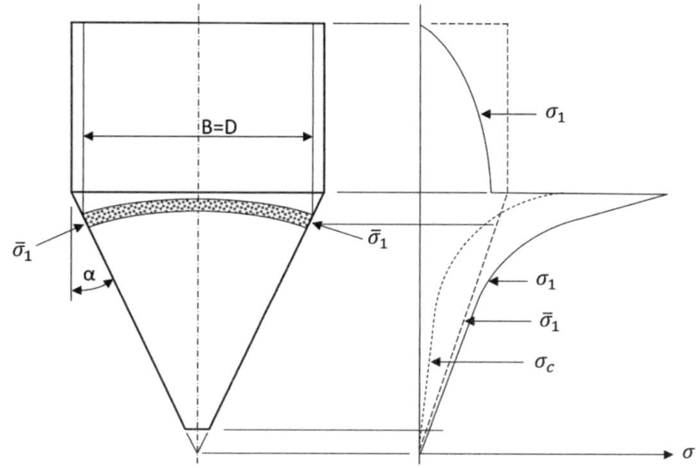

Fig. 4.33 Stress distributions during flow showing arching high up in hopper

opening necessary which can cause problems in obtaining uniform draw down when feeding along the slot. An alternative is to employ a multi-outlet plane-flow hopper as indicated in Fig. 4.35. An additional feature of using several outlets is the provision of a back-up facility, should problems occur. However, for correct operation it is necessary, under normal conditions, to operate all outlets simultaneously to prevent funnel-flow occurring.

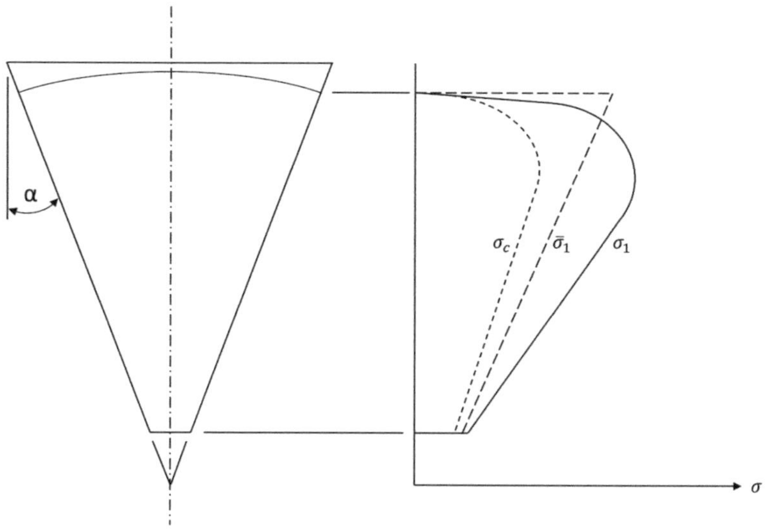

Fig. 4.34 Stress conditions in hopper without surcharge

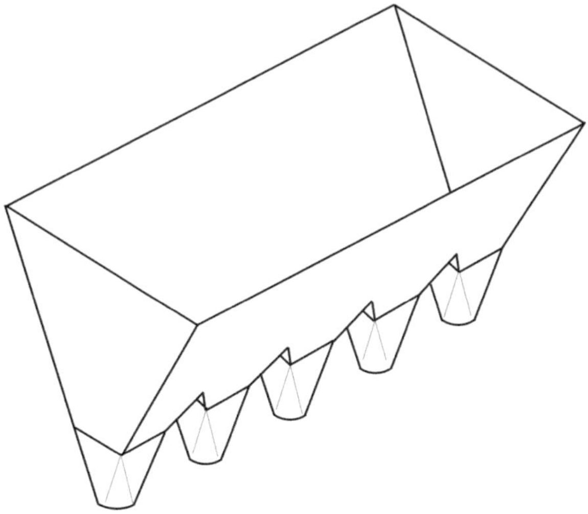

Fig. 4.35 Multi-outlet plane flow hopper

Case Study—Handling of Ammonium Nitrate

Ammonium Nitrate Prill, commonly known as "*Nitropril*", is a classical example of a bulk material that displays the caking characteristics that are described in the foregoing discussion. It is widely used as an explosive at mine sites. Nitropril is quite hydroscopic and temperature sensitive having a phase change of typically 38 °C. Often, as in the case cited [23], the manufacturing plant is located at or near the seaboard requiring storage in large bulk sheds that are equipped with air conditioning for temperature and humidity control. The Nitropril is transported by rail and/or road to the mine sites, which are, almost invariably located inland, where summer temperatures of 40 °C and higher are common.

Figure 4.36b shows the storage bin at the processing plant located near the mine site. As can be seen, the bin is a classical mass-flow, axi-symmetric design comprising a hopper and upper, vertical storage bin section. In the light of the preceding discussion, it is very evident that the bin is not suitable for efficient storage of the Nitropril. This is due to the high consolidation stresses experienced at, and in the vicinity of, the bin and hopper transition resulting in arching and blockage problems as previously discussed. Figure 4.36a shows a sample of agglomerate Nitropril that was dislodged from the bin/hopper transition region.

4.6.5 Agglomeration Problems

The importance of integrating the flow properties of bulk solids with the design of the handling plant to maintain continuous, trouble-free operations cannot be too strongly stressed. When this does not occur, the costs in downtime due to flow blockages

(a) Example of caking of ammonium nitrate

(b) Ammonium nitrate bin with surcharge

Fig. 4.36 Caking problem in ammonium nitrate storage bin

can be very significant. The following case study of coal agglomeration problems experienced in the boiler coal feed systems of a thermal power stations is cited.

Figure 4.37 shows the coal feed system comprising a mass-flow storage hopper and standpipe for transferring coal to a gravimetric weigh feeder. Not shown and not relevant to this discussion is the discharge from the feeder to the coal pulverising mill and the conveying of the pulverised coal to the boiler. The coal feed rate to the boiler is in the order of 70–80 t/h, which, it may be surmised, that this was the basis for the weigh feeder selection for which the feed intake via the standpipe was nominally 600 mm in diameter. By default, it appears that the 600 mm dictated the selection of the 600 mm diameter standpipe and hence the outlet diameter of the mass-flow hopper.

Based on measured flow properties of coal a much larger opening would be required if the moisture content of the coal happened to be at or near the critical strength conditions. This would be exacerbated by the isolating valve being located in the 600 mm standpipe, as shown in Fig. 4.37a.

Figure 4.37b and c refers to another installation which mirrors that of Fig. 4.37a. In this case the rear inspection door of the feeder has been opened to show a large oval shaped block of agglomerated coal in which the longitudinal dimension is large enough to cause a flow feed blockage.

By way of example, two possible modifications are discussed as illustrated in Figs. 4.38 and 4.39.

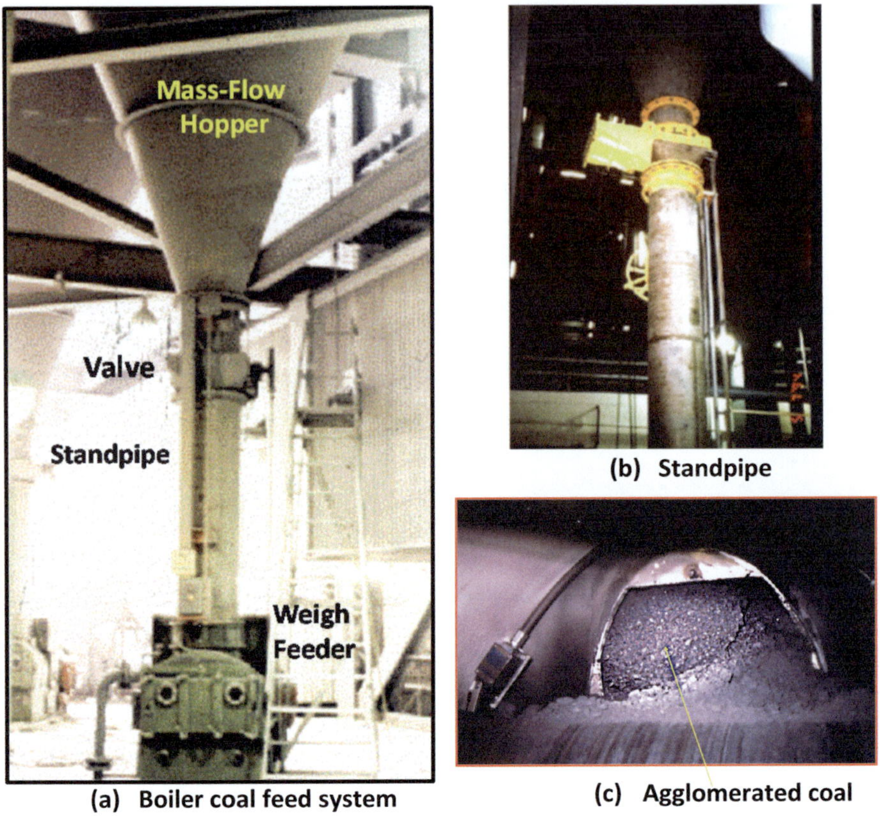

Fig. 4.37 Boiler coal/feed system agglomeration problem

(i) ***Example Modification***

It is quite apparent that the hopper opening dimension of B = 600 mm based on the standpipe diameter is not satisfactory for efficient mass-flow without blockages. The required opening dimension is $B \geq B_{CR}$, where B_{CR} is the minimum arching dimension defined by Eq. (4.4) determined from the flow properties tests performed on the coal that is to be utilised in the boiler feed system. Account is taken of the coal moisture content for maximum bulk strength and periods of likely prolonged undisturbed storage. Hence a larger control valve will be required and installed in the hopper which is modified to accommodate the valve as illustrated in Fig. 4.38.

When the valve is opened and flow commences, there will be dilation of the coal which will have some effect on reducing the bulk strength as the coal enters the standpipe but by how much is uncertain. In any case, as illustrated by Fig. 4.37c, the 600 mm diameter of the standpipe is too small to prevent potential blockages due to agglomeration. The standpipe diameter may be estimated using

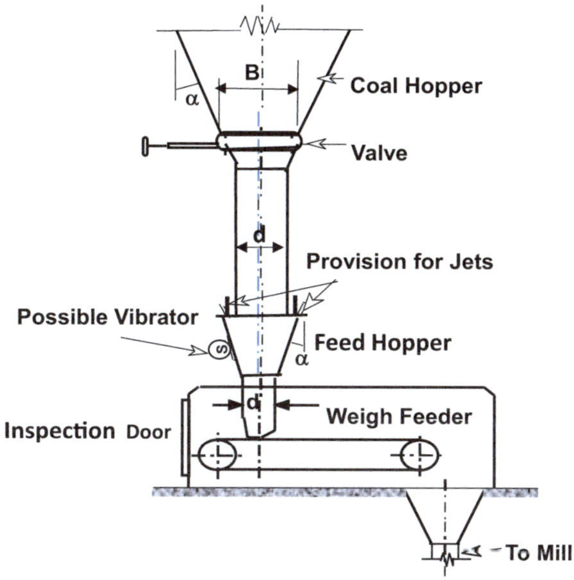

Fig. 4.38 Possible modification

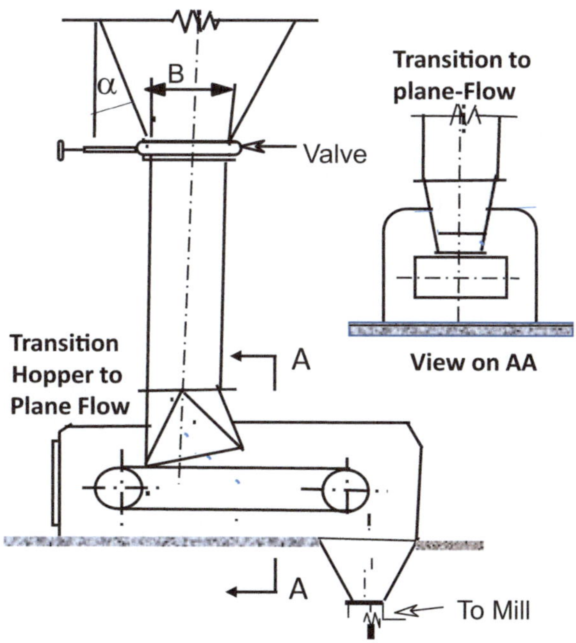

Fig. 4.39 Possible alternative modification

$$d \geq 11.0\frac{\tau_o}{\gamma} \tag{4.45}$$

where

$\tau_o =$ pipe wall and bulk solids cohesive shear stress (kPa)
$\gamma =$ bulk specific weight (kN/m^3).

The aim of this arrangement is to prevent any agglomeration problems occurring in the standpipe but rather to use the feed hopper as a possible collection point where air jets and/or vibration are employed if required. In the case of the vibrator, it should be cycled on and off to avoid blockage problems during discharge. The main objective of the feed hopper is to converge the flow in the 600 mm feed intake to the weigh feeder.

(ii) ***Possible Alternative Modification***

Figure 4.39 illustrates a possible alternative modification incorporating a transition plane-flow feed hopper to gain some possible advantage of the feed slot width being narrower than the circular opening diameter of the axisymmetric feed hopper of the previous example of Fig. 4.38.

4.6.6 Bulk Solids with Appreciable Bulk Elasticity or Springiness

While all bulk solids exhibit friction and cohesive characteristics, most solids have high bulk rigidity. However, there are some bulk solids which exhibit significant bulk elasticity and *springiness*. Such bulk solids include wood chips, chopped-up film, cotton flock, wool, polymer foam and orange peel. As discussed by Johanson [24], such materials have *elastic wind-up* potential that can cause extreme bridging problems.

Severe problems have been experienced with wood chips [24]. In one case, chips measuring approximately 54 mm were able to flow through a 250 mm diameter opening in a mass-flow cone, even after standing at rest for 24 h, provided no elastic energy was stored. When placed in an 11.5 m diameter funnel flow bin, the wood chips arched over a 4.5 m diameter vibrating bin discharger. It is apparent that the bin vibrator caused a significant amount of elastic energy to be stored, the strength increasing approximately linearly with consolidation pressure.

The obvious solution to bin design for materials exhibiting high bulk elasticity or springiness is to ensure that the consolidation pressures are kept within acceptable limits. There are two possibilities for over-pressures to occur in bins.

(i) Initial Filling Pressures

Here there is little solids movement. In the case of mass-flow, movement of the solids can be achieved by initiating flow during the filling process. In this way the over pressure associated with static filling can be eliminated, thus avoiding elastic energy being stored.

(ii) Head of Solids in Vertical Channel

The head of solids in a vertical channel may cause elastic energy to be stored causing arching. The problem may be reduced by employing rough walls in the cylinder. One practical solution involves the use of ledges or shelves as illustrated in Fig. 4.40. It is also good practice, in the case of mass-flow bins, to use "squat" bin proportions by limiting the height of the cylinder.

The limitations in over pressure can be determined from the critical arching dimension given by Eq. (4.4). At the critical condition, Fig. 4.41, the stress acting on the arch $\overline{\sigma}_1 = \sigma_c$ is the unconfined yield strength.

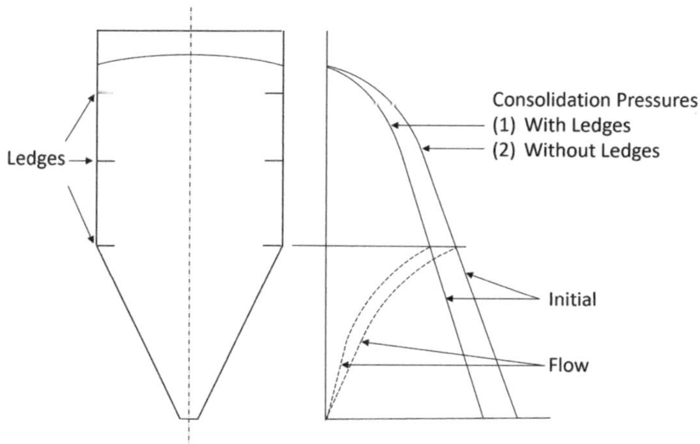

Fig. 4.40 Use of ledges to control surcharge loads

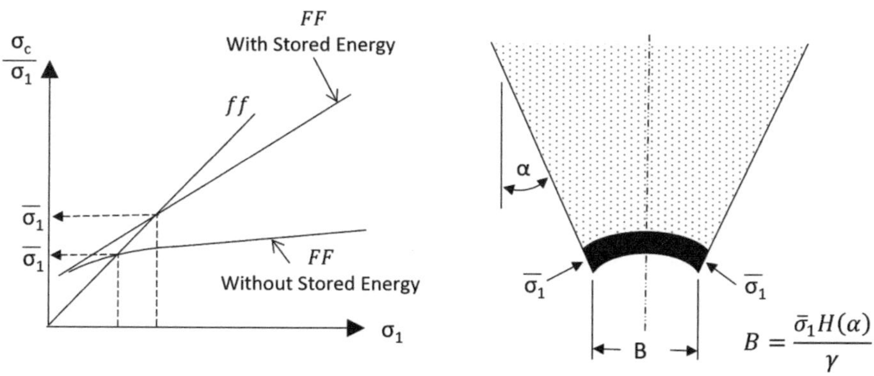

Fig. 4.41 Arching conditions in a hopper when elastic bulk solid is stored

For a given consolidation head or pressure σ_1 the value of σ_c and hence the arching dimension B will be much larger if there is elastic energy stored than if this is not the case. Note also that B becomes larger as the bulk density decreases. This is an added problem when handling materials such as wood chips.

4.7 Use of Inserts to Control Bin Flow Patterns

Inserts may be used in bins to achieve a desirable flow pattern. Often, they are used in the form of a *retro-fit* device to correct an undesirable flow pattern which is giving rise to either flow blockages or dynamic wall loads of large magnitude in silo walls. Knowledge of the flow properties of the bulk solid, together with an understanding of the conditions controlling bin flow patterns will permit the application of innovative design in the use of inserts for a specific purpose.

4.7.1 The Inverted Cone Insert

Johanson [25] proposed the use of inverted conical-shaped inserts to correct the flow in hoppers as illustrated in Fig. 4.42. In effect, the insert converts a conical hopper into a pseudo plane-flow hopper in which the slot is equivalent to a circular annulus. As indicated by the mass-flow limits shown in Fig. 4.1, the hopper half-angle for a plane-flow hopper is about 8° to 10° larger than that of the equivalent conical hopper. It is important to correctly place the insert. This requires consideration of the wall friction angle φ, insert cone angle α_c the insert base radius R and the spacing W, the latter being sufficient to permit flow without arching. Procedures for designing and correctly placing inserts are presented in Johanson's paper.

The application of inverted, conical-type inserts to in-bin blending has received limited attention and it is a subject requiring further research. From a practical point of view, a knowledge of the flow properties of the bulk solid and the boundary friction characteristics of the hopper and inserts can be used to advantage in correcting segregation problems. For instance, if segregation has occurred in a bin with fines congregating in the central region and coarse particles in the outer region, the selection of the appropriate lining surface for the insert as well as its slope angle is important. Since the boundary friction angle for fine particles is higher than for coarse particles due to the particle size and surface roughness interaction, the surface of the insert needs to be smoother than that of the hopper for similar slope angles so that the fines can flow more freely and mix with the coarser particles.

An overriding consideration in the use of inserts in bins for flow correction concerns the magnitude of the draw-down forces that can be generated. Since the inserts must be supported from the hopper walls, it needs to be noted that for the design of the supports, loads greater than hydrostatic can occur. For this reason,

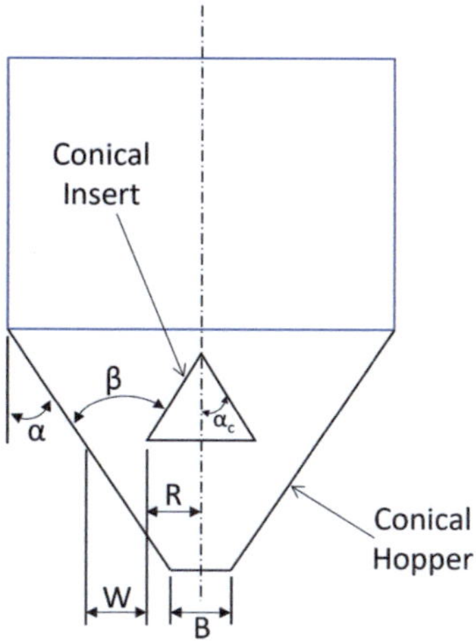

Fig. 4.42 Insert to correct hopper flow pattern

the placing of inserts in bins must be carefully considered in the light of alternative solutions.

The principle of the insert is also embodied in the vibrating *live bottom* bin or pile discharger as illustrated in Fig. 4.43. In this case, the gyratory motion of the insert or baffle assists in promoting flow. It is important that the vibrator be cycled rather than run continuously to avoid compaction of the bulk solid in the bin (or stockpile) if the capacity of the conveyor transporting the material from the bin does not match the feed rate.

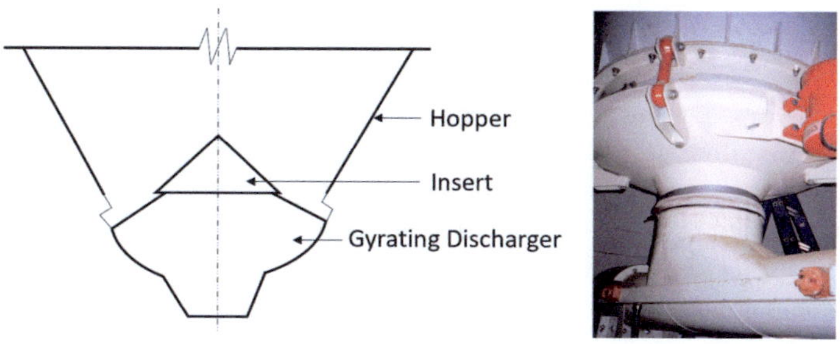

Fig. 4.43 *Live Bottom* bin

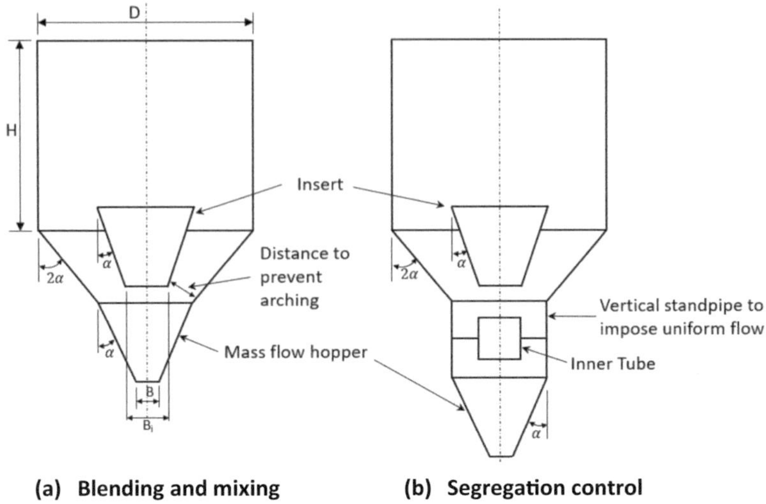

Fig. 4.44 Binsert™ for promoting mass-flow and bin blending

4.7.2 Cone–in Cone Binsert™

The foregoing problems may be solved by means of the cone-in-cone Binsert™ invented by Johanson [10]. Alternative arrangements of this device are shown in Fig. 4.44.

Essentially the half-angle α of the inner cone is determined by the mass-flow design requirement as previously outlined. The half-angle of the outer cone is then set at 2α the hopper lining materials for the inner and outer cones being the same α. Looked at in another way, suppose a cylindrical bin with conical hopper has a hopper half angle 2α and no inner cone is fitted, then quite clearly the bin will operate under funnel-flow. By fitting the inner cone with half angle α, the flow changes to mass-flow. This illustrates the effectiveness of the Binsert™ in correcting problems due to funnel-flow as well as allowing large storage capacities to be achieved with reduced head room than is possible with a conventional mass-flow bin. The device is much more effective than the inverted cone insert.

The principal advantage of the cone-in-cone insert is in blending and anti-segregation control.

4.7.3 Plane-Flow Hopper in Hopper Concept

The concept of the hopper-in-hopper insert may be successfully applied to plane flow bins with wedged-shaped, funnel-flow hoppers. While plane flow offers the advantage of a half-angle being 8 to 10 degrees larger for the equivalent half-angle

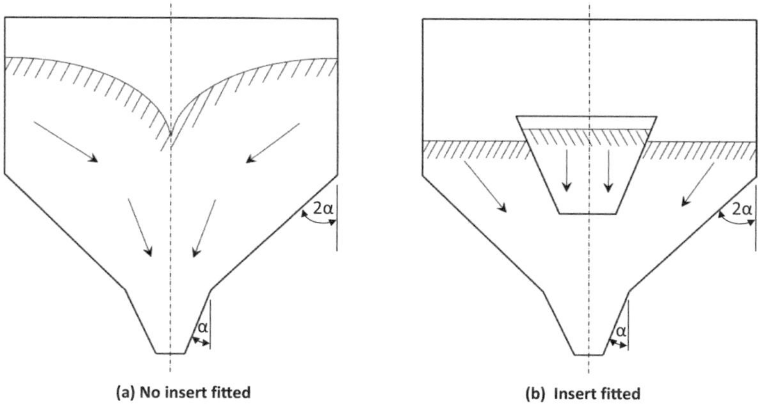

Fig. 4.45 Flow patterns in plane flow bin

for conical hoppers, occasions such as space limitations, particularly restricted head room, necessitate the employment of funnel-flow and larger half-angles as illustrated in Fig. 4.45a. The conversion to reliable mass-flow may be achieved by the incorporation of a plane-flow hopper insert as shown in Fig. 4.45b. The hopper half-angle applies to the insert, while the angle 2α applies to the hopper.

The advantage of the hopper-in-hopper insert was successfully applied to the solution of an industrial handling problem in the food industry. The application was linked to the process of mixing and transport of the powder ingredients for a soft drink processing application.

The flow properties of the powder were determined, and these were used to select the appropriate half-angles and opening dimensions. The required half-angle was $\alpha = 22°$ so this became the angle for the insert, while the hopper was designed with the half angle $2\alpha = 44°$. Tests conducted on a pilot-scale hopper confirmed these results. The final design is shown in Fig. 4.46.

The bin incorporates a specially designed screw feeder to promote uniform draw-down of powder from the upper hopper followed by controlled transfer to the lower mass-flow hopper. It is important that controlled discharge and good mixing be maintained.

4.7.4 Case Study—Coke Bin

To illustrate the application of inserts to correct the flow pattern in a bin, a case of a 4 m square by 15 m tall flat bottom silo constructed of concrete is considered. The silo stored coke and discharges centrally, the flow being controlled by a vibratory feeder. Segregation problems are being experienced. The silo operates under funnel-flow and no doubt has significant *dead* capacity in addition to the segregation problem.

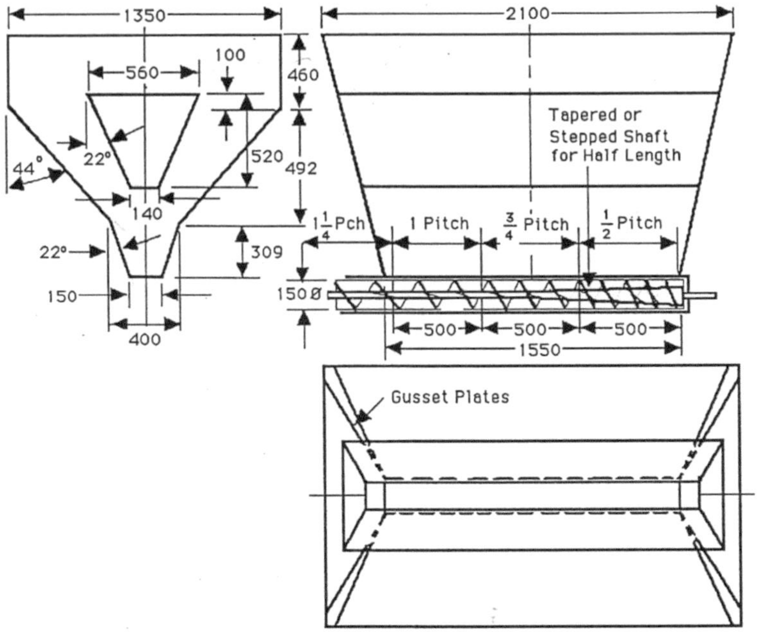

Fig. 4.46 Blending bin final design—project for a soft drink manufacturer

Three possible solutions are reviewed in Fig. 4.47. It is very desirable that the silo be converted to mass-flow and the arrangement in Fig. 4.47a is the simplest solution. The second option of Fig. 4.47b incorporates a pyramidal hopper insert; it would produce good mixing but may be more complex than is necessary. Alternatively, a conical hopper insert may be used. The third option involves an inverted cone insert which still requires a hopper to be fitted within the bin; this arrangement is shown in Fig. 4.47c. The hopper half-angle in this case would need to be fixed by the plane flow limits. The hoppers in each case would need to be lined with a low friction material such as Stainless-Steel type 304 with 2B finish or an ultra-high molecular weight polyethylene material (such as Tivar 88). It is essential that the flow properties of the coke be determined for the range of moisture contents and residence times expected to occur in practice. In this way the correct geometry for the hoppers and inserts can be determined.

It must be noted that the solutions proposed above do not reduce the capacity of the bin through the installation of the hoppers. This is because the bin as it is at present has reduced live capacity due to the *dead* storage. Conversion to mass-flow guarantees the bin will completely empty with good re-mixing performance and with the desired *first-in, last-out* flow pattern.

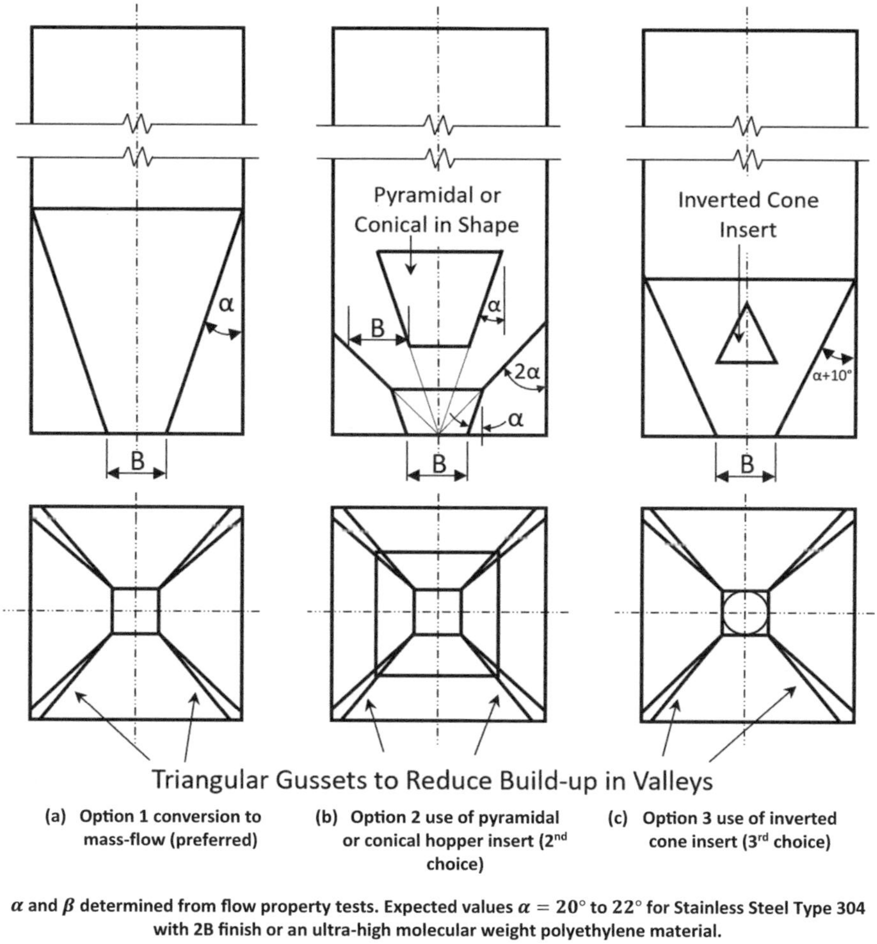

Fig. 4.47 Proposed modifications to coke bin

4.8 Concluding Remarks

This chapter has presented the basic background procedures for bin design with the appropriate case studies to demonstrate the variety of applications that are typically required by industry.

References

1. Jenike, A.W., *Storage and Flow of Solids.* Utah Agric. Exp. St. Bull, 1964. **123**.
2. Jenike, A.W. and T. Leser, *A Flow - No Flow Criterion in the Gravity Flow of Powders in Converging Channels*, in *Proc. 4th Intl. Congress on Rheology*. 1963.
3. Arnold, P.C., A.G. McLean, and A.W. Roberts, *Bulk Solids: Storage, Flow and Handling.* 2nd ed. 1979: The University of Newcastle Research Associates (TUNRA).
4. Guo, J. and U.o. Newcastle, *Investigation of arching behaviour under surcharge pressure in mass-flow bins and stress states at hopper/feeder interface.* 2014.
5. Guo, J., A.W. Roberts, and J.D. Prigge, *The Investigation of Arch Model Acting in Mass-Flow Hoppers.* Advanced Materials Research, 2012. **508**: p. 135–140.
6. Arnold, P.C. and A.G. McLean, *An Analytical Solution for the Stress Function at the Wall of a Converging Channel.* Powder Tech., 1976. **13**: p. 255.
7. Enstad, G., *On the theory of arching in mass flow hoppers.* Chemical Engineering Science, 1975. **30**(10): p. 1273–1283.
8. Moore, B.A. and P.C. Arnold, *An Alternative Presentation of the Design Parameters for Mass-Flow Hopper.* Powder Tech., 1985. **42**(1): p. 79–89.
9. Johanson, J.R., *Method of Calculating Rate of Discharge from Hoppers and Bins.* Trans. Soc. Min. Engrs., AIME, 1965. **232**: p. 69–80.
10. Johanson, J.R., *The Placement of Inserts to Correct Flow in Bins.* Powder Tech., 1967/68. **1**: p. 328–333.
11. Johanson, J.R. and T.A. Royal, *Measuring and Use of Wear properties for Predicting Life of Bulk Materials Handling Equipment.* Bulk Solids Handling, 1982. **2**(3): p. 517–523.
12. Roberts, A.W., *Friction, Adhesion and Wear in Bulk Materials Handling*, in *Keynote Paper, Proc. AntiWear 88, Inst. of Metals, Instn. Mech. Engrs.* 1988: The Royal Society London.
13. Roberts, A.W., M. Ooms, and S.J. Wiche, *Concepts of Boundary Friction, Adhesion and Wear in Bulk Solids Handling Operations.* Intl. Jnl. of Bulk Solids Handling, 1988. **10**(2).
14. Roberts, A.W., *Tribology on Bulk Solids Handling* Transactions, Institution of Engineers, Australia, Mechanical Engineering, 1991. **ME16**(3).
15. Roberts, A.W., *Bin and Feeder Design Criteria for Anti-Segregation and Blending*, in *Proc. Conference on Cost Effective Bulk Materials Handling, Mech 91 Congress, Instn. of Engrs Aust.* 1991: Sydney.
16. McLean, A.G., *Flow Rates Simple Bulk Solids from Mass-Flow Bins.* 1979, The University of Wollongong, Australia.
17. Arnold, P.C. and Z.H. Gu, *The Effect of Permeability on the Flowrate of Bulk Solids from Mass-Flow Bins.* The Intl. Jnl. of Powder Handling and Processing, 1990. **2**(3): p. 229–238.
18. Arnold, P.C., Z.H. Gu, and A.G. McLean, *On the Flowrate of Bulk Solids from Mass-Flow Bins*, in *Proc. 2nd World Congress Particle Technology*. 1990: Kyoto, Japan. p. 2–9.
19. Roberts, A.W., M. Ooms, and K.S. Manjunath, *Feeder Loads and Power Requirements in the Controlled Gravity Flow of Bulk Solids from Mass Flow Bins.* Mechanical Engineering Transactions, The Instn. of Engrs. Australia, 1984. **ME9**(1): p. 49–61.
20. Manjunath, K.S. and A.W. Roberts, *Wall Pressure-Feeder Load Interactions in Mass-Flow Hopper/Feeder Combinations -Part 1* Bulk Solids Handling, 1986. **6**(4): p. 769–775.
21. Manjunath, K.S. and A.W. Roberts, *Wall Pressure-Feeder Load Interactions in Mass-Flow Hopper/Feeder Combinations—Part 2.* Bulk Solids Handling, 1986. **6**(5): p. 903–911.
22. Jenike, A.W., *Quantitative Design of Mass Flow Bins.* Powder Tech., 1967. **1**: p. 237.
23. Roberts, A.W., O.J. Scott, and W. Tiedemann, *Bulk Handling of Ammonium Nitrate: Flow Properties and Storage Bin Design Criteria*, in *International Powder and Bulk Solids Conference.* 1979: Philadelphia, U.S.A.
24. Johanson, J.R. and B.D. Cox, *Bin and Feeder Design for Wood Chips and Other Springy Bulk Solids*, in *14th Annual Powder and Bulk Solids Conference.* 1989: Chicago, U.S.A. p. 464–474.
25. Johanson, J.R., The use of flow-corrective inserts in bins. Journal of Engineering for Industry, 1966. 88(2): p. 224–230.

Chapter 5
Design and Performance of Funnel-Flow and Expanded-Flow Bins and Gravity Reclaim Stockpiles

5.1 Introductory Remarks

This Chapter is devoted to the design of funnel-flow and expanded-flow bins and gravity reclaim stockpiles. Procedures for determining the geometrical shape of ratholes during funnel-flow in bins and gravity reclaim stockpiles are presented. Once the relevant flow properties of the bulk solid have been determined, the critical rathole diameters, draw-down and expansion angles of the ratholes may be determined. Using techniques based on solid geometry, the live capacity of the bin or stockpile may be estimated.

To overcome known limitations in the procedures for funnel-flow design, the new hoop stress theory developed by the authors Roberts et al. [1] is reviewed and its application to funnel-flow design is described. To demonstrate the influence of bin and stockpile geometry on draw-down and live capacity performance, results of experiments conducted on laboratory scale bin and stockpile models are presented. Case studies of large-scale industrial stockpiles are included to demonstrate the validity of the new design theory and test procedures which are shown to provide much improved predictions of stockpile draw-down performance.

5.2 Review of Funnel-Flow in Bins and Stockpiles

5.2.1 Basic Concepts

As explained in previous Chapters, the formation of stable pipes or "ratholes", by which name they are commonly known, is associated with funnel-flow, which is depicted in Fig. 5.1. In this case flow occurs by material sloughing off the top surface of the bin and flowing down the central flow channel which forms above the opening. Flow will continue until the level of the bulk solid in the bin drops an amount H_D

A. W. Roberts, *The Theory and Practice of Bulk Materials Handling*,
https://doi.org/10.1007/978-3-032-19702-3_5

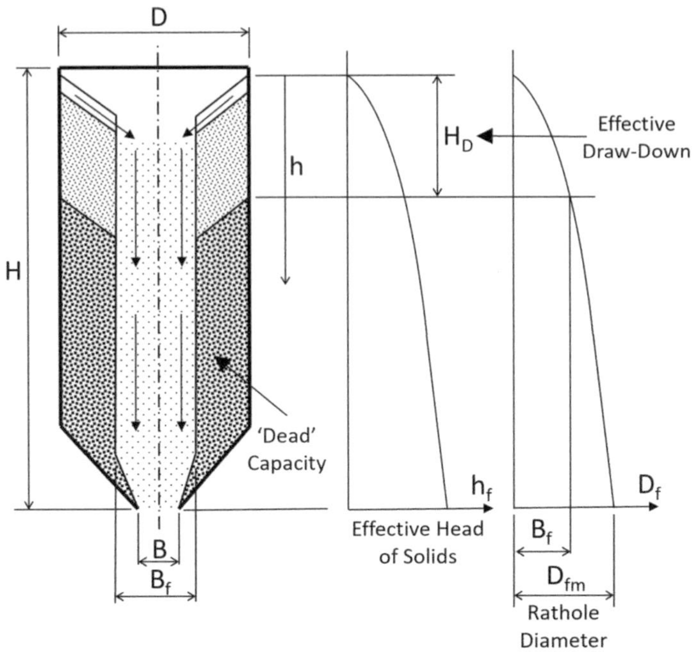

Fig. 5.1 Funnel-flow

equal to the draw-down. At this level, the bulk strength of the contained material is sufficient to sustain a stable rathole of diameter D_f as illustrated in Fig. 5.1.

At the bin outlet the rathole diameter $D_f = B_f$ will be slightly larger than the hopper opening B. Above the opening the rathole walls defined by the diameter D_f are steeply inclined and in many cases approximately vertical. Once the level defined by H_D is reached, there is no further flow and the material below this level represents 'dead' storage. For complete draw-down, B_f should be at least equal to D_{fm}, which is the rathole diameter calculated for the base of the bin. However, for cohesive bulk solids, D_{fm} is often several metres rendering funnel-flow impracticable. Expanded-flow, as illustrated Fig. 5.2, is then a better proposition.

5.2.2 *Gravity Reclaim Stockpiles*

Gravity reclaim stockpiles, as illustrated in Figs. 5.3 and 5.4 are widely used in the mining and mineral processing industries to store cohesive bulk solids in large quantities. In view of the overall height of stockpiles, typically 20 to 40 m, the consolidation stresses can be quite high and, correspondingly, the critical pipe or rathole dimensions quite large. Hence the reclaim hopper and feeder dimensions, from a practical point of view, become prohibitively too large.

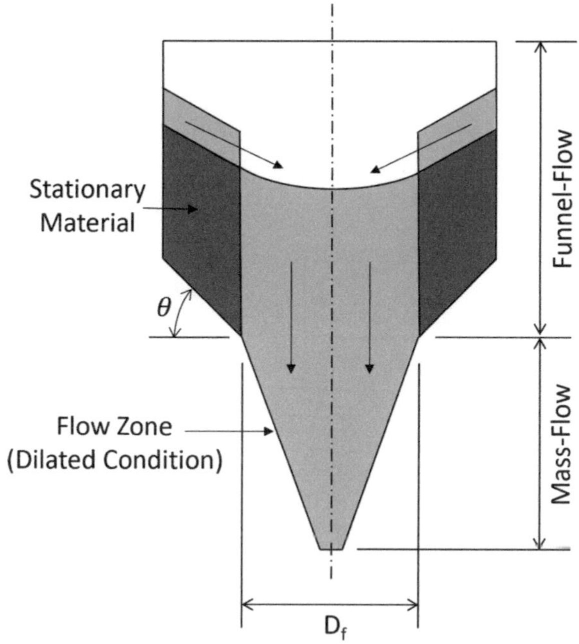

Fig. 5.2 Expanded flow bin

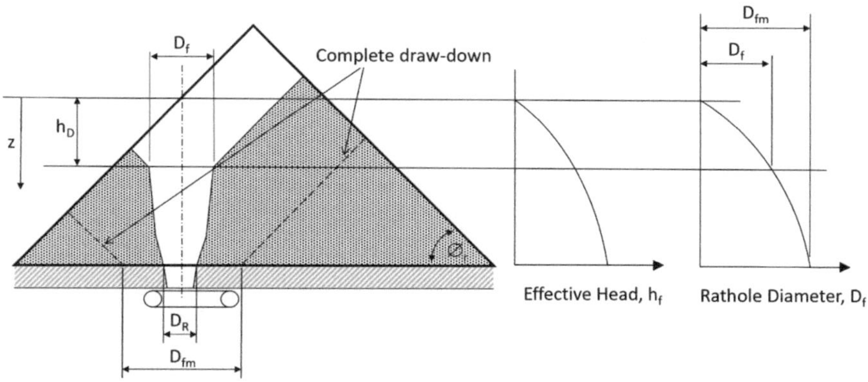

Fig. 5.3 Gravity reclaim stockpile—single outlet

Since complete draw-down is impractical, it is accepted that ratholes need to form above the reclaim openings. It then becomes necessary to employ multi outlets as illustrated by the twin outlet stockpile of Fig. 5.4 where the draw-down and live capacity may be optimised by selecting the appropriate spacing distance 'X'.

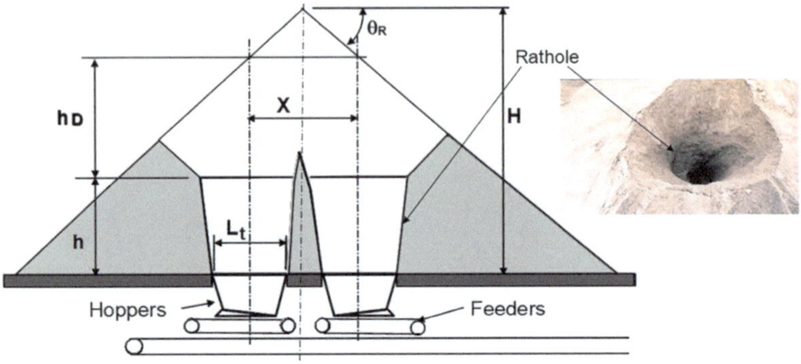

Fig. 5.4 Gravity reclaim stockpile—twin outlets

Gravity reclaim stockpiles operate under expanded flow, with funnel-flow in the upper main storage and discharge via mass-flow hoppers and feeders as shown in Figs. 5.3 and 5.4. The mass-flow hoppers have two important functions:

(I) Flow control at the required mass-flow discharge rate.
(II) Feeder load control through the establishment of a passive stress field in the hopper. This is discussed in more detail in Chap. 6 on Wall Loads and Chap. 11 on Feeder Design.

5.3 Review of Jenike Funnel-Flow Theory

5.3.1 Rathole Geometry

Jenike assumed that ratholes are circular in shape with vertical walls and that an axi-symmetric arched stress field exists in the flow channel. This is shown in Fig. 5.5. He further assumed that the stability of a rathole depended on the critical arching dimension of the bulk solid discharging by mass-flow in the flow channel. The ability of a bulk solid to sustain a stable pipe or rathole depends on its unconfined yield strength σ_c which, in turn, depends on the major consolidation stress, σ_1, of the stored bulk solid. The incipient failure state is defined by the condition $\sigma_c = \bar{\sigma}_1$, where $\bar{\sigma}_1$ is the stress acting at the surface of the rathole. The stress state is depicted in the Mohr diagram of Fig. 5.6.

For mass-flow in the flow channel, as illustrated in Fig. 5.5, the arch is assumed to be of uniform thickness for which the shear stress t is given by

$$\tau = \frac{\gamma D}{4} \tag{5.1}$$

where

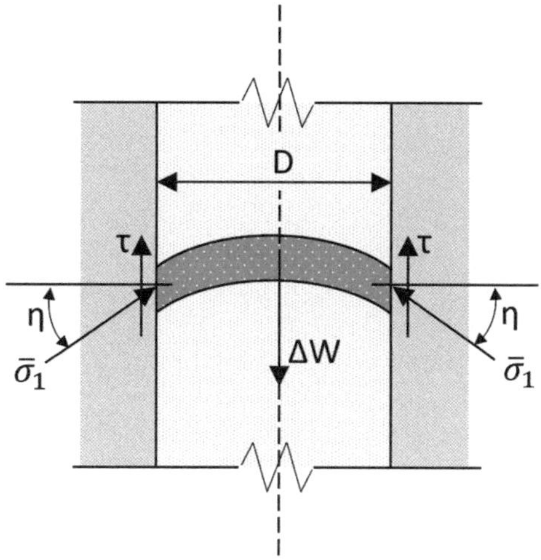

Fig. 5.5 Jenike arched model

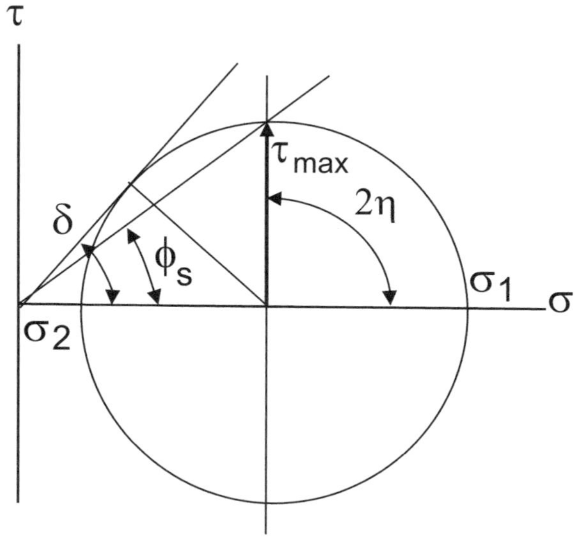

Fig. 5.6 Mohr stress diagram

$\gamma = \rho g =$ bulk specific weight

$\rho =$ bulk density

The stress conditions for failure and flow in the flow channel are depicted in the Mohr stress diagram of Fig. 5.6. It is noted that failure and flow in the flow channel corresponds to the maximum shear stress $\tau = \tau_{max}$ for the hopper half-angle $\alpha = 0°$. That is, the arch thickness parameters $H(\alpha) = 2$ for $\alpha = 0$ as given by Eqs. 4.5 and 4.29 and Fig. 4.10 of Chap. 4.

The angle φ_s defines the equivalent friction angle at the surface of the flow channel corresponding to the effective angle of internal friction δ. The friction coefficient is given by

$$\mu = \tan\varphi_s = \sin\delta \tag{5.2}$$

The angle $\eta = 45°$ defines the stress $\bar{\sigma}_1$ acting at the boundary of the arch.
Jenike derived the following expression for the critical pipe or rathole diameter,

$$D_f = \frac{\sigma_c G(\varphi_t)}{\gamma} \tag{5.3}$$

Where

$\gamma = \rho g =$ bulk specific weight
$\rho =$ bulk density
$\sigma_c =$ $\bar{\sigma}_1$
$\varphi_t =$ static angle of internal friction.

For convenience, $G(\varphi_t)$ may be represented by the empirical equation

$$G(\varphi_t) = 4.3\tan\varphi_t \tag{5.4}$$

A plot of $G(\varphi_t)$ is given in Fig. 5.7.

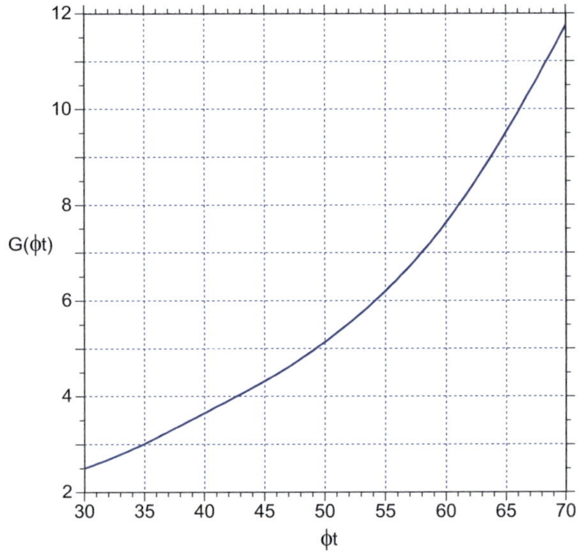

Fig. 5.7 Function $G(\varphi_t)$

5.3.2 Influence of Bin Outlet Geometries

Jenike showed that a pipe or rathole which forms during funnel-flow is controlled by the shape and dimensions of the bin outlet as shown in Fig. 5.8. For circular outlets, the rathole forms across the opening. For square and rectangular outlets, the critical rathole dimension D_f corresponds to the diagonal of the opening as indicated. In the case of symmetrical outlets, such as the circle and rectangle, the rathole dimension D_f is normally greater than the critical arching dimension B_{CR} so that flow blockages due to the formation of cohesive arches are unlikely to occur. However, in the case of the rectangular opening, in addition to the determination of D_f, it is also necessary to ensure that the slot width B is greater than the critical arching dimension B_{CR}. Jenike noted that the flow channel generally breaks away at a half angle of 30° from the sides of a rectangular opening. To ensure that a stable arch does not form, he recommended that a half-angle less than 30° should be used. From mass-flow design, the flow factor corresponding to 30° is $ff = 1.7$ and the arch thickness parameter is $H(\alpha) = 1.15$. The dimension B is determined as follows:

(i) Draw the flow factor $ff = 1.7$ on the Flow Function graph (instantaneous or time whichever is relevant to the design) and determine the value of σ_1 at the point of intersection.
(ii) Compute the dimension B from the following equation:

$$B = \frac{1.15\,\bar{\sigma}_1}{\rho g} \tag{5.5}$$

While the function $G(\varphi_t)$ is based on the undisturbed time storage internal friction angle φ_t, in cases where this parameter is not known, the kinematic instantaneous internal friction angle φ_t may be substituted. Normally these two friction angles compare quite closely.

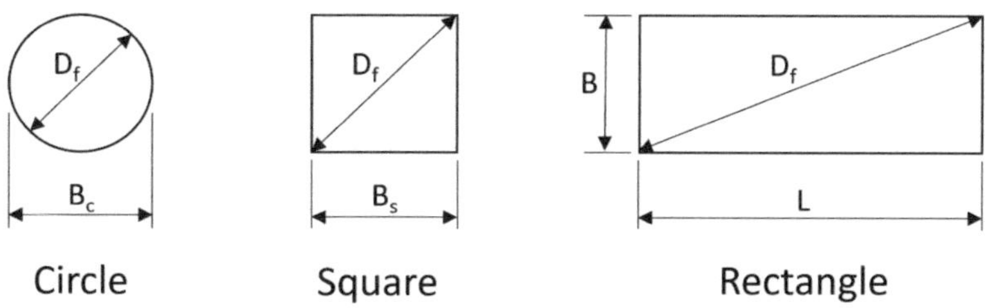

Fig. 5.8 Outlet geometries of funnel flow bins

5.3.3 Flow Properties and Rathole Diameter Predictions

While the flow properties for mass-flow design are determined for the low range of major consolidation stress, typically $0 < \sigma_1 < 20\,kPa$, for funnel-flow design the consolidation stresses are much higher. For example, in the case of a 40 m high gravity reclaim iron ore stockpile, the consolidation stresses approaching a mega-Pascal (MPa) may be experienced. Hence, the need for an extended range of flow property tests to be performed. For the discussion that follows, a set of flow property tests are presented in conceptual form in Fig. 5.9.

The prediction of rathole diameters that have been developed are described as "lower bound" and "upper bound" as now described:

(a) Lower Bound Solution

Noting that the discharge from a funnel-flow bin is characterised by mass-flow in the pipe or rat-hole flow channel, Jenike derived piping flow factors ff_p to determine the relationship between the stress $\bar{\sigma}_1$ and the major consolidation stress σ_1 such that $\bar{\sigma}_1 = \frac{\sigma_1}{ff_p}$. Plots of ff_p as functions of the static, φ_t, and effective angles of internal friction δ are given by Jenike [2]. As shown in Fig. 5.9, the ff_p flow factor line is drawn on the flow property Flow Function graph and the intersection with the instantaneous FF defines the major consolidation stress σ_1. The shortcoming of Jenike's original method is that it ignores the surcharge pressure acting in the stored bulk solid and, therefore, under-estimates rathole diameters. Quite possibly, this result which is referred to as the "lower bound", may only be approached if the bin is being filled and emptied at the same time. It is more of academic interest rather than of any practical value for bin and stockpile design.

(b) Upper Bound Solution

Recognising the deficiencies of the lower bound solution as being inadequate for funnel-flow design, the procedures were modified to take account of the surcharge pressure when calculating the major consolidation stress, σ_1. The unconfined yield strength, σ_c, is obtained directly from the time flow function FF_t as indicated in Fig. 5.9. This method is known as the Upper Bound solution. In general, this revised method has been found to be rather too conservative.

The above two solutions are shown in red on the flow property graphs of Fig. 5.9.

5.4 Modified Theory for Funnel-Flow Design

Based on industrial experience from design and operational feedback associated with large stockpiles, it had become apparent that the procedures for funnel-flow design were rather too conservative. This was largely due to the adoption of the upper bound consolidation stress as the basis for the draw-down live capacity calculations. This led to the new "Hoop Stress Theory", developed by Roberts et al. [1], which has been shown to give improved performance predictions for funnel-flow design.

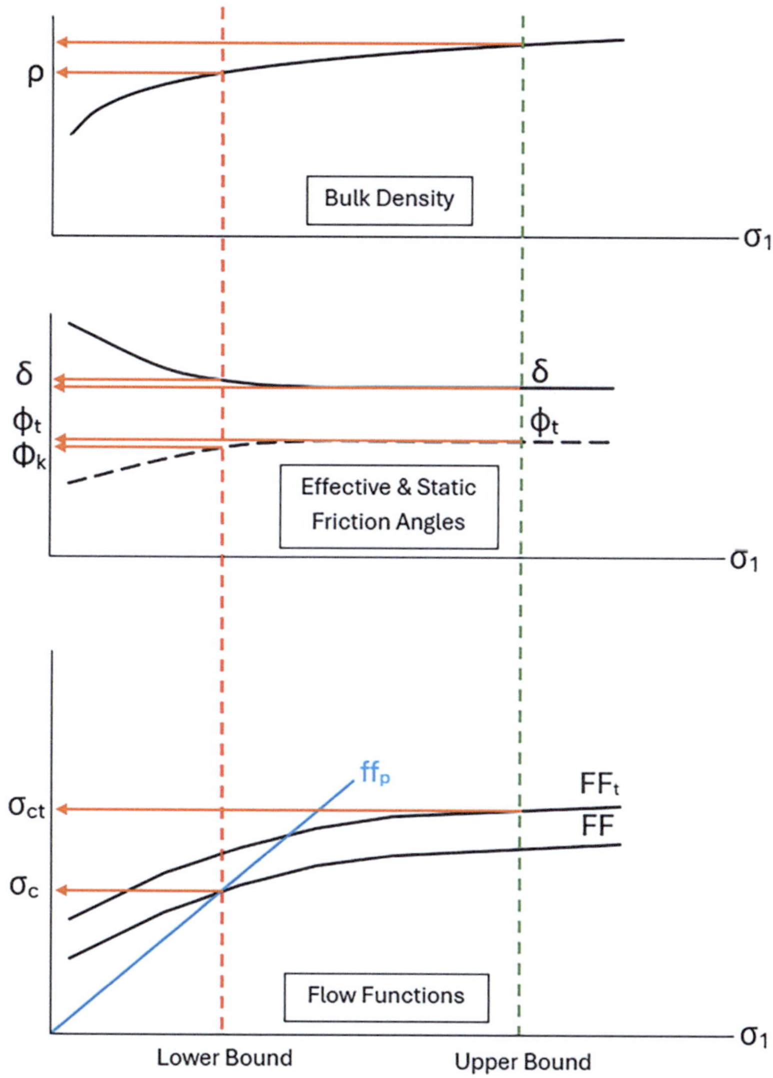

Fig. 5.9 Extended flow property graphs incorporating lower and upper bound design lines

5.4.1 *Hoop Stress Theory*

Observations of the sloughing action from the free surface during funnel-flow show that the slope angle of the free surface is not constant but steepens towards the central flow channel. The analysis assumes the hoop strength of the bulk solid is the dominating condition for rathole stability.

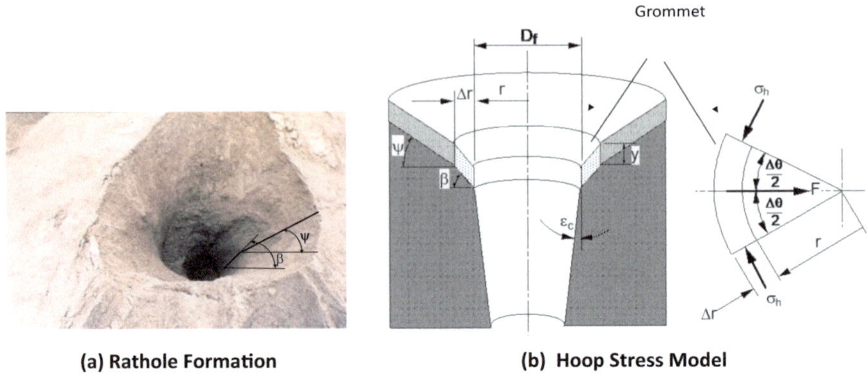

Fig. 5.10 Hoop stress model for funnel-flow

As may be observed from the photograph of Fig. 5.10a, the free surface is assumed to slope at an angle ψ, with the slope increasing to β in the region of the rathole. Figure 5.10b represents the hoop stress model adopted to describe the sloughing action. The rathole is formed when the hoop strength of the "donut-shaped" grommet is just on the verge of failing. The angle ψ is normally in the range $\varphi_t < \psi < \delta$. The angle β for incipient flow is assumed to be

$$\beta = \frac{\pi}{4} + \frac{\varphi_t}{2} \tag{5.6}$$

For the critical condition, the hoop stress is just equal to the unconfined circumferential yield strength $\sigma_{c\theta}$, that is, $\sigma_h = \sigma_{c\theta}$. The critical rathole dimension is given by

$$D_f = \frac{\sigma_{c\theta} G(\beta)}{\gamma} \tag{5.7}$$

where

$$G(\beta) = \frac{2(1 + \mu \tan\beta)}{} \tan\beta - \mu \tag{5.8}$$

Noting that $\mu = \tan\varphi_t$, it may be shown that $G(\beta)$ becomes

$$G(\beta) = 2\tan\beta \tag{5.9}$$

The form of Eq. (5.9) for $G(\beta)$ is similar to that of Eq. (5.4) for $G(\sigma_t)$. Both $G(\beta)$ and $G(\sigma_t)$ are plotted in Fig. 5.11. The determination of the critical rathole diameter D_f using Eq. (5.3) requires the Flow Function FF_θ for the third or circumferential direction to be known. This is shown in Fig. 5.12. FF_θ defines the unconfined yield strength $\sigma_{c\theta}$ in the circumferential direction as a function of the major consolidation stress σ_1.

Normally flow property tests are based on two-dimensional consolidation stress conditions in which the unconfined yield strength σ_c corresponds to the plane defined by the major and minor principal stresses σ_1 and σ_2. It has been shown that the unconfined yield strength, $\sigma_{c\theta}$, in the third or circumferential direction may be approximated quite closely to the value corresponding to the mean stress determined using the conventional 2-dimensional direct shear test. The mean stress is defined by

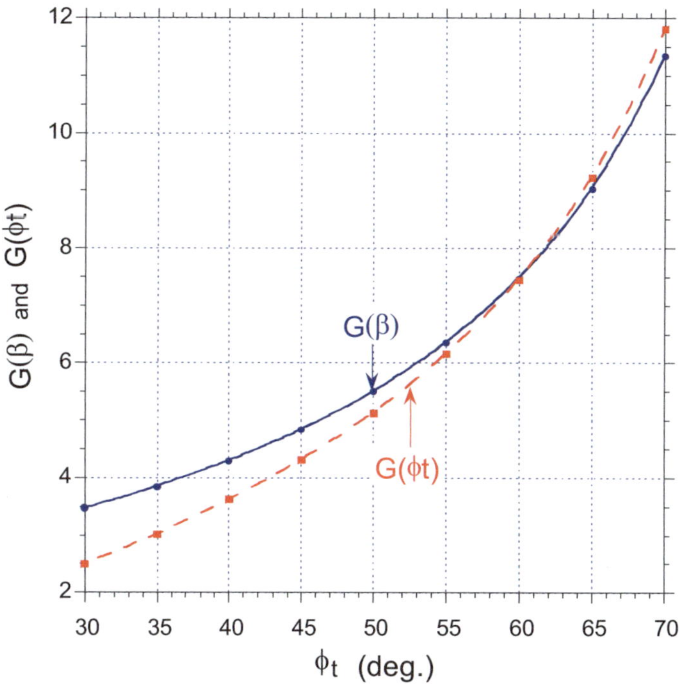

Fig. 5.11 Functions $G(\beta)$ and $G(\varphi_t)$

Fig. 5.12 3-Dimensional consolidation stresses

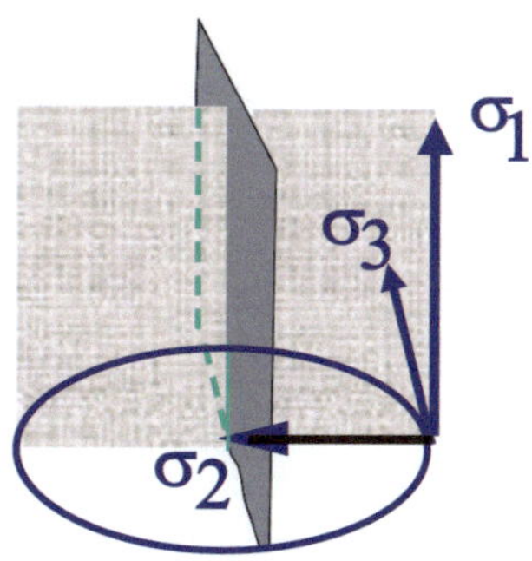

$$\sigma_m = \frac{\sigma_1}{1 + \sin\delta} \tag{5.10}$$

The flow property graphs of Fig. 5.9 may now be updated to include the mean value as shown in Fig. 5.13.

However, the flow function FF_θ defining $\sigma_{c\theta}$ as a function of σ_1 may be obtained from flow property measurements using the Jenike shear cell as illustrated in Fig. 5.14. The procedure is to consolidate each test sample in the shear cell in the normal way to establish the two-dimensional consolidation stress state for axi-symmetry. Following consolidation, the shear cell is then turned through 90° and relocated in the shear tester and the sample is sheared.

Figure 5.15 shows the Flow Functions FF and FF_θ for axi-symmetry and for the circumferential direction respectively for a sample of iron ore at a moisture content of 8.2%. The predicted FF_θ defined by the mean stress is also shown. The two FF_θ's are virtually identical. Sometimes the 90° shear tests give flow functions slightly

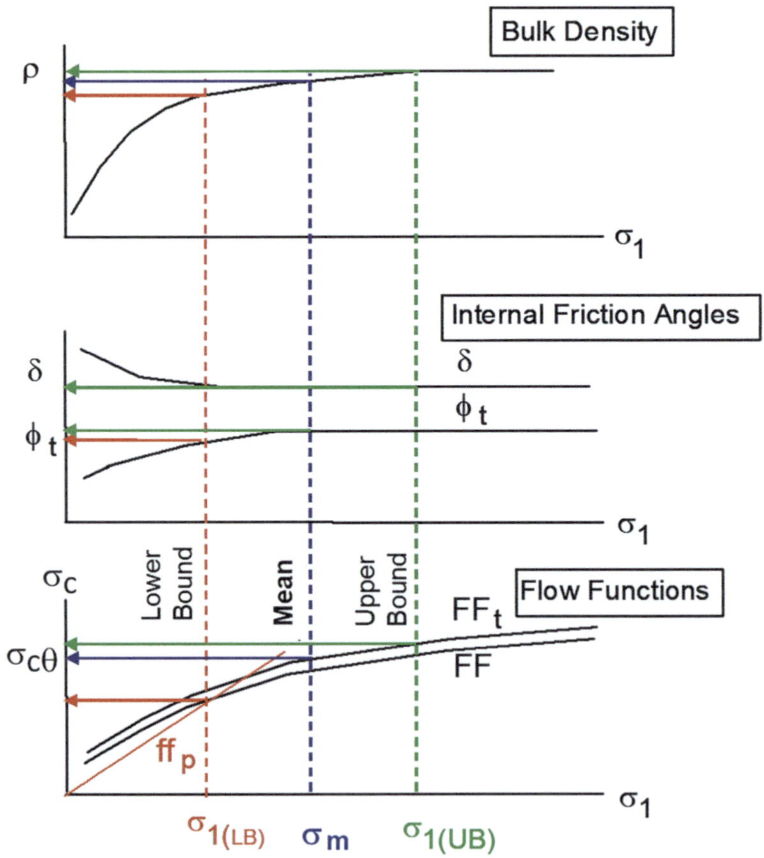

Fig. 5.13 Flow property graphs incorporating lower, mean and upper bound design lines

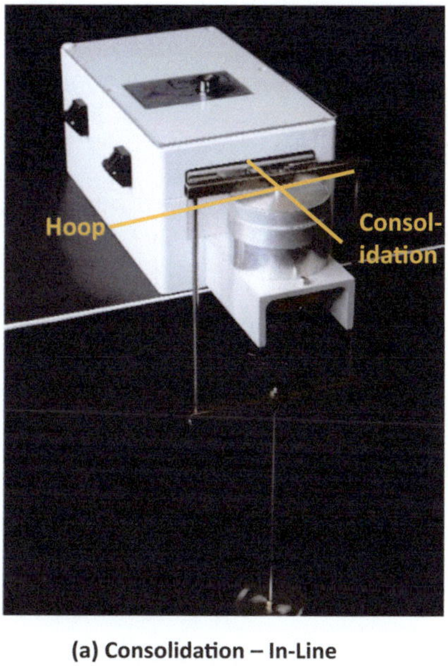

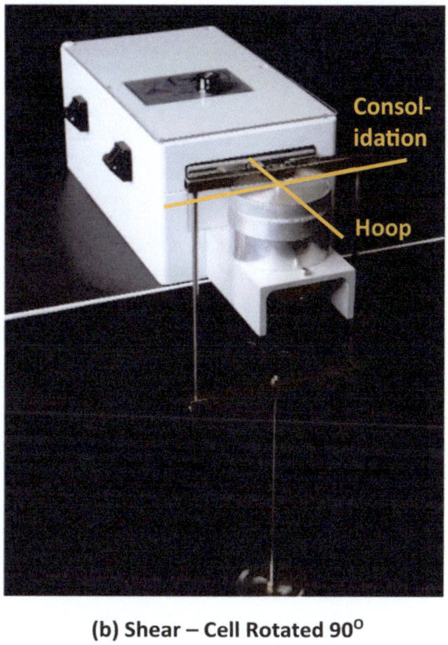

Fig. 5.14 2-Dimension shear test

lower than the predicted based on the mean stress. In other cases, slightly higher flow functions are measured. An example of the latter is shown in Fig. 5.16 which refers to a gold ore at 8.0% moisture content. Overall, the unconfined yield strength based on the mean consolidation stress is regarded as a satisfactory predictor when 90° shear test results are not available.

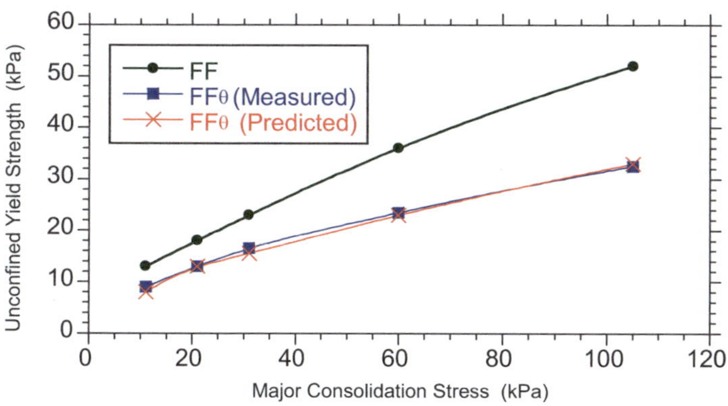

Fig. 5.15 Flow functions for iron ore

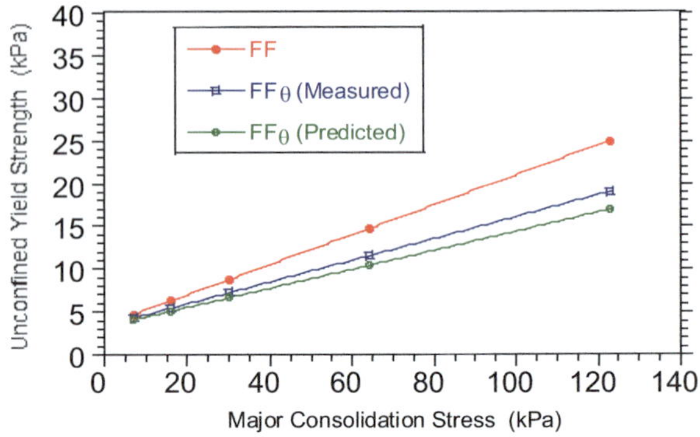

Fig. 5.16 Flow functions for gold ore

5.5 Rathole Geometry

5.5.1 Rathole Walls Expansion Angles

Consistent with Jenike's funnel-flow theory, it is noted that mass-flow occurs in the flow channel in which the friction angle at the boundary surface depends on the effective angle of internal friction δ. It is assumed that failure and, hence, flow at the boundary occurs when the shear stress is a maximum. Referring to the Mohr Circle diagram of Fig. 5.6, it readily follows that the internal friction angle φ_s at the boundary is defined by

$$\varphi_s = \tan^{-1}(\sin\delta) \tag{5.11}$$

Applying the mass-flow limits for the values of φ_s given by Eq. (5.11), the rathole slope angles are plotted in Fig. 5.17. Curve 1, defining ε_c for axi-symmetric flow, is the dominant analytical limit for mass-flow in a circular rathole. For long rectangular openings, the side wall slope angles ε_p are less precise. Curve 2 defines the lower bound design limit for ε_p for plane-flow, while Line 3 for $\varepsilon_p = 30°$ represents the likely upper bound limit. Jenike postulated that along the side walls of rectangular openings, ratholes initially slope away with a half slope angle of 30°. Once the rathole becomes circular in cross-section, the angle ε_c then controls the diverging slope of the rathole. While values of ε_p may fall within the range bounded by Curve 2 and Line 3, in many cases, particularly for long slotted openings, larger values of ε_p up to the 30° nominal limit are more likely to occur.

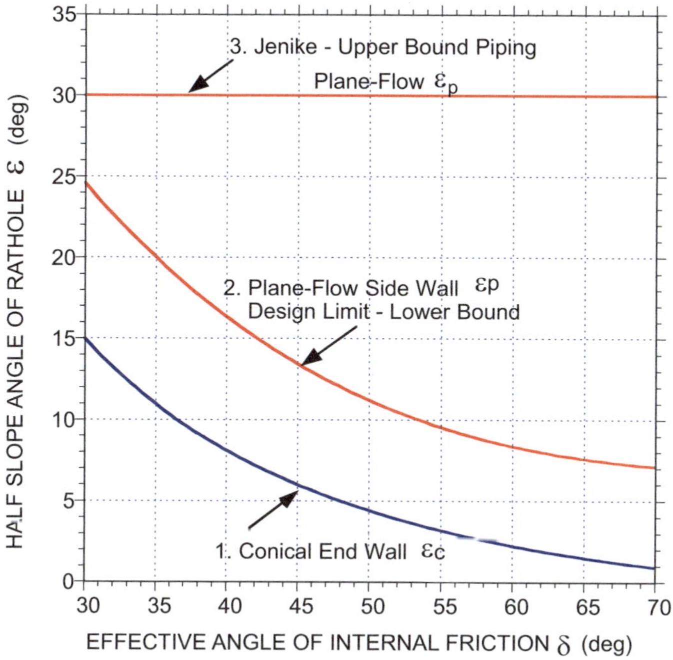

Fig. 5.17 Limiting half slope angles

5.5.2 Approximation of Rathole and Hopper Interface Geometry

The general case of the rathole interface with a rectangular opening associated with a reclaim hopper is considered as illustrated in Fig. 5.18. The rathole forms at the top of the hopper across the diagonal D_R of the rectangle, such that

$$D_R = \sqrt{{L_T}^2 + {D_T}^2} \tag{5.12}$$

$$\frac{D_R}{L_T} = \sqrt{1 + \left(\frac{D_T}{L_T}\right)^2} \tag{5.13}$$

where

L_T = length of hopper transition
D_T = width of hopper transition

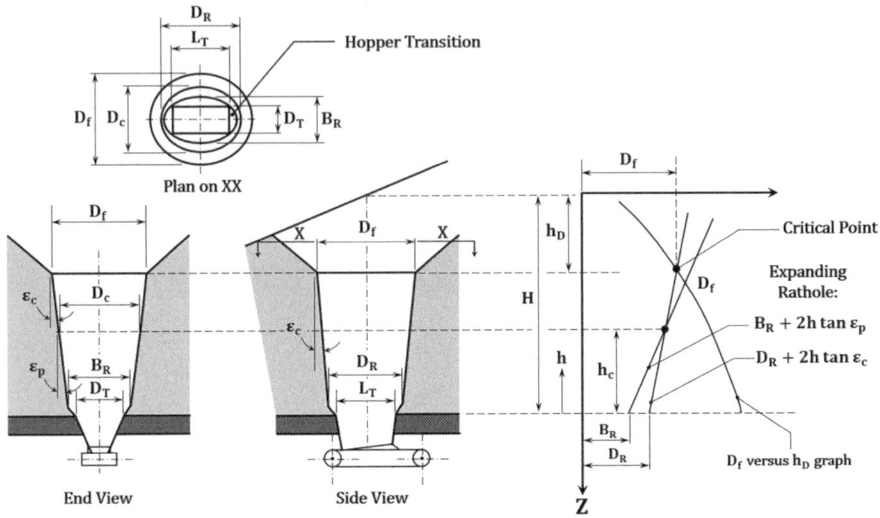

Fig. 5.18 Rathole and reclaim hopper geometry

While the actual shape of the rathole is quite complex with the walls slightly curved rather than straight, a close approximation to the shape may be made using the following assumptions:

(a) Lower Rathole and Hopper Interface Zone

Immediately above the rectangular opening corresponding to the hopper transition, the rathole is approximately elliptical in shape with D_R the major axis and B_R the minor axis. The determination of B_R is based on the following assumptions:

- For rectangular openings in which $L_T \gg D_T$, Fig. 5.18, the flow channel opens out at a half-angle of 30° along the sides, and at an average angle of 5° at the end.
- For a square opening, the flow channel opens out at an average angle of 7° along sides.
- For $D \ll L_h$, transitional arrangements apply.

On this basis, the following relationship is proposed:

$$\frac{B_R}{D_T} = \frac{L_T}{D_T}\left[1 + C_R\left\{1 - e^{\frac{(L_T/D_T)-1}{C_L}}\right\}\right]\left[\sqrt{1 + \left(\frac{D_T}{L_T}\right)^2} - 1\right] + 1 \qquad (5.14)$$

where

$C_R = 5.6$
$C_L = 8$

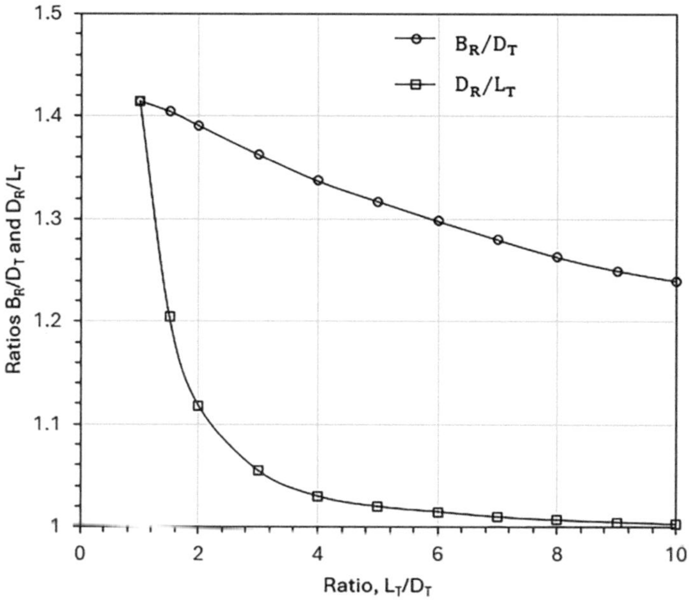

Fig. 5.19 $\frac{D_R}{L_T}$ and $\frac{B_R}{D_T}$ as functions of $\frac{L_T}{D_T}$

Figure 5.19 shows the plot of $\frac{D_R}{L_T}$ (Eq. (5.13)) and $\frac{B_R}{D_T}$ (Eq. (5.14)) as functions of $\frac{L_T}{D_T}$.

(b) The ratio of the area of the approximated ellipse to the area of the rectangular opening at the hopper transition is

$$A_r = \frac{\frac{\pi}{4} D_R B_R}{L_T D_T} = \frac{\pi}{4} \frac{B_R}{D_T} \sqrt{1 + \left(\frac{D_T}{L_T}\right)^2} \geq 1.0 \tag{5.15}$$

The plot of A_r versus $\frac{L_T}{D_T}$ is shown in Fig. 5.20.

(c) The rathole opens out from the elliptical shape at the bottom to a circular shape of diameter D_f (the critical rathole diameter) at the top corresponding to the draw-down h_D.

(d) The end walls of the rathole are sloped at an angle ε_c which is equal to or less than the limiting conical piping angle ψ_c.

(e) The side walls of the rathole are sloped at an angle ε_p which is equal to or less than the limiting plane-flow piping angle ψ_p. Normally, $\varepsilon_p < \psi_p$ since the slope of the side walls is controlled by the rathole becoming circular at the draw-down transition.

(f) Above the draw-down level, the crater sloughs off at an angle θ which normally ranges between the static angle of internal friction φ_t and the effective angle of

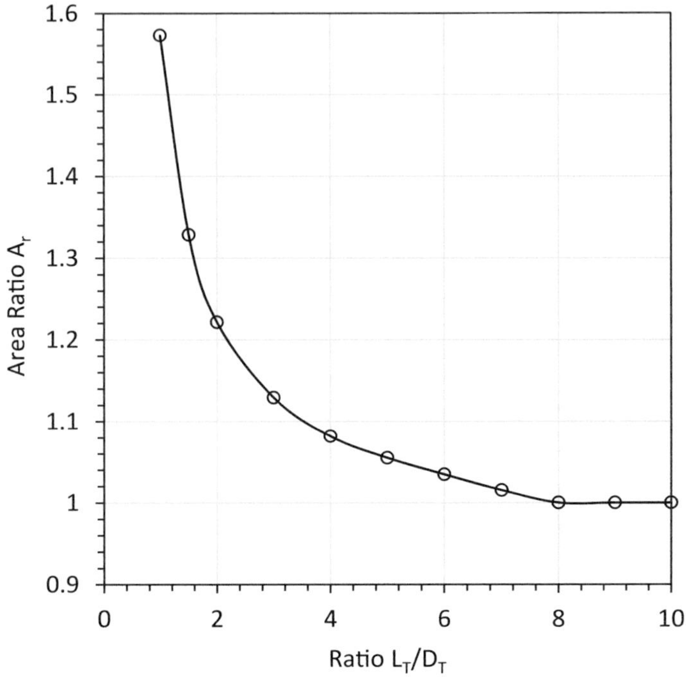

Fig. 5.20 Variation of area ratio with $\frac{L_T}{D_T}$

internal friction δ. Referring to Fig. 5.18, the limiting slope will normally be $\theta_e = \delta$ for the end slope. The slope of the sides is such that $\varphi_t \leq \theta_s \leq \delta$.

5.6 Draw-Down and Live Capacity in Funnel-Flow Storage Bins and Stockpiles

As is well known, funnel-flow is characterised by material sloughing off the top surface and flowing down the central flow channel which forms above the opening. It is a "first in, last out" flow pattern with flow continuing until the level of the bulk solid in the bin drops an amount h_D equal to the draw-down. At this level, the bulk strength of the contained material is sufficient to sustain a stable pipe or 'rathole' of diameter D_f. The bulk solid remaining in the bin represents "dead" capacity.

5.6.1 *Funnel-Flow Bins*

Figure 5.21 shows, by solid lines, a flat bottom bin of diameter D_1 and fill height H with a circular outlet opening of diameter B. Because the consolidation stresses and corresponding bulk strength of the stored material, and hence draw-down, is a function of the bin diameter, for illustration purposes, a bin of larger diameter D_2, depicted by the dotted lines, is superimposed on the first bin as shown. Just above the outlet the flow channel opens out to a slightly larger dimension D_R and then diverges at an angle ε which depends on the effective angle of internal friction δ in accordance with Fig. 5.17. For a circular outlet, the slope angle is $\varepsilon = \varepsilon_c$ which, for the normal values of occurring in practice, indicated the flow channel slopes at a steep angle. Once the diameter of the flow channel equals the critical rathole diameter flow ceases and a stable rathole remains.

The major consolidation stress distributions expressed as effective head of solids h_f for the two bins are determined and the corresponding critical rathole diameters D_f are calculated in accordance with the established procedures. The h_f and D_f graphs are shown in Fig. 5.21. Discharge takes place until the diameter of the flow channel equals the critical rathole diameter at the draw-down level h_D. The draw-down level is defined by the point of intersection of the rathole expansion line and the D_f curve as illustrated in Fig. 5.21.

The major consolidation stress σ_1 of the stored bulk solid is approximated by the vertical pressure determined by the Janssen equation. It is usually expressed as the effective head of solids h_f. That is,

$$h_f = \frac{\sigma_1}{\gamma} = \frac{D}{4\,\mu\,k_j}\left(1 - e^{-4\,\mu\,k_j\,\frac{z}{D}}\right) \tag{5.16}$$

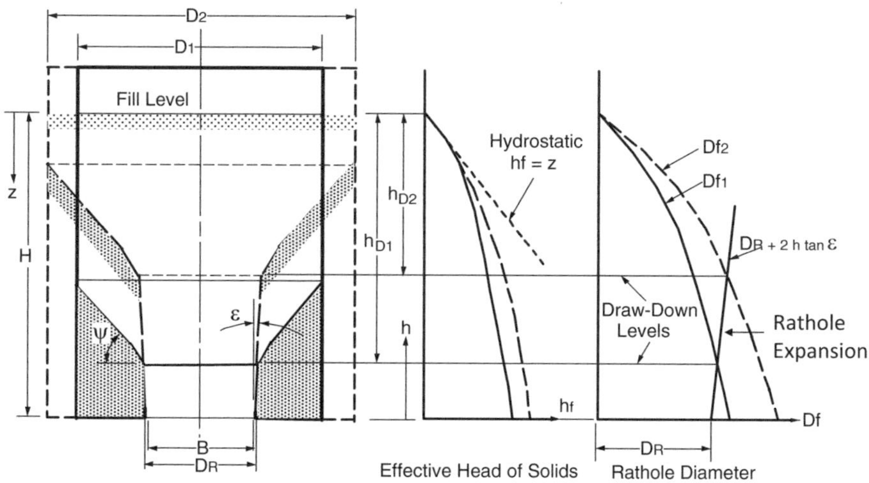

Fig. 5.21 Funnel-flow bin draw-down performance

where

$\sigma_1 =$	major consolidation stress
$\gamma = \rho g =$	bulk specific weight
$\rho =$	bulk density
$g =$	gravity
$D =$	bin diameter
$k_j =$	ratio of normal to average vertical pressure
$\mu = \tan\varphi =$	wall friction coefficient
$\varphi =$	wall friction angle
$z =$	depth below the free surface

It is usual to assume $k_j = 0.4$. The effective head h_f is illustrated in Fig. 5.21 for the two values of D, namely D_1 by the solid line and D_2 by the chain dotted line.

It is useful to examine the slope of the h_f curve. From Eq. (5.16), it follows that

$$\frac{dh_f}{dz} = e^{-4\,\mu\, k_j \frac{z}{D}} \tag{5.17}$$

When $z = 0$, the slope of the h_f curve is unity and $\frac{d\sigma_1}{dz} = \gamma$. That is, the effective head equals the actual head, $h_f = z$. Also, as D becomes larger, $\frac{z}{D}$ approaches zero and the consolidation stress approaches the hydrostatic stress, that is $\sigma_1 = \gamma z$.

It is evident from Eq. (5.16) and Fig. 5.21 that the consolidation stresses expressed as the effective head h_f increase with bin diameter, which for a given opening dimension B and corresponding effective opening D_R, results in a reduction in the draw-down. While the reduced draw-down with increase in bin diameter leads to a reduction in the amount of bulk material discharging from the actual flow channel, this is usually more than compensated for by the greater amount of bulk material flowing over the surface as a result of the sloughing action. Nonetheless, the reclaim efficiency for a given height of storage H decreases with increase in bin diameter. This also means that the amount of 'dead' material increases with increase in bin diameter.

5.6.2 Model Bin Studies

A series of experiments were performed on small scale model bins to confirm the foregoing observations. The experimental work involved a series of tests carried out on four flat bottom funnel-flow bins using iron ore as the bulk material. The diameters of the four bins were 0.15, 0.2, 0.25 and 0.3 m. In all cases the outlet diameter was 0.09 m and the fill height 0.3 m. One test bin is shown in Fig. 5.22.

The results are presented in Figs. 5.22 and 5.23. Figure 5.23a shows that the reclaim increased from 11.8 kg for the 0.15 m bin to 17.2 kg for the 0.3 m diameter bin. Thus, a doubling of the bin diameter has yielded an increase of 46% in the reclaim or live capacity. However, this is at the expense of a decrease in reclaim efficiency from 87% to 36% as the bin diameter increased from 0.15 to 0.3 m.

Fig. 5.22 Model funnel-flow bin—test handling iron ore

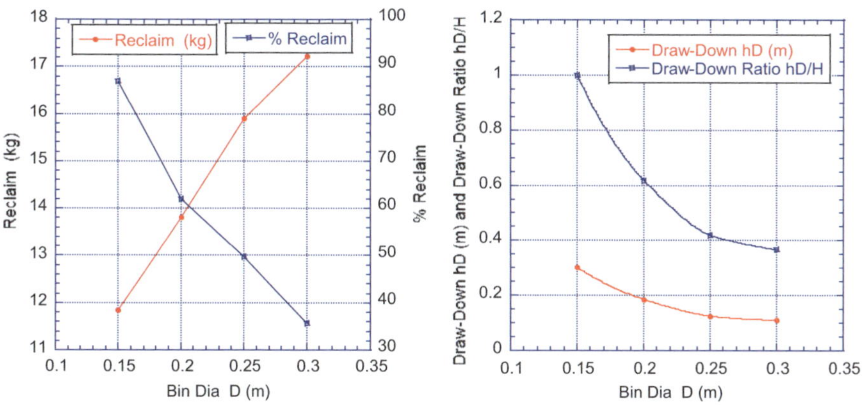

(a) Reclaim performance **(b) Draw-down performance**

Fig. 5.23 Model bin reclaim and draw-down performance

Figure 5.23b shows the corresponding decrease in draw-down performance as the bin diameter increased from 0.15 to 0.3 m. These results are to be expected from an examination of Fig. 5.21 which shows the increase in height of the intersection point of the rathole expansion graph with the D_f graph that accompanies an increase in bin diameter.

5.6.3 Stockpile Reclaim—Model Studies

The draw-down characteristics of gravity reclaim stockpiles are now considered. Figure 5.24 shows a conical stockpile with an effective central reclaim outlet of diameter D_R. To demonstrate the influence of stockpile height on the draw-down and live reclaim, Fig. 5.24 depicts two heights H_1 and H_2.

It is important to note that in funnel-flow, the datum for the consolidation stress is the top free surface. Hence a marginal gain in draw-down and live capacity through increased height is more than offset by a decrease in reclaim efficiency. This is demonstrated by tests conducted on small scale iron ore conical stockpiles. These were performed on model stockpiles of height 0.2, 0.25 and 0.3 m.

The test results are plotted in Fig. 5.25a and b. Figure 5.25a shows that the reclaim increased from 5.9 to 8.3 kg as the stockpile height increased from 0.2 to 0.3 m. However, there was a corresponding decrease in reclaim efficiency from 18.1 to 7.7%. Figure 5.25b shows the increase in draw-down h_D and increase in h_D/H ratio that accompanied the increased in stockpile height, but this does little to influence the live capacity.

5.7 Expanded Flow Bin Design Example

The design of the expanded-flow bin shown in Fig. 5.26 handling steaming coal at a moisture content of 22.2% is discussed. The flow properties of the steaming coal are presented in Chap. 4, Sect. 4.4.2. An overview of the bin design is summarised as follows:

(a) Upper Funnel Flow Section

The upper funnel-flow bin has a 10 m diameter constructed of mild steel with an effective storage height of 10 m. From the flow-properties of the steaming coal the following parameters are relevant:

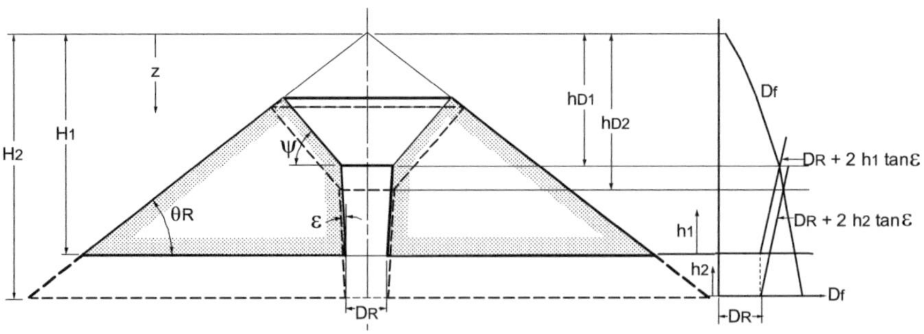

Fig. 5.24 Conical stockpile showing influence of height on reclaim performance

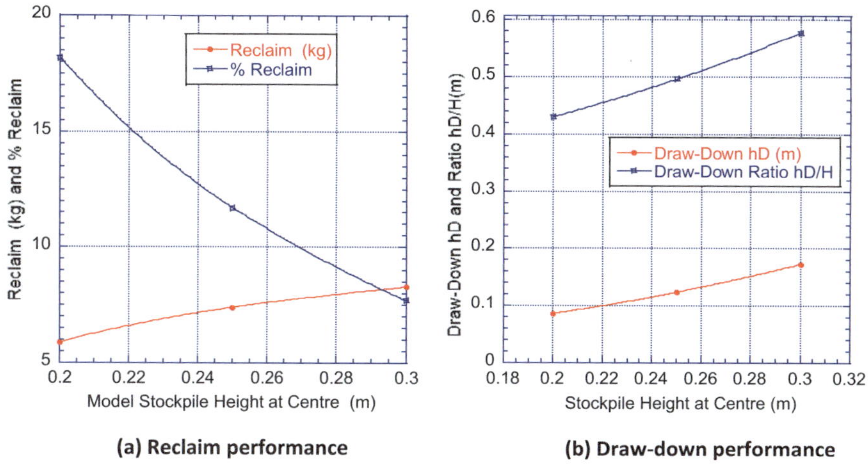

Fig. 5.25 Model iron ore stockpile reclaim and draw-down performance

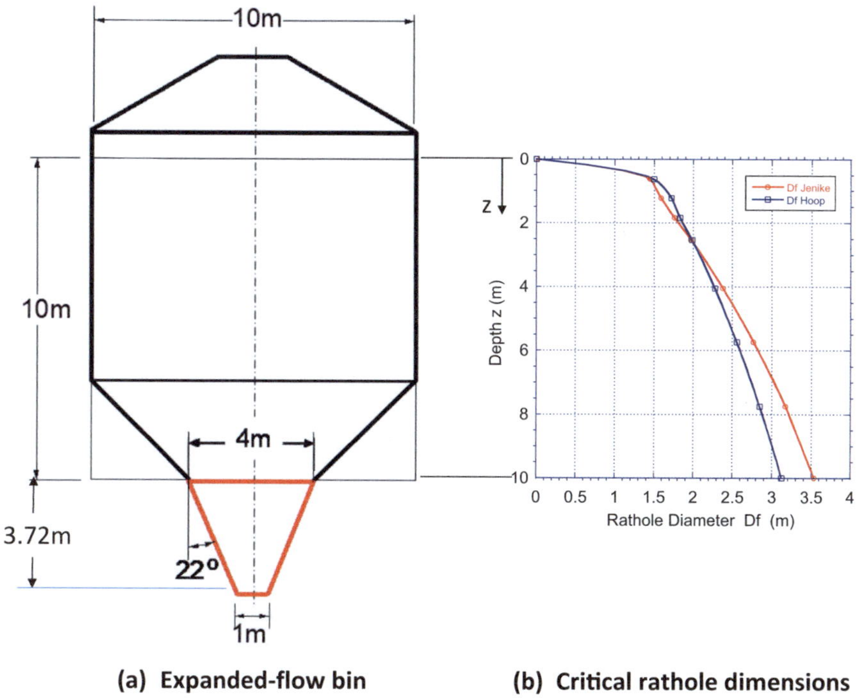

Fig. 5.26 Expanded-flow bin design example

Bulk density of the coal $\rho = 0.9$ t/m^3; Wall friction angle $\varphi = 30°$; Effective angle of internal friction varying from, $\delta = 51°$ to $44°$; Static angle of internal friction $\varphi_t = 38°$

(i) The D_f versus depth h graph is determined as follows:

- Applying Eq. (5.16), that is Janssen's equation, determine the effective head h_f as a function of depth z measured below the top surface.
- For each value of h_f determine the corresponding values of the unconfined yield strength σ_c or σ_{ct} (whichever is relevant) from the Flow Function graph.
- Selecting the corresponding static value φ_t of the static angle of internal friction, calculate the D_f using Jenike's Eq. (5.3) which is repeated below for convenience:

  $D_f = \frac{\sigma_c G(\varphi_t)}{\gamma}$, Eq. (5.3) repeated, where $G(\varphi_t) = 4.3 \tan \varphi_t$

(ii) The D_f versus depth z graph, Hoop Stress Method:

- The procedure is similar to the above with the following changes:
  - The major consolidation stress σ_1 replaced by the mean stress σ_m
    Where $\sigma_m = \frac{\sigma_1}{1 + \sin\delta}$, Eq. (5.10) repeated.
- The critical rathole diameter is determined using:

  $D_f = \frac{\sigma_{c\theta} G(\beta)}{\gamma}$, Eq. (5.7) repeated and $\beta = \frac{\pi}{4} + \frac{\varphi_t}{2}$, Eq. (5.6) repeated,

Where $\sigma_{c\theta}$ = unconfined yield strength corresponding to the mean stress σ_m

(b) Mass-Flow Hopper

Figure 5.26b shows that the two D_f plots, the slightly more conservative Jenike D_f values and the revised hoop stress D_f values are relatively close.

The critical outlet diameters are 3.5 m and 3.1 m, respectively. Adopting 4 m as the opening diameter assures the bin will empty by funnel flow. Based on the mass-flow hopper design example in Chap. 4, the chosen hopper to achieve uniform, controlled discharge is as shown in Fig. 5.26a.

5.8 Stockpile Consolidation Stresses and Loads

For gravity reclaim stockpile design, a knowledge of the magnitude and distribution of the consolidation stresses is necessary for the determination of the rathole geometries and live capacities as well as the base loads generated by the stored mass. The latter is required for the design of the reclaim vaults or tunnels, along with the load determination necessary for the design of the reclaim hoppers and feeders. It is recognised that stockpile load analysis is a complex problem largely due to the variations in the storage and flow properties of the bulk solids under variable climatic conditions along with the specified "fill and reclaim" operational cycles.

Typically, industrial stockpiles, particularly those associated with the mining and mineral processing industries, range from 20 to 50 m in height. The highest in the experience of the TUNRA Bulk Solids group of the University of Newcastle is 70 m, a copper ore stockpile located in Irian Jaya. On such a scale, the consequences of failure may well be catastrophic, so the temptation is to err on the conservative side in the design. Yet the cost of being too conservative is difficult to justify.

A review of the salient aspects of the stress and load analysis is presented.

5.8.1 From Sand Heaps to Stockpiles

It is reasonable to assume that the vertical pressure acting at each base location is directly related to the head of bulk solids at that location. That is, $p_v = \gamma h$, where $\gamma = \rho g =$ bulk specific weight and $h =$ height of the pile. In effect the loading is often referred to as "hydrostatic" and the base load pressure profile is the mirror image of the stockpile as illustrated by the red line in Fig. 5.27. The maximum pressure occurs at the centre under the apex of the pile.

Of particular relevance to the stockpile load distribution problem is the research conducted by physicists in the 1990's focusing on heaps of sand. It had been observed that when sand is poured onto a flat surface to form a conical pile the pressure in the centre under the apex actually decreases, as illustrated in Fig. 5.27. This has become widely known as the "M distribution". The complexity of this deceivingly simple problem promoted significant research interest in attempting to provide an explanation for this so-called dip in the central pressure. The answer came from the research of Wittmer et al. [3, 4] through their development and application of the "Fixed Principal Theorem" (FPA).

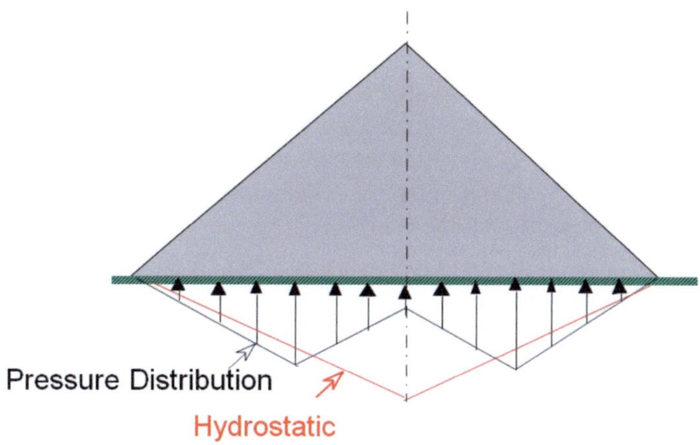

Fig. 5.27 Stockpile base load showing the "M" distribution

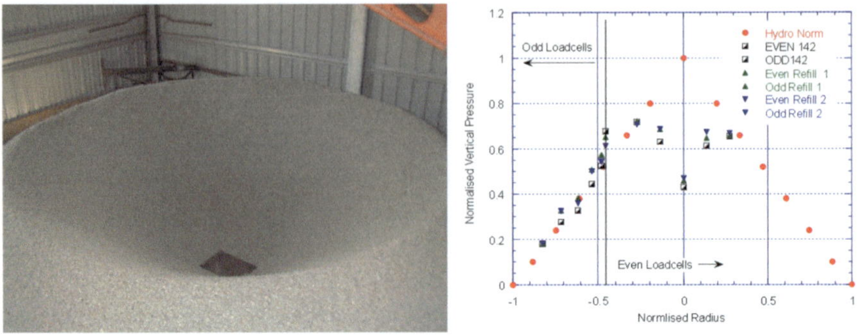

(a) Pilot scale stockpile test rig **(b) Sample base pressure load distributions**

Fig. 5.28 Test results for gravel stockpile showing "M" distribution [6]

This research aroused the interest of the science writer Watson, [5] who was obviously impressed by the complexity of this seemingly simple problem when he wrote:

"the humble sand pile is to granular mechanics as Fermat's Last Theorem was to number theory: a tantalising simple problem that stubbornly eludes solution'.

Follow up research was performed by McBride [6] who conducted experiments on a 2 m high conical stockpile storing and discharging gravel as illustrated in Fig. 5.28a. A sample of his test results, shown in Fig. 5.28b, displays evidence of the "dip" in the central pressure thus confirming the presence of the "M Distribution". Also shown in red, for reference purposes, is the hydrostatic pressure distribution which is presented in normalised form along with the various pressure measurements.

5.8.2 Stockpile Surcharge Design Loads

While accepting the importance of the "M" base load distribution, its application to the design and operation of large-scale industrial stockpiles is not yet fully understood. Nonetheless, it is important that the complexities involved in stockpile design be identified and dealt with appropriately. A cautious approach is required.

Some points to note are as follows:

(i) Variability of the Bulk Solid and Consolidation—Sand is a cohesionless, free flowing material, whereas practically all mineral ores are cohesive. In mining operations, the bulk solid is quite heterogeneous, comprising a wide range of particle shapes and particle sizes from large rocks to a few microns in a random packing array. The random behaviour is influenced by the loading arrangement and consequent segregation that may result. The bulk solid is mostly cohesive, its strength varying with moisture, temperature and prolonged periods of undisturbed consolidation. Under extreme rainfall conditions, stockpile slumping may occur giving rise to complex, variable base loading conditions.

(ii) There is always the possibility of ground settlement relative to the top of a reclaim vault or tunnel and this can induce very high surcharge pressures on the tunnel roof. Furthermore, the purpose of a stockpile is to store a bulk solid and then reclaim it by either mechanical means or by gravity flow. In the case of the latter, reclaim hoppers and feeders are employed as in Fig. 5.29. The flow patterns induced by the reclaim process can cause high pressure concentrations due to the draw-down forces. This is illustrated in Fig. 5.30.

The use of mass-flow reclaim hoppers interfaced with the feeders is important from the point of view of achieving reliable feed, and for controlling the loads on the feeders and the corresponding drive torques and powers. This is explained by reference to Fig. 5.31 along with Fig. 5.30. Due to the arched stress field conditions in the hopper, illustrated in Fig. 5.30, after feeding has been initiated, even if the feeder is then stopped with the stockpile still relatively full, the load V_f on the feeder is independent of the surcharge head. The load V_f is much lower than the initial load V_i which occurs when the crater is filled from the empty condition.

The initial load V_i on the feeder is more difficult to predict and may vary considerably from when the stockpile is filled for the first time to when a pre-formed rathole or flow-channel is filled from the empty condition after the stockpile has been in use. While the very conservative approach is to assume that the surcharge pressure p_s is the hydrostatic pressure, this will normally result in an over-design which cannot be justified on economic grounds. If a pre-formed flow channel exists, then this channel will act as a pseudo bin or silo in which the shear stresses generated at the boundaries will provide support for the load generated by the bulk solid and reduce the surcharge pressure p_s. Even when a stockpile is filled for the first time, there is likely to be an

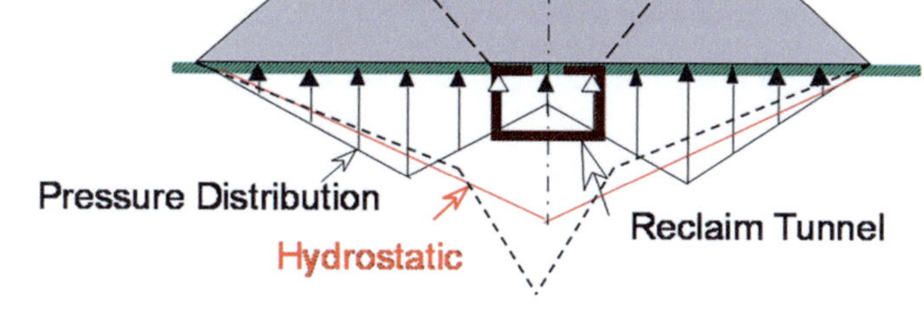

Fig. 5.29 Gravity reclaim stockpile showing influence of ground settlement

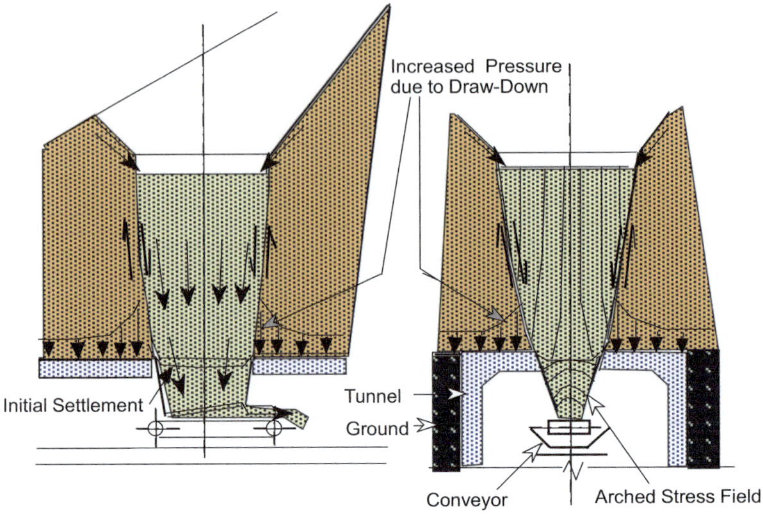

Fig. 5.30 Influence of draw-down loads in flow channel

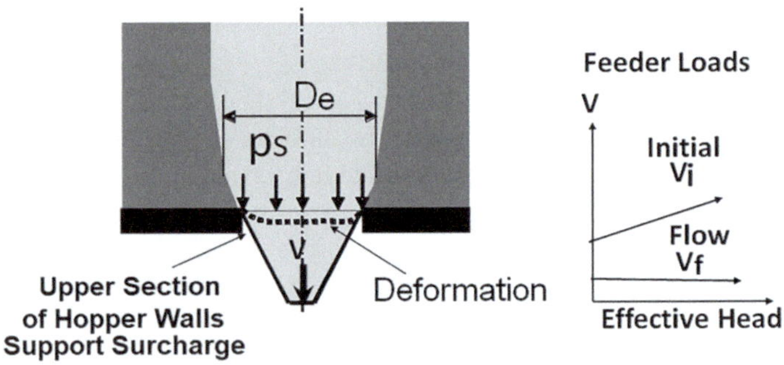

Fig. 5.31 Hopper surcharge pressure

initial settlement of the bulk solid at the transition level of the hopper and stockpile as illustrated in Fig. 5.29. This will help to define the flow channel and reduce, at least partially, the surcharge pressure p_s.

Summarising, the following remarks and recommendations are presented:

(a) Uniformly consolidated Stockpile—"Hydrostatic" pressure

$$p_s = \gamma h_s \tag{5.18}$$

where

$\gamma = \rho g =$ Bulk specific weight

That is, the effective head is equal to the actual head. This equation may be applied to the determination of the base load pressures but is rather too conservative for the determination of the rathole dimensions as well as the feeder loads.

(b) Rankine Pressure

$$p_s = \gamma h_s \cos \theta_R \tag{5.19}$$

where

$\theta_R =$ Angle of repose

Less conservative modification to upper bound solution.

(c) Hoop Stress

As described in this chapter, this method utilises the consolidation condition described by the mean stress $\sigma_m = \frac{\sigma_1}{1+\sin\delta}$ where $\sigma_1 =$ major consolidation stress and $\delta =$ effective angle of internal friction. It is recommended for its application in the rathole diameter calculations.

(d) Preformed Rathole

$$p_s = \frac{\gamma D_e}{4k \sin \delta} \tag{5.20}$$

where

$D_e =$ Effective rathole diameter at hopper interface
$k = 0.4 =$ pressure ratio;
$\delta =$ Effective angle of internal friction.

This method is based on the asymptotic value of Janssen's equation. It is recommended for reclaim hopper and feeder load calculations.

5.9 Industrial Stockpile Design and Performance—Case Studies

5.9.1 Example 1. Channar Iron Ore Stockpile

A case study example that demonstrates the developments in gravity discharge stockpile design is the Channar iron ore stockpile in the Pilbara area of Western Australia. This case study is based on the original project work undertaken by TUNRA Bulk Solids of the University of Newcastle [7] in the latter part of the 1980's. The stockpile is of conical form with two reclaim hoppers and feeders, the geometry being as shown in Fig. 5.32. The height of the stockpile when fully loaded is 32 m.

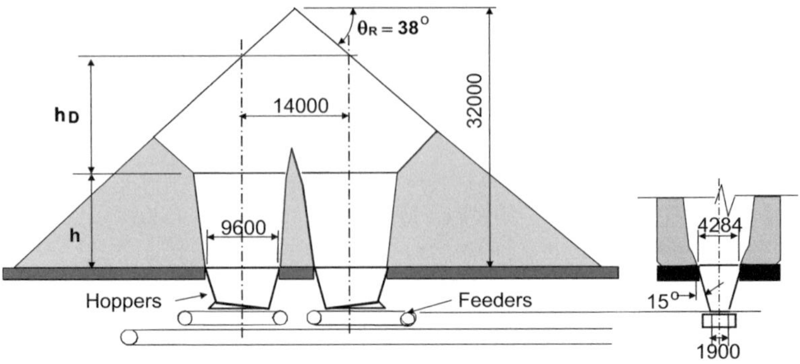

Fig. 5.32 Channar iron ore stockpiles

Commencing as a "greenfield" mine site development, it was necessary to conduct flow property tests on bore samples of iron ore extracted from the new site. These flow properties were used to design the storage and handling plant notably, in this case, the primary stockpile of Fig. 5.32. As an aid to the design, an experimental simulation was performed on a 50:1 scale model using the actual iron ore sample with the top particle size scaled down. Employing the dynamic simulation modelling procedure as described by Roberts [8], the stockpile consolidation conditions were based on the non-dimensional stress parameter $N_{\sigma 1}$ given by

$$N_{\sigma 1} = \frac{\sigma_1}{\rho g B} \tag{5.21}$$

where

$\sigma_1 =$ major consolidation stress
$\rho =$ bulk density
$B =$ characteristic dimension such as reclaim hopper transition width

Thus, the required consolidation stress for the model tests is given by

$$\sigma_{1m} = \sigma_{1p} \frac{\rho_m}{\rho_p} \frac{B_m}{B_p} \tag{5.22}$$

where subscript "m" denotes the model and subscript "p" the full-size prototype.

The appropriate values of the major consolidation stress, bulk density and angle of internal friction were obtained from the flow property tests. An example of the model simulation tests is shown in Fig. 5.35b. The results of the experimental simulation gave a live capacity of 28% confirming the results that had been predicted using analytical procedures.

In order to compare the results of the original design with the more recent design methodology presented in this Chapter, the original flow property test results have been used to compute the upper bound and mean or average rathole dimensions as

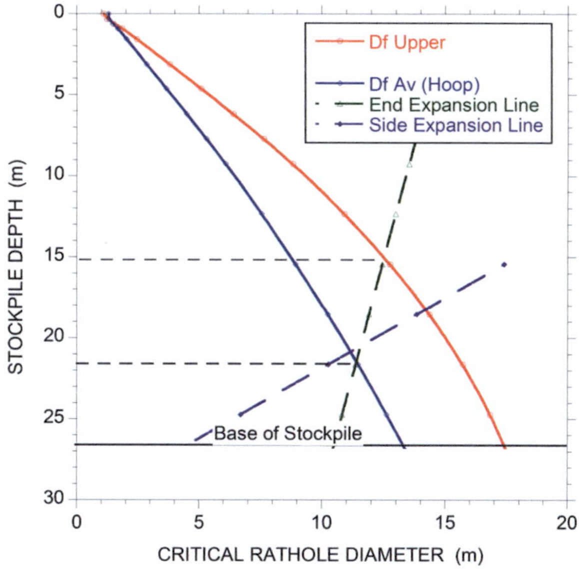

Fig. 5.33 Critical rathole dimensions and expansion lines

functions of the centre line depth above each reclaim hopper opening. The results are plotted in Fig. 5.33. Also plotted are the rathole side and end expansion lines.

As previously discussed, the end slope ε_c dominates the draw-down and rathole geometry. Because of the rectangular proportions of the hopper transition length L_T and width D_T in this case, the angle ε_p for the side expansion line is taken to be 30°. The draw-down and live capacity results are presented in Table 5.1. The live capacities have been computed using a geometrical modelling package, the cross-sectional views for the average or mean consolidation and upper bound consolidation cases are shown in Fig. 5.34.

The mean or average case based on the hoop stress theory gave a 28% live capacity which is in full agreement with the value that was predicted. The calculated upper bound case gave a much-reduced draw-down with a predicted live capacity of only 16% as indicated by Fig. 5.34 and Table 5.1.

A pictorial overview of this case study is presented in Fig. 5.35. Figure 5.35a shows the 50:1 scale model which gave 28% measured live capacity, the model studies being performed prior to the stockpile construction and plant commissioning. Figure 5.35b

Table 5.1 Draw-down and live capacities for iron ore stockpile

Consolidation condition	Draw-down h_D (m)	Total filled volume (m^3)	Reclaimed volume (m^3)	% Reclaim (%)
Average or mean case	21.7	56,215	15,741	28
Upper bound case	15	56,215	8,994	16

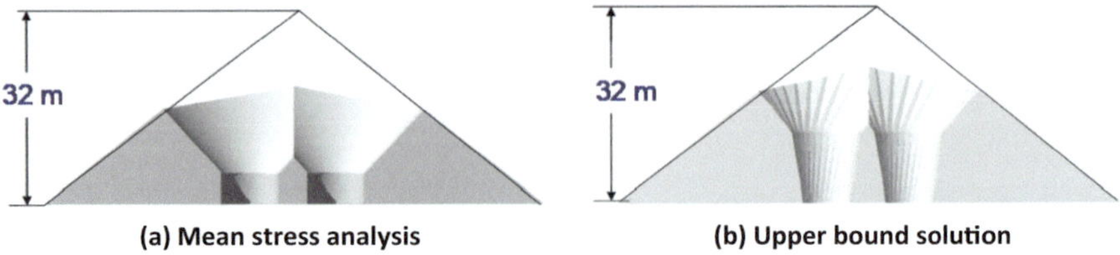

Fig. 5.34 Comparison with mean and upper bound stress solutions

Fig. 5.35 Comparing predicted and actual draw-down of Channar Iron Ore stockpile

shows a computed generated model based on the hoop stress theory, while Fig. 5.35c and d show the actual stockpile in operation during the commissioning stage in 1989 where live capacities ranging from 28% to 30% were achieved. The comparison between the experimental simulation, the computer simulation and the actual full scale stockpile draw-down and live capacity performances are most encouraging.

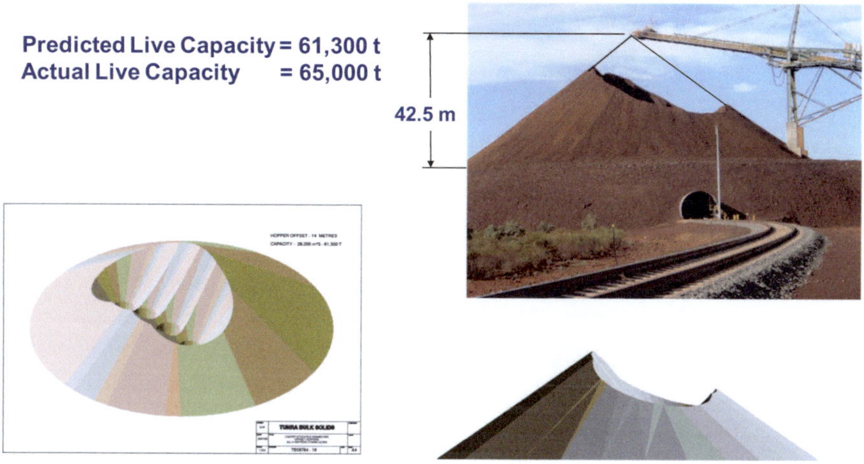

Fig. 5.36 42.5 m high iron ore stockpile [9]

5.9.2 Example 2. 42.5 m High Iron Ore Stockpile

A more recent installation is the 42.5 m high iron ore stockpile presented in Fig. 5.36 which shows the stockpile in operation, together with the three-dimensional simulations which formed the basis for the final design. The simulations utilised the measured flow properties of the iron ore.

In this case four wedge-shaped reclaim hoppers are employed, the hoppers being arranged in line and offset 14 m to one side of the centre line. The reason for this is to reduce the consolidation pressures due to the reduced head to gain an increase in reclaim and, hence, live capacity. The simulations gave a predicted capacity of 29,200 m^3 or 61,300 t. The actual live capacity is 65,000 t.

5.9.3 Example 3. Increasing the Height of the Reclaim Hoppers

The iron ore stockpile of Fig. 5.37 has two reclaim openings, each 2 m square and spaced 14 m apart. The original arrangement is shown in Fig. 5.37a for which the height of the pile is 23.5 m. The live capacity is 1096 m^3, only 4.92% of the stored volume. In the modified version of Fig. 5.37b, the reclaim hoppers are extended 1.5 m above the ground level enabling the opening dimension to be increased to 4 m square. Despite the reduction in storage height from 23.5 to 22 m, the live capacity is increased to 2802 m^3 which is 15.34% of the stored volume, giving an overall gain of 156%.

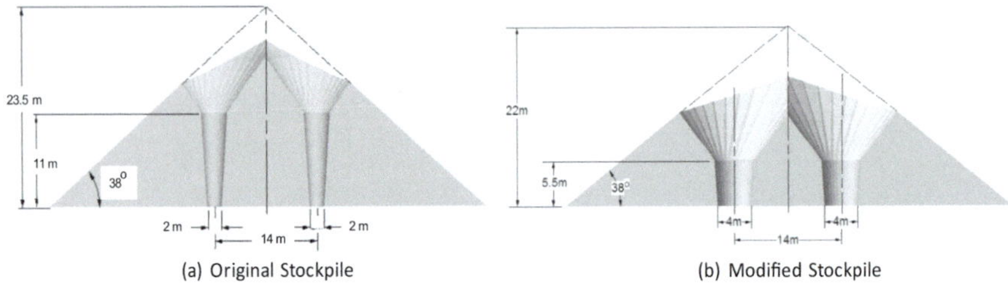

Fig. 5.37 Increasing height of reclaim hoppers in an iron ore stockpile

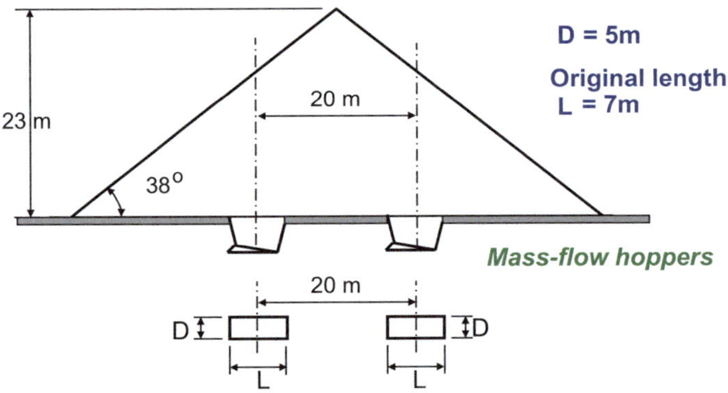

Fig. 5.38 Iron ore stockpile with 2 fixed spacing opening slots

5.9.4 *Example 4. Increasing Reclaim Slot Length for Increased Capacity*

The iron ore stockpile shown in Fig. 5.38 has two reclaim hoppers spaced 20 m apart. The transition dimensions of the hoppers are length L = 7 m and width D = 5 m. The aim is to increase the reclaim capacity by increasing L while maintaining the width D = 5 m and 20 m spacing. Figure 5.39 shows draw-down predictions for slot lengths ranging from 7 to 11.7 m. Figures 5.40 and 5.41 show the percentage reclaim and live capacity as functions of slot length, respectively.

The results show that increasing L_T from 7 to 10.6 m, an increase of 51%, results in an increased reclaim from approximately 15 to 59%.

5.9.5 *Concluding Remarks*

This chapter has been devoted to the design of funnel-flow and expanded-flow bins and gravity reclaim stockpiles. Procedures for determining the geometrical shapes of

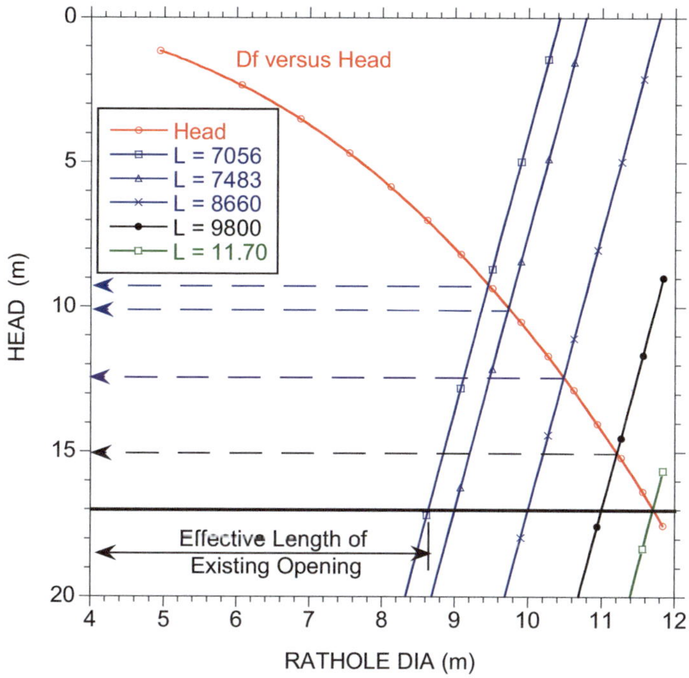

Fig. 5.39 Rathole diameter and draw-down determination

Fig. 5.40 Draw-down and % reclaim

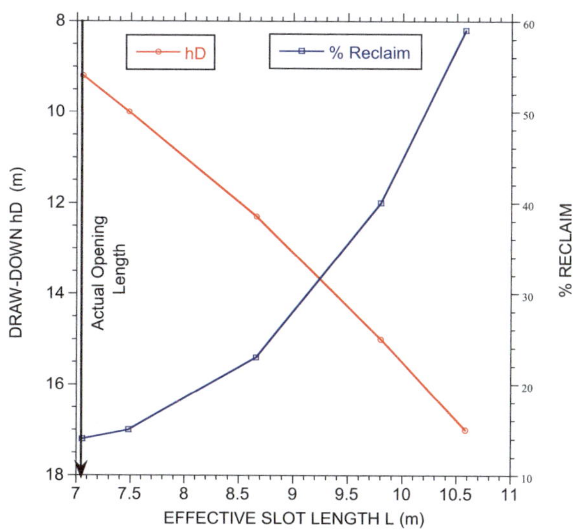

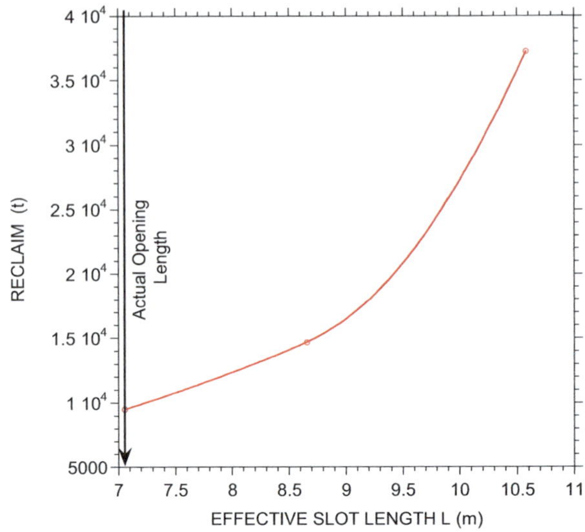

Fig. 5.41 Live capacity

ratholes during funnel-flow in bins and gravity reclaim stockpiles have been reviewed. The importance of the relevant flow properties of the bulk solids to the optimum design and performance of funnel-flow bins and stockpiles is strongly stressed. This is linked to the determination of the critical rathole diameters to maximise the draw-down, coupled with the positioning of the reclaim hoppers and feeders aimed at maximising the live capacity. Using techniques based on solid geometry, the live capacity of the bin or stockpile may be readily estimated.

To overcome known limitations in the design procedures, the advantages of the new hoop stress theory, which is linked to the mean consolidation stresses, has been shown to give storage and live capacity predictions which match quite well with results from actual industrial stockpiles. This is demonstrated by the case study examples which clearly show the validity of the new design theory and test procedures in providing capacity predictions.

References

1. Roberts, A.W., S.J. Wiche, and T. Krull, *Review of funnel-flow theory in relation to draw-down and live capacity of gravity reclaim stockpiles.* Bulk Solids Powder Sci. Technol., 2008. **3**: p. 97–107.
2. Jenike, A.W., *Storage and Flow of Solids. Bull. 123.* The Univ of Utah, Engng Exp Station, USA, 1964.
3. Wittmer, J.P., et al., *An explanation for the central stress minimum in sand piles.* Nature, 1996. **382**(6589): p. 336–338.
4. Wittmer, J.P., M.E. Cates, and P. Claudia, *Stress propagation and arching in static sandpiles.* Journal de Physique II, 1997. **7**(1): p. 39–80.
5. Watson, A., *Searching for the sand-pile pressure dip.* Science, 1996. **273**(5275): p. 579–580.

6. McBride, W. and U.o. Newcastle, *Mechanics of bulk solid stockpiles*. 2001.
7. Roberts, A.W. and L.H. Teo, *Design considerations for maximum reclaim capacity of conical stockpiles*. 1990. **10**: p. 9–15.
8. Roberts, A.W., *Characterisation for Hopper and Stockpile Design*, in *Characterisation of Bulk Solids*. 2005. p. 85–131.
9. Wiche, S., James, G, *Conceptual design of stockpile at Mesa A mine*, in *Australian Bulk Handling Review* 2010. p. 70–72.

Chapter 6
Bin Wall Loads

6.1 Introductory Remarks

In bin design, the prediction of bin wall loads continues to be a subject of some considerable complexity. This fact is highlighted by the number of structural failures of bins and silos that have occurred over the past hundred and more years, largely due to the lack of understanding, at both the design and operational stages, of the influences of the loading and discharge flow patterns on the bin walls. Despite the widely varying approaches to the analysis of bin wall loads, it is clear that the loads are directly related to the flow pattern developed in the bin. In the case of mass-flow bins, the discharge flow patterns are generally well defined and the corresponding wall loads are predictable and reproducible. However, in funnel-flow bins the flow patterns are more difficult to ascertain, especially if the bins have multiple outlet points, the bin loading is not central and the discharge is eccentric. Unless there are compelling reasons to do otherwise, bin shapes should be kept simple and symmetric.

6.1.1 Foundations of Silo Load Research

As described in Chap. 1, the foundation research into bin and silo loads commenced in the latter 1880's spanning some 25 years into the early 1900's. The motivation for this research was the growing need to store and handle bulk materials, typically grain, in large quantities. Of particular note is the research of Isaac Roberts linked to the design and construction of a large grain silo in the Port of Liverpool in the north of England. Roberts's research was published in the proceedings of the Royal Society of London in 1884 [1]. The research of the German Engineer, H.A. Janssen, followed 10 years later with the publication of his epic research paper and silo load equation in 1895 [2]. Both the work of Isaac Roberts and that of Janssen were confined to tall,

A. W. Roberts, *The Theory and Practice of Bulk Materials Handling*,
https://doi.org/10.1007/978-3-032-19702-3_6

flat-bottom, cylindrical silos for the initial filling and static storage conditions. They did not study the silo's loads when the silos were discharging.

Undoubtedly, the work of Janssen, particularly his silo load equation is, by far, the most widely known of the early silo research emanating from the early foundation period. The centenary of Janssen's work was celebrated in the conference "100 Years of Janssen" held in Nuremburg, Germany, in 1995, Roberts [3]. The lesser-known contributions of the other researchers included studies of silo loads under dynamic, discharge flow conditions. Of particular note is the research of the Canadian engineer J.A. Jamieson, [4] who demonstrated the influence of over-pressures on silo walls during symmetrical discharge and the highly significant load variations due to eccentric discharge.

6.1.2 Follow-On Developments

It was not until the mid-1960's that the fundamental research into bin loads began to take place. Undoubtedly the research conducted by Jenike, together with Johanson, into the storage and modes of flow of bulk solids in mass-flow and funnel-flow bins provided the stimulus for the significant developments in bin load research that followed. By way of example, a selection of papers published over the period mid 1960's to latter 1970's is cited; Jenike et al. [5–7], Walker [8], Walters [9, 10], Clague [11], Arnold et al. [12]. Developments have continued, including the established bin and silo load codes such as the Australian Standard AS3774-96 [13] and the more recent Euro Code [14]. The principal focus of this Chapter is to review the basic, *classical* fundamentals of wall load analysis noting that this knowledge underpins the more recent, widespread applications of numerical procedures such as Discrete Element Method (DEM) Analysis.

6.2 Mass-Flow Bins

The stress fields and normal wall pressures occurring in mass-flow bins for the initial filling and flow conditions are shown in Fig. 6.1. When a bin is initially filled from the empty condition, a peaked stress field occurs as in Fig. 6.1a; the major principal pressure is almost vertical. When flow occurs, the stress field in the hopper switches to an arch stress field, the switch travelling up the hopper becoming locked in at the transition as in Fig. 6.1b. In the arched stress field, the load is transmitted to the wall of the hopper with the major principal stress acting more in a horizontal direction.

Above the hopper, that is in the cylinder, the peaked stress field remains, although imperfections in the cylinder wall which give rise to localised flow convergences cause over-pressures to occur in the cylinder. Imperfections in bin walls that give rise to over-pressures may result from manufacturing defects such as weld projections or plate shrinkages in steel bins, or from construction issues like deformation of

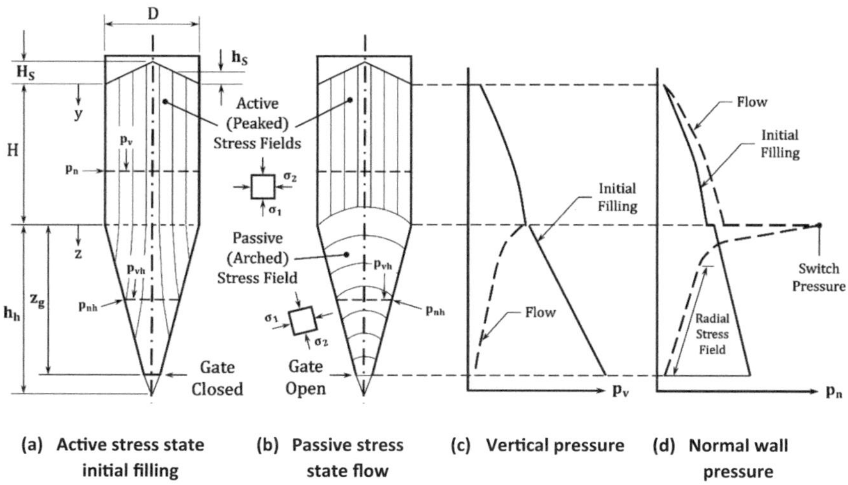

Fig. 6.1 Pressures acting in mass-flow bins

formwork in concrete bins. Jenike has used strain-energy methods to analyse these over-pressures during flow in the cylinder. In view of the difficulty of using the strain energy method, design codes generally employ over-pressure factors to account for flow conditions in the cylinder. This is indicated by the upper bound Pn curve for the cylinder in Fig. 6.1b.

6.3 Mass-Flow—Cylinder Pressures

The cylinder refers to the upper, parallel section of the bin. The initial pressures are those occurring when the bin is filled from the empty condition.

6.3.1 *Initial Filling Case*

At a given depth y from the actual datum (Fig. 6.1), the average vertical pressure Pv and normal wall pressure Pn are given by the Janssen equation with an initial surcharge term included. Referring to Fig. 6.1, it can be established that the differential equation for the vertical pressure is,

$$\frac{dP_v}{dy} + \frac{\mu k P_v}{r_c} = \gamma \tag{6.1}$$

Solving Eq. (6.1) and noting that when y = 0 the average vertical surcharge pressure is Pvo, the following equation is obtained:

$$P_v = \frac{\gamma r_c}{\mu k}\left[1 - e^{-\left(\frac{\mu k y}{r_c}\right)}\right] + P_{vo} e^{-\left(\frac{\mu k y}{r_c}\right)} \tag{6.2}$$

where

$r_c =$	characteristic radius dimension of cylindrical container.
$\gamma = \rho g =$	Bulk specific weight.
$\rho =$	bulk material density.
$g =$	acceleration due to gravity.
$\mu = \tan \phi_w =$	coefficient of wall friction.
$\phi_w =$	wall friction angle.
$P_{vo} =$	average surcharge pressure due to natural surcharge of material following filling.
$k = \frac{P_{ni}}{P_v}$	lateral pressure ratio.

The normal pressure P_{ni} acting on the wall is given by

$$P_{ni} = kP_v \tag{6.3}$$

Hence from (6.2) and (6.3),

$$P_{ni} = \frac{\gamma r_c}{\mu}\left[1 - e^{-\left(\frac{\mu k y}{r_c}\right)}\right] + P_{no} e^{-\left(\frac{\mu k y}{r_c}\right)} \tag{6.4}$$

where

$$P_{no} = kP_{vo} \tag{6.5}$$

The initial pressure curves are shown in Fig. 6.1c and d.

6.3.2 Characteristic Radius Dimension

The following definitions are from the Australia Standard AS3774-96 [13].

(i) For a round or square container:

rc = D/4

where

D = bin diameter or width.

(ii) For a rectangular container of length L and width D, see Table 6.1.

Table 6.1 Characteristic radii for rectangular containers

L/D	For short side	For long side
1.5	0.27D	0.30D
2.0	0.30D	0.33D
4.0	0.33D	0.40D
5.0	0.35D	0.50D

6.3.3 Pressure Ratio k

Referring to Eq. (6.2), the coefficient k relates the normal stress or pressure P_n to the average vertical pressure Pv for the cross-section. That is,

$$k = \frac{P_n}{P_v} \tag{6.6}$$

The value of k varies depending on the cylinder geometry. In accordance with AS3774, k is such that

$$k = \frac{1 + \sin^2\delta - 2\sqrt{\sin^2\delta - \mu^2\cos^2\delta}}{4\mu^2 + \cos^2\delta} \geq 0.35 \tag{6.7}$$

where

$\delta =$ Effective angle of internal friction.
$\mu =$ Coefficient of friction for bulk solid on cylinder wall.

Equation (6.7) implies that k shall be the larger of the value determined by Eq. (6.7) or 0.35. However, for almost all practical values of δ and μ, Eq. (6.7) gives values of $k < 0.35$, so that $k = 0.35$ becomes the lower limit. In the absence of more detailed information the value $k = 0.4$ is often used.

6.3.4 Initial Surcharge Pressure

The initial surcharge pressure Pvo may be estimated by:

$$P_{vo} = \frac{\gamma r_c}{\mu k}\left[1 - e^{-\left(\frac{\mu k h_s}{r_c}\right)}\right] \tag{6.8}$$

However, it is appropriate to determine Pvo from

$$p_{vo} = \gamma h_s \tag{6.9}$$

where

$h_s =$ effective surcharge

The corresponding value of P_{no} is

$$P_{no} = kP_{vo} = k\gamma h_s \tag{6.10}$$

h_s depends on the bin shape and manner in which the bin is loaded.

Assuming central loading, then for an axi-symmetric bin, a conical surcharge is assumed. For a plane-flow bin in which the length is greater than the width, an approximate triangular shape would occur. With these two limits:

$$h_s = \frac{H_s}{m_s + 2} \tag{6.11}$$

where

$m_s = 0$ for triangular surcharge
$m_s = 1$ for conical surcharge.
$H_s =$ actual surcharge head.

The surcharge head is given by

$$H = \frac{D\tan\theta_r}{2} \tag{6.12}$$

where

$D =$ diameter, or width of bin
$\theta_r =$ angle of repose.

6.3.5 Mass-Flow Cylinder Pressures—Discharge Flow Case

With the flow of bulk solids along the internal walls of the cylinder, any protrusions of the wall surface, as illustrated in Fig. 6.2, can give rise to localised stress concentrations as shown. The protrusions may arise as a result of manufacturing or construction imperfections such as localised buckling of the cylinder walls, undressed welds, or ledges arising from ill-fitting liner plates. For bins of reinforced concrete construction, protrusions leading to stress concentrations can arise from imperfections in the formwork. Since such localised imperfections and stress concentrations cannot be known at the design stage, the approach is to assume that they can occur anywhere and calculate the locus of all possible overpressures as illustrated by the dashed lines in the cylinder section of Fig. 6.1d and in Fig. 6.2.

A fundamental analysis of the overpressures, based on strain energy theory, was undertaken by Jenike et al. [5]. However, it has been widely accepted that the overpressures can be adequately described as a function of Janssen's equation. For

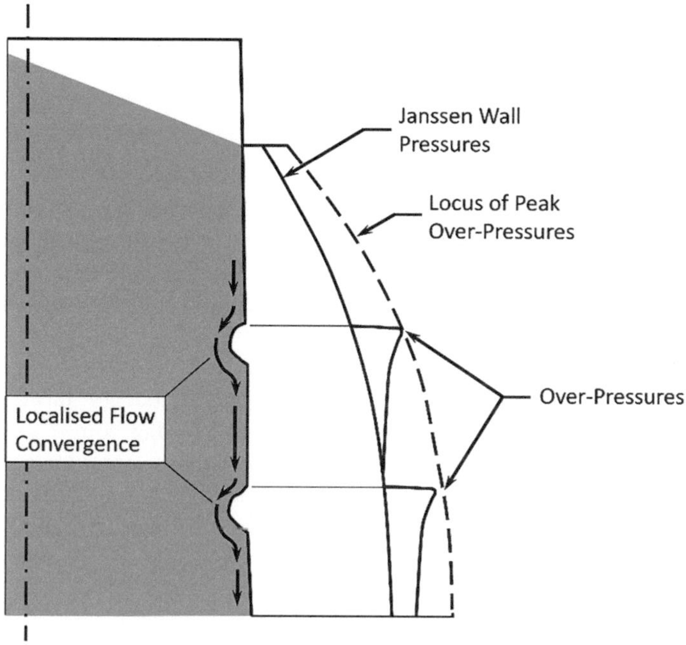

Fig. 6.2 Flow imperfections and over-pressures

example, AS3774-1996 has recommended that the flow pressures in the cylinder be obtained by multiplying the Janssen pressure p_{ni} by c_{nf}, that is:

$$p_{nf} = c_{nf} p_{ni} \tag{6.13}$$

where c_{nf} is the larger of;

$$c_{nf} = \left[7.6\left(\frac{h_b}{d_c}\right)^n - 6.4\right]c_c \tag{6.14}$$

and

$$c_{nf} = 1.2c_c \tag{6.15}$$

$h_c + h_s =$ total effective height of stored solid
$c_c =$ 1 for axi-symmetric flow.
$c_c =$ 1.2 for planar flow $n = 0.06$
$d_c =$ D
$P_{ni} =$ Janssen pressure given by Eq. (6.4)

The flow pressures for the cylinder are shown in Fig. 6.1d.

6.4 Mass-Flow Hopper Stress Fields and Wall Load Analyses

6.4.1 *The Horizontal Slice Stress Field Model*

As is evident from the theory of Jenike, along with the methodology for mass-flow hopper design presented in Chap. 4, the arched stress field model describing the passive stress state is the basic, fundamental design model. While the arched stress field model may also be applied to the determination of hopper wall loads, it has been found that the horizontal *slice* model, illustrated in Fig. 6.3, is widely used.

The advantage of the slice model is that it may be applied to mass-flow hopper design for both the active or initial, stress field state as occurs during filling of the hopper and the passive or arched stress field state during hopper discharge. A key parameter in the hopper wall load determinations is the pressure ratio k_h. The horizontal slice method allows adjustments to be made to the selection of the most appropriate pressure ratio value defined to suit a particular design requirement.

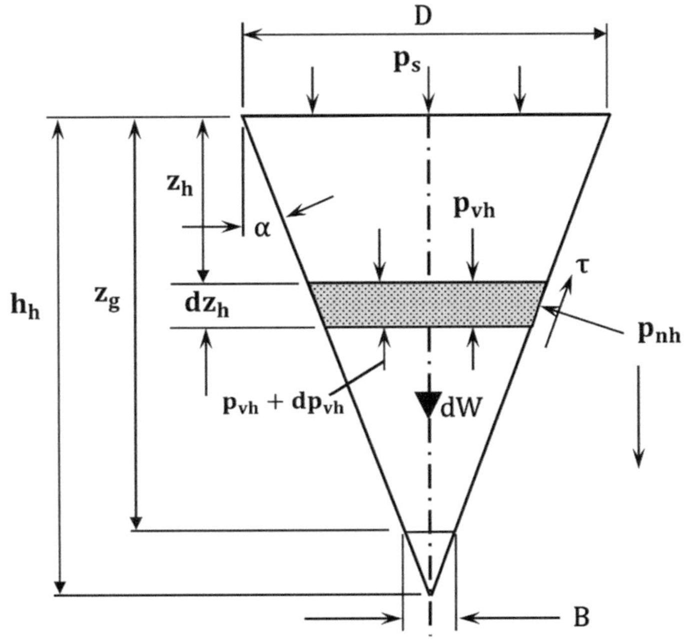

Fig. 6.3 Equilibrium of element in hopper

6.4.2 *The Horizontal Slice Passive Stress Field*

Referring to the mass-flow hopper model of Fig. 6.3, the parameter k_h relates the normal pressure P_n acting on the hopper wall at depth z_h to the average vertical pressure p_{vh} acting across the slice. That is,

$$k_h = \frac{P_{nh}}{P_{vh}} \tag{6.16}$$

From an equilibrium analysis, the average vertical pressure acting in the hopper at an arbitrary depth zh below the transition is described by the following differential equation:

$$\frac{dP_{vh}}{dz_h} + \frac{jP_{vh}}{(h_h - z_h)} = \gamma \tag{6.17}$$

Solution of this equation leads to

$$P_{vh} = \gamma\left[\frac{h_h - z_h}{j - 1}\right] + \left[P_s - \frac{\gamma h_h}{(j-1)}\right)\left(\frac{h_h - z_h}{h_h}\right)^j \tag{6.18}$$

The normal pressure on the hopper wall is

$$P_{nh} = k_h P_{vh} \tag{6.19}$$

$$j = (m+1)\left[k_h\left(1 + \frac{\tan \varnothing_w}{\tan \propto}\right) - 1\right] \tag{6.20}$$

where

$j =$	stress field parameter
$\propto =$	hopper half-angle.
$P_s =$	surcharge pressure at datum transition.
$h_h =$	distance from apex to transition.
$\gamma = \rho g =$	bulk specific weight, where $\rho =$ bulk density.
$\varnothing_w =$	wall friction angle.
$m =$	symmetry factor.
$= 1$	for axi-symmetric or conical hoppers.
$= 0$	for plane-flow hoppers.

It is noted that Eqs. (6.18), (6.19) and (6.20) apply for both the active, or filling condition, and passive, or discharge flow condition subject to the appropriate values of "j" and "k_h" being used. Where necessary for clarity purposes, the additional subscripts "i" and "f" are incorporated. That is, j_i, j_f, k_{hi} and k_{hf}.

6.4.3 Mass-Flow Hopper Initial Filling Case—Limiting k_h Value

For the initial filling, active stress field case, j cannot be negative, leading to the minimum value of $j = 0$ for which Eq. (6.18) reduces to the simple hydrostatic form with the vertical pressure P_{vi} being given by

$$P_{vh} = P_s + \gamma z_h \tag{6.21}$$

The corresponding normal pressure P_{nh} acting on the hopper walls is given by

$$P_{nh} = k_h P_{vh} = k_{hi}(P_s + \gamma z_h) \tag{6.22}$$

and from (6.20) with $j = 0$,

$$k_{hi} = \frac{\tan \alpha}{\tan \alpha + \tan \varnothing_w} \tag{6.23}$$

6.5 Mass-Flow Hopper Passive Stress Field Pressure Ratios

The application of the horizontal slice model to describe the passive stress field in a mass-flow hopper is highly dependent on the selection of the appropriate pressure ratio k_{hf}. The various methods that may be applied are discussed:

6.5.1 Modified Jenike Based, Mass-Flow Arched Stress Field Method

For reference purposes, the hopper pressure ratio k_{hf} based on Jenike's classical arched stress field model presented in Chap. 4 is reviewed. The relevant mass-flow equations are those derived from the work of Enstad [15] and Arnold et al. [16]. For convenience, the equations are reproduced here.

The pressure ratio, labelled k_{hfm}, is presented as the ratio of two, non-dimensional parameters, one for the normal wall stress or pressure and the other for the vertical pressure as shown in Eq. (6.24) below:

$$k_{hfm} = \frac{\frac{\sigma_w}{\gamma B}}{q\left(\frac{4}{\pi}\right)^m} \quad (6.24)$$

where

$$\frac{\sigma_w}{\gamma B} = \frac{Y(1 + \sin\delta\cos2\eta)}{2(X - 1)\sin\alpha} \quad (6.25)$$

$$q = \left(\frac{\pi}{3}\right)^m \frac{1}{4\tan\alpha}\left\{2\left(\frac{\sigma_w}{\gamma B}\right)(\tan\alpha + \tan\phi_w) - \frac{1}{1+m}\right\} \quad (6.26)$$

$$\eta = \frac{1}{2}\left[\phi_w + \sin^{-1}\left(\frac{\sin\phi_w}{\sin\delta}\right)\right] \quad (6.27)$$

$$X - \frac{2^m\sin\delta}{1 - \sin\delta}\left[\frac{\sin(2\eta + \alpha)}{\sin\alpha} + 1\right] \quad (6.28)$$

$$Y = \frac{[2(1 - \cos(\eta + \alpha))]^m(\eta + \alpha)^{1-m}\sin\alpha + \sin\eta\sin^{1+m}(\eta + \alpha)}{(1 - \sin\delta)\sin^{2+m}(\eta + \alpha)} \quad (6.29)$$

$\varnothing_w = \tan^{-1}(\mu) =$ wall friction angle
$\sigma_w = P_{nhf} =$ normal pressure at wall
$\delta =$ effective angle of internal friction.
$m = 1$ for axi-symmetric flow.
$m =$ 0 for plane-flow.

In Eq. (6.29) the plane-flow numerator term $(\eta + \alpha)$ must be in radians.

In general, the modified Jenike method tends to over-estimate the switch pressure (Fig. 6.1).

6.5.2 *Theoretical Estimates of k_{hf}*

(a) ***Stress Fields Model***

The analytical study of the stress fields in mass-flow hoppers undertaken by Roberts [17] has provided the basis for the derivation of appropriate, theoretical values of k_{hf} which have an application in hopper wall load studies. The model, upon which this analytical work is based, is shown in Fig. 6.4. The aim is to determine the stress fields and wall loads in the horizontal slice model that correspond to the fundamental parameters of the arched model. The parameters embodied in the slice model apply to the coordinate dimension "z" below the hopper and cylinder transition.

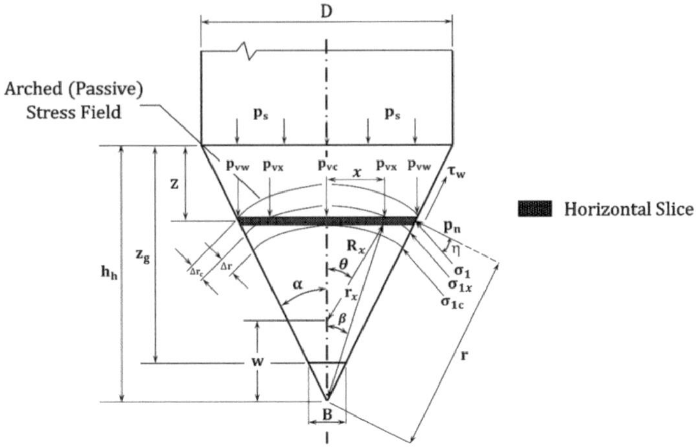

Fig. 6.4 Horizontal and arched stress field model

In accordance with Jenikes' radial stress theory, the major principal consolidation stress "σ_1" is directly proportional to the radial distance "r" from the point of intersection of the hopper walls as shown. The slope of the hopper walls is defined by the hopper half-angle "α". For the coordinate depth "z", the major consolidation stress "σ_1" intersects the wall at the angle "η" with respect to the normal stress "P_n". At the centreline of the hopper "σ_1" is horizontal. This pattern is repeated for the various circular arcs of the stress field.

For the horizontal slice model, the major consolidation stress decreases from a maximum value "σ_1" at the wall to a minimum value "σ_{1c}" at the centre intersection point. For the arbitrary location defined by horizontal distance "x" and corresponding angle "θ" for which the arch radius of curvature is "R_x", the corresponding major consolidation stress is "σ_{1x}". The corresponding vertical pressures are "P_{vw}" at the wall, "P_{vc}" at the centre and "P_{vx}" at the arbitrary intersection point.

b. ***Variation of Vertical Pressure Across the Horizontal Slice***

At an arbitrary position, x, corresponding to the arc radius R_x and angle θ, the vertical pressure P_{vx} as a function of σ_{1x} is given by

$$P_{vx} = \sigma_1\left(1 - \frac{\Delta r}{r}\right)\left(\frac{1 - \sin\delta\cos 2\theta}{1 + \sin\delta}\right) \tag{6.30}$$

where

$$\frac{\Delta r}{r} = 1 - \frac{\cos\alpha\sin(\alpha + \eta)}{\sin\alpha\cos\theta + \sin\eta} = \frac{\sin(\alpha + \eta)}{\sin\theta}\left(\frac{\sin(\alpha - \beta)}{\cos\beta\sin\alpha} - 1\right) + 1 \tag{6.31}$$

and

$$\tan\beta = \frac{\sin\alpha\sin\theta}{\sin\alpha\cos\theta + \sin\eta} \tag{6.32}$$

The mean or average pressure for the slice is given, conceptually, by

$$P_v = \frac{1}{\alpha + \eta}\int_0^{\alpha+\eta} P_{vx}\,d\theta \tag{6.33}$$

where

$$P_{vx} = f(x, \theta, \sigma_{1x}) \tag{6.34}$$

(c) ***Approximate Pressure Ratio*** k_{hf}

Noting that the direction of σ_1 is tangential to the arc and assuming the magnitude of σ_1 can be approximated using the mean value of Δr, an approximate closed form solution for the pressure ratio k_{hf} may be obtained as follows:

$$k_{hf} = \frac{(1 + \sin\delta\cos2\eta)}{C_s(1 - 0.5\sin\delta\sin2(\alpha + \eta)]} \tag{6.35}$$

where

$$C_s = 0.5 + \left(\frac{\cos\alpha\sin(\alpha + \eta)}{2(\sin\alpha + \sin\eta)}\right) \tag{6.36}$$

6.5.3 *Horizontal Slice Maximum and Minimum Pressure Ratio Bounds*

Figure 6.5a shows the upper bound defined by the vertical stress "P_{vw}" at the wall and lower bound "P_{vc}" at the centre. The corresponding Mohr Circle stress fields are shown, respectively, by C1 and C2, in Fig. 6.5b.

(a) **Stress Condition at the Hopper Wall**

Referring to Fig. 6.4, the angle "η" defining the direction of the major consolidation stress "σ_{1x}" with the normal stress or pressure "P_n" is given by

$$\eta = 0.5\left(\phi_w + \sin^{-1}\left(\frac{\sin\phi_w}{\sin\delta}\right)\right) \tag{6.37}$$

From the Mohr Circle diagram C1, the normal stress or pressure is given by

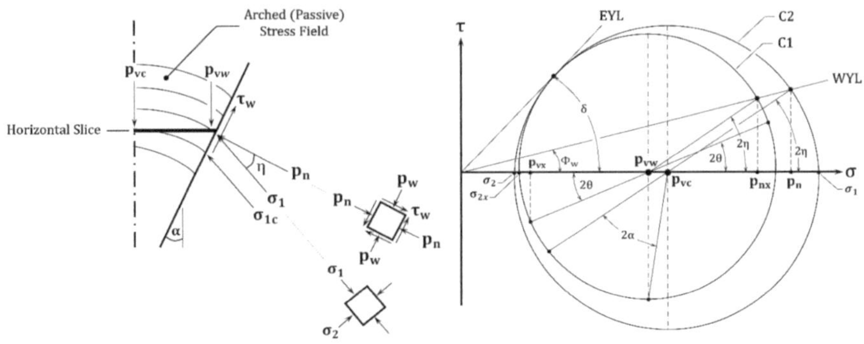

(a) Pressure ratio bounds **(b) Mohr circle stress fields**

Fig. 6.5 Horizontal slice pressure ratio bounds

$$P_n = \frac{\sigma_1}{1 + \sin\delta}(1 + \sin\delta\cos2\eta) \tag{6.38}$$

and the corresponding vertical stress or pressure P_{vw}, shown in Fig. 6.5 acting at the wall is given by

$$P_{vw} = \frac{\sigma_1}{1 + \sin\delta}(1 - \sin\delta\cos2(\alpha + \eta)) \tag{6.39}$$

Hence the value of the pressure ratio acting at the wall is

$$k_{bf} = \frac{P_n}{P_{vw}} = \frac{1 + \sin\delta\cos2\eta}{1 - \sin\delta\cos2(\alpha + \eta)} \tag{6.40}$$

AS3774-1996 k_{hf} Values

While noting that the k_{hf} ratio is theoretically correct at the hopper walls, the Australian Standard adopted Eq. (6.40) as the average pressure ratio for hopper wall load determination. The reasons for this assumption were based on practical industrial case studies and experience. An example explaining this decision is include in the case study example in Sect. 6.8.

(b) **Stress Condition at the Centre**

Referring to Fig. 6.4, the stress condition at the centre is described by the Mohr Circle C2 for which the major consolidation stress is "σ_{1c}". At the centre, it acts in a horizontal direction while the minor principal stress is vertical corresponding to the vertical stress or pressure P_{vc}. On the assumption that the effective friction and "δ" does not vary across the horizontal slice, it is assumed that $\sigma_{2c} \approx \sigma_2$. Hence $P_{vc} = \sigma_2$, which, from Fig. 6.4, σ_1 is constant around the arch and not across the slice. Also, the vertical pressure P_{vhf} cannot be constant across the slice. The net result is that the k_{hf} value given by Eq. (6.40) tends to under-estimate the normal pressures in the region of the transition in some cases. For this reason, the modified value derived by Roberts [17, 18] and summarised below is recommended.

The lower bound value of the vertical pressure is given by

$$P_{vc} = \frac{\sigma_1}{1 + \sin\delta}(1 - \sin\delta) \tag{6.41}$$

The upper bound value of the vertical pressure is given in Eq. (6.39) and the normal pressure is given by Eq. (6.38). Assuming, for simplicity that the average vertical pressure acting on the slice is the mean of the upper and lower bound values, the modified pressure ratio is

$$k_{hf} = \frac{2(1 + \sin\delta\cos 2\eta)}{2 - \sin\delta(1 + \cos 2(\alpha + \eta))} \tag{6.42}$$

6.5.4 Summary of Passive Pressure Ratios

See Fig. 6.6.

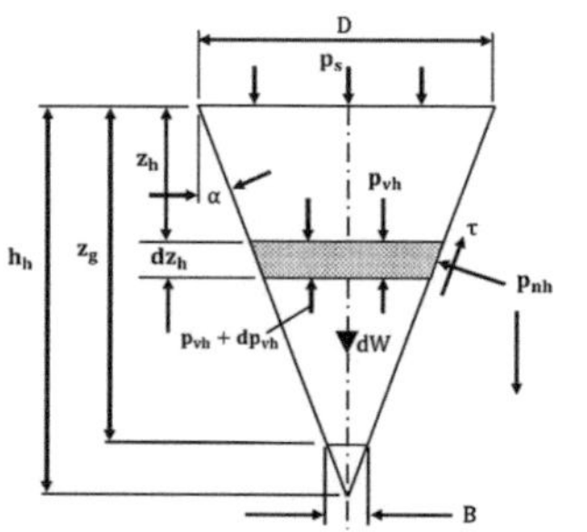

B	1.0 m
D	4.0 m
z_g	2.82 m
α	28 °
ϕ	26 °
δ	53 °

Name	k_{hf} Values	
	Equation	k_{hf}
AS3774 - k_{hfa}	6.38	1.0497
Modified k_{hfm}	6.4	1.825
Closed Form	6.33 - 34	2.485
Jenike (Modified)	6.23	2.622

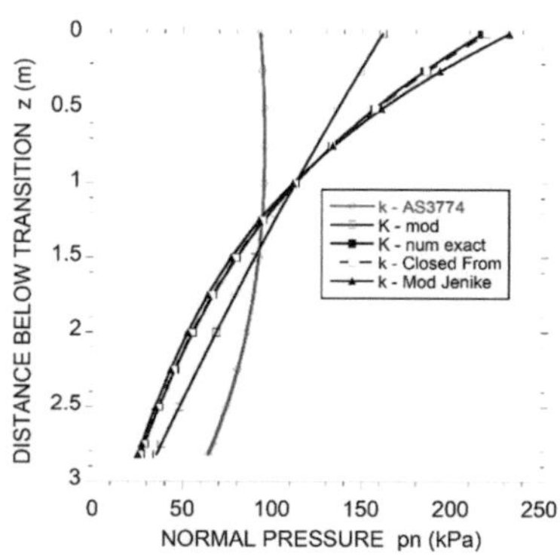

Fig. 6.6 Example plane-flow hopper normal loads summarising pressure ratios

6.6 Surcharge Pressure p_S Acting on Hopper

The computation of the surcharge pressure P_s acting on a mass-flow hopper at the transition datum level depends on the stress field established in the surcharge load. The following cases are considered:

6.6.1 *Mass-Flow and Expanded Flow Bins*

Referring to Fig. 6.7, the stress field in the upper cylindrical section is peaked and the surcharge pressure is given by the Janssen equation:

$$P_s = \frac{\gamma r_c}{\mu k}\left[1 - e^{-(\mu k h_c / r_c)}\right] + P_{vo} e^{-(\mu k h_c / r_c)} \tag{6.43}$$

where

$$P_{vo} = \gamma h_s \tag{6.44}$$

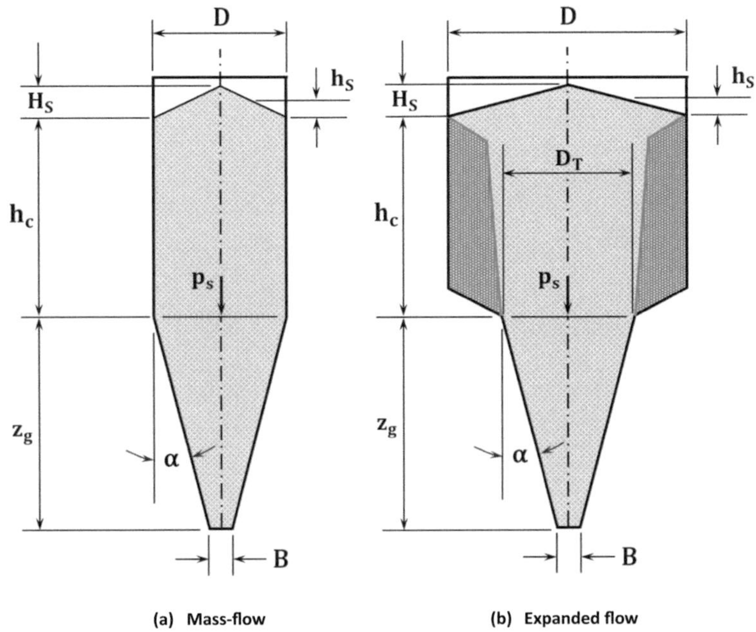

Fig. 6.7 Mass-flow and expanded-flow bins

$h_c =$ height of material in contact with cylinder walls
$k \geq 0.35$ (Often assumed as 0.4)

Other variables as previously defined.

Note: The surcharge pressure computed from Eq. (6.35) is used when determining *both* the initial and flow pressures acting in the hopper

6.6.2 Gravity Reclaim Stockpile

The use of mass-flow reclaim hoppers and feeders under stockpiles is illustrated in Fig. 6.8. The determination of surcharge head and pressure in the case of stockpiles is somewhat uncertain owing to the significant variations that can occur in the consolidation conditions existing within the stored bulk solid. The state of consolidation of the bulk solid is influenced by such factors as.

- Stockpile management and loading history
- Loading and unloading cycle times and length of undisturbed storage time

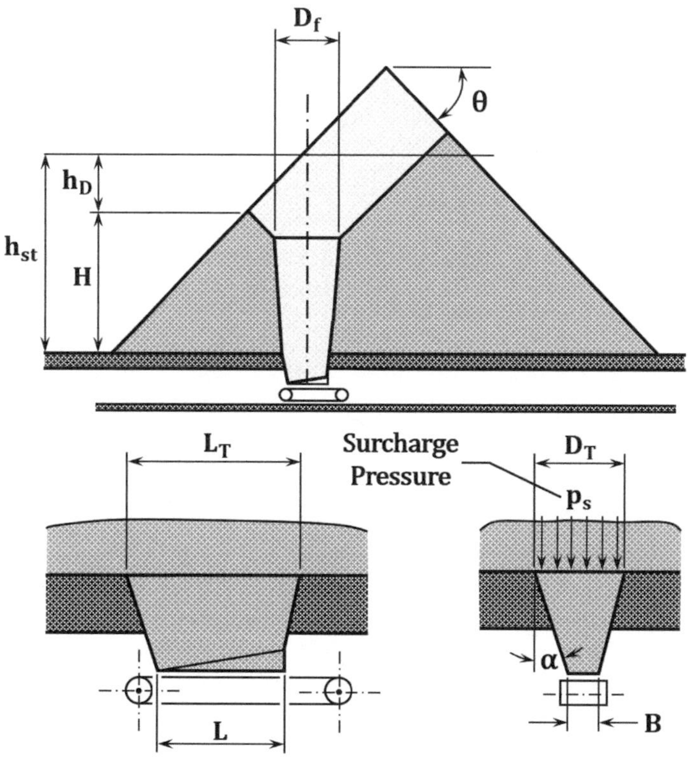

Fig. 6.8 Gravity reclaim stockpile

- Variations in moisture content
- Degree of segregation
- Variations in the quality of bulk solid over long periods of time
- Compaction effects of heavy mobile equipment that may operate on the stockpile surface.

The following three cases are considered:

(a) **Case 1—Freshly Formed or Uniformly Consolidated Stockpile in Which Withdrawal Occurs at Regular Intervals.**

$$P_s = \gamma h_{st} \tag{6.45}$$

That is the effective head is equal to the actual head. This is the most conservative solution.

(b) **Case 2—Less Conservative Solution**

In some cases, a less conservative solution may be applied through the use of the Rankine pressure or head. That is

$$P_s = \gamma h_{st} \cos\phi_r \tag{6.46}$$

where

$\phi_r =$ Angle of repose

(c) **Case 3—Preformed Rathole**

Where a stockpile is well consolidated and a stable, preformed rathole exists, the rathole acts as a pseudo bin. The consolidation pressure may be estimated using Eq. (6.43) where

$$r_c = \frac{D_f}{4} \tag{6.47}$$

$$H = h_{st},\ h_s = h_D \tag{6.48}$$

$$\mu = \sin\delta \tag{6.49}$$

Note: The surcharge pressures computed for the cases described are used when determining *both* the initial and flow pressures acting in the hopper.

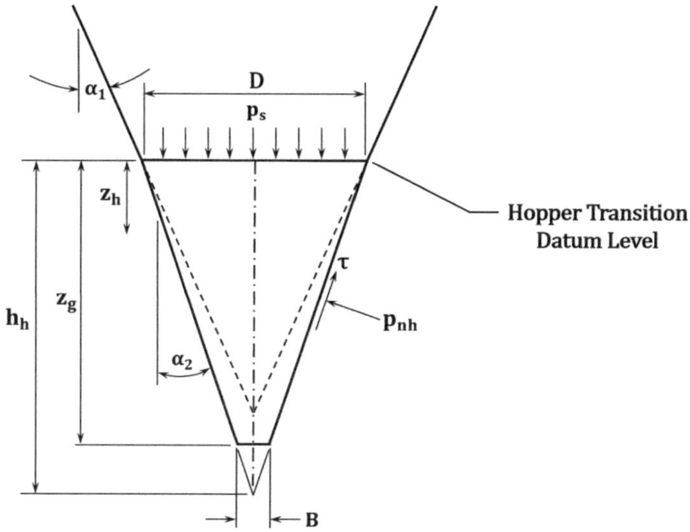

Fig. 6.9 Change in hopper slope and/or lining surface

6.6.3 Change in Slope and/or Lining in Mass-Flow Hopper

Sometimes either or both the hopper slope and lining surface are changed as illustrated in Fig. 6.9. In such cases a peaked or *active* stress field exists for the initial case and an arched or *passive* stress field for the flow case. Hence the surcharge pressures P for both the *active* and *passive* cases at the transition will, in turn, need to be applied.

6.7 Mass-Flow Bin Wall Load Design Examples

6.7.1 Example 1

The case of the mass-flow bin of Fig. 6.10 is considered. The bin is to handle coal. The following information is given:

$\phi_w = 30°$ for cylinder; $\phi_w = 20°$ for hopper, bulk density $\rho = 1$ t/m^3, $\delta = 50°$

The wall pressures have been determined using the procedures described herein. The pressures are plotted in Fig. 6.11. The relevant parameters are:

For Cylinder:$k = 0.35$, $c_{nf} = 1.387$

For Hopper:$k_{hi} = 0.526$, $k_{hf} = 2.46$ (based on Eq. (6.35)),$j = 7.36$

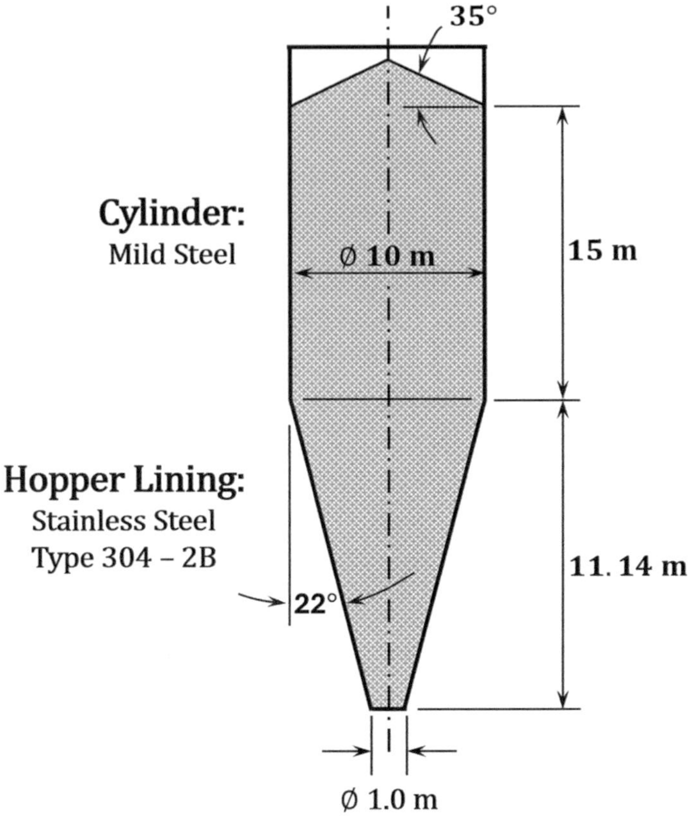

Fig. 6.10 Mass-flow bin for coal

6.7.2 Example 2

Coal Bin with two Hopper Sections as Shown in Figs. 6.12, 6.13.

6.8 In-Plane Forces

At a given location in the bin, the design loads for that location are selected to be the larger of the initial and flow loadings. The loads which act on the bin walls are shown in Fig. 6.1, the shear stresses τ are assumed to be related to the normal pressure P_n by:

$$\tau = P_n \tan\phi_w \tag{6.50}$$

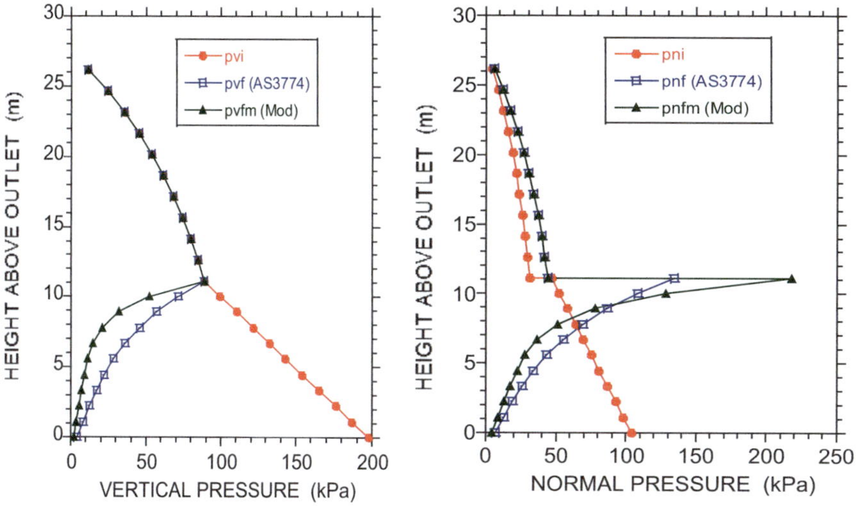

Fig. 6.11 Vertical and normal wall pressures for bin in Fig. 6.10

Fig. 6.12 Mass-flow bin with 2 hopper sections

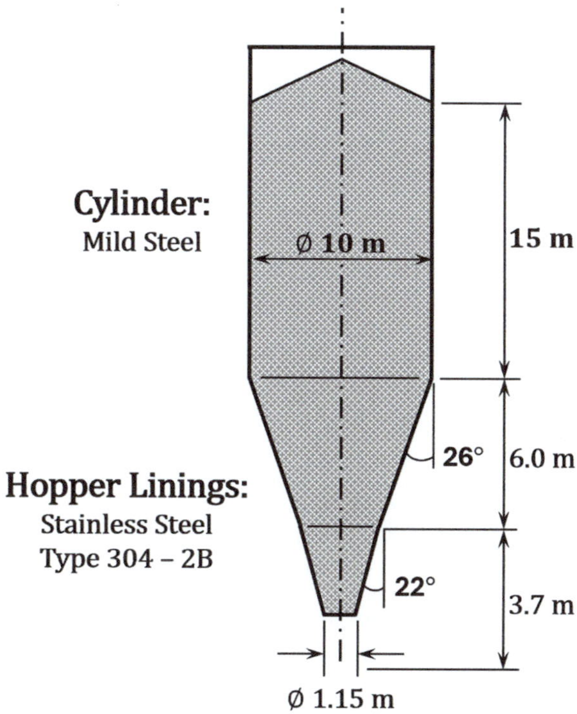

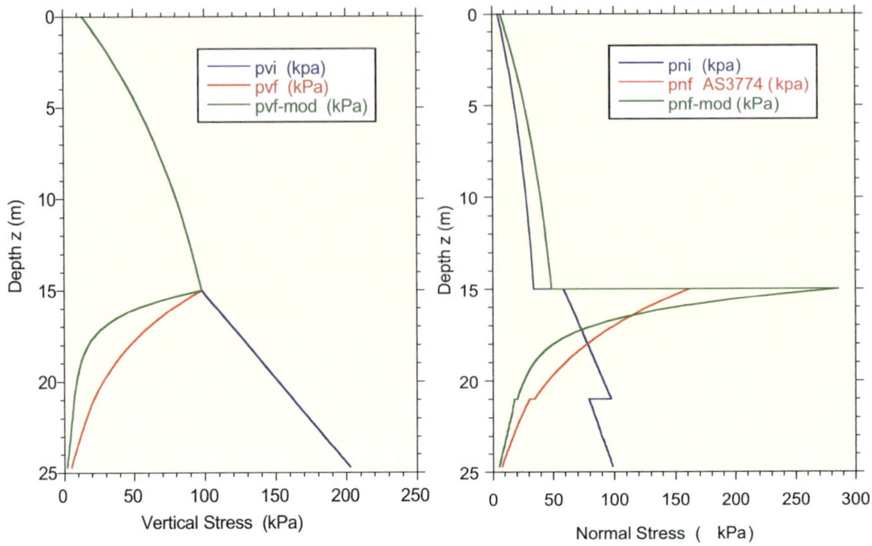

Fig. 6.13 Vertical and normal wall pressures for bin in Fig. 6.12

The in-plane loads are given by

$$V = \int \tau dz \tag{6.51}$$

The in-plane forces in the cylindrical section of the silo walls are obtained by integrating the shear force along the silo wall. For a silo of circular cross-section, the initial or static in-plane force per unit circumference in kN/m is given by

$$N_{zi} = \frac{\gamma D}{4}\left[y + \left(\frac{D}{4\mu k} - h_s\right)\left(e^{-4\mu ky/D}\right) - 1\right)\right] \tag{6.52}$$

where

$\mu =$ upper bound value of wall friction angle
$k =$ ratio of lateral to vertical pressure.
$z =$ depth below surface.

The flow values of the in-plane force are computed from

$$N_{zf} = c_{qf} N_{zi} \tag{6.53}$$

where c_{qf} is the traction flow multiplier, $c_{qf} = 1.2$ for cylinder and $c_{qf} = 1.4$ for plane-flow containers.

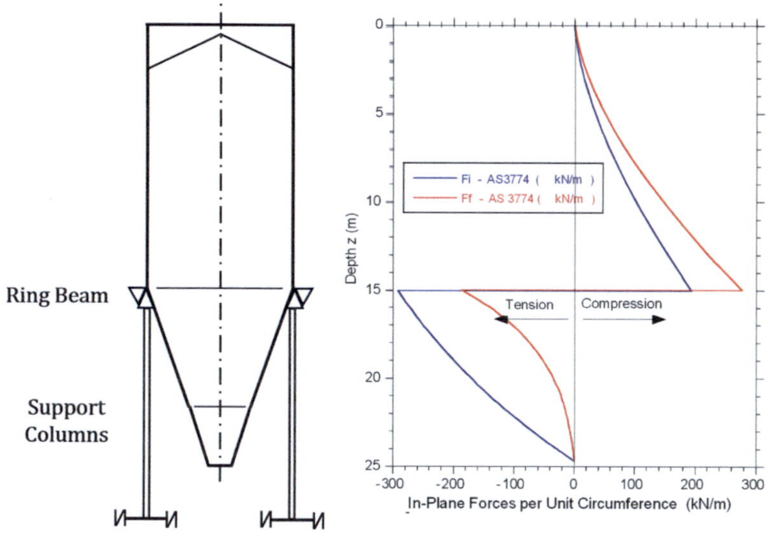

Fig. 6.14 In-plane forces for bin supported at ring beam

In the hopper walls, the accumulated effects of the shear stresses and the normal stresses acting on the hopper walls give rise to both in-plane forces and forces perpendicular to the hopper walls respectively. The magnitude of these forces change at discontinuities in the hopper walls. These forces in the hopper walls may be expressed per unit circumference (Fig. 6.14).

6.9 Influence of Bulk Solids Parameters on Wall Loads

In general, the wall loads are maximised if the design values in Table 6.2 are used.

6.10 Symmetrical Funnel-Flow Bins

The methods for predicting the wall loads in funnel-flow bins of symmetrical, cylindrical and rectangular profiles commonly adopt a similar approach to that of mass-flow as outlined in Sect. 6.3. The Janssen method is applied to predict the static and initial filling loads on the vertical walls, with the application of over-pressure factors to account for the increased wall loads under dynamic or flow conditions. The theories assume a single symmetric outlet. Such an outlet is the only one which allows the flow pattern in the bin to be predicted with any degree of certainty. It should be appreciated that even with such an arrangement there can be no guarantee that the flow pattern will remain symmetrical.

Table 6.2 Design values for wall load calculation

	Friction value	
	Internal	Wall
	δ	Φ_w
Cylinders:		
Normal Wall Pressure	Independent	Minimum
In-plane wall forces	Independent	Maximum
Hopper:		
Normal wall pressures	Maximum	Cylinder—minimum
Shear stress on wall	Maximum	Hopper—maximum
In-plane wall forces		Hopper—maximum
Bulk Density:		
For all predictions use maximum value		

In symmetric funnel-flow bins, an effective transition controls the flow channel as indicated in Fig. 6.15. AS3774-1996 gives the lower bound of the effective transition as

$$h_t = 0.4d_c \tan\delta \tag{6.54}$$

where

$d_c = D =$ Bin diameter or width
$\delta =$ Effective angle of internal friction.

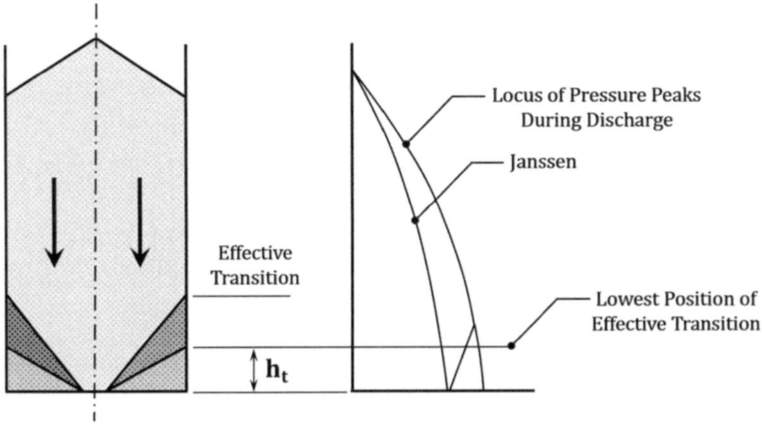

Fig. 6.15 Pressures in funnel-flow bin

Above the effective transition the flow occurs along the walls similar to that occurring in a tall mass-flow bin. An over-pressure occurs at the effective transition similar to the switch pressure in mass-flow. Since the effective transition is not stationary but tends to move during the discharge process, the procedure for design involves defining the position at which the effective transition is likely to occur. In the upper cylindrical section of the bin, imperfections in the walls can give rise to increased pressures. The flow pressures, depicted in Fig. 6.15, are normally computed by the application of over-pressure factors on Janssen. As described in Sect. 6.3.

6.11 Loadings Due to Eccentric Discharge

The analysis of the pressures in bins with non-symmetric flow patterns due to eccentric outlets is a subject of considerable complexity. The walls of bins having a single eccentric outlet or bins with multiple outlets are subject to bending stresses in addition to hoop stresses. It is possible for bending stresses in the order of three times the hoop stresses to be experienced [19, 20]. In general, eccentric loading should be avoided if possible.

6.11.1 Eccentric Discharge Examples

Typical examples of eccentric discharge are illustrated in Fig. 6.16. The flow channel, often described as a *rathole*, behaves as a pseudo mass-flow bin where the discharge is governed by the internal friction of the bulk solid. Depending on the position of the eccentric outlet, the flow channel may be formed completely within the bulk solid without contact with the bin wall. However, as is often the case, the flow channel also intersects the wall, thus mobilising the so-called wall friction in the wall contact zone. Off-loading chutes, illustrated in Fig. 6.16b are a typical example.

The typical form of the cross-sectional stress profile due to eccentricity during discharge is illustrated in Fig. 6.16c. It is well established that the wall pressures "P" due to the static stored bulk solid are proportional to the radius "R", while the reduced pressures "P_o" of the discharging bulk solid in the flow channel are proportional to the radius "r" as illustrated in Fig. 6.16 c. Research has shown that over-pressure, *switch* pressures, occur at the locations of the bin and flow channel as illustrated in Figs. 6.16 c, 6.17.

Apart from the non-uniform pressure distribution, the internal surface of the wall above the outlet is subjected to considerable impact and abrasive wear which further exacerbates the wall bending and buckling. Examples of industrial off-loading chute applications are illustrated in Fig. 6.16 a and b [21].

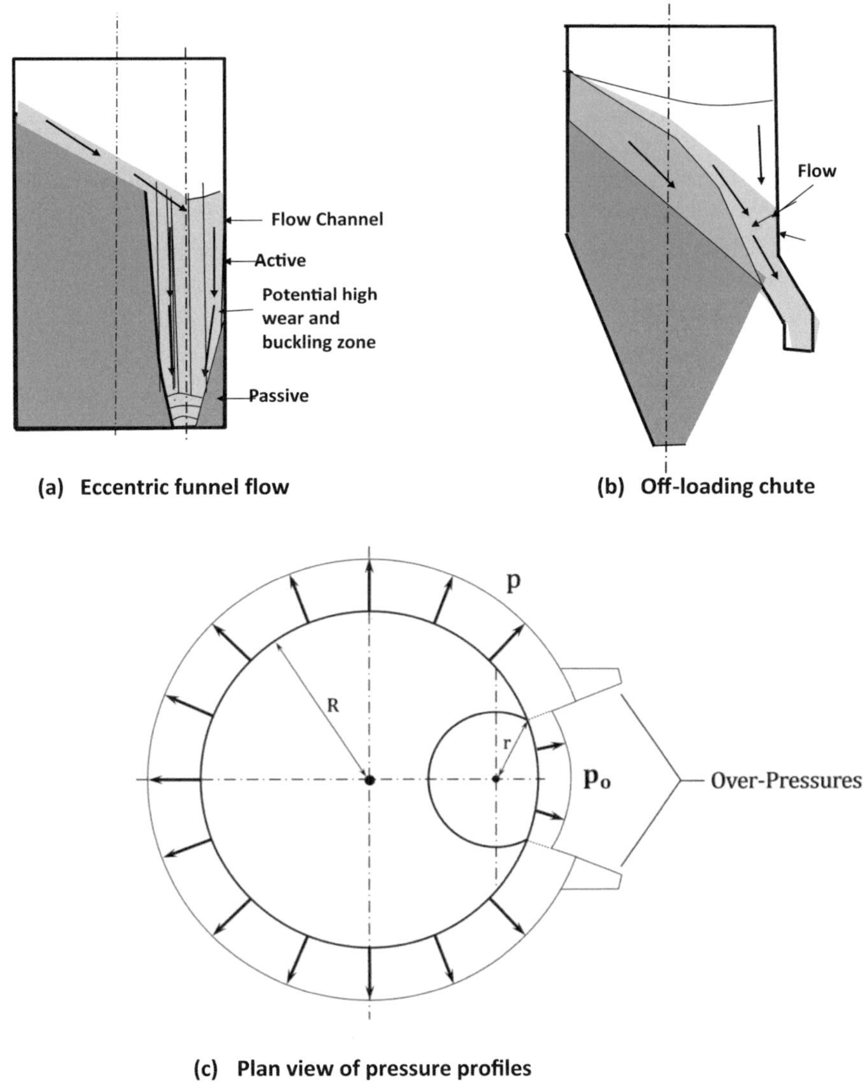

Fig. 6.16 Typical bin eccentric discharge arrangements

6.11.2 Buckling of Thin Wall Steel Bins Resulting from Eccentric Discharge

With the knowledge that eccentric loadings induce bending in the walls of bins or silos, there is a need to consider the combined effect of bend and hoop stresses. Focusing on bins and silos of steel construction it is important that the overpressure factors also include variations in wall thickness. By way of background, it is useful to

(a) Coal storage bin (b) Grain storage silos

Fig. 6.17 Off-loading chute examples

review the early introductory work of Jenike [5] who developed moment relationships for steel bins subject to eccentric loading.

The case of a cylindrical steel bin with a single eccentric outlet as illustrated by the cross-sectional view in Fig. 6.18 is considered. The bin radius is denoted by "R" and the flow channel radius, which intersects the wall, is denoted by "r". The static pressure due to the stored bulk solid is denoted by "P" and the flow pressure for the discharge flow by "P_o". The load eccentricity is represented by "e".

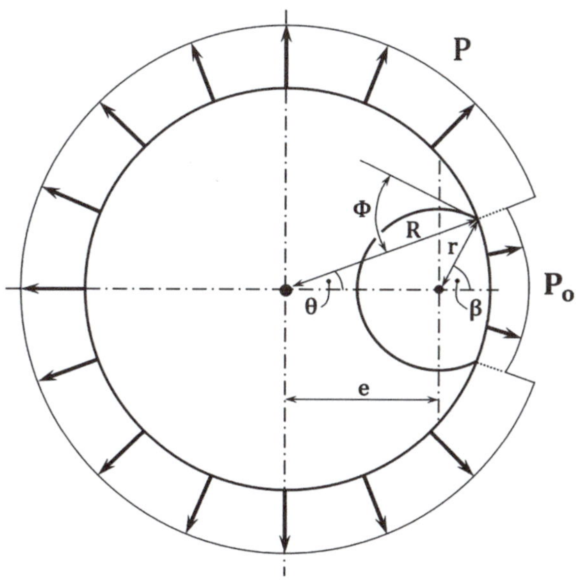

Fig. 6.18 Plan view of pressure profiles

A direct proportionality between the pressures "P" and "P_o" and their corresponding radii is assumed. That is

$$\frac{P_o}{P} = \frac{r}{R} \tag{6.55}$$

The flow channel intersects the wall at an angle β as indicated in Fig. 6.18. The angle ∅ defines the tangent of the flow channel at the wall intersection as illustrated. Jenike assumes that ∅ is the wall friction angle between the bulk solid and the bin wall. From the geometry of Fig. 6.18 and using the moment equation, the moment "M" per unit length due to eccentric loading is taken to be

$$M = K_m R^2 P \tag{6.56}$$

$$K_m = \frac{\sin^2\theta\tan\theta}{\pi}\left(1 - \frac{\sin\theta}{\cos(\varnothing - \theta)}\right) \tag{6.57}$$

where

$R =$ bin radius
$r =$ flow channel radius.
$\theta =$ eccentricity angle.
$\phi =$ wall friction.
$P =$ bin load pressure.
$P_o =$ flow channel pressure.

The factor K_m is plotted in Fig. 6.19 for the value of f = 30°. Key points of the K_m graph are determined from the accompanying $\Delta_2 k_m$ graph which is a plot of the second derivative $\frac{d^2K_m}{d\theta^2}$ of the slope of the K_m graph.

While Fig. 6.19 indicates a maximum value of K_m when $\theta = 46°$, in view of the geometrical constraints of the K_m this maximum value of θ is not likely. Jenike analysed the conditions that could occur and suggests $\theta = 21°$ which corresponds approximately to the inflection point of Fig. 6.19. Larger values of θ may be possible in certain cases.

For an unstiffened steel silo, Jenike postulated that the ratio of bending to hoop stress can be expressed by

$$\frac{\sigma_b}{\sigma_{hoop}} = \frac{6K_m R\left(1 - \upsilon^2\right)}{t} \tag{6.58}$$

where

$t =$ Wall thickness
$\upsilon =$ Poisson's Ratio.

For $\upsilon = 0.3$ and using the value of k_m corresponding to $\theta = 21°$, Eq. (6.58) becomes

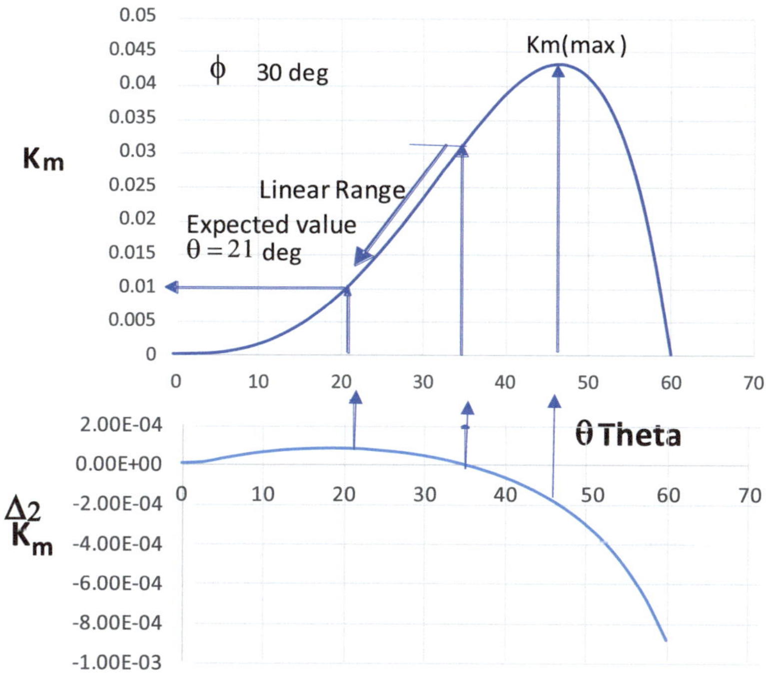

Fig. 6.19 Bending moment factor K_m variation with theta (θ)

$$\frac{\sigma_b}{\sigma_{hoop}} = \frac{0.055R}{t} \tag{6.59}$$

Values of $\frac{\sigma_b}{\sigma_{hoop}}$ based on (6.59) are shown in Fig. 6.20. As indicated, the bending effects are significant, particularly at low shell thicknesses and larger bin radii or diameters.

Since the total stress is the sum of the hoop plus bending stresses, it is possible to devise a maximum stress over-pressure factor c_m for steel bins based on Eq. (6.60). The form of expression is

$$c_m = c_o\left(1 + \frac{c_1 R}{t}\right) \tag{6.60}$$

where c_1 is the eccentricity factor. For conditions relating to Eq. (6.59) $c_1 = 0.055$.

Because some localised buckling or yielding leading to a re-distribution of stress in steel bins can be tolerated, the factor c_o is introduced where $c_o < 1$. While further analysis is required to establish appropriate values of c_o, the degree of stiffening and actual static pressure distribution accounting for the surcharge and flow channel developed will have an influence. It needs to be noted that c_m must always be larger

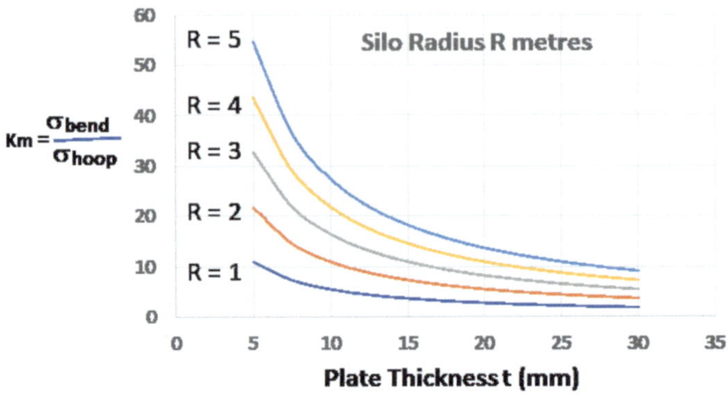

Fig. 6.20 $\frac{\sigma_b}{\sigma_{hoop}}$ for unstiffened steel silos based on Eq. (6.59)

than the values specified for symmetrical funnel-flow. The factor c_m is applied to the static pressure to give an equivalent pressure for hoop stress analysis.

The foregoing discussion is, to a large extent, somewhat conceptual requiring further research.

6.11.3 *Industrial Example of Bin Crease*

An example of denting or creasing in a steel bin handling coal in a power station is shown in Fig. 6.21. The problem arose as a result of an interfacing problem at the hopper outlet and the inappropriate coal feeder which, induced eccentric funnel-flow, thus preventing the desired mass-flow with the hopper outlet fully active. This case study is discussed in more detail in Chap. 10 which is devoted to the feeding of bulk solids.

6.11.4 *Bending and Hoop Stresses in Reinforced Concrete Bins*

For reinforced concrete bins and silos subjected to combined bending and hoop stresses, a similar procedure may be applied [19]. In this case the moment of resistance of the concrete wall is taken into account. Consider the case of a concrete silo designed for hoop tension with reinforcement steel being located nominally at the centre of the walls. If the wall thickness relative to the curvature is of low significance, as is usually the case, then the ratio of bending to hoop tension is given by

Fig. 6.21 Creasing in a coal hopper at a power station

$$\frac{\sigma_b}{\sigma_{hoop}} = \frac{K_m R}{t_c} \tag{6.61}$$

t_c is the effective thickness or moment arm of resistance for reinforced concrete wall. The ratio of total to hoop is given by Eq. (6.62).

$$\frac{\sigma_{total}}{\sigma_{hoop}} = 1 + \frac{K_m R}{t_c} \tag{6.62}$$

Equation (6.62) neglects the influence of vertical reinforcement in reducing the lateral bending effects.

6.12 Design Approach to Eccentric Discharge

Following on from the previous section, a more general approach to the design of bins and silos with eccentric discharge is outlined. The discussion is, in effect, a review of the procedures presented in AS3774-1996 [13]. The design pressures shall depend on the eccentricity of the outlet, e_o, relative to the dimension of the container, d_c. An eccentricity less than 0.1 d_c shall be treated as no effective eccentricity. The relevant details are given in Fig. 6.22.

Points on the wall on the same side as the outlet shall be termed the *NEAR SIDE*, while those on the wall remote from the outlet shall be termed the *FAR SIDE.*

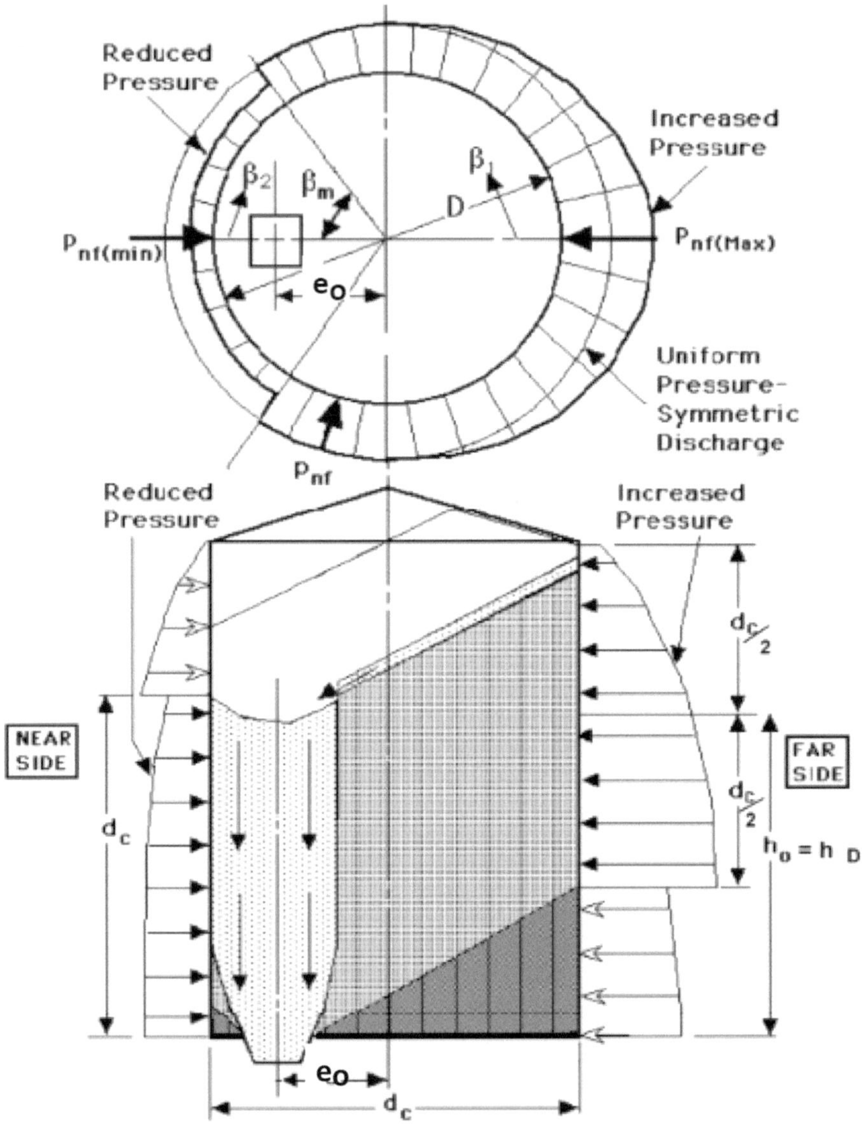

Fig. 6.22 Eccentric flow pressure distribution for single eccentric outlet

6.12.1 *Pressure Increases*

The pressure increase on the *FAR SIDE* of the silo, see Fig. 6.22, is found to act over a height d_c and is centred a distance h_D above the silo base. The value of h_D is

$$h_D = (0.5d_c + e_o)\tan\delta \tag{6.63}$$

where e_o is the eccentricity of the discharge outlet. The maximum pressure increase, defined to be $P_{ef,max}$ by AS3774-1996, acts on the eccentricity axis and on the *FAR SIDE* (Fig. 6.22) and is denoted by

$$P_{ef,max} = P_{nf}\left(\frac{e_o}{d_c} - 0.1\right) \geq 0.0 \tag{6.64}$$

where P_{nf} is the normal pressure acting on the wall during material flow. For a circular container the pressure increase variation is:

For β Between 90° and 270°:

$$P_{ef} = P_{ef,max}(-\cos\beta) \tag{6.65}$$

For β Between −90° and 90°:

$$P_{ef} = 0 \tag{6.66}$$

where:

$\beta =$ circumferential coordinate measured from the eccentricity axis on the *NEAR SIDE* of the silo

$P_{ef} =$ normal pressure increase due to eccentric flow.

6.12.2 *Pressure Reductions*

The maximum pressure reduction, defined to be "$P_{ef,\,red}$" by AS3774-1996, acts on the *NEAR SIDE*, (Fig. 6.22) and is denoted by

$$P_{ef,red} = 1.5P_{nf}\left(\frac{e_o}{d_c} - 0.1\right) \tag{6.67}$$

where P_{nf} is the normal pressure acting on the wall during material flow. The pressure reduction given by Eq. (6.67) is taken to be constant at a particular height above the silo base, and, for a circular container, acts over an angle of β_e either side of the silo eccentricity axis. The value of β_e is taken to be

$$\beta_e = 105 - \frac{150e_o}{d_c} \tag{6.68}$$

where β_e is the circumferential range of pressure reduction expressed in degrees.

6.13 Multi-Outlet Bins

The complexity of the wall loads in bins and silos with multiple outlets requires particular attention. This is illustrated in Fig. 6.23, which shows, by way of example, the seven outlets and feeder arrangement in a large coal bin. While the outlet and feeder pattern is symmetrical, the circumferential spacing of the outlets and feeders will give rise to the deformation bending pattern indicated by the dotted profile in Fig. 6.23. In any case, there is no guarantee that all feeders will operate at the same time. There is always the distinct possibility that, at some stage, one or more feeders will be switched off leading to more significant eccentric loads and bending effects. A case study based on this example is presented in Sect. 6.14.

As a further example, Fig. 6.24 shows the catastrophic failure of a bin handling raw sugar. The bin, which was in new state of construction when the failure occurred, comprised two rows of four feeders as shown. While the actual reasons for the failure occurring are not known, it is almost most certain it was due to the eccentric loadings resulting from the way the bin discharge outlets were operated.

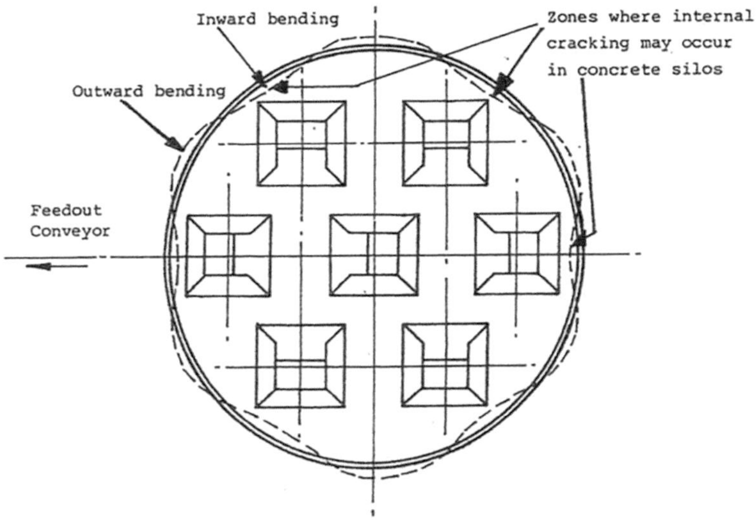

Fig. 6.23 Multi-outlet coal storage bin

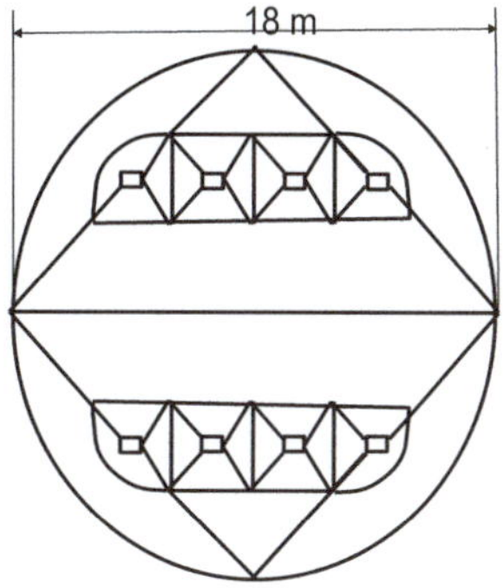

Fig. 6.24 Collapsed raw sugar bin

6.14 Multi-Outlet, Reinforced Concrete Coal Silos—Case Study

This case study concerns the severe cracking experienced in the walls of the two 3500 tonne storage capacity, identical, concrete coal storage silos shown in Fig. 6.25a. The vibratory feeders discharge coal onto the conveyor belt located centrally as indicated. Inspection of the internal surfaces of the silo walls revealed the presence of cracks as illustrated in Fig. 6.25b, necessitating retrofitting the silos with steel bands as shown in Fig. 6.25 a and d. The silos are 15 m diameter and 25.4 m high from the top of the flat floor. The silo discharge is via seven outlets and vibratory feeders arranged as shown in Figs. 6.23 and 6.25 c with six outlets symmetrically spaced around the periphery and one located centrally. Key issues are discussed in the following subsections.

6.14.1 Silo Walls

The concrete walls of the silos are 250 mm thick with both vertical and horizontal reinforcement bars nominally located centrally within the walls. Little or no allowance was made for eccentricities in loadings and associated bending. Inspection of the internal surfaces of the silo walls revealed that vertical cracks had occurred as illustrated in Fig. 6.25b. The cracks commenced approximately 3 m above the silo floor and extended upwards to about 10 m. It was noted that vertical cracks occurred in a pattern of 5–6 cracks above each outlet. While the feeders are located symmetrically, the reclaim flow channel linked to each feeder created eccentricities due to the stationary zones as illustrated in Fig. 6.25c. Furthermore, the eccentricities were exacerbated when all feeders were not operating. On closer inspection of the silo wall, similar cracks, although not as severe, were noticed on the outside but were

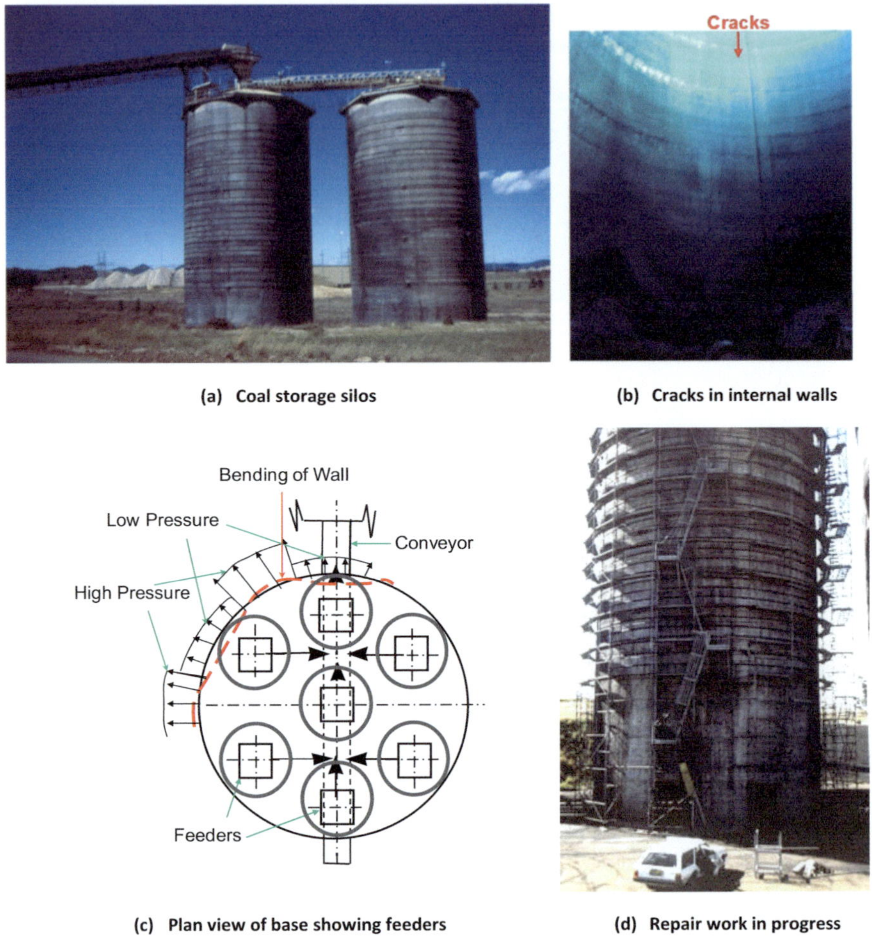

(a) Coal storage silos (b) Cracks in internal walls

(c) Plan view of base showing feeders (d) Repair work in progress

Fig. 6.25 Two 3500 tonne coal silos showing repair work

located between each of the 6 outlets. The second silo was also examined and found to display the cracking problem.

6.14.2 Measurement of Hoop and Bending Strains and Loads

To gain insight into the cracking problem, strain gauges were attached to the hoop reinforcement bars as illustrated in Fig. 6.26.

The concrete at the selected locations was chipped away to allow the gauges to be attached. They were then sealed over. Theoretical analysis predicted that highest bending would occur at locations approximately 7 m from the silo floor. Three rows

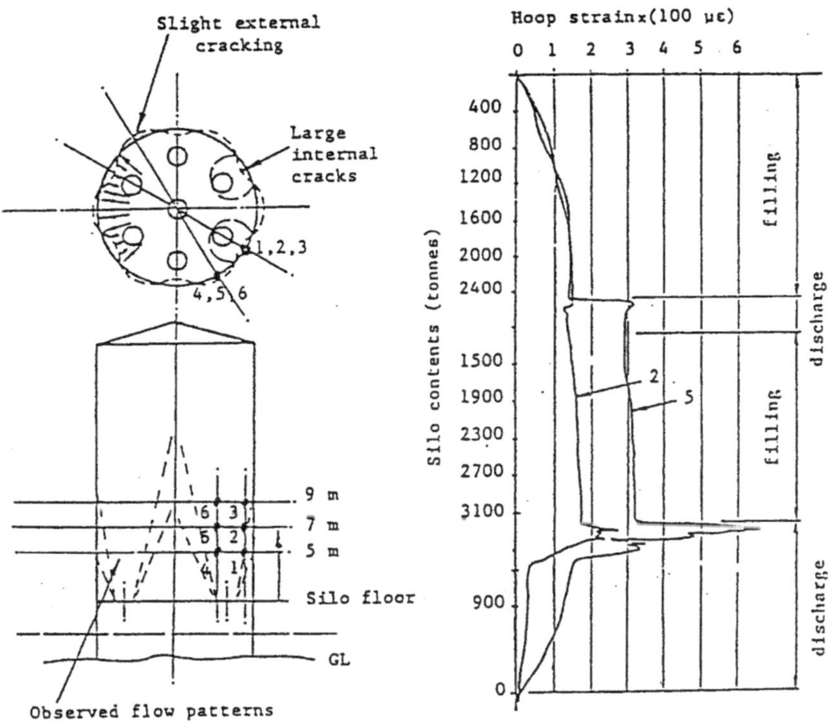

(a) Measurement locations

(b) Typical measurements at 7 m above bin floor

Fig. 6.26 Wall loads and strain measurements in 3500 tonne coal silo

of gauges were installed at 5 m, 7 m and 9 m above the silo floor and arranged directly above an outlet and between two outlets as shown in Fig. 6.26a. The strains in the hoop reinforcement at these points were continuously and simultaneously recorded and plotted on a chart recorder during filling and emptying of the silo. A typical recording is shown in Fig. 6.26b for strain measurements at points 7 m above the bin floor. The dynamic effects are shown quite clearly by the bending caused when all feeders are operated at the same time. The maintenance of a *static* bending moment is clearly evident after discharge is stopped. This is due to the uneven flow pressures set up within the silo during discharge. Further filling of the silo simply increased the hoop stress at both points of inward and outward bending.

Full scale tests were performed with various combinations operating to generate different combinations of eccentric discharge patterns. With all feeders operating, the hoop and total stresses (hoop plus bending) for a range of operating conditions are presented in Table 6.3 for a typical set of results. Also shown for comparison purposes are calculated results based on $\theta = 31°$ which was consistent with the mode of operation of the silo. The calculated results accounted for the combined concrete

Table 6.3 Measured and calculated hoop and bending stress ratios

Gauge location (refer to Fig. 6.26)	Ratio $\frac{\sigma_{total}}{\sigma_{hoop}}$	
	Measured	Calculated
4	2.61	2.90
5	3.71	2.90
6	2.81	2.90

and hoop reinforcement under outward bending (tension) and applied a 0.75 factor for inward bending where steel reinforcement is less effective in compression. The measured values correspond to locations 4, 5, 6 of Fig. 6.26, these locations being in the region of maximum bending moments.

With the silo full and all feeders operating, ratios of total to hoop stresses as high as 4.4 for the wall between outlets and 0.75 for inward bending were determined. These correspond, respectively, to calculated values of 3.2 and 1.0. With only one feeder operating, the corresponding stress ratios are 1.7 for the wall between outlets (outward bending) and 0.1 for the wall directly above an outlet (inward bending). It is evident that the eccentricity influence is less severe in this case.

While the foregoing results show some variations between the calculated and measured values, it needs to be noted that the measured values are obtained from the silo operating with the walls already cracked. Furthermore, with the strain gauges being located on the outside surface of the reinforcing bars, the measurements would include the bending stresses in the bars themselves. However, the results indicate the general order of magnitude of the total stress in relation to the hoop stress as predicted by the method of analysis used. It needs to be noted that the calculated results are based on Eq. (6.62).

6.14.3 *Improvement of Bin Live Capacity*

Apart from the silo wall cracking problem, the seven feeder reclaim configuration of Figs. 6.23 and 6.25 c created substantial dead zones and associated reduced live capacity. Based on the measured flow properties of the coal, the decision was made to modify each of the two coal silos by installing a mass- flow hopper above the central outlet and vibratory feeder as illustrated in Fig. 6.27a. The hoppers were manufactured from type 304 Stainless Steel with a 2B finish, Fig. 6.27b shows one of the mass-flow hoppers prior to its installation in the silo. The increased live capacity of the silos was quite significant.

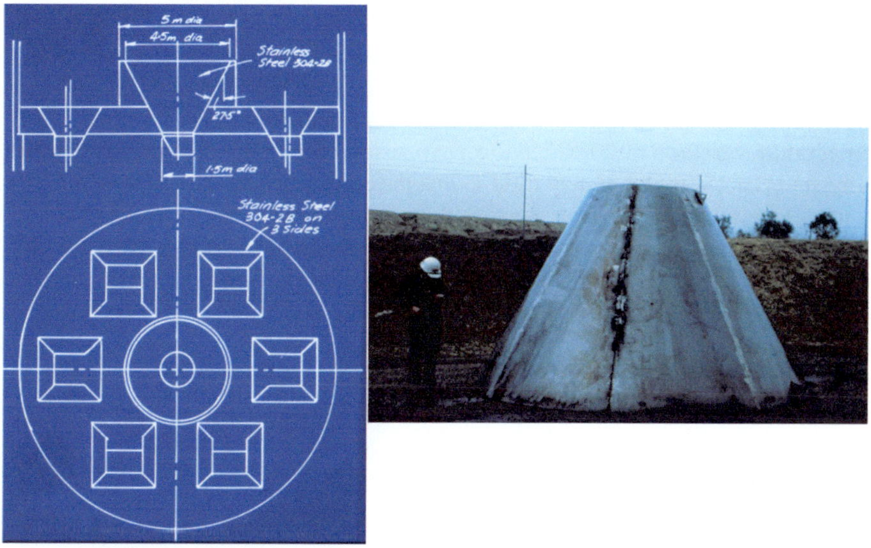

(a) Modifications made to each silo (b) Stainless steel mass-flow hopper

Fig. 6.27 Modifications to coal silos to improve live capacity

6.15 Influence of Temperature and Moisture on Bin and Silo Loads

There are additional factors, such as those linked to environmental conditions along with variations in the properties of the stored bulk solids, that have an influence on silo loadings and structural integrity. The following examples are deemed to be of particular importance:

6.15.1 Temperature Variations

The various methods used to analyse the wall pressures in steel silos generally assume the silo to be a rigid structure. This is not likely to be the case. Daily, weekly and monthly variations in atmospheric temperatures would cause the shell of a silo to expand and contract with some settling of stored material during expansion with little recovery, if any, during contraction. This can be a problem in the case of silos storing bulk materials over long periods of time without discharge.

Measurements of daily temperature fluctuations and pressures generated against the walls of a 15 m diameter steel storage bin in Queensland, Australia, were observed during on site tests. Pressure measurements indicated quite clearly a significant increase in pressures during the night period when temperatures were lower than during the day. In another case a large steel wheat silo collapsed in the early hours of the morning during the early winter period, the silo having been filled initially, during the summer. Apart from the daily temperature fluctuations, the overall changes from summer to winter would have a significant effect. For the silo in question, a daily temperature variation of 20 °C led to an expansion of 3.3 mm in 15 m. If, hypothetically, the wheat was completely rigid so that the contraction of the silo was prevented as the temperature dropped, an increase in wall pressure of 18 kPa would occur, this being 41% of the maximum static wall pressure. However, the elasticity of the wheat relative to the steel needs to be considered. The modulus of elasticity of wheat varies significantly from one wheat variety to another [22]. Based on the properties of individual grains, a typical elasticity modulus for wheat is $E_w = 2.07x10^9 kPa$ which when compared with the modulus for steel $E_s = 207x10^9 kPa$ gives

$$\frac{E_w}{E_s} = 0.01 = 1\% \tag{6.69}$$

While this is in itself insignificant, the cumulative effect of expansion and contraction over a period of days and weeks is not insignificant. For example, for the field study quoted, extrapolation of the data for a two-month period indicated a cumulative increase in pressure of 11 kPa or 25% of the maximum static wall pressure. A further matter requiring consideration is the orientation of the silo relative to its position with respect to the sun. It is possible that non uniform circumferential heating of the silo walls will occur leading to eccentric loading and the associated structural problems.

6.15.2 Moisture Variations in Grain

a. **Free Expansion Due to Moisture Increase**

Grain, such as wheat, when subjected to increases in moisture due to changes in humidity conditions will undergo an increase in volume. This is illustrated in Table 6.4 which shows the effects of wheat grains after prolonged exposure to high humidity conditions, the results being based on samples conditioned in a humidity cabinet. As indicated the moisture content increased from 9.54% to 20.07% after 4 days, the corresponding increase in volume due to swelling being 17.8%. For moisture changes of lower order, the degree of swelling is correspondingly lower.

b. **Constrained Expansion**

By way of example, in a grain storage bin or silo, the volume increases due to swelling of the grain is constrained by the rigidity of the bin. To gauge the possible volume expansion under such constrained conditions, tests were conducted on wheat samples contained in cylindrical cells with rigid bottoms and side walls. A loaded piston was located on top of each sample, the piston being free to move in the vertical direction.

Table 6.4 Free expansion of humidified wheat

Sample	Storage time (hrs)	Moisture content (% w.b.)		Volume Increase (%)
		Initial	Final	
1	1	6.32	7.38	0.62
2	4	4.91	8.65	4.74
3	5	10.09	13.83	6.05
4	24	11.16	16.58	14.52
5	29	9.51	19.26	16.4
6	95	9.54	20.07	17.79

The moisture content of the wheat in the various cells was varied and movement of the piston due to swelling of the wheat over varying periods of time was recorded. A typical set of results is shown in Fig. 6.28 which applied to a piston pressure of 28 kPa. The results showed that increasing the moisture content from 9.54% to 20.07% over four days caused the wheat to swell by 17.8% in volume. For smaller moisture increases, the degree of swelling was proportionally lower.

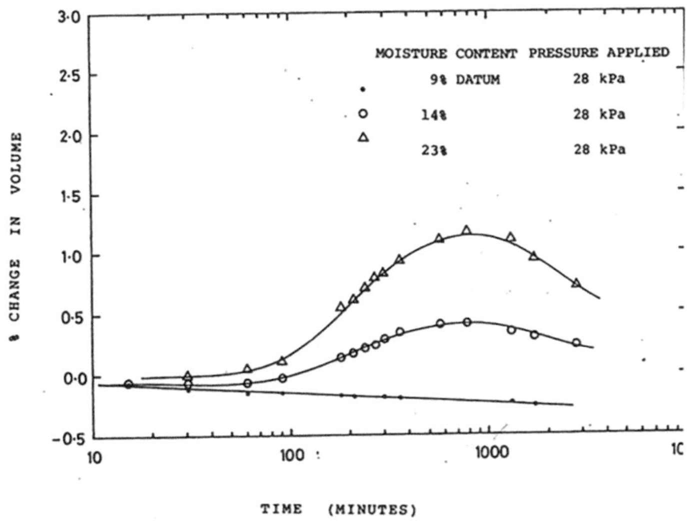

Fig. 6.28 Expansion of wheat due to swelling under constrained storage pressure of 28 kPa [22]

6.16 Case Study—Failure of Grain Silo

As an illustration of the influence of variations of ambient temperature and grain moisture, the case of the failed grain storage silo shown in Fig. 6.29a is reviewed. The silo, which was one of four similar silos located in southern NSW of Australia, was 15 m in diameter and 2 5m tall as illustrated in Fig. 6.29b. Two aeration ducts were incorporated in the silo base to allow air circulation of the stored grain for temperature control and associated conditioning of the stored grain. The aim was to prevent insect infestation during prolonged periods of undisturbed storage.

The silo, along with its neighbours, was constructed of steel plates, mainly by bolted connections plate to plate to form the cylindrical shape, and by the vertical rib members to provided the height support. In effect the silo had limited circumferential stiffnes as is evident from the failure by *splitting open* as depicted in Fig. 6.29a.

6.16.1 *Factors Contributing to the Failure—Ambient Temperature Variations*

Clearly the daily and monthly variations in the ambient temperatures and associated expansion and contractions of the silo walls led to fatigue degradation of the steel structure. The fact the collapse occurred in the early hours of the morning during

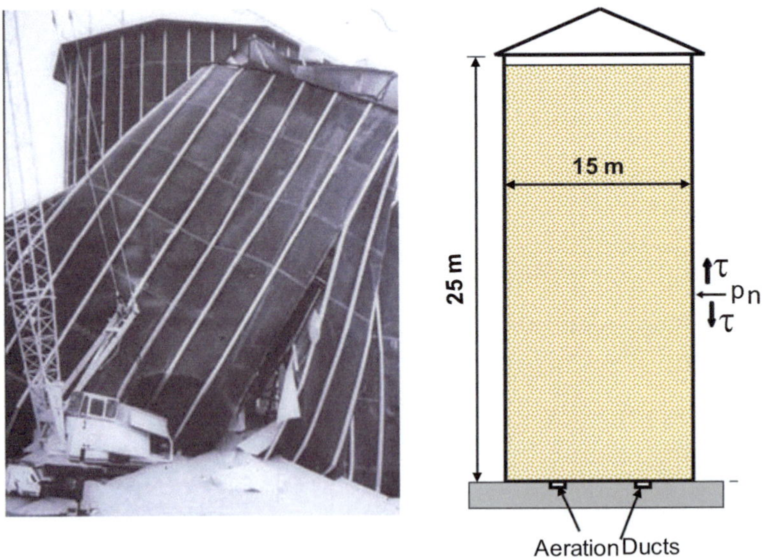

(a) Collapsed grain silo

(b) Silo geometry

Fig. 6.29 Failed grain silo

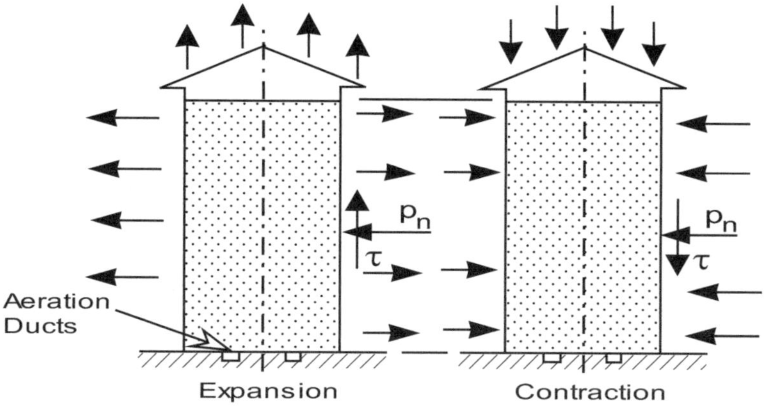

Fig. 6.30 Expansion and contraction of a silo

the onset of winter is an indication of fatigue failure. Furthermore, the fact that the silo was located on the end with limited protection provided by the neighbouring silos suggests uneven expansion and contraction of the silo walls due to the relative exposure to sun. This is a possible explanation as to why the failure occurred on the exposed side of the silo wall as shown in Fig. 6.29a.

In addition to lateral expansion, a bin or silo will expand upwardly in the longitudinal direction, the amount of longitudinal expansion, for most bin aspect ratios, being of the same order as the lateral expansion. This is illustrated in Fig. 6.30. When the bin cools, not only does it contract laterally, but it also contracts longitudinally. As the bin walls '*pull down*' due to this latter action, there is a strong likelihood that the shear stress at the wall due to wall friction will change direction from up to down. This reverse friction is the '*piston effect*' which can result in wall pressures greatly higher than the Janssen and hydrostatic pressures. Pressures due to reverse friction can significantly add to those induced by the lateral contraction of the silo.

It is also possible that the silo walls may have had a "*denting*" or "*creasing*" problem as discussed in Sect. 0. However, the overriding evidence pointed to an increase in moisture content of the stored grain resulting from an operational mishap which is explained in the following sections.

6.16.2 Variation in Grain Moisture

It was established that the automatic control system connected to the air circulating fans was not operating leaving the fans running during a period of high humidity, possibly during a period of rainfall. This led to an increase in the moisture content of the grain in the lower section of the silo to increase from 9% as stored to 13%. Laboratory triaxial tests performed on wheat showed that for increases of moisture content over 4 days from approximately 9% (w.b.) to 13% (w.b.) resulted in **nine**

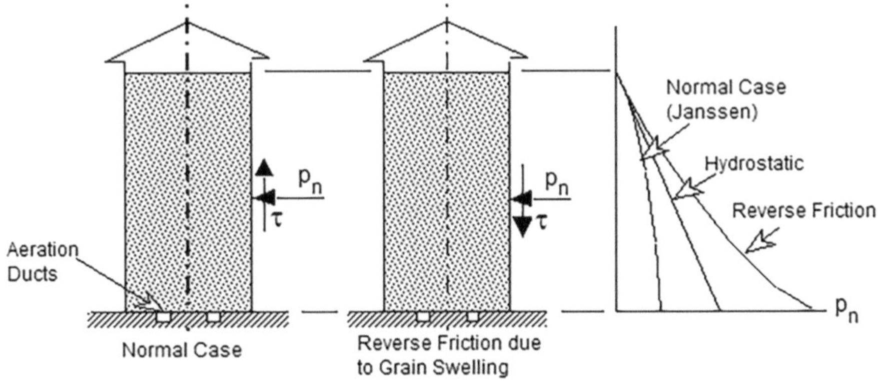

Fig. 6.31 Silo wall pressures under both normal and reverse friction conditions

times increase in the lateral pressure due to swelling of the wheat. In the case of the silo of Fig. 6.29a, the grain swelling resulted in reverse friction on the silo walls leading to a substantial increase in lateral pressures on the silo walls as illustrated in Fig. 6.31.

Noting that the moisture transmitted to the grain is via the aeration ducts, there is almost certainly a non-uniform moisture distribution throughout the stored grain mass. This may be explained by the so-called boundary layer effect governing the moisture distribution as illustrated in Fig. 6.32a and b. This will influence the pressure and load distribution on the silo walls. Denoting the boundary layer by the parameter "X", the predicted silo wall loads for the range of "X" values from 2 to 7 are illustrated in Fig. 6.32b. For no increase in moisture of the grain mass, the wall loads are predicted by Janssen's equation. As the moisture increases, the degree of grain swelling and associated reverse friction increases to the maximum level as shown in Fig. 6.32b.

6.17 Use of Anti-Dynamic Tremmie Tube to Control Silo Wall Pressures

The case of tall silos handling free flowing, cohesionless bulk solids, notably grain, under symmetrical funnel-flow, as discussed in Sect. 6.10, is considered. In most cases, the silos have flat bottoms with the outlet centrally located resulting in the effective transition to occur well down near the silo base. The discharge is nominally by mass-flow with flow along the walls over the substantial area of the silo height. Owing to the inevitable wall surface imperfections and convergences causing dynamic overpressures, the loads on the silo walls are significantly higher than those of the static filling case.

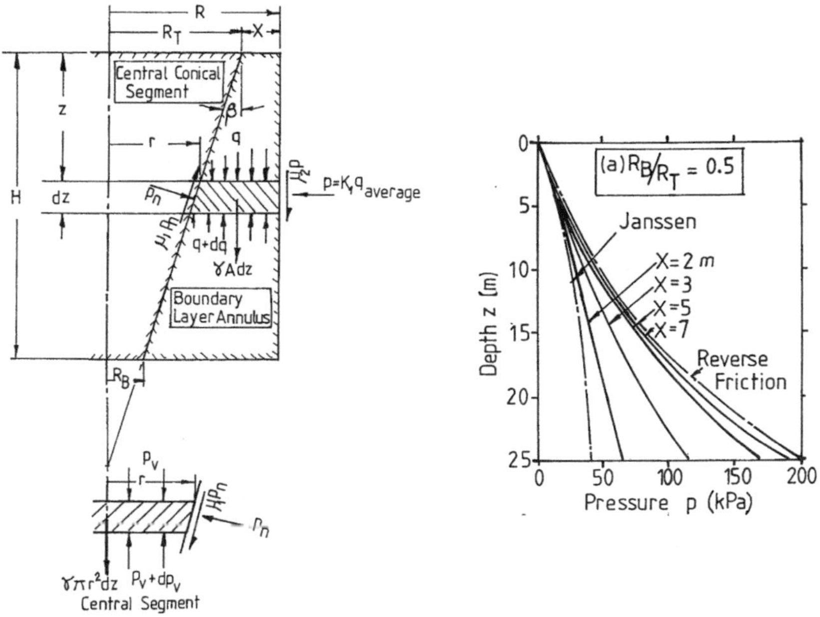

(a) **Boundary layer model** (b) **Normal pressure variations as function of X**

Fig. 6.32 Influence of boundary layer on silo wall pressure distributions [22]

6.17.1 The Anti-Dynamic Tube of Reimbert

As shown by Reimbert [23], it is possible, by the use of anti-dynamic tubes, to control the flow pattern so that funnel-flow always occurs without flow along the walls and dynamic overpressure. In this way, the wall pressures never exceed the static values. Reimbert's anti-dynamic tube, placed centrally in a symmetrical silo, extends almost the whole height and has a series of feed holes or ports to allow grain to enter the tube at various levels.

6.17.2 The Tremmie Tube

A variation of the Reimbert tube is the Tremmie tube which has no feed holes or ports in the walls and extends slightly less than half the height of the silo. Research using this type of anti-dynamic tube was conducted by Ooms and Roberts [24] at the University of Newcastle, Australia. The opportunity to study this problem arose from an industrial project concerning an installation of grain silos at a grain terminal facility. A tenth scale model, measuring 1.2 m diameter and 3.5 m tall, of the actual silo, constructed of *Perspex* is shown in Fig. 6.33.

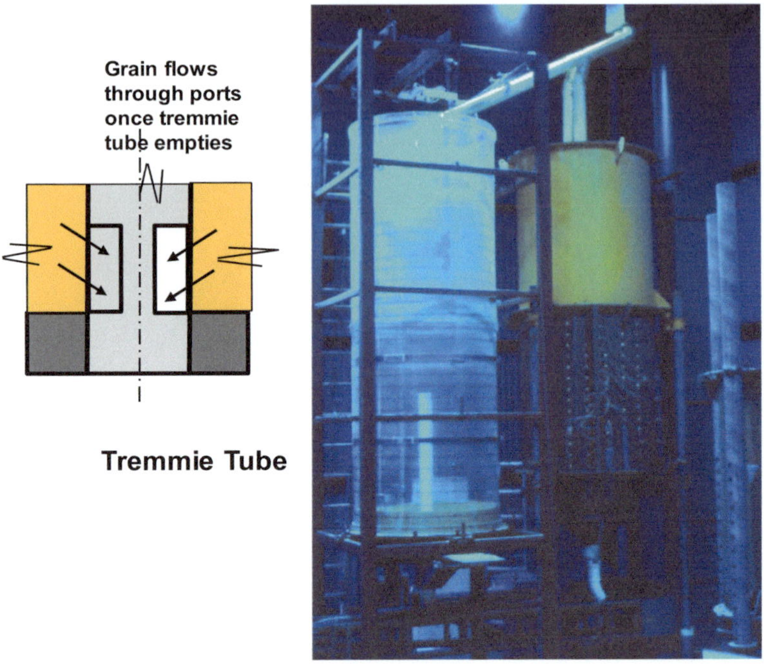

Fig. 6.33 Anti-dynamic Tremmie tube

The tube is self-choking, not requiring valves or gates to control the flow. Discharge flow of grain from the region outside the Tremmie tube does not occur until the tube empties. In effect, the Tremmie tube divides a tall funnel-flow silo into two squat silos in series. The arrangement ensures that at no time does the effective transition intersect the walls of the silo during discharge. Grain from the top half of the silo discharges first via the Tremmie tube to the silo outlet. Once the Tremmie tube empties, discharge from the lower part of the silo commences as grain is then directed to the silo outlet via lateral flow through the outlet ports as illustrated in Fig. 6.33. The inside diameter of the Tremmie tube is based on the required silo opening dimension. The tube height may be selected to be a little less than half the silo height. For the test silo, $H_T = 0.41$ H was chosen.

The overriding aim of the Tremmie tube system is to ensure that the normal pressure on the silo walls never exceeds the static or initial filling condition. This is illustrated by the set of results presented in Fig. 6.34 which compares the results with the tube fitted with those with no tube. As the results show, the silo with the Tremmie tube ensures that the normal pressures do not exceed the initial or static pressures. Without the tube, the normal pressures are orders of magnitude higher.

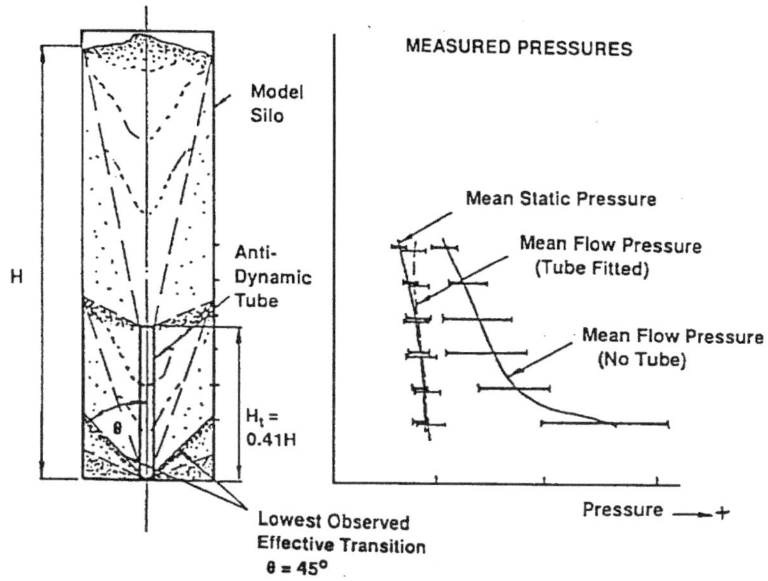

Fig. 6.34 Influence of anti-dynamic, tremmie tube in controlling silo wall pressures[24]

6.17.3 The Anti-Dynamic Tremmie Tube Applied to Eccentric Discharge

The anti-dynamic Tremmie tube described above has been successfully used to control the flow in silos having eccentric and multiple discharge points. Without the tube in place, the walls of such silos are subject to significant bending stresses in addition to the hoop stresses (Fig. 6.35).

6.17.4 Important Application Matters Requiring Consideration

While anti-dynamic Tremmie tubes clearly have a few advantages as demonstrated, their use in solving bin flow problems is limited. The following points need to be taken into consideration:

- They are limited to cohesionless, free flowing bulk solids such as grain
- The draw-down loads due to material flow are quite substantial causing high loads on the silo base which must be accounted for in the design
- It is essential that the tubes remain in the centre of the silo without lateral bending
- The bulk solid loads on structural members providing lateral support of the tubes must be assessed as part of the design process.

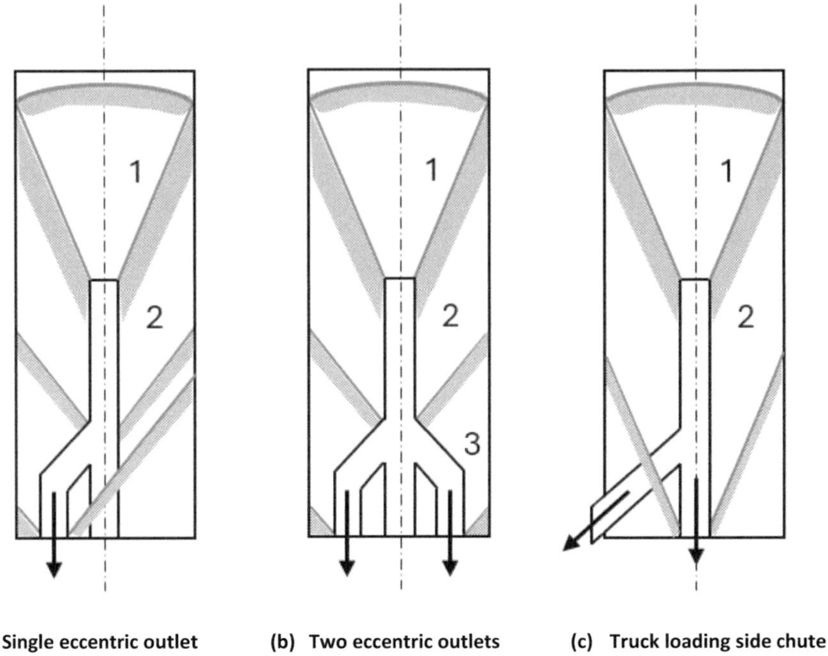

Fig. 6.35 Example of the tremmie tube to overcome eccentric discharge

6.18 Closing Comments

There is no doubt that the prediction of loads acting on bins and silos continue to be a subject of considerable complexity. As illustrated by the case study examples presented in this Chapter, the number of silo and bin failures that have occurred in past years has been largely due to the lack of understanding, at both the design and operational stages, of the influences of the loading and discharge flow patterns on the bin walls. Despite the widely varying approaches to the analysis of bin and silo wall loads, the loads are directly related to the flow pattern developed in the bin or silo. As is well known, in the case of mass-flow bins, the discharge flow patterns are generally well defined and the corresponding wall loads are predictable and reproducible. However, in funnel-flow bins the flow patterns are more difficult to ascertain, especially if the bins have multiple outlet points, the bin loading is not central, and the discharge is eccentric. Unless there are compelling reasons to do otherwise, bin shapes should be kept simple and symmetric.

However, this is not always achievable placing greater emphasis on the need for ongoing research linked with the application of simulation packages notably Discrete Element Modelling (DEM) coupled with, as required, Finite Element Analysis (FEA) and/or Computational Fluid Dynamics (CFD). While accepting the role of DEM in its capability to provide solutions to bin and silo design, it is important to understand that it is a *means to an end, but not an end in itself*. The fundamentals of bulk solids handling embraced by the *classical mechanics* approach must not be overlooked. This is the overriding philosophy of the subject of wall loads presented in this Chapter along with the content of all other chapters of the book.

References

1. Roberts, I., *Determination of the vertical and lateral pressures of granular substances. Part I - Wheat and peas.* Royal Society of London. Proceedings., 1884. **36**: p. 225–240.
2. Janssen, H.A., *The Measurement of Pressures in Grain Silos.* Z. Ver. Dtsch. Ing. Beih, 1895. **35**: p. 1045-1049.
3. Roberts, A.W. *100 Years of Janssen.* 1995.
4. Jamieson, J.A., *Grain Pressures in Deep Bins.* Trans. Can. Soc. Civ. Eng., 1903. **17**(554–654).
5. Jenike, A.W., *Load Assumptions and Distributions in Silo Design*, in *Conference on Construction of Concrete Silos.* 1977: Oslo, Norway.
6. Jenike, A.W., J.R. Johanson, and J.W. Carson, *Bin Loads - Part 3: Mass-Flow Bins.* Jnl. of Engng for Industry, Trans. A.S.M.E., Series B, 1973. **95**(1).
7. Jenike, A.W., J.R. Johanson, and J.W. Carson, *Bin Load - Part 4: Funnel-Flow Bins.* Jnl. of Engng for Industry, Trans. A.S.M.E., Series B, 1973. **95**(1).
8. Walker, D.M., *An Approximate Theory for Pressure and Arching in Hoppers.* Chem. Engng. Sci., 1966. **21**: p. 975.
9. Walters, J.K., *Theoretical Analysis of Stresses in Axially-symmetric Hoppers and Bunkers.* Chem. Eng. Science, 1973. **28**: p. 779.
10. Walters, J.K., *Theoretical Analysis of Stresses in Silos With Vertical Walls.* Chem. Eng. Science, 1973. **28**: p. 779.
11. Clague, K., *The Effects of Stresses in Bunkers.* 1973, Univ. of Nottingham.
12. Arnold, P.C. and A.W. Roberts, *A useful Procedure for Predicting Stresses at the Walls of Mass Flow Bins*, in *Symposium on "Solids Flow and Handling' AIChE, 80th National Meeting.* 1975: Boston.
13. AS3774–1990, A.S., *Loads on Bulk Solids Containers.* Standards Association of Australia, 1990.
14. EN 1091–4, *Eurocode 1 - Action on Structures. Part 4: Silos and tanks.* European Committee for Standardization, Central Secretariat, Brussels, Belgium, 2005.
15. Enstad, G., *On the theory of arching in mass flow hoppers.* Chemical Engineering Science, 1975. **30**(10): p. 1273-1283.
16. Arnold, P.C., A.G. McLean, and A.W. Roberts, *Bulk Solids: Storage, Flow and Handling.* 2nd ed. 1979: The University of Newcastle Research Associates (TUNRA).
17. Roberts, A.W., *Basic principles of bulk solids storage, flow and handling.* 1993: Institute for Bulk Materials Handling Research.
18. Roberts, A., *Review of mass-flow hopper design with respect to stress fields and surcharge loads.* Particuology, 2010. **8**: p. 591-594.
19. Roberts, A.W. and M. Ooms, *Wall Loads in Large Steel and Concrete Bins and Silos due to Eccentric Draw-down and other Factors*, in *Proc. Int. Conf. Design of Silos for Strength and Flow.* 1983: Stratford-Upton-Avon, U.K.
20. Ooms, M. and A.W. Roberts, *An Overview of Silo Wall Loads with Special Emphasis on the Flow Pattern and Wall Pressure Interrelationship*, in *Proc. Int. Conference on Particulate and Multi-Phase Processes.* 1985: Miami Beach, U.S.A.
21. Ooi, J. and J. Rotter, *Elastic and plastic predictions of the storing pressures in conical hoppers.* Transactions of the Institution of Engineers, Australia. Mechanical engineering, 1989. **14**(3): p. 165–169.

22. Roberts, A.W. *Some aspects of grain silo wall pressure research: influence of moisture content on loads generated and control of pressures in tall multi-outlet silos.* in *Proceedings of the Powder and Bulk Solids Conference, Rosemont, Illinois.* 1988.
23. Reimbert, M.L. and A.M. Reimbert, *Silos: Theory and Practice, by Marcel L. Reimbert & Andre M. Reimbert*. 1976: Trans Tech Publications.
24. Ooms, M. and A. Roberts, *The reduction and control of flow pressures in cracked grain silos.* Bulk Solids Handling, 1985. **5**(5): p. 1009-1016.

Chapter 7
'Silo Quaking' Dynamic Characteristics in Bins and Silos

7.1 Introductory Remarks

A recurring problem in the operation of bins and silos is dynamic loads that arise during gravity discharge of the stored bulk material. The discharge flow is characteristically, a cyclic or pulsating type flow phenomenon which is described by the title of *silo quaking*. The pulsating flow generally arises as a result of changes in density during flow and by varying degrees of mobilisation of the internal friction and boundary wall friction and associated *slip stick* characteristics. Silo quaking is known to occur in bins and silos of widely varying geometries including tall and squat mass-flow and funnel-flow bins, as well as expanded-flow and multi-outlet bins. The pulsating load problem is more pronounced at low flow rates where the period of pulsing may be from a few seconds to many seconds or even minutes. The consequence of the pulsating loads may range from nuisance values arising from the transmission of shock waves through the ground to disturb neighbouring areas, to structural fatigue failure when the natural frequencies of the silo and structure are excited by the flow pulses. The combination of cyclical shock loads associated with silo quaking, together with the static and flow loads, have an important influence on the fatigue life of the bin or silo [1].

The subject of pulsating flow of bulk solids in bins and silos has been studied by several authors, a selection of papers on this subject being given in References [2–19]. The material presented in this chapter is mainly based on the research and industrial case studies performed by Roberts et al. [9–12]. A variation of the silo quaking problem is *silo music* and *silo honking* as has been reported by Tejchman and Gudehus [13, 14]. The *honking* phenomenon is known to occur in tall, aluminium silos which store plastic powders. In this case the higher frequency components of the flow pulsations can give rise to loud, periodic, *fog horn* type sounds that have a distinct nuisance effect.

The discussion that follows has two main objectives. Firstly, to review relevant research into the *silo quaking* problem as it occurs in mass-flow, funnel-flow,

A. W. Roberts, *The Theory and Practice of Bulk Materials Handling*,
https://doi.org/10.1007/978-3-032-19702-3_7

expanded-flow, intermediate flow and multi-outlet bins. Secondly, to illustrate various practical manifestations of silo quaking problems by reference to selected case study examples.

7.2 Tall Mass-Flow Silos

Silo quaking is known to occur in tall mass-flow silos in which the height of fill is above the critical height H_{cr} measured above the hopper and cylinder transition, where $H_{cr} \approx 0.8D$ to $1.0D$ for which D is the cylinder diameter. The problem drew the attention of Roberts et al. [9–12] who conducted a program of research involving a pilot scale study of this dynamic load problem. An overview of this research is now presented.

7.2.1 Pilot Scale Mass-Flow Silo Research

The characteristics of the quaking are demonstrated in tests performed on the bin shown in Fig. 7.1 a and b. The cylindrical section of the test bin is constructed of mild steel, the dimensions being 1.2 m diameter and height 3.5 m. The bin is fitted with a mass-flow hopper constructed of stainless steel. Fourteen load cells, which measured normal and shear stress simultaneously, were fitted into the bin walls. To preserve the smooth, internal concave surface of the silo wall, 14 sections of the wall were removed to allow the specially designed load cells to be manufactured and then refitted. The cells were designed by Professor V. Askegaarde of the Technical University of Denmark.

A selection of test results using wheat is presented, the focus being on the loads acting in the cylinder. Referring to Fig. 7.1b, the load cell locations at which the measurements were taken, along with the three head heights of wheat in the cylinder are provided in Table 7.1.

7.2.1.1 Filling of Bin from Empty Condition and Load Settlement

As the bin is filled from the empty condition, normal pressures and shear stresses were recorded as the fill level became in contact with each load cell in turn. A typical set of results for load cell 14 corresponding to the fill ratio $H/D = 2.9$ is shown in Fig. 7.2. A dynamic *stick–slip*, load pulse effect is depicted in both normal pressure and shear records during filling and continues after the bin is filled. The load pulses become more pronounced in the case of the shear records as illustrated.

It is evident that during undisturbed storage, the stored mass approaches, asymptotically, its critical state consolidation condition by a pulse-like settling action. The phenomena depicted in Fig. 7.2 was observed at all locations, with pulses at each

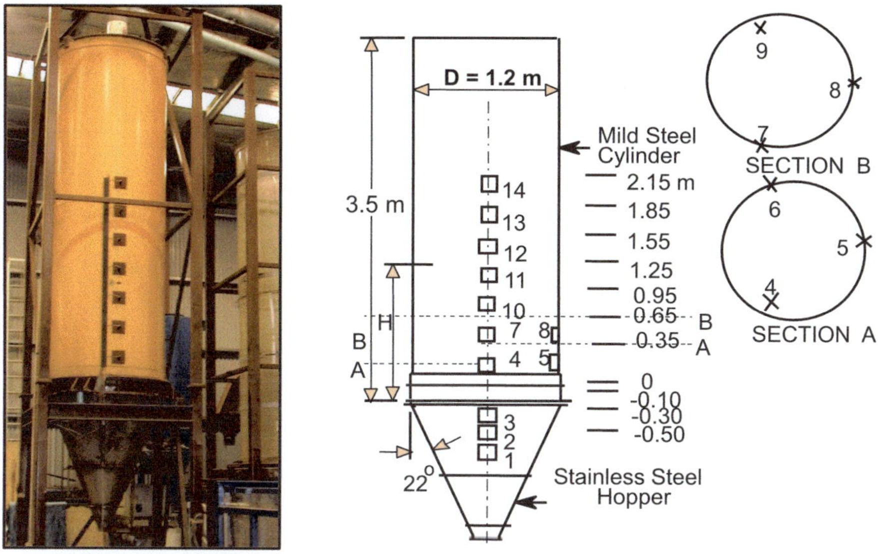

(a) Photograph of test bin

(b) Test bin showing load cell locations

Fig. 7.1 Mass-flow test bin

Table 7.1 Location of measurements and head height of wheat in cylinder

Load cells	5	7	10	12	13	14
Head H (m)	3.44		2.41		1.1	
H/D	2.9		2.0		0.9	

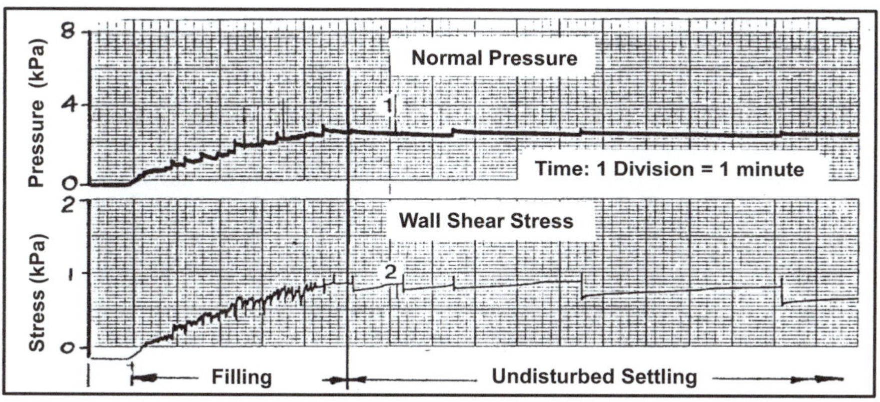

Fig. 7.2 Bin filling and undisturbed settling

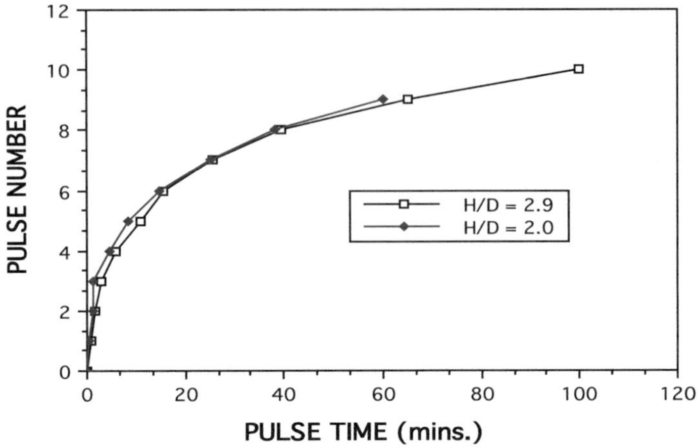

Fig. 7.3 Undisturbed settling pulse motion

location occurring at the same time intervals. Records of the undisturbed settling were monitored over long periods of time. The same effect was observed for the three H/D ratios examined. As the settling time increased, the pulse period also increased in an exponential manner as illustrated in Fig. 7.3. This also shows that the settling characteristics for $H/D = 2.9$ and $H/D = 2.0$ are consistent.

7.2.1.2 Flow Patterns and Wall Pressures During Bin Discharge

Figure 7.4 shows, schematically, the velocity profiles during the bin discharge. The flow convergence in the hopper is quite pronounced in the centre of the hopper as shown. The commencement of the flow convergence begins to occur in the cylindrical section at the critical height H_{cr} above the hopper and cylinder transition. Above the critical height, plug type flow occurs with the velocity profile substantially uniform over the cross-section. Below the critical level in the region of the transition, the flow starts to converge due to the influence of the hopper, and the velocity profile is no longer uniform.

The velocity profile is further developed in the hopper as shown. As the flow pressures form in the hopper, dilation of the bulk solid occurs. As a result of this dilation, it is possible that the vertical supporting pressures decrease slightly thus reducing the support given to the plug of bulk solid in the cylinder. This causes the plug to drop momentarily giving rise to a load pulse. The cycle is then repeated.

Examples of the wall pressure and shear stress records at locations 5 and 14 in the cylinder for the fill ratio $H/D = 2.9$ are shown in Fig. 7.5. As shown, the pulsing was quite pronounced at location 14 well up the cylinder, the pulse period averaging approximately 5 secs. The pulsing amplitude decreased down the cylinder

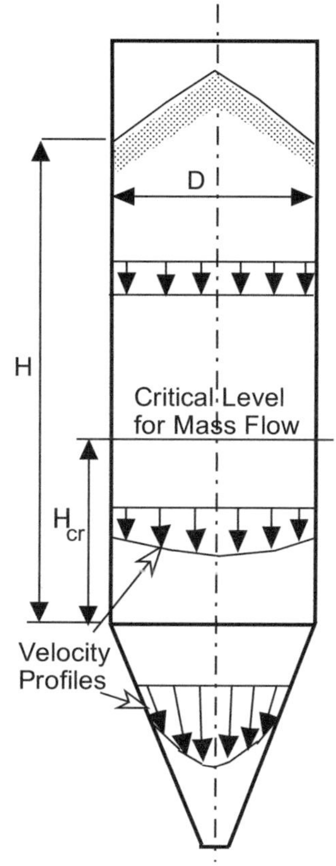

Fig. 7.4 Velocity profiles

with virtually no pulsing being evident at location 5 near the transition. These results support the conclusion that silo quaking in mass-flow is predominately a plug-flow phenomenon.

7.2.1.3 Effect of Surcharge Head

Similar dynamic effects were experienced for the case of $H/D = 2.0$, but when the head was reduced to $H/D = 0.9$ there was no evidence of any pulsations during discharge. This confirms the conclusion that self-excited pulsating flows in mass-flow bins, which give rise to the *silo quaking* phenomenon, are associated with surcharge heads greater than a minimum value of nominally $H/D \approx 1.0$.

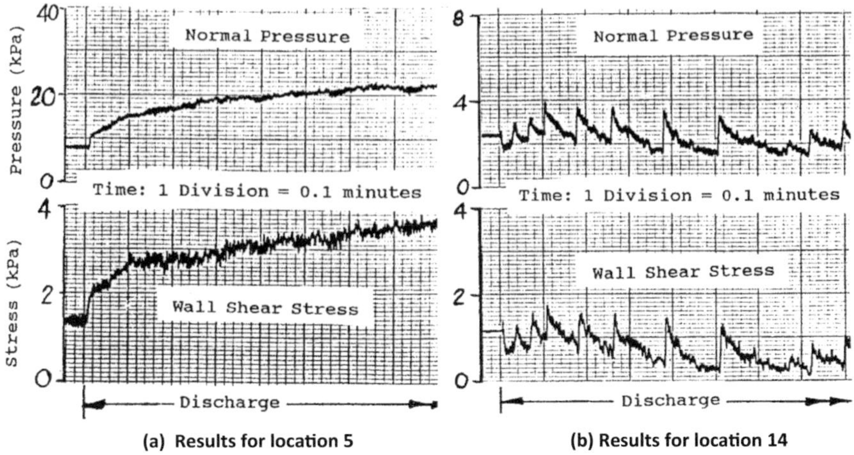

Fig. 7.5 Pressure and shear stresses during emptying (H/D = 2.9)

7.2.1.4 Wall Friction

The normal and shear stress records permit the wall friction to be determined. Values obtained from the test results ranged from 12 to 14° which are significantly lower than the friction angles of 18 degrees measured by the Jenike shear tester. The latter value represents static friction whereas the values determined from the silo tests represent kinetic friction. It is also possible that some rolling as well as sliding of the granular particles may have occurred which could account for the lower angles of kinetic friction. These results highlight the importance of accurately determining wall friction.

The subject of wall friction is discussed in [16] where a linear action friction and wear testing machine is described. In this apparatus the rubbing speed between the bulk solid and the sample lining surface can be varied to simulate more closely the conditions pertaining to hoppers. Hence kinetic friction is measured as distinct from static or limiting friction as determined by the Jenike shear tester.

7.3 Dynamic Loads in Tall, Cylindrical Mass-Flow and Funnel-Flow Silos

The findings of the foregoing study which was based on the tall mass-flow pilot scale bin, is also applicable to tall funnel-flow bins and silos. While it was ecognized by Roberts [12] that multiple shock planes may exist, the assumption of an averaged single shock plane model has been shown to provide a good practical predictor of loads in industrial silos experiencing quaking problems.

7.3.1 Silo Discharge Flow Model

The system model upon which the analysis is based is shown in Fig. 7.6a and b. Referring to Fig. 7.6a, the case of a single pulse occurring on a plane located at height H_{sh} above the cylinder and hopper transition is considered. Above this plane, the bulk solid is considered to behave as a plug which imparts step loads due to slip-stick motion which may arise from variations in bulk density due to dilation, together with variations in internal and wall or boundary friction as the friction changes from limiting or static to kinetic values.

After each shock occurs, the dynamic wave will decay with time according to the bulk elastic properties and damping characteristics of the consolidated bulk solid. It is assumed that the amplitude of the shock pressure increase, ΔP_{wy} normal to the wall decays exponentially with distance measured on each side of the shock plane. Since the discharge flow of bulk solids normally involves several shock planes at various heights in the bin, the locus of the peak pressure amplitude is assumed to be bounded by an increasing exponential curve as illustrated in Fig. 7.6a.

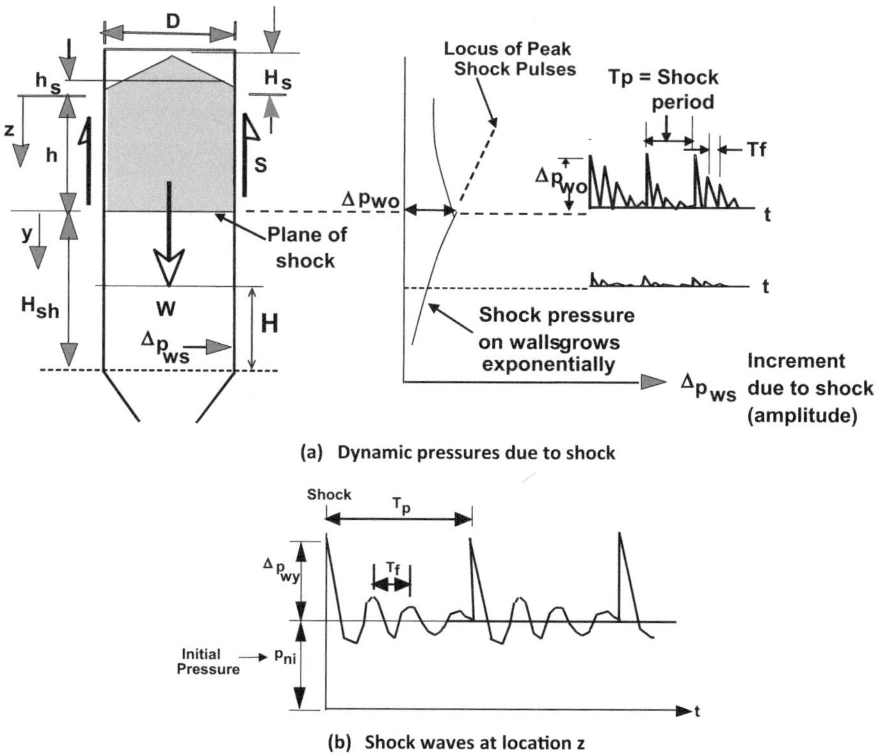

Fig. 7.6 Dynamic pressures due to shock

As illustrated in Fig. 7.6b, silo quaking is a dominant low frequency, dynamic load effect with pulses occurring at extended time periods T_p. The higher frequency components, illustrated by T_f, are of lower magnitude and would have a lesser influence on the quaking phenomena.

7.3.2 Dynamic Loads Resulting from Quaking

The maximum increment in the average vertical pressure, ΔP_{vo}, due to the shock load is

$$\Delta P_{vo} = k_d \gamma \left[\left(1 - e^{-\mu K h/R}\right) \left(\frac{R}{\mu K} - h_s \right) + h_s \right] \tag{7.1}$$

The corresponding maximum increment in lateral wall pressure due to shock load is

$$\Delta P_{wo} = K \Delta P_{vo} \tag{7.2}$$

where the effective radius is given by Eq. (7.3).

$$R = \frac{D}{2(1+m)} \tag{7.3}$$

and effective surcharge head is calculated using Eq. (7.4).

$$h_s = \frac{H_s}{m+2} \tag{7.4}$$

The above variables are defined as follows:

$\gamma = \rho g =$	bulk specific weight.
$\rho =$	bulk density.
$h =$	head.
$K =$	ratio of lateral to vertical pressure.
$H_s =$	actual surcharge head.
$D =$	diameter or width of flow channel.
$m = 0$	for rectangular silo.
$m = 1$	for circular silo.
$\mu = \tan\varnothing =$	coefficient of wall friction.
$\varnothing =$	wall friction angle.
$k_d =$	dynamic load factor.

Based on the results of the pilot scale tests, an exponential distribution for the stress or pressure amplitude is assumed. The variation in the maximum amplitude of the pulse pressure ΔP_{wy} on each side of the shock surface is given by Eq. (7.5).

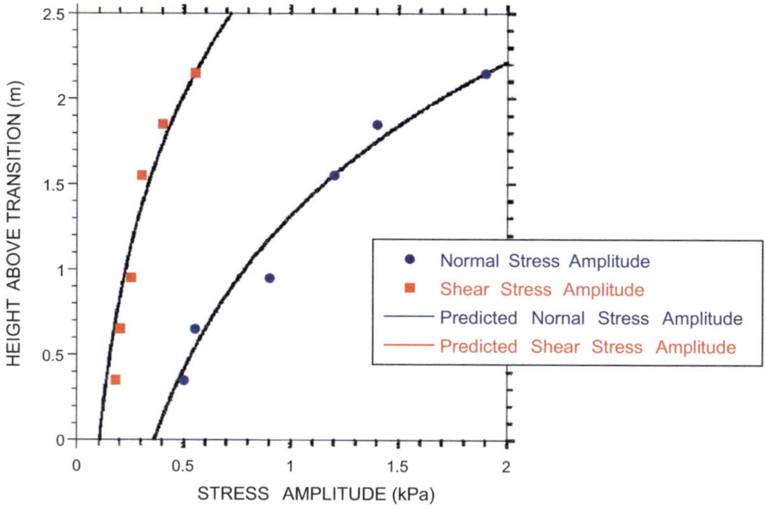

Fig. 7.7 Predicted and measured normal and shear shock load amplitudes for pilot scale silo

$$\Delta \mathrm{P}_{wy} = \Delta P_{wo} e^{-\mu K y/R} \tag{7.5}$$

As y →h, it is assumed that $\Delta \mathrm{P}_{wy}$ *tails off* to approach the flow pressure at the top surface in the cylinder. It is noted that $\Delta \mathrm{P}_{wy}$ is the additional wall pressure applied to the initial or static pressure.

If the shock load is suddenly applied at each pulse period, then $k_d = 1.0$. Under more difficult conditions, such as in the case of cohesive bulk materials when there is a delay for the pulse to be initiated, $k_d > 1$.

7.3.2.1 Example: Comparison of Predicted and Measured Wall Pressure Amplitude

Figure 7.7 shows the measured values (plotted points) and calculated (full lines) wall pressure amplitudes for the wheat silo of Fig. 7.1 a. The calculated values are based on Eqs. (7.1) to (7.3) using the following values: $H_{sh} = 2.14$ m (location 14 of Fig. 7.1a), $\varnothing = 30°$, $\rho = 0.85t/m^3$, $D = 1.2$ m, overall fill height above transition, $H_{sh} + h = 3.0$ m, $k = 0.4$, and $k_r = 0.29$ = ratio of shear to normal stress amplitude.

7.3.3 Pulse Period

Pulsating flow has been studied by several authors [2–6]. For instance, Shinohara et al. [2] recognised the existence of a dynamic arch and studied the flow of both sand and glass beads from a PVC bin of $H/D = 3$. It was noted that the location of the arch was higher for glass beads than for sand, this being due to the difference in the void fraction of the particle. The authors derived an expression to predict the frequencies of the pulsating flows. Firewicz [6] proposed an improved solution to that presented by Shinohara et al.

However, the work of Shinohara et al. and Firewicz refers to the flow from the hopper and the influence of this pulsating flow on the behaviour in the main body of the bin is difficult to predict. The problem is made more complicated by the variability in bin shapes, variability in the properties of the bulk solid, particularly when mined products such as coal are handled where there are a wide range of particle sizes, moisture contents and impurities such as clay.

The theory of Roberts, [10], has been shown to be a good predictor of the pulse frequency and period arising from practical industrial cases of silo quaking. Roberts' equation for the determination of pulse period is as follows:

$$T = \left(t_0 + \frac{v}{a}\right) + \sqrt{\frac{v}{a}\left(2t_0 + \frac{v}{a}\right)} \tag{7.6}$$

where $a = \frac{\Delta P_{vo}}{\rho(h+h_s)}$, $v = \frac{Q}{\rho A}$ and $t_0 = \frac{\Delta \varepsilon_y}{v}$

$a =$ acceleration of upper mass during pulse motion
$v =$ average velocity of the bulk solid in the cylinder during discharge.
$Q =$ discharge rate.
$t_0 =$ time for motion of upper mass to be initiated.
$\Delta\varepsilon_y =$ dynamic displacement of consolidated mass in vertical direction.

In many cases the average velocity is quite low. For example, consider the discharge of a bulk solid of bulk density $\rho = 1$ t/m^3 from a tall, 10 m diameter cylindrical silo fitted with a conical hopper. The discharge rate is 100 t/h. The average velocity in the cylindrical section of the silo can be calculated as $v = 0.0004$ m/s $= 0.40$ mm/s. It is also necessary to gain some appreciation of the magnitude of the accelerations occurring during pulsating flow. In general, the pulse accelerations lie in the range $0.1\ \text{g} \le a \le \text{g}$. It follows that for most practical cases, the ratio $\frac{v}{a} \approx 0$. Hence $T = t_0$. That is,

$$T = \frac{\Delta\varepsilon_y}{v} \tag{7.7}$$

The parameter $\Delta\varepsilon_y$ is dependent on the average particle size, void ratio and properties of the boundary surface of the flow channel. A plot of pulse period versus average discharge velocity in the cylindrical section of a silo is presented in Fig. 7.8.

Roberts [11] also conducted wall friction tests for sand on Perspex using the Jenike shear tester for a range of shear strain rates from 0.004 to 0.085 mm/s. Stick–slip was quite pronounced with the period decreasing with both increase in strain rate or shear velocity and decrease in normal stress. The relationship of $T = \frac{De_y}{v}$ was confirmed. It was also shown that

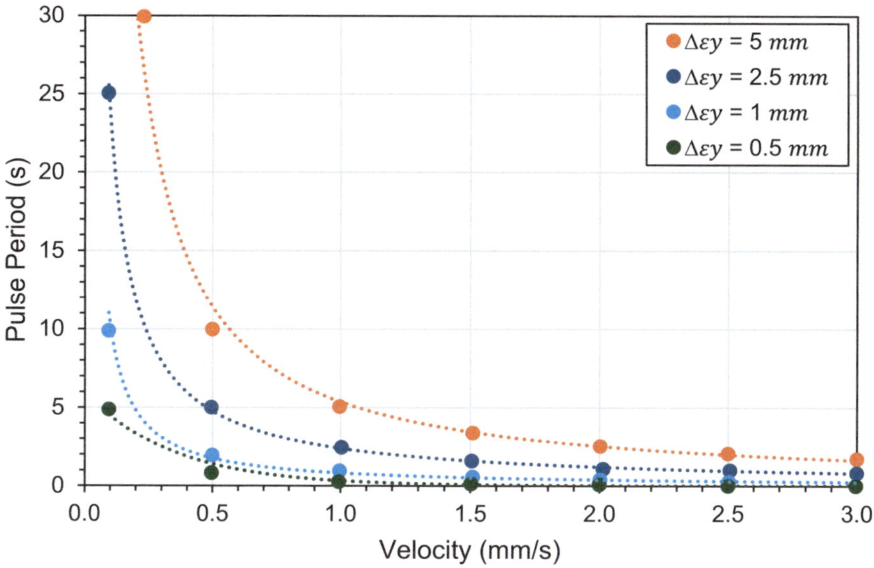

Fig. 7.8 Pulse period versus velocity

$$\Delta\varepsilon_y = \frac{\sigma_n}{K_T} \tag{7.8}$$

where K_T = contact stiffness $\left(\frac{kPa}{mm}\right)$ and σ_n = normal stress (kPa). The value of K_T = 333 for sand on perspex, but it is believed that this value was influenced, to some extent, by the stiffness of the load cell of the shear tester.

7.4 Stress Wave Propagation in Tall Silos

7.4.1 The Contribution of Wensrich

Following on the earlier work of Roberts, a comprehensive fundamental study of the silo quaking problem in tall mass-flow silos was performed by Wensrich [12, 15, 16]. He examined stick–slip motion of bulk solids discharging by gravity and used continuum models employing hypo-plasticity theory to analyse the wave motion. His models clearly show the exponential growth of various wave amplitudes within the silo.

To confirm the analytical predictions, Wensrich conducted experimental studies using a small scale *silo* simulation model. The model, together with typical test results, are shown in Fig. 7.9. The model comprised a Perspex tube, 90 mm diameter and approximately 1 m long. A plunger or piston was used to control the downward

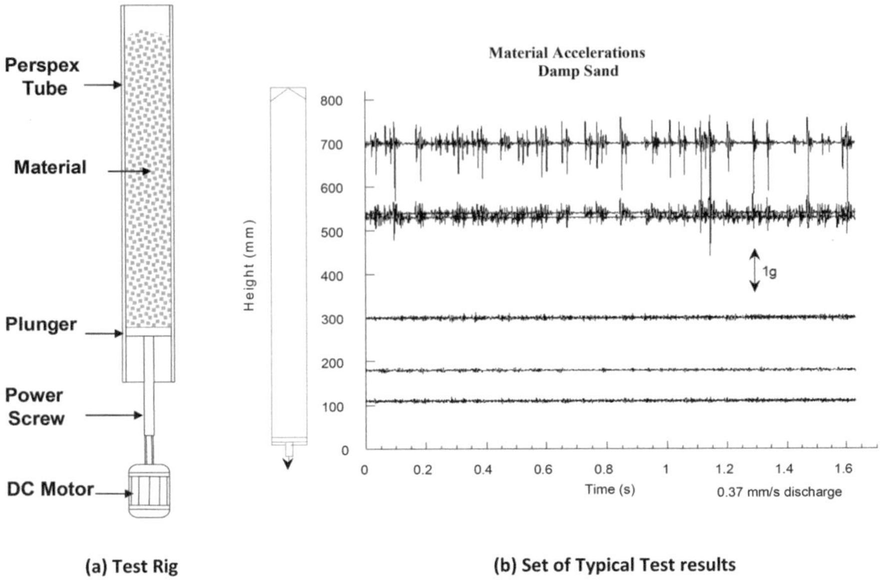

Fig. 7.9 Experiments performed by Wensrich [15]

movement of the bulk material. The piston is retracted at pre-determined velocities by means of the DC motor connected to the piston with a screw. Sand at approximately 0.5% moisture content was chosen as the bulk material. Tiny accelerometers were employed to measure the induced accelerations in the flowing sand.

By way of example, Fig. 7.9 b presents the accelerations within the material at various heights for the discharge velocity of 0.37 mm/s. The results clearly show that the amplitudes of the accelerations increase exponentially with height up the tube owing to the stick–slip effect. Towards the bottom of the tube the accelerations are practically zero. The results confirm the predictions of Roberts [10] as summarised in Sect. 7.3.

7.4.2 Stress Wave Attenuation

To attenuate the growth in shock wave amplitude up the silo, Roberts and Wensrich [19] conducted experiments using a hopper insert, as depicted in Fig. 7.10a. In effect, the insert divides a tall mass-flow silo into two squat mass-flow silos. As the results of Fig. 7.10b show, the acceleration amplitude above the insert is reduced by approximately 50%.

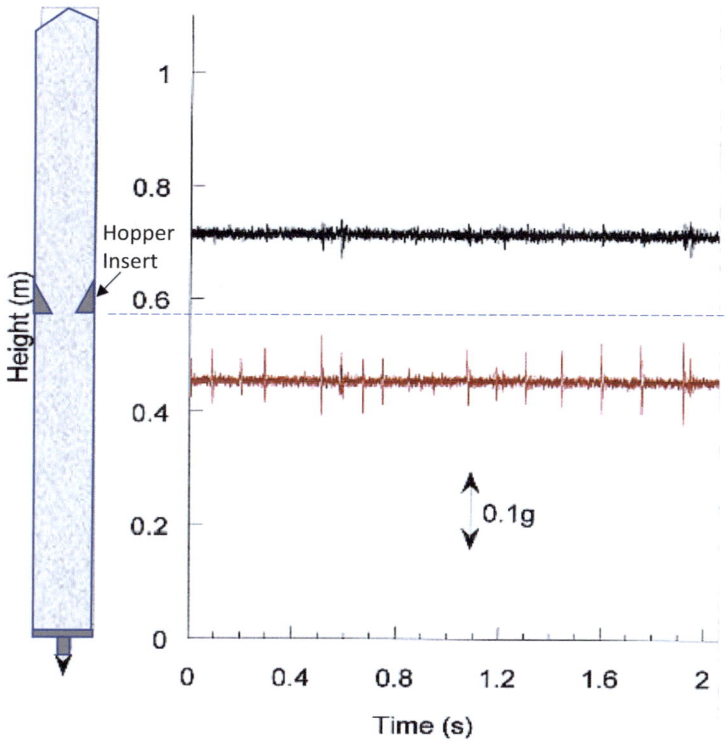

(a) Measured accelerations above and below inserts

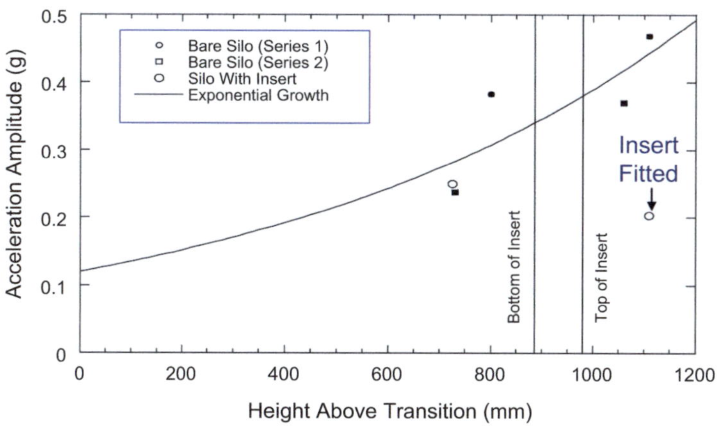

(b) Amplitude of acceleration above and below inserts

Fig. 7.10 Influence of hopper insert in reducing acceleration during discharge

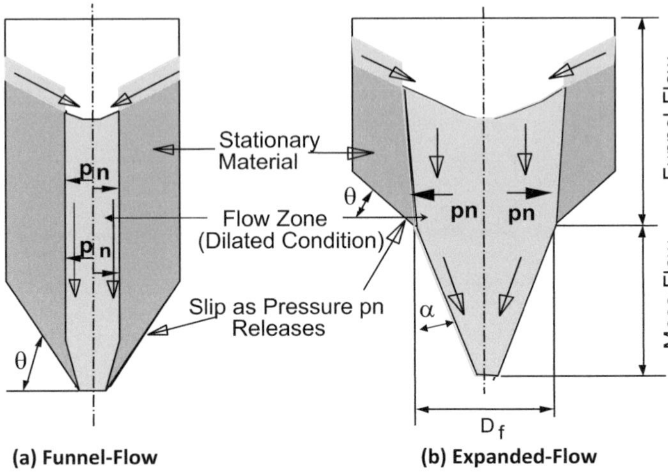

Fig. 7.11 Funnel-Flow and expanded-flow

7.5 Mass and Funnel Flow Bins of Squat Geometry

7.5.1 Funnel and Expanded-Flow Bins

Pulsating flow and dynamic loads may occur in mass-flow, funnel-flow and expanded-flow bins of squat proportions as illustrated in Fig. 7.11. In such cases, it is the hopper that influences the pulsating load. During funnel-flow with no flow along the walls, as depicted in Fig. 7.11a, dilation of the bulk solid occurs as it expands in the flow channel. As a result, some reduction in the radial support given to the stationary material may occur. If the hopper is steeply sloped, say [$\theta \geq \delta$], then the stationary mass may slip momentarily causing the pressure in the flow channel to increase because of the *squeezing* action. The cycle then repeats. A similar behaviour may occur in expanded flow bins, such as the bin depicted in Fig. 7.11b.

7.5.2 Intermediate-Flow

Benink [17] identified an additional flow regime, namely, intermediate-flow, which occurs between mass-flow and funnel-flow. Intermediate flow is illustrated in Fig. 7.12. Whereas the radial stress theory of Jenike [18] ignores the surcharge head, Benink has shown that the surcharge head has a significant influence on the flow pattern generated.

For mass-flow to occur in the hopper, the surcharge head of material in the cylinder (Fig. 7.11) must be greater than the critical head H_{cr}. For squat bins in which $H < H_{cr}$ and for which the hoppers are close to the borderline between mass and funnel-flow,

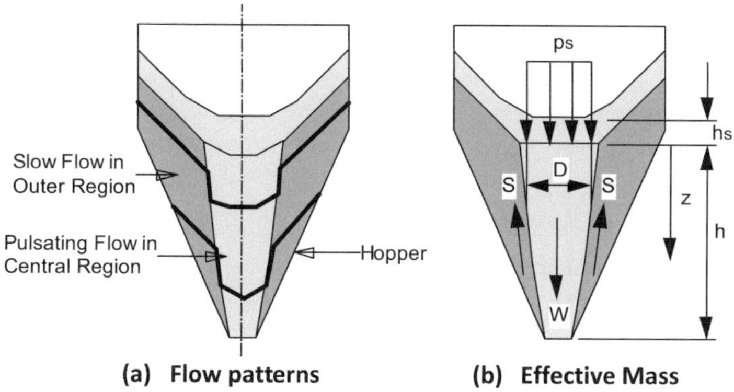

Fig. 7.12 Intermediate-flow illustrating pulsating flow regime

discharge will be characterised by intermediate-flow. Such flow is defined by rapid flow in the central region of the hopper, and a slower flow in the outer regions as illustrated in Fig. 7.12a. Funnel-flow is really the special case of intermediate-flow in which the outer region is stationary.

Based on observed flow patterns in model and full-size bins, an estimate of the dynamic loads may be made. The effective dynamic mass of bulk solids in the central flow channel of the hopper section is the total mass in this section less the frictional support due to shear at the boundaries between the central and outer flow channels. This is depicted in Fig. 7.12b.

Here the analysis is similar to that presented in Sect. 7.2 for tall bins, except in this case the flow pulsations are generated in the lower hopper section. The pulsating mass is as indicated in Fig. 7.12b with the substitution of $D_f = D =$ width or diameter of flow channel, and $\mu =$ coefficient of internal friction. Normally

$$\tan \varnothing_t \geq \mu \geq \sin\delta \tag{7.9}$$

where

$\varnothing_t =$ static angle of internal friction
$\delta =$ effective angle of internal friction defining upper bound of shear stress.

7.6 Tall Coal Bins with Chisel Hoppers—Case Study

An installation comprising three reinforced concrete coal silos experienced severe silo quaking problems. Each silo, illustrated schematically in Fig. 7.13, stands 58 m above ground level, with a height above the base of 52 m, an internal diameter of 21.4 m, and a wall thickness of 0.3 m. Adjacent silos are interconnected by concrete along their vertical lines of intersection.

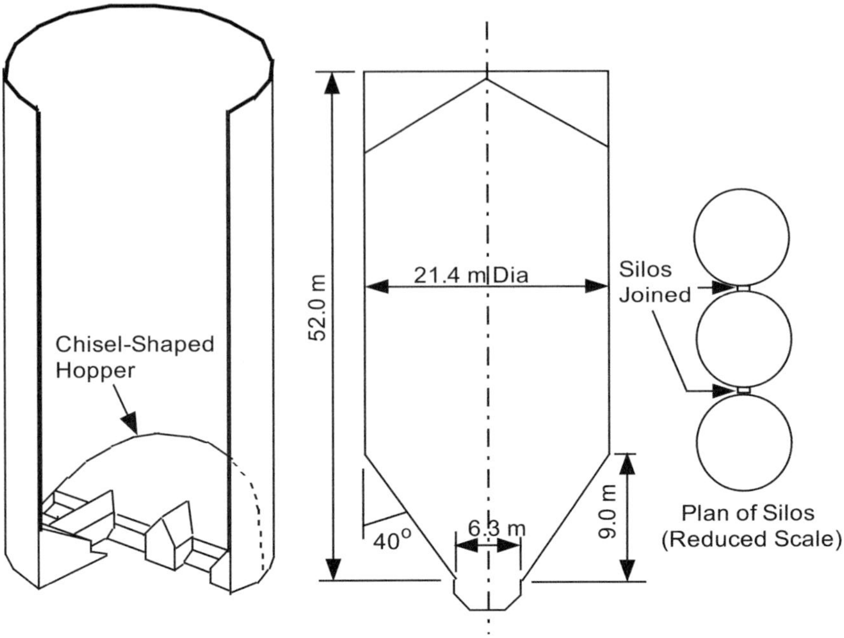

Fig. 7.13 Tall coal silos

The silos are equipped with chisel-shaped, plane-flow hoppers lined with carbon steel plate. The hopper half-angle, α, is 40° relative to the vertical. Each hopper divides into three outlets, with the lower sections lined in stainless steel. The maximum discharge rate is 2,700 tonnes per hour, achieved through vibratory feeders.

During operation, severe shock loading was reported, with a shock period of 3 to 5 s, most pronounced when the silos were substantially full. After several years of service, significant structural damage occurred, including sections of concrete being dislodged.

7.6.1 Computation of Shock Loads

Based on the silo diameter of 21.4 m, the shock plane is assumed to be located at a height of 26 m above the base of the hopper. The head of coal above the shock plane is $h = 22$ m. Using the procedures presented in Sect. 7.2, the estimated amplitude of the dynamic pressure is $\Delta Pwo = 56kPa$.

Equation (7.3) has been used to compute the locus of the shock pressure amplitude, the results being illustrated in Fig. 7.14. Also shown are the static and flow pressures. In view of the curved transition of the chisel-shaped hopper with the cylinder, the pressure profiles will vary around the periphery of the bin. Figure 7.14 applies to the

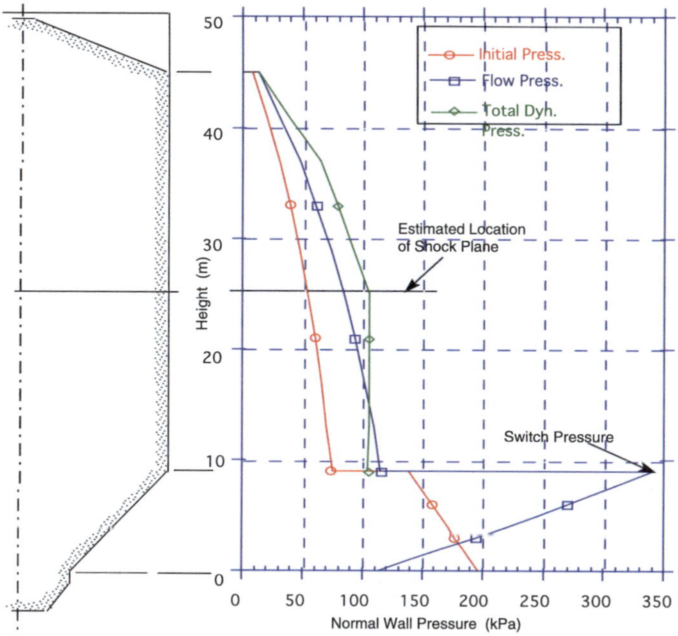

Fig. 7.14 Pressure distributions for bin

side of the bin where the chisel-shaped hopper has its highest intersection point with the cylinder wall.

7.6.2 *Estimation of Pulse Period*

The application of Eq. (7.7) gives the following predicted values of the pulse period:

For discharge rate $Q =$ 2700 *t/h*, $T = 2.4\,secs$
For discharge rate $Q =$ 1500 *t/h*, $T = 4.4\,secs$

These values, which are based on $\Delta\varepsilon y = 5$ mm for the coal, are in general agreement with the approximate values of 3 to 5 secs as reported for the actual silo.

7.6.3 *Vibration of Structure*

A critical factor in the operation of the silos is the influence of the dynamic characteristics of the overall structure. Noting that the silos were supported on columns on a base, which, in turn, was supported on piles, a simplified dynamic model of the silos is shown schematically in Fig. 7.15. There will be vertical and lateral stiffness due

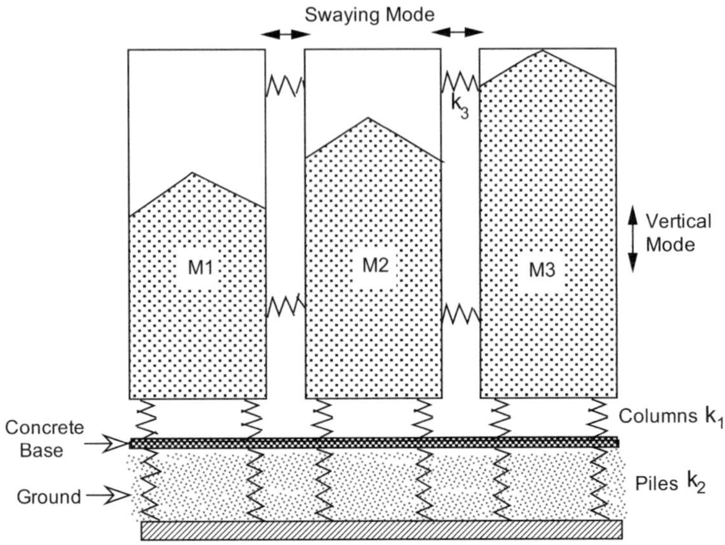

Fig. 7.15 Simplified dynamic model of silos

to the columns and piles as well as vertical and lateral stiffness due to the concrete connecting adjacent silos. In view of the significant variation in the silo mass from the full to the empty condition, there is a significant variation in the natural frequencies.

The natural frequency for the fundamental mode is given by

$$f_n = \frac{1}{2\pi}\sqrt{\frac{k}{m}} \tag{7.10}$$

Noting the variation in mass, the natural frequency of each silo will change as follows:

$$\frac{f_{nfull}}{f_{nempty}} = 0.56 \tag{7.11}$$

The natural frequency of the full silo is only 0.56 of the empty silo. While the various stiffnesses of the structures are not known, it is possible that the load pulse frequency of 3 to 5 s could excite one of the silo modes.

As each silo fills and empties, there will be different mass contents and, hence, different frequencies for adjacent silos. As a result, there will be dynamic coupling between adjacent silos which could impose significant loads on the concrete connecting these silos. The modes of vibration, while complex, would involve a combination of vertical and sideways swaying motion, the latter induced by non-symmetrical loadings of coal in the silos as well as variations in ground stiffness in the zone of the supporting piles.

7.7 200 Tonne Wheat Conditioning Silo—Case Study

7.7.1 Problem Description

This case study is concerned with the *silo quaking* problem being experienced with the 200-tonne wheat conditioning silo shown in Fig. 7.16.

Wheat is dosed with water prior to being fed into the top of the silo, the objective being to bring the wheat to a uniform moisture content of around 16% prior to discharge for the subsequent milling process. The magnitude of the shock loads was quite severe, particularly when the silo was full or near full. Observation showed the shock loads were transmitted through the ground into the neighbouring concrete silo structure leading to some cracking.

Calculations showed that the vertical dynamic load for the full silo when discharging grain amounted to 30% of the full static load, thereby increasing the total vertical load to 130% of the static load. The load analysis indicated that if the silo is operated at reduced head or fill, the shock loads are significantly reduced. For instance, at 60% fill height, the dynamic load is 46% of its maximum value, whereas at 50% fill height, the dynamic load is only 21% of its maximum value. Obviously,

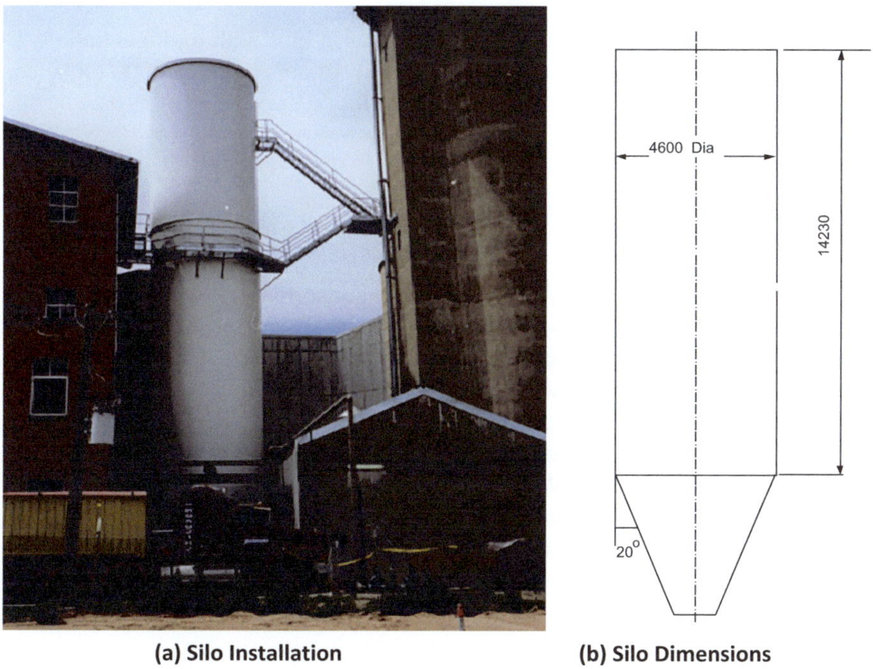

Fig. 7.16 200 tonne wheat conditioning silo

operating at 50% fill height would be the preferred option provided the residence time of the wheat in the silo is still sufficient to achieve good conditioning.

The vibration frequency of the silo structure due to the vertical motion and swaying varies with the degree of fill. The fundamental natural frequency of the silo when 100% full is estimated to be 40% of the corresponding frequency when the silo is empty.

7.7.2 *Recommended Solution*

The proposed solution for the silo is to incorporate a hopper insert as described in Sect. 4.2 and shown in Fig. 7.17. It needs to be noted that the insert will cause an increased normal, switch type pressure on the silo wall and consequently, the silo wall may need to be strengthened in this region. This is illustrated by the stiffening ring shown in Fig. 7.17.

7.8 6000 Tonne Multi-Outlet Coal Bin—Case Study

Silo-quaking problems have been known to occur in bins with multiple outlets. By way of illustration, the case study referred to in [10] concerning a 6000 tonne coal bin is reviewed. The bin is illustrated in Fig. 7.18.

7.8.1 *Problem Description*

The bin is of reinforced concrete construction supported on concrete columns as illustrated in Fig. 7.18b. Coal is discharged by means of seven vibratory feeders onto a centrally located conveyor belt as shown in Fig. 7.18c. When the bin was full or near full, severe shock loads were observed at approximately 3 s intervals during discharge. The discharge rate from each feeder was in the order of 300 t/h. When the level in the bin had dropped to approximately half the height, the shock loads had diminished significantly. With all the outlets operating, the effective transition was well down towards the bottom of the bin walls and the critical head H_m was of the same order as the bin diameter and greater than D_f. Substantial flow occurred along the walls, and since the reclaim hoppers were at a critical slope for mass and funnel-flow as determined by flow property tests, the conditions for *silo quaking* to occur were highly likely.

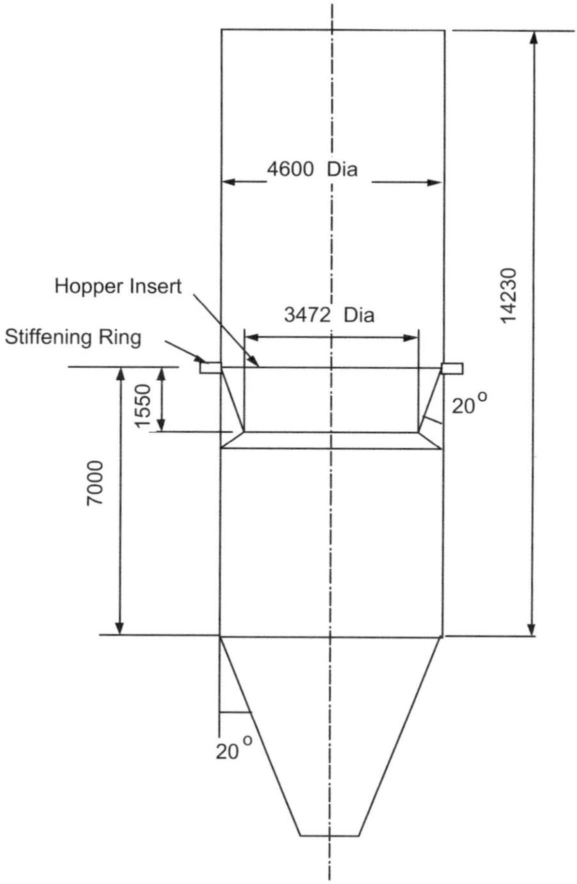

Fig. 7.17 Modified silo showing hopper insert

7.8.2 Field Trials

Confirmation of the mechanism of silo quaking was obtained in field trials conducted on the bin. Strain gauge load sensors and accelerometers were installed on selected columns. In one series of tests the three feeders along the centre line parallel with the reclaim conveyor were operated, while the four outer feeders were not operated. This arrangement is shown in Fig. 7.19. This induced funnel-flow in a wedged-shaped pattern as indicated in Fig. 7.19, with the effective transition occurring well up the bin walls, that is $H_m < H_{cr}(= D_F)$ or $H_m \ll D$. The same was true when only the central feeder (feeder 1) was operated. In this latter case the stationary material in the bin formed a conical shape. Under these conditions, the motion down the walls was greatly restricted and, as a result, the load pulsations were barely perceptible.

In a second set of trials, the three central feeders were left stationary, while the four outer feeders were operated. This arrangement is shown in Fig. 7.20. This gave rise to the triangular prism shaped dead region in the central region, with substantial

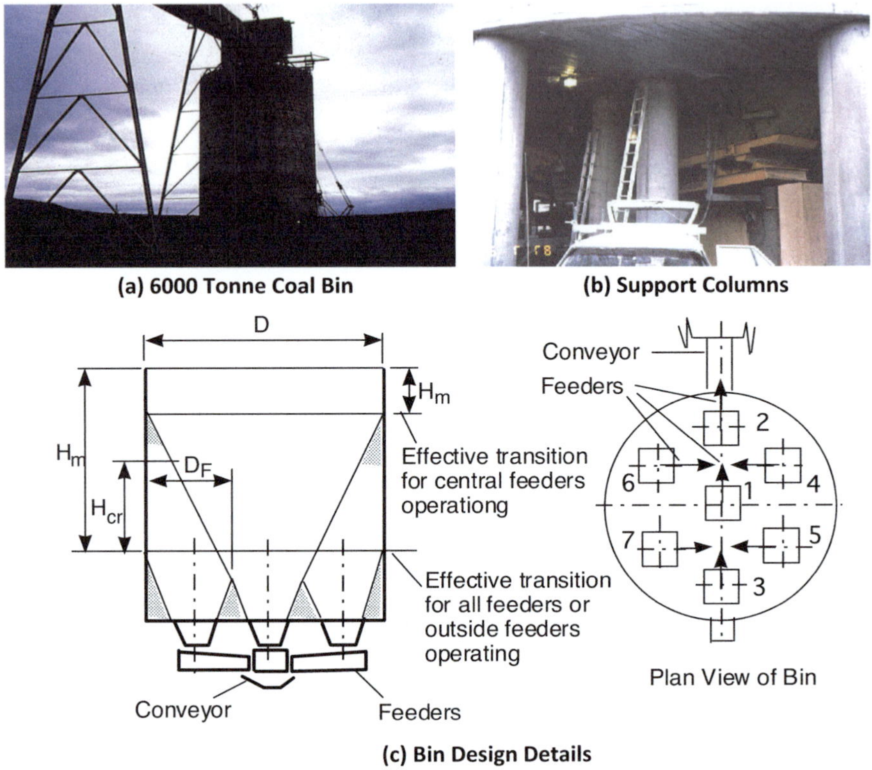

Fig. 7.18 6000 Tonne Multi-outlet coal bin

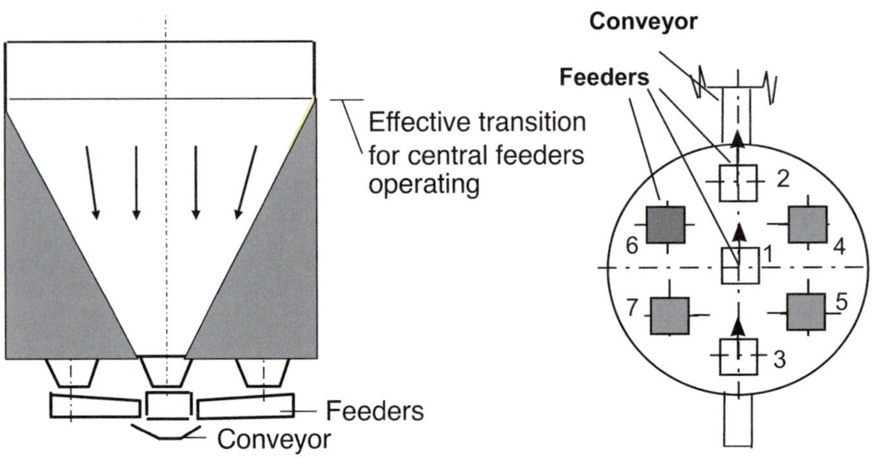

Fig. 7.19 Test Arrangement No. 1—Feeder 1 or Feeders 1, 2 & 3 operating

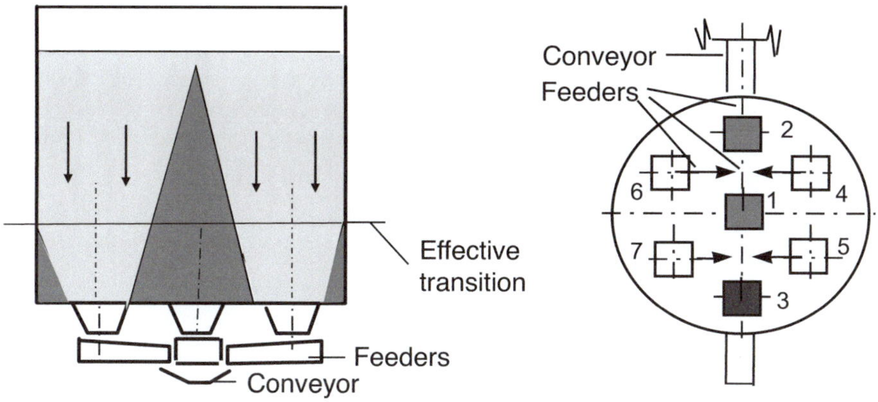

Fig. 7.20 Test arrangement No. 27 Feeders 4, 5, 6 & 7 operating

mass-flow along the walls. The load pulsations were just as severe in this case as was the case with all feeders operating.

7.8.3 Column Strain Measurements

The foregoing observations were quantified by the column dynamic strain measurements as functions of the bin level as illustrated in Fig. 7.21. These results clearly demonstrate that in order to avoid shock loadings, the effective transition must be high up in the bin to limit the height of coal moving against the wall. That is, $H_m \leq D_f$, or $H_m \ll D$ as indicated in Fig. 7.18c.

From an operational point of view, when the silo is full or near full, it is better to operate the discharge through the three centre line feeders with the four outer feeders shut down. Once the level approaches 50%, the four outer feeders can then

Fig. 7.21 Dynamic strain test results

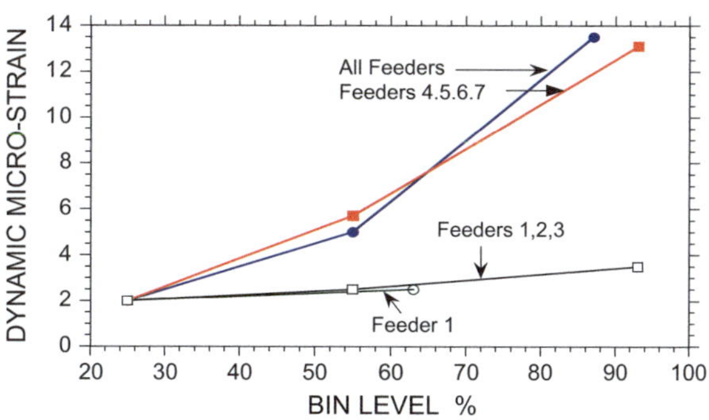

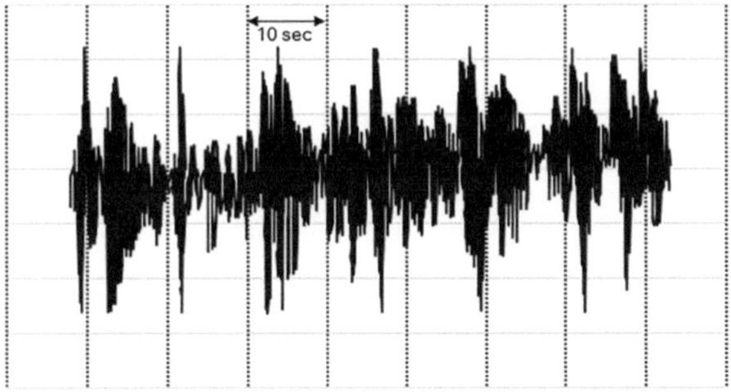

Fig. 7.22 Dynamic strain for feeder

be brought on-line to complete the discharge. However, the problem with this arrangement is that the initial discharge rate with only the three centre line feeders operating may be insufficient to meet the plant requirements. The strategy then, to overcome the dynamic *quaking* problem would be to operate the silo at reduced fill levels approaching 50%.

A sample of the dynamic strain record for the case when the four outer feeders were operating is shown in Fig. 7.22. In this case the dynamic strain is plotted as a function of time thus enabling the pulse period to be determined.

7.8.4 Dynamic Stability of the Silo Structure

A critical factor in the operation of *quaking* silos is the dynamic stability of the overall structure. The silo in question is supported on columns from a concrete base which, in turn, is supported on piles as illustrated in Fig. 7.23. In view of the significant decrease in total mass of the silo from the full to empty condition, there will be a corresponding increase in the natural frequencies of the silo during the emptying process. The modes of vibration involve combinations of vertical, swaying and twisting modes which are induced as a result of non-symmetrical loading of the silo, the pulsating flow during discharge and variations in the ground stiffness in the zone of the supporting piles.

7.9 Expanded-Flow Bin with Twin Hoppers—Case Study

Figure 7.24 shows a 2800 tonne coal bin of steel construction, which operates under expanded flow. The lower chisel shaped hopper splits into two outlets from which flow is controlled by vibratory feeders. The maximum discharge rate is 2500 tonne/hour.

Severe silo quaking was observed and unlike the previous case study examples where the quaking diminished significantly when the bins discharged half their

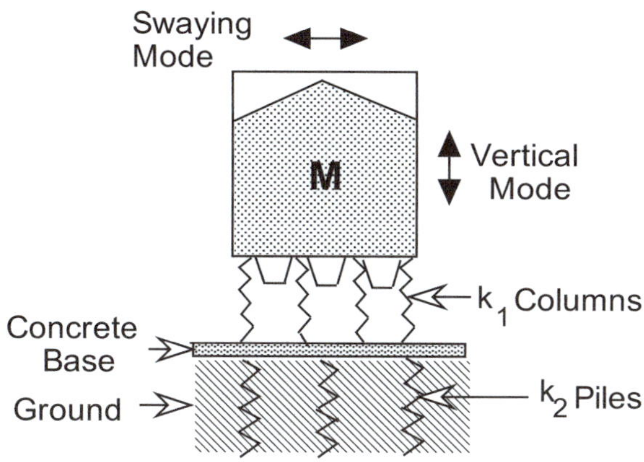

Fig. 7.23 Silo dynamic model

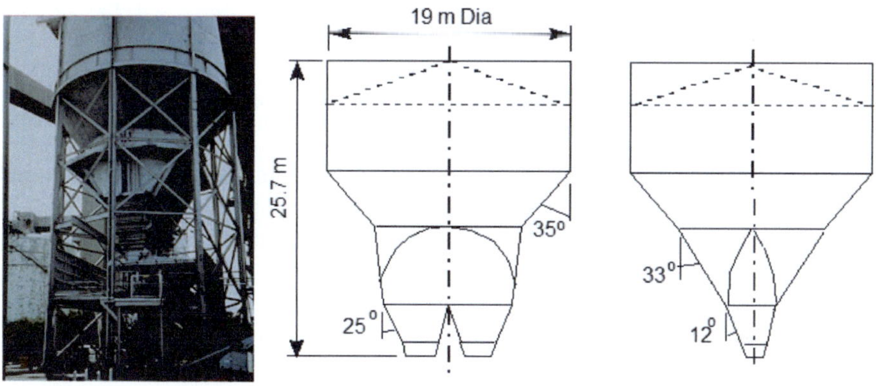

Fig. 7.24 Expanded flow bin with twin intermediate-flow hoppers

contents, in this case the shock loads were still quite substantial even when the bin level had dropped below 50%. The load pulsations were more pronounced when wet coal was handled.

7.9.1 Dynamic Loads and Stresses

Using a scale model of the coal bin constructed of clear plastic confirmed that the pulsating loads were due to the flow patterns developed in the lower chisel-shaped hoppers. Flow was characterised by rapid flow in the central region and slow flow in the outer region. Flow instabilities were observed at the interface of the flow zones giving rise to stick–slip type motion. The pulsating flow was most pronounced when moist, cohesive coal was used; when dry coal was used, the problem was not

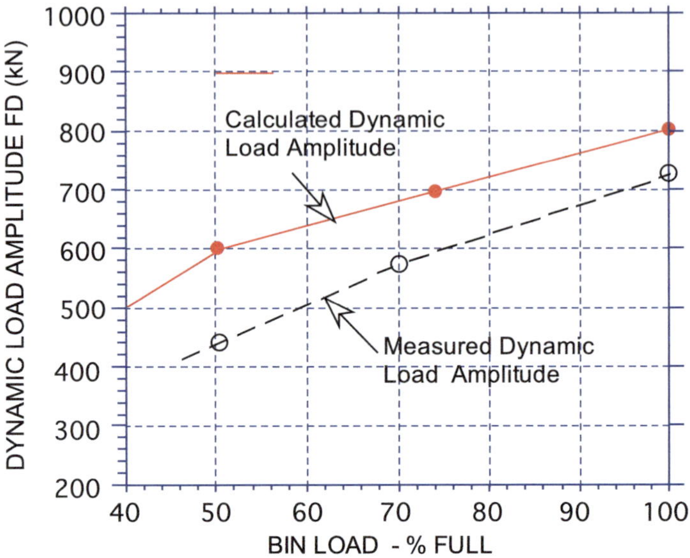

Fig. 7.25 Estimated and measured dynamic loads

observed. It is quite apparent that the curved interface geometry of the chisel-shaped hopper, combined with the large plane-flow hopper half-angle and higher friction of the moist coal on the hopper surface, render the hopper marginal for mass-flow. In fact, the flow is more characteristic of intermediate-flow as defined by Benink [17]. This flow mode is illustrated in Fig. 7.12.

The dynamic loads due to pulsing were computed for various levels of fill, these values being converted in the estimated dynamic stresses in the bin support columns. Strain gauges were mounted on the support columns of the full scale bin and the dynamic strains were recorded during several loading cycles. Figure 7.25 shows the predicted and measured results. The measured dynamic loads are slightly lower than the predicted values, but the comparison is considered to be quite reasonable.

7.9.2 *Natural Frequencies of the Bin and Structure*

It is contended that flow instabilities giving rise to pulsating loads or *silo quaking* are present in most bins or silos; whether the dynamic loads are experienced depends on the stiffness of the structure as a whole. In the case of the bin in question, the structure is quite flexible when the support columns and wooden piles are taken into consideration. The vibration modes of interest are the vertical and lateral swaying modes as shown in Fig. 7.26.

The dynamic motion is a combination of both modes. However, in view of the observed non-symmetrical loading of the bin, it is more likely that the swaying mode

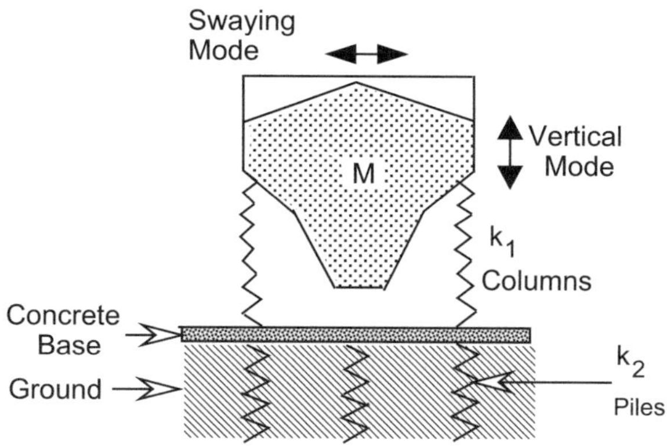

Fig. 7.26 Vibration modes

Table 7.2 Natural frequencies of bin

Mode	Bin load	Natural frequency (Hz)	Ratio f_{full}/f_{empty}
Vertical	Full	3.4	0.23
	Empty	14.78	
Lateral	Full	1.82	0.27
	Empty	6.84	

dominates. The fundamental natural frequencies were determined for the vertical and swaying modes both for the full and empty conditions. The results are presented in Table 7.2.

7.10 Concluding Remarks

This chapter has examined the phenomenon of self-excited dynamic loads in bins, often referred to as the *silo quaking* problem. Mechanisms under which pulsating flow and dynamic loads may occur in tall mass-flow and funnel-flow bins and silos, and in squat funnel-flow, expanded-flow, intermediate-flow, and multi-outlet bins have been described. Based on laboratory and field studies, theories have been presented to describe the characteristics of the dynamic loads that are generated during discharge. The shock load amplitude is shown to increase exponentially with height measured above the hopper and cylinder transition. The pulse period is found to be inversely proportional to the average velocity of the discharging bulk solid in the cylindrical section of the silo.

Studies of stick–slip motion provide a great insight into the *quaking* phenomenon. Current research on this topic involves the application of lumped mass models and continuum models based on hypo-plasticity theory to describe dynamic flow behaviour. For tall mass-flow silos, the results confirm the exponential growth of the shock pressures from the lower to the upper regions of the cylinder.

It is believed that flow pulsations occur in almost all bins. Whether the corresponding dynamic loads induced are observable depends on the degree of severity of the pulsations and on the natural frequencies of the bin and supporting structure as well as the geotechnical characteristics of the ground providing the foundation support for the structure as a whole. It is important that bin designers become fully aware of the *silo quaking* phenomenon and ensure that the natural frequencies of the overall bin structure avoid these frequencies due to the flow pulsations.

References

1. Roberts, A.W., *Review of the "Silo quaking" Problems in Bins of Various Geometrical Shapes and Flow Patterns.* Task Quarterly, 2003. **7**(4).
2. Shinohara, K., et al., *Mechanism of Gravity Flow of Particles from a Hopper.* Industrial & Engineering Chemistry Process Design and Development, 1968. **7**(3): p. 378-383.
3. Kmita, J., *Silo as a System of Self-Induced Vibration.* Journal of Structural Engineering-asce - J STRUCT ENG-ASCE, 1985. **111**.
4. Moriyama, R., *The effect of filling methods on the wall pressure near the transition in a bin.* Bulk solids handling, 1985. **5**: p. 603-609.
5. Moriyama, R. and G. Jimbo, *Reduction of pulsating wall pressure near the transition point in a bin.* Bulk Solids Handl.;(Germany, Federal Republic of), 1988. **8**(4).
6. Firewicz, H., *Kinematics of the gravity flow of granules.* Aufbereitungs-Technik, 1988. **2**: p. 61-70.
7. Kmita, J., *An experimental analysis of internal silo loads.* Bulk solids handling, 1991. **11**(2): p. 459-468.
8. Kmita, J., *A study on the self-induced vibration in silos.* Bulk Solids Handling, 1992. **12**: p. 25-25.
9. Roberts, A. and S. Wiche, *Silo quaking—a pulsating load problem during discharge in bins and silos.* IMechE Paper, 1991(C418/048).
10. Roberts, A., *Shock loads in silos - the" silo quaking" problem.* Bulk Solids Handling(Switzerland), 1996. **16**(1): p. 59–70.
11. Roberts, A.W., *Flow pulsations in the gravity flow of bulk solids and the influence on silo loads.* Chemeca 99 (27th: 1999: Newcastle, Australia), 1999: p. 142–158.
12. Roberts, A.W. and C.M. Wensrich, *Flow dynamics or 'quaking' in gravity discharge from silos.* Chemical Engineering Science, 2002. **57**(2): p. 295-305.
13. Gudehus, G. and J. Tejchman, *Silo music and silo quake.* Silos-Forschung und Praxis Tagung, 1992. **92**: p. 103-110.
14. Tejchman, J. and G. Gudehus, *Silo-music and silo-quake experiments and a numerical Cosserat approach.* Powder Technology, 1993. **76**(2): p. 201-212.
15. Wensrich, C.M., *Analytical and numerical modelling of quaking in tall silos.* 2025, Open Research Newcastle.
16. Wensrich, C.M., *Shock Waves During Silo Quaking*, in *Proc. 7th International Conference on Bulk Materials, Storage, Handling and Transportation.* 2001: Newcastle, Australia.
17. Benink, E.J., *Flow and stress analysis of cohesionless bulk materials in silos related to codes.* 1990.
18. Jenike, A.W., *Gravity Flow of Bulk Solids.* Utah Agric. Exp. St. Bull, 1961. **108**.
19. Roberts, A.W.W., C.M.;, *Silo Quaking Problem and Hopper Discharge System of Wheat Conditioning Silo.* 2000: Internal Report, TUNRA Bulk Solids, The University of Newcastle, Australia.

Chapter 8
Gate Loads and Load Settlement

8.1 Introductory Remarks

Storage bins for bulk solids handling operations rely on flow isolating gates to prevent outflow of bulk solids during initial bin filling from the empty condition, as well as during periods of stoppages in the plant operation. While in some cases slide gates may be employed in a flow throttling mode to adjust the outflow, this practice is not recommended in view of the negative influence this may have on the distribution of bin wall loads due to resulting eccentricities in the bulk solids discharge flow patterns. For efficient flow control the gate must be fully opened during the plant operation and a flow rate controlling device such as a belt or vibratory feeder being employed to achieve the required discharge feed rate.

The focus of this Chapter considers the design of isolating gates in large-load out bins such as those employed in the iron ore and coal industries for the filling of bulk rail wagons while the train is in continual motion. A typical train load-out bin facility is illustrated in Fig. 8.1 [1]. The bin incorporates a hopper which utilises a slide-type flow isolating gate in conjunction with a swing action, inclined chute fitted with a clam shell flow trimming gate. The discharge control chute is fitted below the flow isolating slide gate. During the filling operation with the slide gate closed, the chute is held in the raised position to allow sufficient head room for the train locomotive to pass underneath. Once this happens, the chute can be lowered to a position just clear of the top of the rail wagons. The clamshell gate is then opened to its controlled operating position and the continuous, sequential filling of the wagons commences. As each wagon is filled the clamshell gate is closed and then opened again when the next empty wagon arrives at the load-out position. The load-out bin operates between "high" and "low" fill levels in conjunction with the bin loading conveyor.

The subject of gate loads is intrinsically linked to the study of feeder loads, a subject which has received significant research attention, as indicated by the selected sample of references [2–6]. In essence, this research has emphasised the importance of treating the mass-flow hopper and feeder as an integral, interactive system in

A. W. Roberts, *The Theory and Practice of Bulk Materials Handling*,
https://doi.org/10.1007/978-3-032-19702-3_8

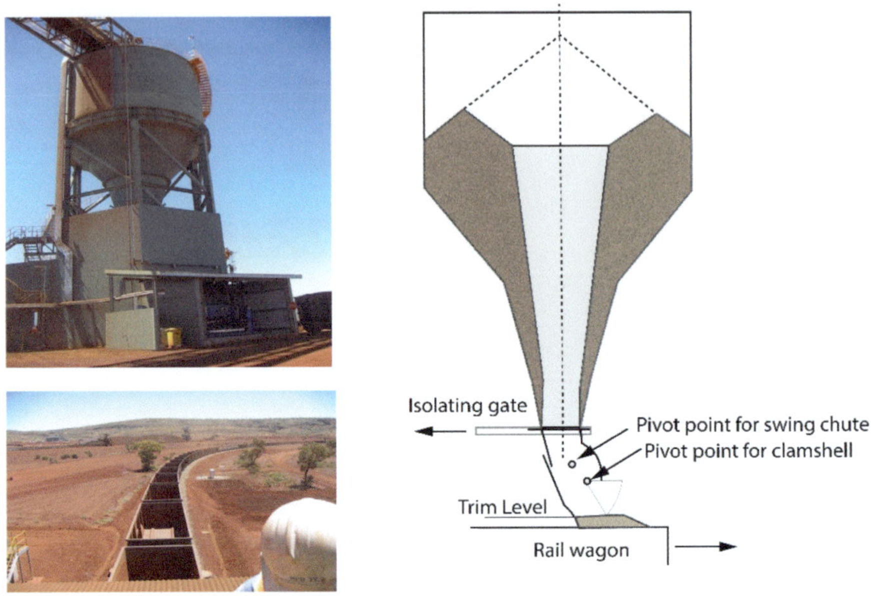

Fig. 8.1 Loadout bin used for train loading [1]

which the bulk solids stress fields in the hopper directly influence the feeder loads and the associated drive torques and powers. It has been well established that as a result of the change in stress states in the feed hopper, the vertical loads acting on a feeder during bin outflow, are significantly lower than the loads acting during the initial filling of the hopper with the feeder not operating. The reduction in loads from filling to running in the order of 80% have been recorded. Feeder design and load analysis dealt with in Chapters 10 and 11.

Gate loads are influenced primarily by the degree of compressibility of the bulk solids and degree of rigidity or stiffness of the gate. The Australian Standard AS 3774–1996 [7], provides some guidance for gate load determination by specifying appropriate so-called "j" values based on notional bulk solid compressibility and gate stiffness levels. In view of the widely varying influence of the "j" factor has on the gate load determination, it is important that the subject of gate loads be investigated. This led to the research of Roberts et al. [8] which is the basis of the material presented in this Chapter, the aims of which are summarised as follows:

(a) Review current theories pertaining to mass-flow bin stress fields and associated loads during filling and discharge;
(b) Outline experimental studies to measure gate loads on a pilot scale laboratory bin;
(c) Test two types of bulk materials having completely different properties and behaviour and corroborate the results with the theoretical results;
(d) Examine the influence of bulk solids settlement following bin loading;

(e) Establish gate load analysis procedures which incorporate improved methods for selection of the most appropriate value of j.

8.2 Review of Mass-Flow Bin and Gate Load Analysis

The gate loads of a mass-flow hopper are derived from wall load theory such as that presented in Chap. 6. For convenience, the relevant sections of the wall load theory are reproduced here.

8.2.1 Stress Fields

The two principal stress fields in a hopper are the *active* stress state as shown in Fig. 8.2a and the *passive* stress state as shown in Fig. 8.2b. Of particular relevance to the analysis of gate loads is the combined stress state called the *active/passive* stress field which can occur when the *passive* state is partly developed as shown in Fig. 8.2c. The corresponding stress fields and wall loads are shown in Fig. 8.2d. The *active* stress state normally accompanies the initial filling of a bin and hopper from the empty condition and is characterised by the major consolidation stress, σ_1, acting substantially in the vertical direction and curving away slightly as the stress direction lines intersect the hopper walls. As the bin filling continues, the surcharge load leads to the stress field switching to a passive or arched stress field as depicted in Fig. 8.2c. This results in an increase in the wall loads and a corresponding decrease in the vertical load on the gate as illustrated in Fig. 8.2d. That is, in the region of the hopper outlet the hopper walls carry a substantial proportion of the vertical load of the bulk solid, thus relieving the load on the gate.

The *pass*ive state of stress occurs when there is any downward movement of the bulk solids such as during filling or during discharge. It can even occur due to undisturbed settlement. In this case the convergence of the downward movement of the bulk solids in the hopper causes an increase in the normal wall loads as the stress field switches from the peaked or *active* stress state to the arched or *passive* state depicted in Fig. 8.2. The major consolidation stress, σ_1, for each arch stress line within the bin hopper is constant in magnitude around the arch but varies in direction. Accordingly, the corresponding normal pressure acting on the bin walls as well as the vertical material stress decrease substantially from the bin hopper transition.

An example of the *active/passive* stress state is load cushioning where the hopper is never emptied completely, so retaining the arched, *passive* stress state when being refilled. This means once the *passive* stress state is established, it remains in a stable condition, even if the discharge is stopped. Another case of the *active/passive* stress state occurs during settlement after filling from the empty condition. The associated downward movement of the bulk solid as it approaches its critical consolidation

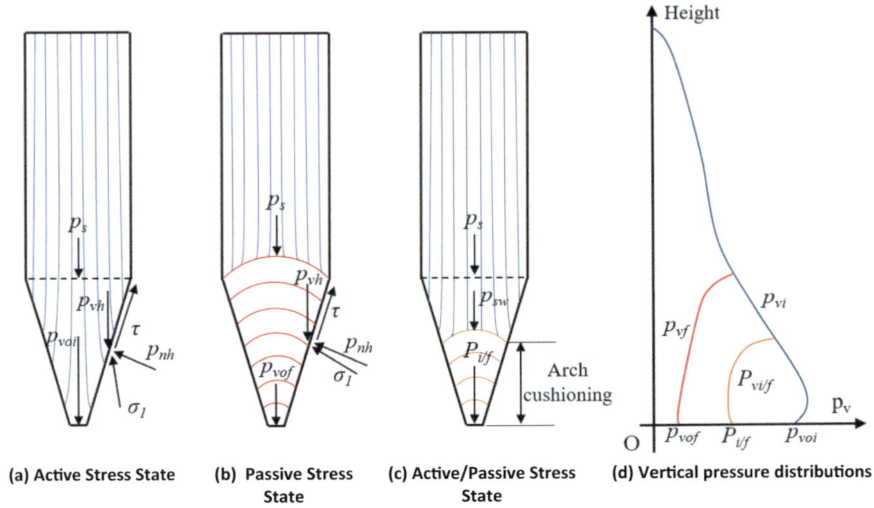

Fig. 8.2 Mass-flow hopper stress fields

condition results in load transfer to the hopper walls and hence, a zone of *passive* state of stress.

As shown in Fig. 8.3d, the pressure distribution is closely associated with the states of the bulk solids inside the bin. Switching from the active to passive stress field can lead to great reductions in the loads around the bin hopper region. Therefore, the stress states are an important property that is employed in feeder load determination and load control, as outlined in the work of Roberts [4].

8.2.2 *Mass-Flow Load Model*

The mass-flow model for the active/passive stress state is based on the wall load theory of Chap. 6 and shown in Fig. 8.3. For convenience, the relevant analysis is reproduced here.

Considering the equilibrium of the model presented in Fig. 8.2, the following differential equation is derived:

$$\frac{dP_{vh}}{dz_h} + \frac{jP_{vh}}{(h_h - z_h)} = \gamma \tag{8.1}$$

The solution to this equation yields the vertical pressure P_{vh}

$$P_{Vh} = \frac{\gamma(h_h - z_h)}{(j-1)} + \left[P_s - \frac{\gamma h_h}{(j-1)}\right]\left[\frac{(h_h - z_h)}{h_h}\right]^j \tag{8.2}$$

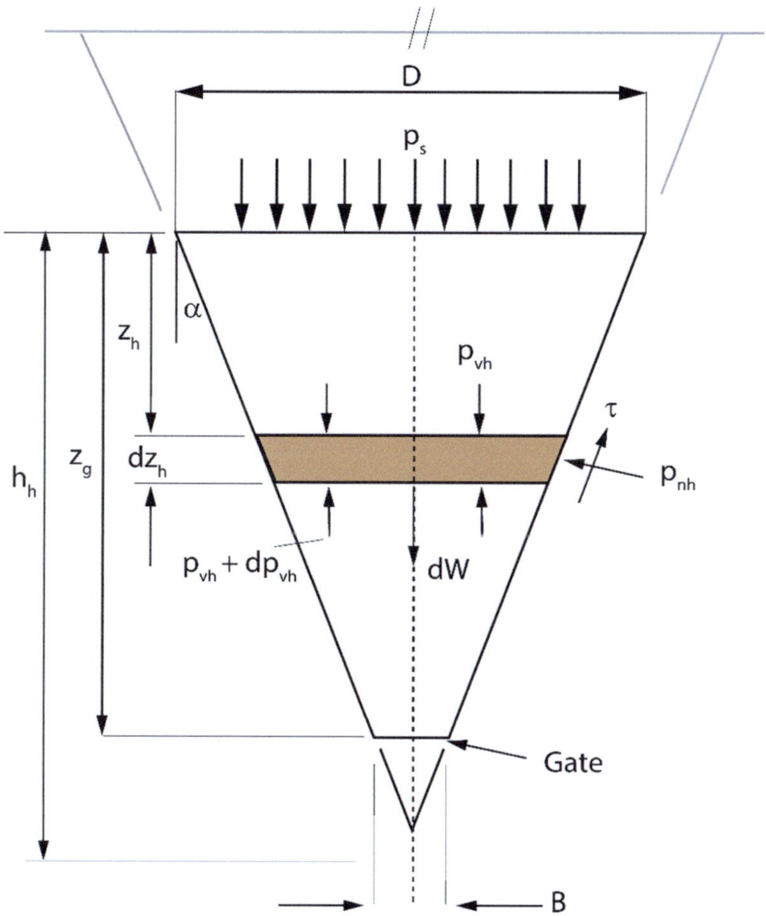

Fig. 8.3 Equilibrium of a slice element in a hopper

where

$$j = (m+1)\left|k_h\left(1+\frac{tan\phi_w}{tan\alpha}\right)-1\right| \tag{8.3}$$

$m =$	hopper flow symmetry factor
$m =$	0 for plane flow or wedged shaped hopper
$m =$	1 for axi-symmetric or conical hopper
$\phi_w =$	wall friction angle.
$\alpha =$	hopper half-angle at outlet.
$P_s =$	surcharge pressure at switch level.
$\gamma = \rho\, g =$	bulk specific weight.
$\rho =$	bulk density.

$k_h = \frac{P_{nh}}{P_{vh}} =$ pressure ratio.
$P_{vh} =$ average vertical pressure acting on slice at depth z_h.
$P_{nh} =$ corresponding normal pressure acting on wall

Setting $z_h = z_g$ in Eq. (8.2), the vertical pressure at the hopper outlet acting on the gate is given by

$$P_{VO} = \frac{\gamma B}{2(j-1)tan\alpha} + \left[P_s - \frac{\gamma D}{2(j-1)tan\alpha}\right]\left[\frac{B}{D}\right]^j \tag{8.4}$$

where
$B =$ hopper diameter at the outlet
$D =$ hopper diameter at the transition.

8.2.3 *Pressure Ratio k_h*

For the flow condition, the pressure follows the arched pressure distribution of the passive stress state. Meanwhile, the pressure ratio is not constant, and it varies across the horizontal slice due to the distribution of the principal consolidation stress within a hopper. However, for practical reasons applicable to the specific case of axi-symmetric mass-flow bins with conical hoppers, the Australian Standard AS 3774–1996 adopted pressure ratio values based on the normal and vertical pressure acting at the hopper walls for which the pressure ratio k_{hf} is given by

$$k_{hf} = \frac{1 + sin\delta cos2\eta}{1 - sin\delta\cos 2(\alpha + \eta)} \tag{8.5}$$

where

$$\eta = 0.5\left[\varnothing_w + sin^{-1}\left(\frac{sin\varnothing_w}{sin\delta}\right)\right] \tag{8.6}$$

δ is the effective angle of internal friction.

However, when applied to mass-flow bins, particularly those with plane-flow, wedged-shaped hoppers, it has been found that this k_{hf} value tends to underestimate wall loads in the upper hopper section and overestimate wall loads in lower hopper region. Considering the symmetry of the pressure distribution for a bulk solid slice, As discussed in Chapter 6, Roberts [6] recommends a modified value of k_{hf} by averaging the pressure across the slice.

$$k_{hf} = \frac{2(1 + sin\delta cos2\eta)}{2 - sin\delta(1 + \cos 2(\alpha + \eta))} \tag{8.7}$$

8.2.4 Selection of J Value

For bin and gate load design, the initial filling and subsequent refilling of the bin are of primary concern. Since the initial active stress state depicted in Fig. 8.2 a is perceived to dominate in this case, the selection of the parameter "j" has a significant influence on the magnitude of the calculated gate loads. If hypothetically, the bulk solid is assumed to be completely incompressible and the gate completely rigid, the value of j = 0 and Eq. (8.4) reduces to the "hydrostatic" load given by

$$P_{vhi} = P_s + \gamma z_h \tag{8.8}$$

with the correspond pressure ratio given by

$$k_{hi} = \frac{\tan\ \alpha}{\tan\ \alpha + \tan\ \varnothing_{\mathrm{w}}} \tag{8.9}$$

The Australian Standard AS 3774–1996 [7] "Loads on Bulk Solids Containers" combined gate loads with feeder loads and specified that the value of j be selected as follows:

(a) j = 0.1 for very incompressible materials stored above stiffly supported feeders;
(b) j = 0.45 for moderately compressible materials stored above flexibly supported feeders;
(c) j = 0.9 for very compressible materials stored above softly supported feeders.

While a gate will generally be relatively stiff, there is considerable uncertainty as to the degree of compressibility of the bulk solid during the bin filling operation. While ignoring the value j = 0, the safe, conservative approach is to select a low value of j, such as $j = 0.1$. However, the design will almost be far too conservative due to the high, unrealistic values of the computed loads. This provided the motivation for the test program that is now described.

8.3 Pilot Scale Test Program

The experimental work program associated with this research involved the pilot scale, mass-flow test bin of Fig. 8.4. Iron ore was the selected bulk solid in view of its relevance to the mining industry. The properties of the iron ore obtained from flow property tests are summarised in Table 8.1. The iron ore was handled in a relatively dry state.

Referring to the test bin, Fig. 8.4 a shows the bin geometrical details while Fig. 8.4 b shows the test set-up with the bin being filled with iron ore at the commencement of a test run. Filling of the bin was accomplished using a loading skip in conjunction with an overhead crane.

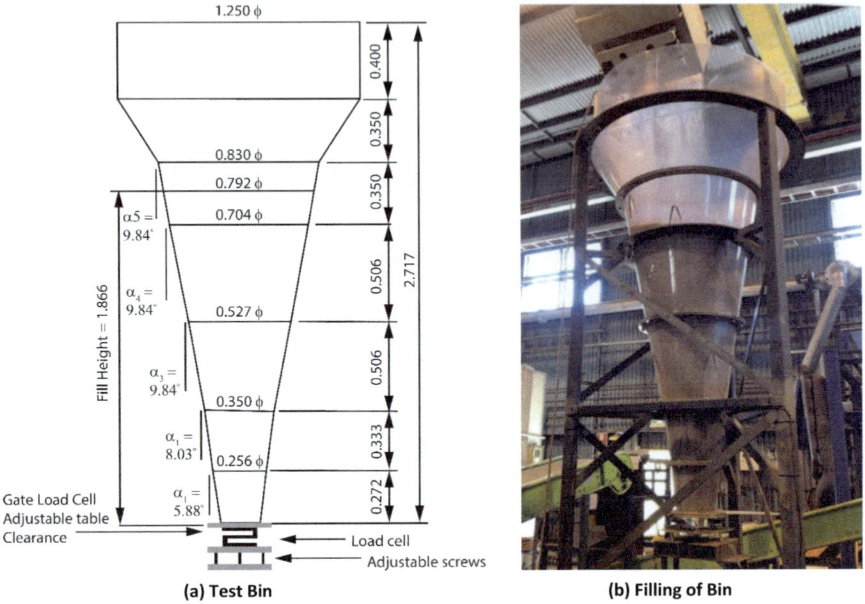

Fig. 8.4 Pilot scale test bin

Table 8.1 Iron Ore test parameters

Average bulk density, ρ	2.03 t/m^3
Wall friction angle, ϕ_w	25°
Effective angle of internal friction, δ	50°
Bin fill rate, Q_m	35 t/h
Total mass, M	0.767 t

The bin is of Perspex construction comprising a five-section, variable geometry mass-flow hopper for which the half-angles are $\alpha_1 = 5.88°$, $\alpha_2 = 8.03°$ and $\alpha_3 = \alpha_4 = \alpha_5 = 9.84°$. In view of the three different hopper angles a sequential approach to the gate load analysis is required. The bin is supported by three columns, each of which is mounted on load cells to enable the bin load to be measured. For the measurements of the gate load, a special load cell was developed comprising a flat plate mounted on three load transducers that, in turn, are attached to a base plate. Adjusting screws are incorporated to enable the clearance between the hopper outlet and gate surface to be set. The gate load cell unit is completely independent of the bin structure. For the loading and load settlement stages of the tests, the gate is set with a minimum clearance of approximately 0.5 mm from the hopper outlet to simulate a stiff gate without bin contact. The quantity of iron ore used in the experiments was 0.767 tonne which corresponded to the fill height of 1.866 m as indicated in Fig. 8.4a.

Filling of the bin was performed using a skip in conjunction with an overhead crane as shown in Fig. 8.4b. While some minor variations in the loading cycle occurred from one test run to the next, the average bin loading rate was, approximately 35 t/h $= 13.3\ m^3/h$.

During each test run, the net bin loads and gate loads were recorded using a data logging system. The loads were recorded from the commencement of the bin loading to the completion of the bin filling operation. The load recording then continued for a period of approximately 2.5 h to measure the changes in the bin and gate loads due to the settlement of the iron ore in the bin.

The final stage of the tests involved progressively lowering the gate to record the decrease in gate load as the clearance between the bin outlet and gate surface increased. The purpose of this final stage of the test was to observe the potential decrease in the gate load due to the stress field change from active to passive field state as a result of the increase in gate clearance, albeit small. This information is relevant to feeder design in view of the importance of the passive stress field in controlling feeder loads and drive powers.

A set of recorded results from the initial filling of the bin is shown in Fig. 8.5. The results illustrate the progressive increase in gate loads and pressures as the bin fills, along with the increase in time and the estimated decrease in impact pressures. The iron ore drop height commences at the loading skip 1 m above maximum fill level.

8.4 Predicted Performance Based on Selection Appropriate *J* Values

As an aid to the review of the measured test results, a selection of relevant predicted performance characteristics of the test bin is presented. The analysis is based on established bin and feeder load theories. The principal focus is on the selection of appropriate *j* values.

8.4.1 *Gate Loads During Bin Filling,* J = *0.45—Active Stress State*

With the gate closed during the filling process, the stress state is deemed to be active, which assumes the vertical pressure distribution is dominant. Based on AS 3774–1996, a *j* value of 0.45 was selected. The computed results are plotted in Fig. 8.6 which shows the increase of the gate load in relation to the fill height. The gate loads increase from 20.23 kg at the height of 0.272 m, that is the top of the lowest hopper Sect. 8.1, to, 85.26 kg at the top level of the fill height of 1.866 m (Fig. 8.4a).

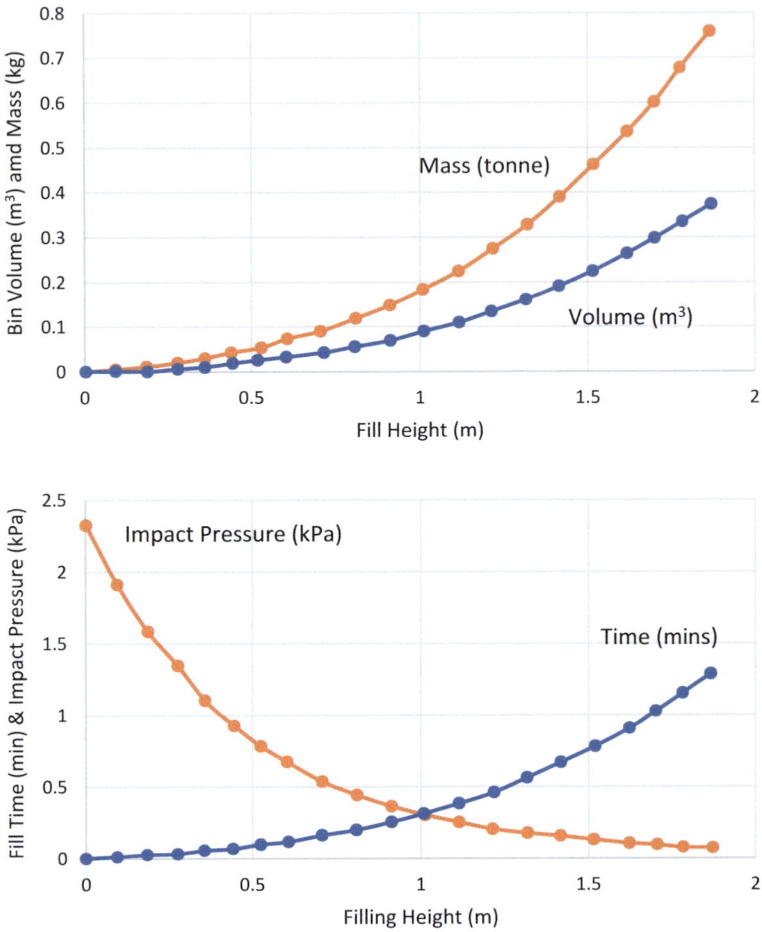

Fig. 8.5 Bin contents, filling time and impacted pressure versus fill height

8.4.2 Influence of J *Values on Gate Loads Based on Full Bin*

The selection of the *j* factor has a significant influence on the gate load determination. As the value of *j* increases the predicted gate load decreases as a result of the stress field approaching the passive state. This is illustrated in Fig. 8.7. For reference purposes, the gate loads associated with the values of $j = 0.1$, 0.45 and 0.9 based on AS 3774–1996 [7] are shown. The corresponding gate loads are, 115 kg, 84kg and 62 kg, respectively.

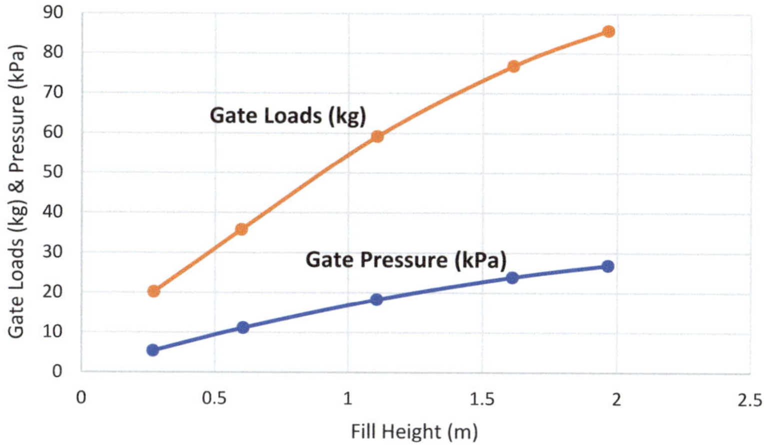

Fig. 8.6 Gate loads as function of fill height

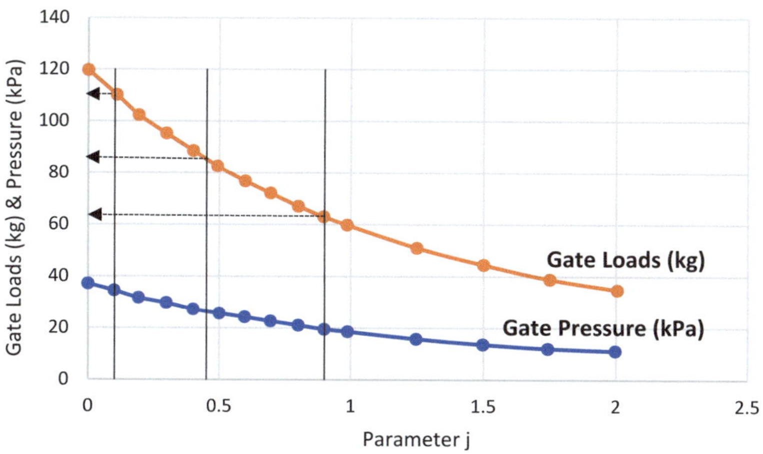

Fig. 8.7 Influence of *j* factor on gate loads—Full bin

8.4.3 Active/Passive Stress States

The case when the stress state switches from an *"active"* state to a fully developed *"passive"* state due to converging flow in the lower part of the hopper is now considered. This case was introduced in Fig. 8.2c and d of Sect. 8.2. It is particularly important in view of the gate load cushioning provided by the *"passive"* stress state. For the bin of Fig. 8.4a, the vertical stress distributions were determined analytically for the lower four sections with the Switch Heights (y_{sw}) of 0.272 m, 0.605 m, 1.111 m and 1.617 m. That is, for each case, the passive stress state was assumed to be

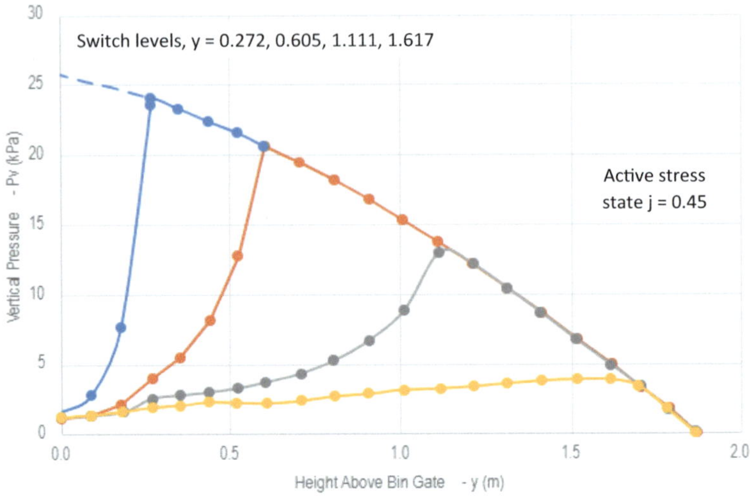

Fig. 8.8 Vertical stress distributions

developed below the switch height, whereas the active stress states, respectively, are maintained above the switch levels.

For the *"active"* case, j is assumed to be 0.45 for which the corresponding calculated values, based on the theory presented in Sect. 8.2, for the four switch levels are 0.22, 0.285, 0.335 and 0 0.335. For the "passive" case, the values of j for the chosen four switch levels are, respectively, $j = 17.123$, 13.145, 10.13 and 10.13. The corresponding *"passive"* pressure ratios are $k_{hf} = 1.89$, 1.76, 1.656 and 1.656. The vertical stress distributions are plotted in Fig. 8.8.

As the graphs indicate, the vertical pressures on the outlet decrease rapidly since the *arched* or *passive* stress field transfers a substantial part of the bin load to the hopper walls. The vertical pressure and potential gate load reduction due to switching from *active* to *passive* is further illustrated in Fig. 8.9 which shows the load under a fully *active* state is 85 kg, decreasing from "active" to passive to 5 kg at the switch level $y_{sw} = 0.22$ m, approaching asymptotically to 4 kg.

8.5 Test Results

Several test runs involving the filling of the bin from an empty condition with the iron ore were conducted, a typical set of the measurements being presented herein.

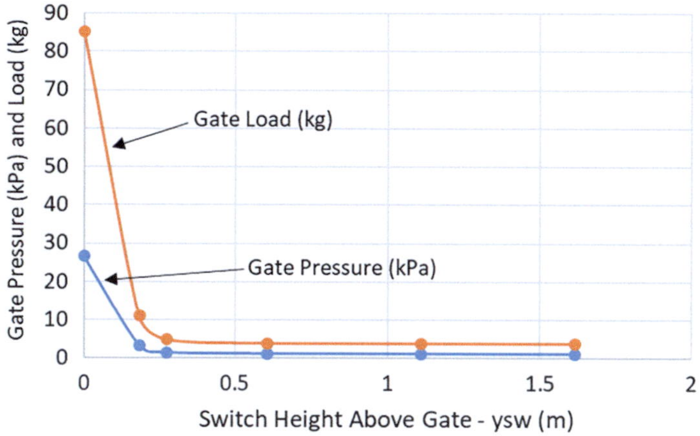

Fig. 8.9 Gate load variation with stress switch heights

8.5.1 *Initial Filling*

The bin filling time was approximately 1.5 min. The gate load increases rapidly for the first few seconds and then gradually to a peak value as the bin fill level is reached. That is, a *step change plus ramp* loading pattern. The gate initially experienced the major portion of the load without the frictional support of the hopper walls. As the filling progressed, the benefit of the increasing contact with the walls and the associated wall support took effect. This demonstrates the benefit of the hopper wall support in controlling the gate loads. After filling has ceased, the recorded gate load was 38 kg, representing 5% of the full load. The corresponding bin load carried by the walls was 730 kg, 95% of the full load of 769 kg (Fig. 8.10).

8.5.2 *Load Settlement*

To investigate the influence of load settlement with undisturbed storage time, the recording of the bin and gate loads continued for a period of, nominally, 2.5 h following the completion of the filling operation. This period was adequate as the gate and bin loads converged, asymptotically, to stable values at the end of the settlement. As shown in Fig. 8.11, the load settlement resulted in a decrease of 8 kg in the gate load and a corresponding increase of 8 kg in the bin load. These load changes are a result of the switching in the stress fields from *active* to the *passive* in the bin. The natural consolidation due to the settlement of the bulk solids and the associated *passive* or *arched* stress field state are sufficient to support a major proportion of the total stored load in the bin. This indicates the effect of load transfer to the hopper walls due to the downward converging creep as the bulk solids approach the critical consolidation conditions.

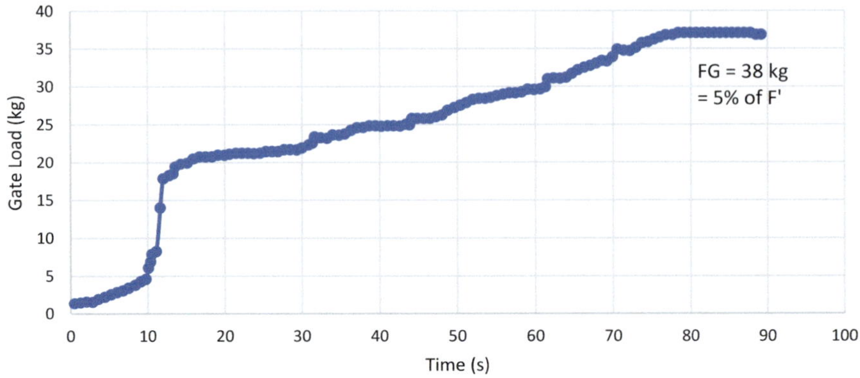

(a) Gate load

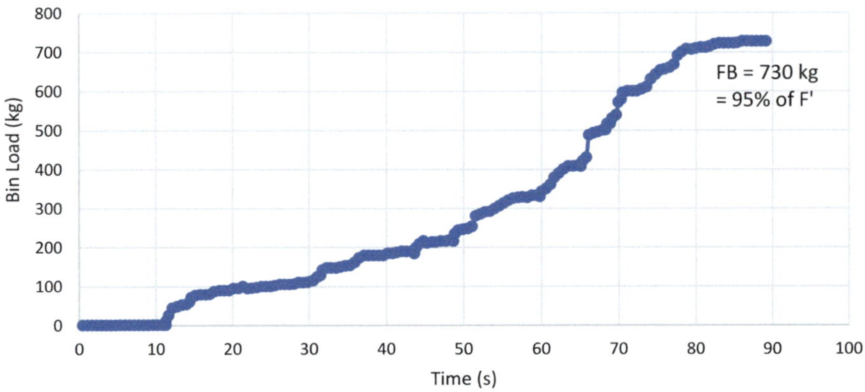

(b) Bin load

Fig. 8.10 Bin and gate load versus time during filling

8.5.3 *The* J *Factor Parameter*

The j factors associated with the gate loads were back calculated for the iron ore from the test results. Referencing Figs. 8.8 and 8.9, the corresponding j factors were determined as shown in Table 8.2.

It can be observed that the computed j values during the filling and settlement are greater than the values recommended in AS 3774–1996 [7]. In the case of the test bin, these computed j values may be regarded as global values based on the adopted theory were only the active stress state conditions are considered. It is quite apparent that the stress is a combination of the active and passive stress states as described in Sect. 8.2.

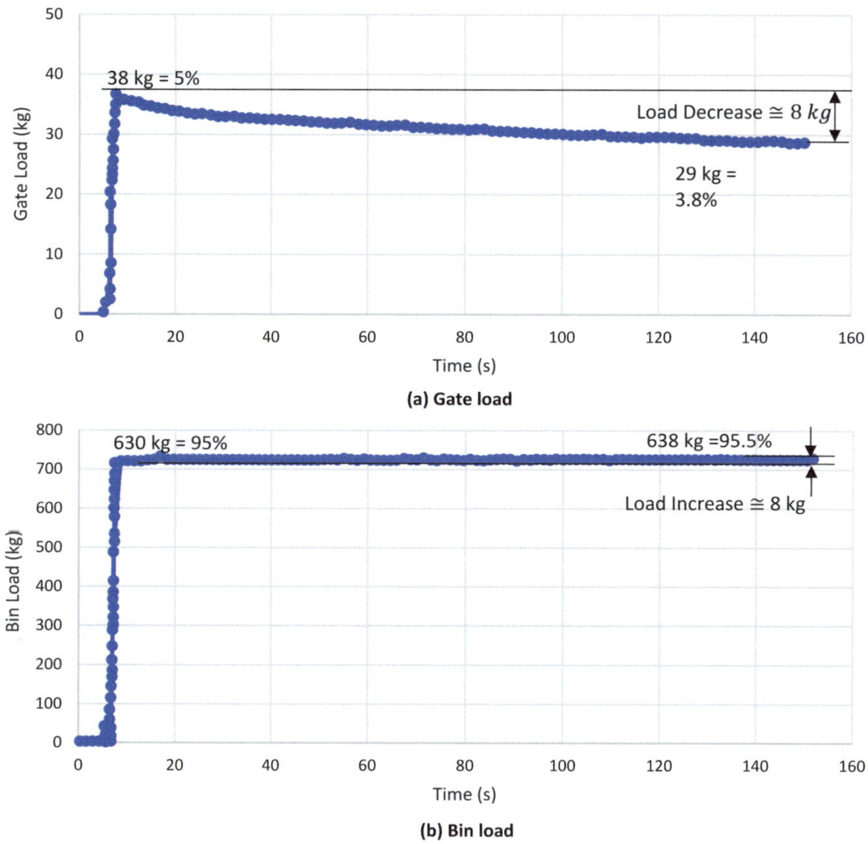

Fig. 8.11 Bin and gate loads as a function of settling time

Table 8.2 Gate loads and j parameters

	Bin filled		
Parameter—j	Gate pressure (kPa)	Gate load (kg)	Method
0.1	34.458	110.345	AS 3774–1996
0.45	26.622	82.255	
0.9	19.698	64.081	
1.8	11.866	38	Measured
Summary	Bin filled 82 $j = 1$		
	Settlement 39 $j = 2$		

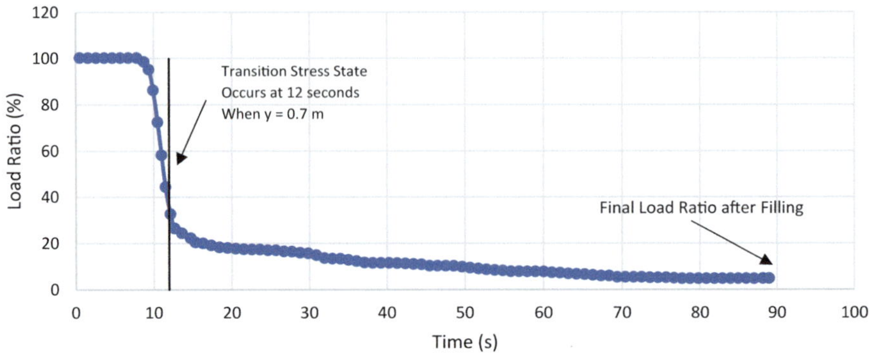

Fig. 8.12 Gate to bin load ratio during filling

8.5.4 *Transition Considerations*

The important influence of the *active* to *passive* change in the stress field in the hopper in controlling the gate load is demonstrated in Fig. 8.13. The graph shows the variation of the gate load as a ratio of the total bin load from the commencement of the filling operation. For the first 8 s the gate carries 100% of the loaded iron ore without bin wall support. The stress field is deemed to be *active.* During the next 4 s as the load transfer to the bin wall commences, the gate load decreases rapidly to 30% as shown. This is due to the transition to the *passage* stress state. Then, for the remainder of the filling operation as the transition to the fully developed *passive* stress test state continues, the gate load further decreases approaching, asymptotically the value of 5% of the total load.

For the pilot scale bin of Fig. 8.4, the stress switch commences after approximately 12 s from the commencement of filling. This corresponds to the fill level of 0.7 m which is approximately 0.1 m above the bottom level of the hopper Section S_3 of Fig. 8.4. This clearly demonstrates the importance of the *passive* stress field in controlling the gate load, as has been discussed. In effect, the natural consolidation due to load settlement of the iron ore in the lower hopper sections of the bin and the associated passive stress field state is sufficient to support the total stored load in the bin without any further increase in the gate load (Fig. 8.12).

8.5.5 *Influence of Gate Clearance*

In the test program, following the nominal 2.5-h load settlement period, the gate was progressively lowered to increase the clearance between the bottom of the hopper outlet and the gate. The results were deemed to reflect the influences that the lowering of the rigidity of the gate structures may introduce on the gate load during the filling and settlement. Figure 8.13 shows the measured gate loads for three separate tests

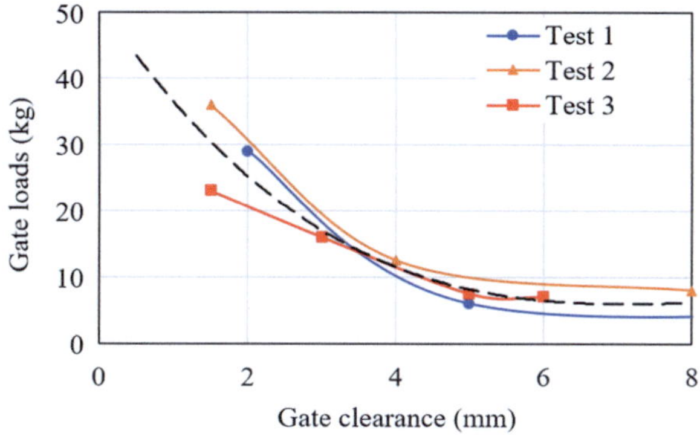

Fig. 8.13 Reduction in gate loads as a function of gate clearance (Iron ore)

using the iron ore in conjunction with a trendline of the best fit. As shown the gate load decreases quickly as the clearance increases in the range from 0 to 4 mm, after that it approaches asymptotically a value of 5 kg as the clearance further increases. It is interesting to compare the results plotted with the predictions presented in Fig. 8.9 where this combination of the *active* and *passive* stress states is considered. From the comparisons, it can be found that the *passive* stress is, to some extent, developed in the hopper and bin with the gate clearances as the measured value of 5 kg is very close to the predicted value of 4 kg. However, these results are an indication of the inferred stiffness on the reduction in gate loads. The switch of the active and passive stress states is not clear and may need to be quantitatively defined.

8.5.6 Cohesive Strength Considerations

An aspect of mass-flow bin design concerns the influence on the gate load determination of the cohesive strength of the bulk solid which is the basis of determining the critical hopper outlet dimension. When the actual opening dimension is greater than the critical outlet dimension, discharge flow can occur. Also loads can be transferred to the gate. Otherwise, a stable arch will form over the outlet without load transfer to the gate and without the possibility for discharge flow. This is a plausible explanation as to why the measured gate loads in the pilot scale tests are lower than those that were predicted. This is an aspect of gate load determination that warrants further investigation. By way of background, the study of stable arch formation in wedged-shaped, plane-flow hoppers handling iron ore was performed by Guo [9, 10]. Also, the work of Roberts [11] is of relevance.

8.6 Dynamic Load Effects

Another aspect of the study of gate loads concerns the dynamic shock and *quaking*, pulse type loads induced during the bin filling as well as during discharge. To illustrate this behaviour a selected sample of the filling operation of the test bin is shown in Fig. 8.14. Figure 8.14a shows the recorded gate and bin loads, while Fig. 8.14b shows the corresponding amplified/frequency records to display the measured amplitude and frequency components of the load measurements. Allowing for the fact that this is a small sample of the overall bin filling operation, there is a "hint" of pulse type dynamic loads bearing a similarity to the load versus frequency characteristics associated with *silo quaking* as discussed in Chap. 7. Comparing the trough-to-peak load amplitudes for the bin and gate, the gate to bin load ratio is approximately 24:1.

This aspect of bin and gate load research requires further investigation.

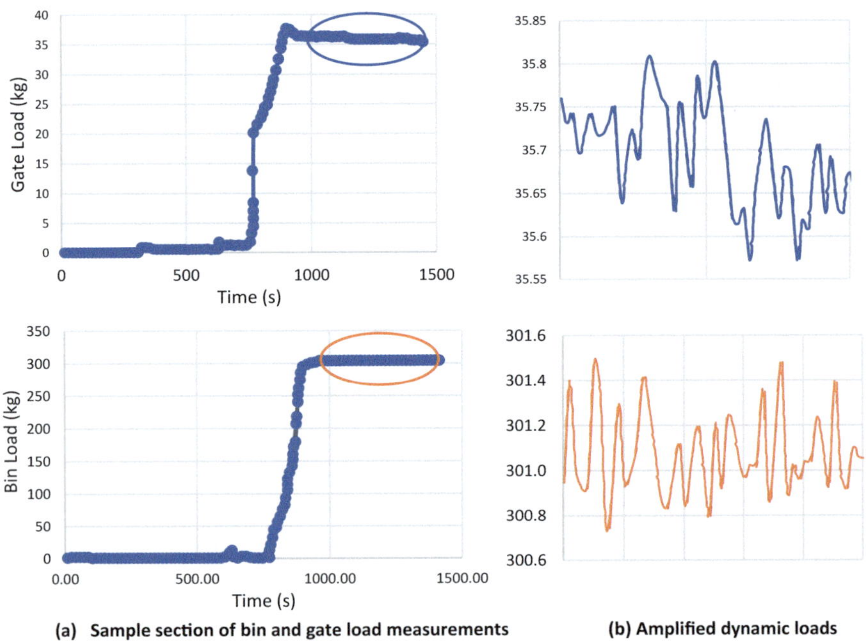

Fig. 8.14 Sample of bin and gate loading records showing dynamic loads

8.7 Concluding Remarks

This chapter has presented both analytical and experimental investigations into the loads acting on isolation gates installed on mass-flow bins. The existing methodology for gate load determination was reviewed and validated against measurements obtained from tests on the pilot-scale mass-flow bin shown in Fig. 8.4.

The principal findings of this study are summarised below:

(a) Gate loads are strongly influenced by the stress conditions within the stored bulk solids during filling and static storage prior to discharge. Loads can range from high values under a fully active stress state to significantly lower values when the stress condition transitions partially or fully to a passive state. The extent of this reduction depends on the compressibility of the bulk solid and the flexibility of the gate and its mounting.

(b) For the iron ore, tests indicated that immediately after filling, the gate load peaked at 38 kg, representing approximately 5% of the total bin load, with 95% of the load carried by the bin walls. After 2.5 h of undisturbed storage, downward creep of the ore reduced the gate load by 8 kg (21%), with a corresponding increase in wall load.

(c) Incremental lowering of the gate to increase clearance demonstrated that a 5 mm clearance (2.5% of the 200 mm opening) reduced the gate load by at least 85% of its initial value, indicating that the stress condition had transitioned to a fully developed passive or arched state. Good correlation was achieved by assuming passive stress formation in the lower bin regions at heights below 0.25 m.

(d) The calculation of gate loads depends on the selection of the j value, which is influenced by hopper geometry, the hopper half-angle α, wall friction angle ϕ_w, and the ratio of normal wall pressure to vertical pressure k_h. The Australian Standard AS 3774–1996 [7] specifies j values of 0.1, 0.45, and 0.9, based on the compressibility of the bulk solid and the stiffness of the feeder or gate. Experimental results from the iron ore tests indicate a global j value of 1.8, which exceeds the AS 3774–1996 recommendations. Since gate load determination is highly sensitive to variations in bulk solid flow properties and bin geometry, it is suggested that AS 3774–1996 be applied with caution. In particular, the j = 0.1 value may be considered to have an overly conservative influence on the load calculations. Current experience suggests that j values in the range of 0.4 to 0.5 provide more realistic estimates, but it is important that each application be assessed individually to account for material and design specifics.

In summary, this chapter has highlighted critical aspects of gate load determination and, more broadly, the fundamental relationship between bulk solid consolidation and stress fields during storage and flow. These topics remain the focus of ongoing research.

References

1. Roberts, A., *Shock loads in an iron ore train loading bin*, in *Structures and Granular Solids*. 2008, CRC Press. p. 79-88.
2. McLean, A., M. AG, and A. PC, *A simplified approach for the evaluation of feeder loads for mass-flow bins*. 1979.
3. Roberts, A., M. Ooms, and K. Manjunath, *Feeder load and power requirements in the controlled gravity flow of bulk solids from mass-flow bins*. Trans. IE Aust., Mechanical Engineering, Vol. ME9, 1984(1): p. 49–61.
4. Roberts, A., *Belt conveyor technology-an overview of feeder design focusing on belt and apron feeders*. Bulk Solids Handling, 2001. **21**(1): p. 13-26.
5. Arnold, P.C., A.G. McLean, and A.W. Roberts, *Bulk Solids: Storage, Flow and Handling*. 2nd ed. 1979: The University of Newcastle Research Associates (TUNRA).
6. Roberts, A.W., *Basic principles of bulk solids storage, flow and handling*. 1993: Institute for Bulk Materials Handling Research.
7. Standard, A.A., *AS-3774*. Loads on bulk containers. Sydney. AS, 1996. **3773**.
8. Roberts, A., et al., *Study of gate loads related to stress states in mass-flow bins during bulk solids filling and settlement*. Advanced Powder Technology, 2021. **32**(3): p. 683-692.
9. Guo, J., A.W. Roberts, and J.-D. Prigge, *Investigation of arching behavior in mass-flow hoppers under surcharge pressures*. 7th International Conference for Conveying and Handling of Particulate Solids, CHoPS 2012, 2012.
10. Guo, J., *Investigation of arching behaviour under surcharge pressure in mass-flow bins and stress states at hopper/feeder interface*. 2025, Open Research Newcastle.
11. Roberts, A.W., *Review of mass-flow hopper design with respect to stress fields and surcharge loads*. Particuology, 2010. **8**(6): p. 591-594.

Chapter 9
Loads on Support Structural Members Buried in Bins and Stockpiles

9.1 Introduction

Large bulk storage facilities such as stockpiles, bulk storage sheds and bins often contain structural members which become partly buried by the bulk solid during the filling process. Typical examples include trestle support legs on stockpile load-out conveyors and support columns in bulk storage sheds, as illustrated in Fig. 9.1. Figure 9.2 shows an industrial bulk fertiliser storage facility. Two views are shown, one with the shed filled to capacity and the other with the shed completely empty. Comparing the two, raises the dilemma as to how much support of the loaded columns is provided by the bulk solids?

The loads on buried structural members are quite complex and depend on several factors such as the methods of filling and reclaiming, load settlement, the flow properties of the bulk solid and any lateral deflection of the structural element. Furthermore, the cumulative, in-plane type compressive loads due to the shear stresses acting along the structural members can contribute to the buckling loads, thus exacerbating the lateral bending. Depending on the method of reclaim employed, the lateral loads may increase further during the unloading process. These various loading conditions are described and design equations are developed. The background theory was developed by Roberts [1] followed by Katterfeld et al. [2], which formed the foundation of the PhD thesis of He [3] and follow-up work by He et al. [4].

A. W. Roberts, *The Theory and Practice of Bulk Materials Handling*,
https://doi.org/10.1007/978-3-032-19702-3_9

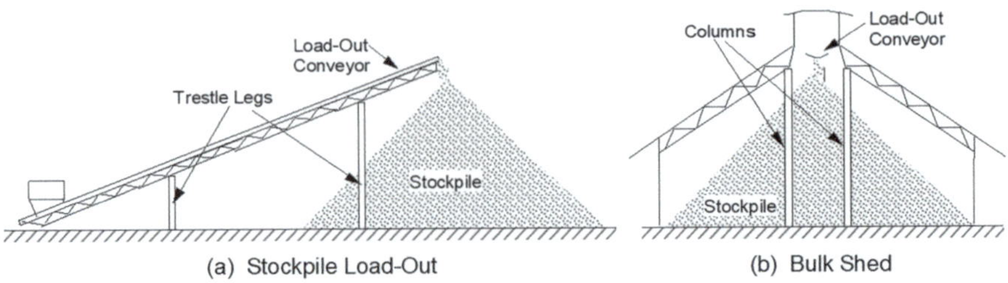

Fig. 9.1 Buried structures in stockpiles

(a) Shed loaded to maximum level **(b) Shed empty showing fully exposed columns**

Fig. 9.2 Bulk storage shed for fertiliser showing loaded and empty conditions

9.2 Loading Conditions on Columns

9.2.1 Loading Conditions on Columns

The loads can be quite variable depending on the following conditions:

- Variations in the flow properties of the material.
- Loading and unloading history and mechanism for loading and unloading.
- Length of time the bulk material remains undisturbed in the stockpile.
- Rigidity of the stockpile floor—whether any settlement has occurred.
- Variations in stress fields from active to passive.

9.2.2 Loading Process—Active Stress State

The column loads are depicted in Fig. 9.3. As the bulk solid is discharged from the load-out conveyor, it flows down the sloping surface of the pile as it is formed, coming to rest with the surface inclined at the angle of repose θ_R of the bulk solid. Within the stored mass, as the depth z increases, the bulk solid continues to move

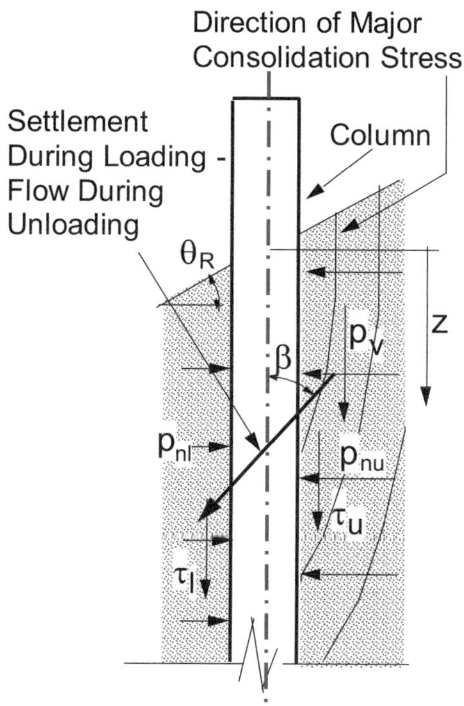

Fig. 9.3 Column loads—active case

by internal shear at a steeper angle. In this case, the slope angle measured from the horizontal approaches that of the effective angle of internal friction δ. The angle β of Fig. 9.3 is such that $\beta = (90^\circ - \delta)$.

Internal creep due to shear continues for some time after the stockpile is loaded. Depending on the stockpile capacity and the properties of the bulk solid, creep may continue for several hours or even longer periods as the bulk solid approaches, asymptotically, its critical consolidation condition. While this movement is not normally observable, being incrementally small as time increases, nonetheless it does occur. The loading condition is said to be *active* since the major principal consolidating stress acts more in the steeply inclined direction as illustrated in Fig. 9.3.

9.2.3 Load Condition—Passive State of Stress

During both storage and reclaim, a *passive* stress state, illustrated in Fig. 9.4, may be established in the region of the columns. In this case, a wedge-shaped flow convergence arises as a result of downward movement or creep of the load due to such causes as

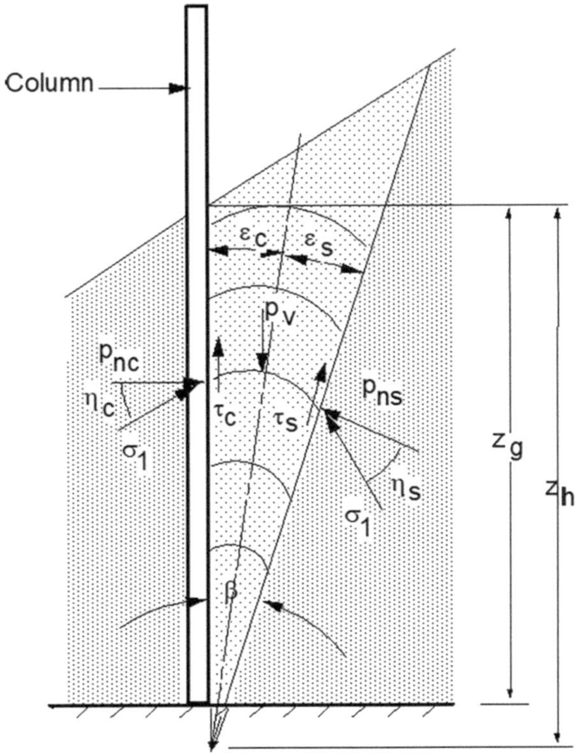

Fig. 9.4 Column loads—Passive case

- Settlement during storage.
- Lateral deflection due to bending of the columns.
- Gravity flow during reclaim.

The downward converging movement against the columns may give rise to an arching condition in which the major consolidating stresses act more in the lateral direction, tangential to the arches.

9.2.4 Loads on Column Faces

The pressures exerted by the bulk solid impose loads on the column faces, as depicted in Fig. 9.5. Referring to Fig. 9.5a, lateral forces F_U act on the leading faces of the columns, with frictional drag forces F_S acting along the sides. A void is created initially behind the lower or rear face. This void may become loosely filled with bulk material, giving rise to a small restraining force F_I acting on the lower face. In the case of cohesive bulk solids, some build-up is likely to occur on the column faces, as illustrated in Fig. 9.5b. This increases the effective areas of the faces. Figure 9.5c shows the loading condition when the rill is at an angle, defined by δ, to the column.

$$F_T = F_U + 2F_S - F_I \tag{9.1}$$

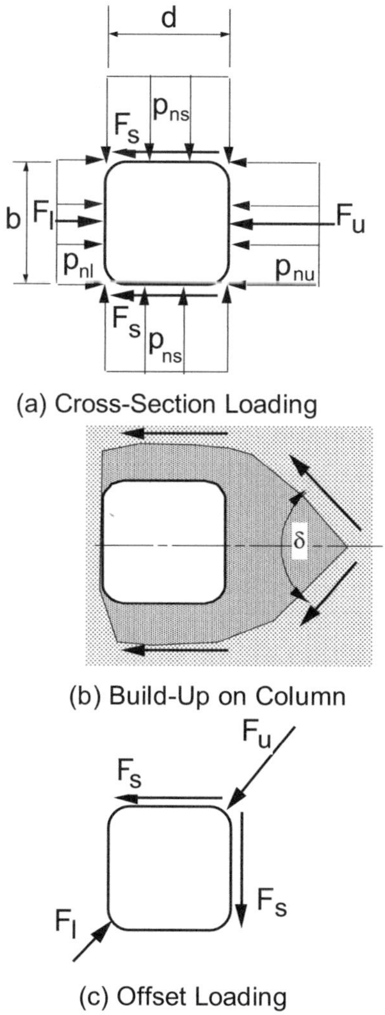

Fig. 9.5 Loads on column faces

9.3 Column Load Model

In the load model for the columns shown in Fig. 9.6, the existence of the failure surface for internal shear defined by the angle β gives rise to the wedge loading condition as illustrated. The centreline of the wedge is defined by the angles α_c and α_s, which depend on the angle of friction at the column surface and the internal shear surface, respectively. It is noted that $\alpha_c + \alpha_s = \beta$. Choosing the coordinates "x" and "y" the equilibrium analysis of the elemental slice of thickness dy is given by the following differential equation:

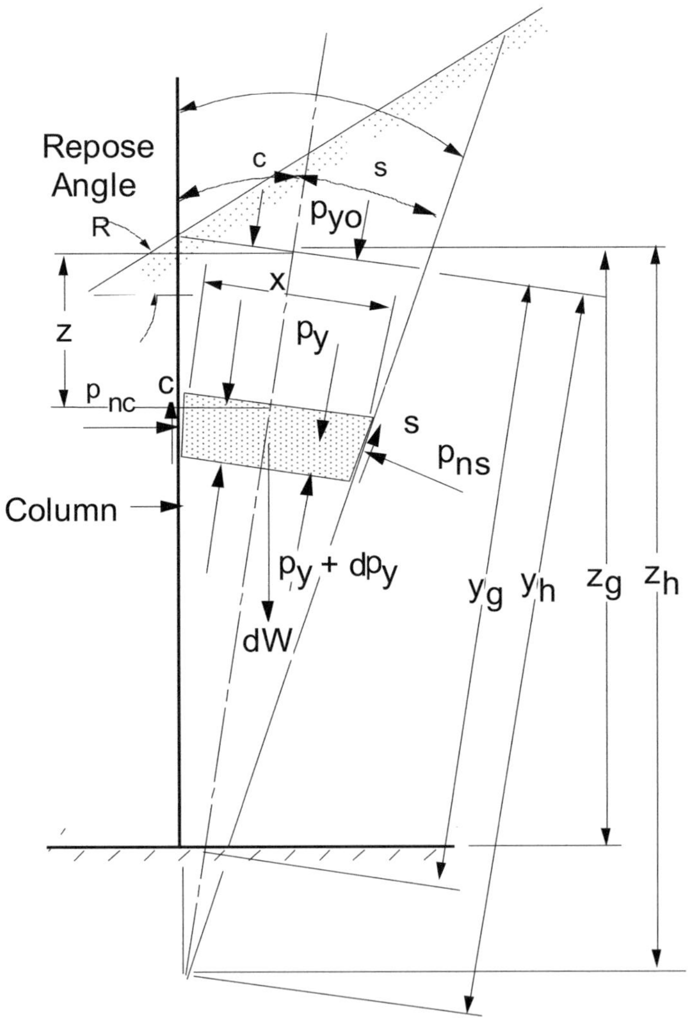

Fig. 9.6 Load analysis model

$$\frac{dP_y}{dy} + \frac{JP_y}{y_h - y} = \gamma_y \tag{9.2}$$

Stress conditions in the wedge of a bulk solid leads to the following equation. The wedge is defined by the angles α_χ and α_σ which depend on the friction at the column surface and for the stress or pressure P_y.

$$P_y = \gamma_y\left[\frac{y_h - y}{j - 1}\right] + \left[\left(P_{yo} - \frac{\gamma_y y_h}{j - 1}\right)\left(\frac{y_h - y}{y_h}\right)^j\right] \tag{9.3}$$

where

$$j = \frac{K_c(\tan\alpha_c + \tan\phi_w) + K_s(\tan\alpha_s + \sin\delta)}{\tan\alpha_c + \tan\alpha_s} - 1 \tag{9.4}$$

$$K_c = \frac{P_{nc}}{P_y};\ K_s = \frac{P_{ns}}{P_y};\ \gamma_y = \gamma\cos\alpha_c$$

γ = bulk specific weight, (kN/m^3); ϕ_w = wall friction angle for column surface

δ = effective angle of internal friction; P_{yo} = surcharge pressure; $y = \frac{z}{\cos\alpha_c}$

9.3.1 Active Case

Based on the discussion in Sect. 9.2, the angle $\beta = (90° - \delta)$. For the active loading condition, it is reasonable to assume that $\alpha_c = \alpha_s = \frac{\beta}{2}$ for which the pressure ratios, $K_c = K_{ca}$ and $K_s = K_{sa}$ are given by

$$K_{ca} = (j + 1)\left[\frac{\tan\alpha_c}{\tan\alpha_c + \tan\phi_w}\right] \tag{9.5}$$

and

$$K_{sa} = (j + 1)\left[\frac{\tan\alpha_s}{\tan\alpha_s + \sin\delta}\right] \tag{9.6}$$

The analysis assumes that on the slip plane the coefficient of internal friction corresponds to the maximum shear stress. Hence $\mu_i = \sin(\delta)$. The parameter 'j' in Eq. (9.5) allows for the degree of compressibility of the bulk solid and the rigidity of the stockpile floor. In this case, it is assumed that $0 \le j \le 0.2$. The maximum limiting case for the pressure P_y occurs when $j = 0$ for which the pressures P_y, P_v, P_{nc} and P_{ns} increase linearly with depth as follows:

$$P_y = P_{vo} + \gamma_s y \tag{9.7}$$

$$P_v = P_{vo} + \gamma z \tag{9.8}$$

$$P_{nc} = K_c(P_{vo} + \gamma_s y) \tag{9.9}$$

$$P_{ns} = K_s(P_{vs} + \gamma_s y) \tag{9.10}$$

9.3.2 Passive Case

In this case the stress field is as shown in Fig. 9.3 with the slope angle β of the failure plane defined as $\beta = 45^\circ - \frac{\delta}{2}$. Equation (9.3) applies in this case with the pressure ratios $K_c = K_{cp}$ and $K_s = K_{sp}$ given by

$$K_c = \frac{1 + \sin\delta\cos(2\eta_c)}{1 - \sin\delta\cos[2(\alpha_c + \eta_c)]} \tag{9.11}$$

$$K_s = \frac{1 + \sin\delta\cos(2\eta_s)}{1 - \sin\delta\cos[2(\alpha_s + \eta_s)]} \tag{9.12}$$

where

$$\eta_c = 0.5\left[\phi_w + \sin^{-1}\left(\frac{\sin\phi_w}{\sin\delta}\right)\right] \tag{9.13}$$

$$\eta_s = 0.5\left[\phi_{ws} + \sin^{-1}\left(\frac{\sin\phi_{ws}}{\sin\delta}\right)\right] \tag{9.14}$$

$$\phi_{ws} = \tan^{-1}(\sin\delta) \tag{9.15}$$

9.3.3 Column Side Load Pressures

The pressures acting on the column side walls denoted by P_{ns} in Fig. 9.5a may be taken to be the same as pressure P_{nu} on the front face. Normally, these side pressures will be due to an active stress state, although both active and passive stress states may be considered.

9.3.4 Support Loads on Rear or Lower Face of Columns

As previously discussed, the support against lateral deflection of the columns is uncertain and, generally, will be of low order, particularly in the case of cohesive bulk solids. The conservative approach is to assume that the pressure P_{hi} (Fig. 9.5a) is zero, which may certainly be the case during the reclaim or unloading process, as illustrated in Fig. 9.7. However, for the case when the stockpile is being filled, it is reasonable to assume that some rear support is achieved through the process of back-filling on the lower or rear side of the column. This loading condition is depicted in Fig. 9.8. It is assumed that an active state of stress exists with the vertical stress P_{vi} and normal stress P_{nl} increasing linearly with depth z as follows:

$$P_{vi} = \gamma z \tag{9.16}$$

and

$$P_{nl} = k\gamma z \tag{9.17}$$

It may be shown that the ratio of normal to vertical stress, k, is given by Eq. (9.18).

$$k = \frac{1 + \sin^2\delta - 2\sqrt{\sin^2\delta - \tan^2\phi_w \cos^2\delta}}{4\tan^2\phi_w + \cos^2\delta} \tag{9.18}$$

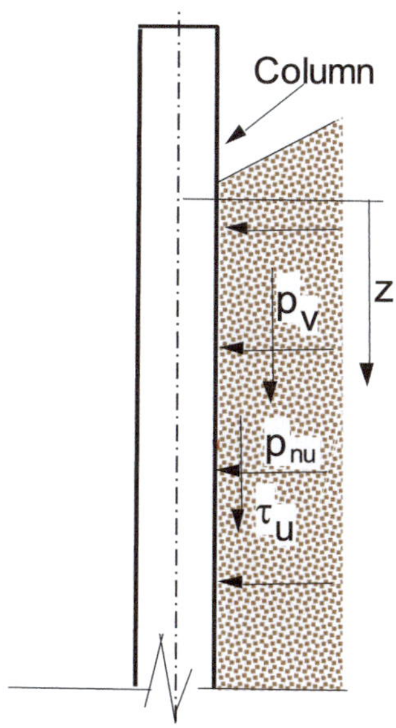

Fig. 9.7 Unloading—no rear support

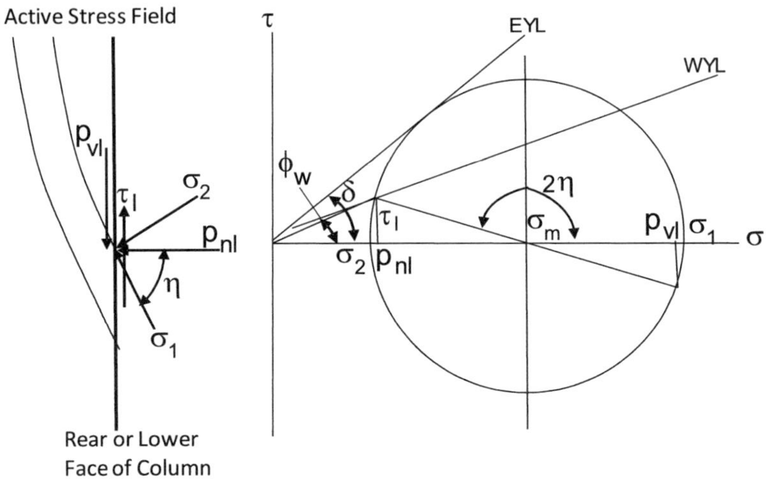

Fig. 9.8 Stress conditions on rear face

To examine the level of support provided by the bulk solid in contact with the rear face of the column, it is useful to compare the pressure ratio k given by Eq. (9.18) with the pressure ratio K_{ca} for the active case as given by Eq. (9.5). The computed results are presented in Fig. 9.9. The lower the value of the ratio of $\frac{k}{K_{ca}} < 1.0$, the less the support provided by the bulk solid acting on the rear face of the column. As the ratio $\frac{k}{K_{ca}} \rightarrow 1.0$, the greater the support and the less the net lateral load on the column. Total support is achieved, theoretically, when $\frac{k}{K_{ca}} = 1.0$. For normal values of the friction angles for contact of the bulk solid with the column surface, this condition is only approached for free-flowing bulk solids of low effective angles of internal friction. It needs to be noted that once a passive state of stress is established on the upper face of the column, the passive stress ratio K_{cp} as defined by Eq. (9.11) will, invariably, be larger than the value of k as defined by Eq. (9.18). Hence, at best, the back filling can only provide partial support against lateral bending of the column.

9.3.5 Total Lateral Forces on Columns

At an arbitrary depth z, the net force per unit depth is given by

$$F_T = F_U + 2F_S - F_l \tag{9.19}$$

Assuming $P_{ns} = P_{nu}$

$$F_T = C_A\, P_{nu}\, (b + 2\, d \tan \phi_w) - P_{nl}\, b\ (\text{kN/m}) \tag{9.20}$$

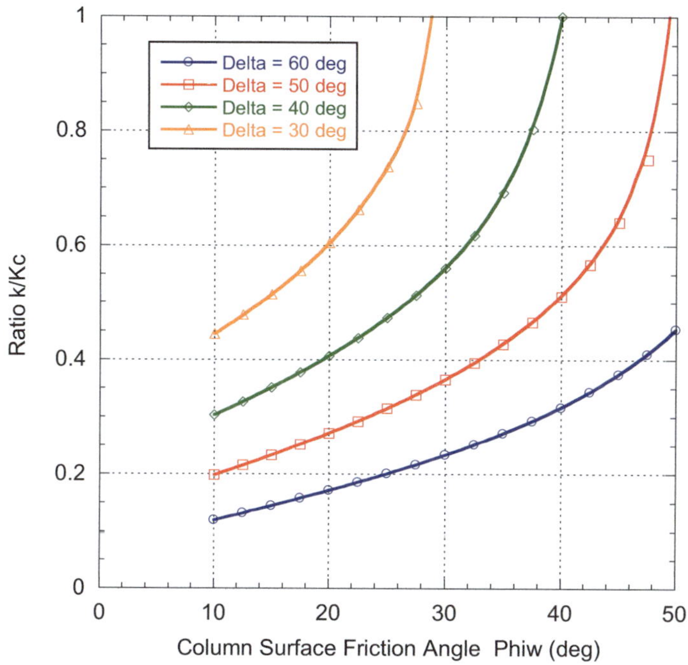

Fig. 9.9 Ratio of k/K_{ca} for active stress states

The coefficient C_A allows for build-up on the columns as in Fig. 9.5b. Normally, $1.5 \leq C_A \leq 2.0$, "b" and "d" are the cross-sectional dimensions of the column.

9.3.6 Column Compressive Loads

The vertical shear stresses acting along the column faces give rise to compressive loads which increase down the column. The compressive force per metre width of column face is given by

$$F_C = b \int_0^z \left((P_{nu} + P_{ni}) + 2P_{ns} \frac{d}{b} \right) \tan \phi_w \, dz \, (kN) \tag{9.21}$$

For the upper and side faces, when $P_{nu} = P_{ns}$; $\frac{d}{b} = 1$; and $P_{ni} = 0$, Eq. (9.21) becomes

$$F_C = 3b \left[\frac{C_1}{2} \left[z_h^2 - (z_h - z)^2 \right] + \frac{C_2}{j+1} \left[z_h^{j+1} - (z_h - z)^{j+1} \right] \right] + F_{zo} \, (kN) \tag{9.22}$$

where

$$C_1 = \frac{\mu K_c \gamma}{j - 1} \tag{9.23}$$

$$C_2 = \frac{\mu K_c}{z_h^j}\left[P_{zo} - \frac{\gamma z_h}{j - 1}\right] \tag{9.24}$$

$$\mu = \tan \phi_w$$

$$F_{zo} = \text{surcharge load on column}$$

When the pressure P_{nl} acting on the rear face is considered; there will be an additional compressive load imposed. This additional load is given by

$$F_{CL} = \mu\, k\, \gamma\, b\, \frac{z^2}{2} \tag{9.25}$$

Hence, the total compressive load is given by the sum of F_C and F_{CL}.

9.4 Case Study Example

To illustrate the application of the foregoing analysis, the case of a storage shed similar to that of Figs. 9.1b and 9.2 is considered. The support columns are 250 mm square tube and the bulk solid is a particular fertiliser for which $\rho = 2.1$ t/m^3, $\delta = 46°$ and $\phi_w = 25°$. When the fertiliser is fully loaded in the shed, the columns are buried to a depth of 13.5 m. For the active case, the parameter "j" of Eq. (9.5) is taken to be 0.2 and the loads on the upper and side faces of the columns are determined for the parameter $C_A = 1.5$ in Eq. (9.20).

The lateral loads per unit height of the columns are plotted in Fig. 9.10 for the active and passive cases. The F_T loads are the computed loads assuming that the pressure on the lower rear face of the columns $P_{nl} = 0$. The F_{Tl} loads allow for some support on the lower rear face. The lateral forces give rise to bending moments in the columns. Figure 9.11 compares the compressive in-plane traction forces in the columns due to the stored fertiliser for the active and passive cases. It is assumed that the friction on all four faces of the column is fully mobilised over the total immersed depth. These compressive forces are in addition to the structural compressive loads due to the roof and load-out conveyor supports, all of which need to be considered in the buckling analysis.

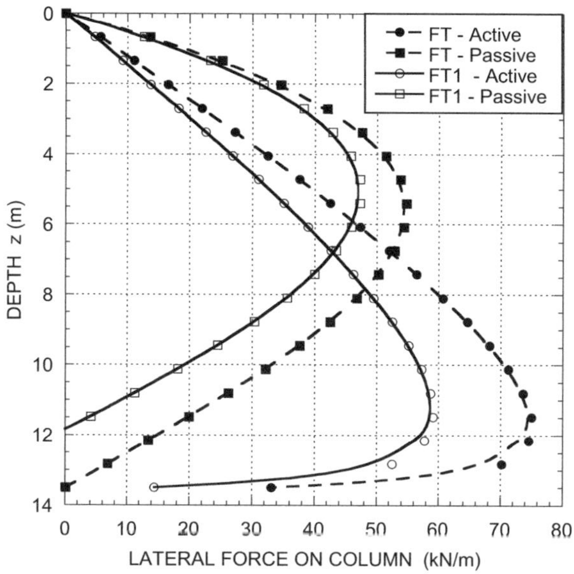

Fig. 9.10 Total distributed lateral forces

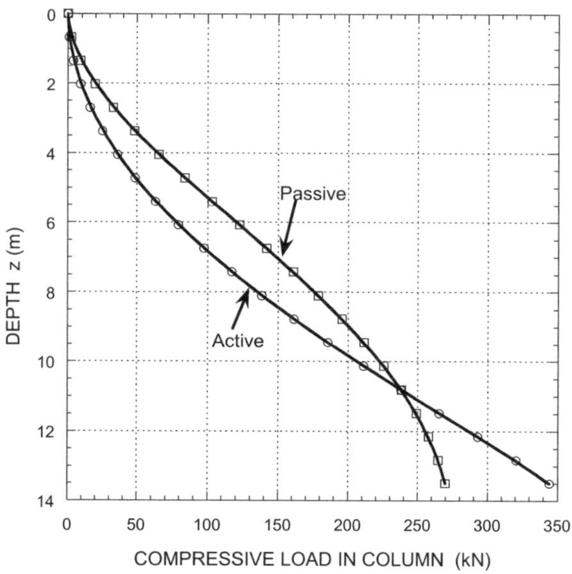

Fig. 9.11 Compressive traction forces

9.5 Experimental Model Studies

Tests were conducted on a small scale, half section conical model stockpile fitted with a column made of a plastic material shown in Fig. 9.12. A Perspex plate was fitted to represent the centreline plane of the stockpile and allowed visual observations of the loaded column to be made. To enable the column loads to be measured during the experiments, the column was calibrated to determine its load versus deflection characteristics under lateral bending conditions. The deflections were to vary linearly with load over the range of loads experienced during the experiments. Both dry and moist sand were used as the test materials. In the case of the dry sand, which was quite free flowing, the back-filling process was quite rapid during the formation of the pile and there were no observable lateral deflections of the column. In this case, the sand provided good support against lateral bending, thus confirming the predictions outlined in Sect. 9.3.4.

However, when moist sand was used, the cohesive characteristics of the sand greatly impeded the back-filling operation leading to little rear support and significant lateral bending as shown in Fig. 9.12b. The calculated loads on the column for the active and passive cases, allowing for rear support, are shown in Fig. 9.12c. The study indicated that the passive load condition is the more likely since the calculated deflection of 13 mm, based on the passive case with some rear column support, compared reasonably well with the measured value of $X = 12$ mm (Fig. 9.12c).

9.6 Tubular Trestle Leg Conveyor Support Structures Buried in Stockpiles

9.6.1 Introductory Remarks

In the mining industries, the storage of bulk solids in large quantities is often achieved through the use of open stockpiles in which the reclaim is by gravity flow through hoppers, feeders and/or chutes onto reclaim conveyors installed below ground level. In many cases, the load-out conveyor supplying the bulk solid to the stockpile is via a boom conveyor supported by a vertical support structure located outside the peripheral area of the stockpile rill zone at ground level. The support structure is not in contact with the stored stockpile material. In other cases, particularly in the coal industry, the load-out conveyor is supported on trestle legs which are substantially buried in the stockpile as illustrated in Fig. 9.13 [5].

For the efficient operation of the storage and load-out system, it is important that the stockpile geometry and the positioning of the reclaim hoppers and feeders be such that the tonnage of bulk solids reclaimed by gravity flow is achieved. Procedures for achieving this objective are presented by Roberts et al. [6, 7]. Reference to Fig. 9.13 shows that the reclaim hoppers, feeders and conveyor are offset an amount defined by “x” from the centre of the stockpile. This is deliberate. Since the reclaim volume

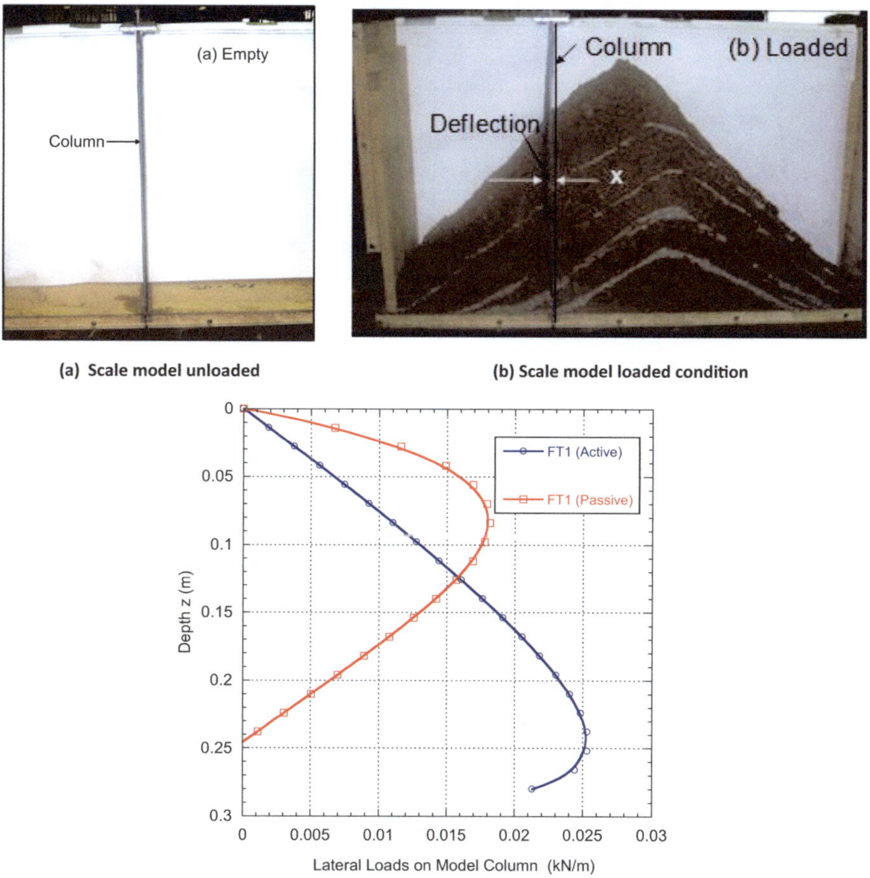

(a) Scale model unloaded **(b) Scale model loaded condition**

(c) Calculated column loads

Fig. 9.12 Scale model stockpile experiments

is controlled by the draw-down H_D measured from the free surface of the pile, it can be shown that the improvement in live capacity is possible over that achieved by locating the reclaim centrally. Furthermore, the offset reclaim assists the layout for the central positioning of the load-out conveyor and its trestle leg supports.

9.6.2 Loading Conditions on Trestle Legs

The loading conditions and associated theory outlined in the previous sections apply to buried structural members with square or rectangular cross-sections. For trestle legs, the same principles are relevant; however, their circular cross-section introduces additional considerations that must be accounted for in the analysis. In general, the

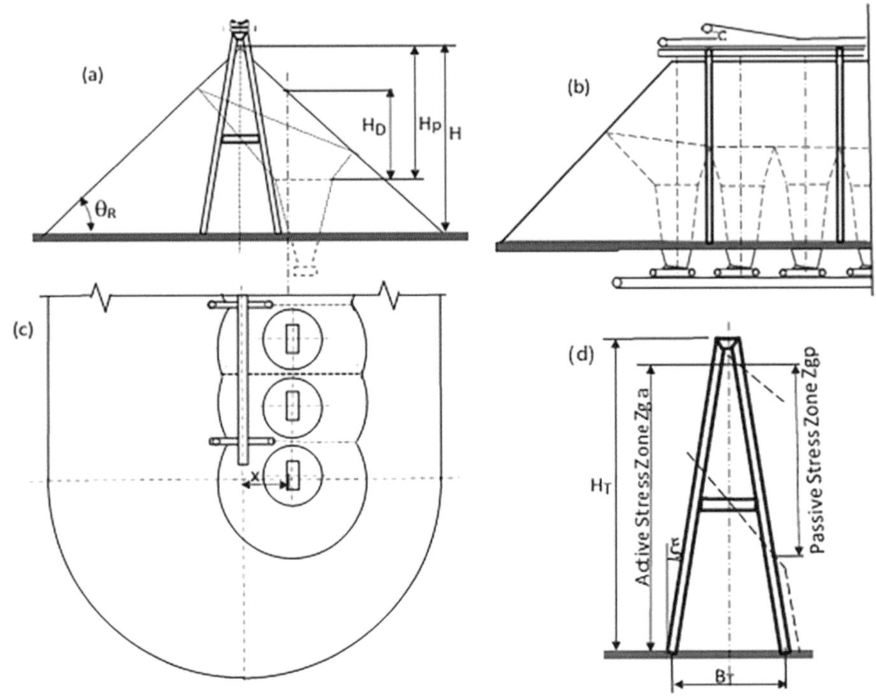

Fig. 9.13 Schematic arrangement of stockpile showing trestle legs, load-out conveyor and reclaim system [5]

principal loads are those due to the variation of the normal pressures with depth during the filling and reclaim processes, and the non-uniformity or eccentricity of these pressures acting around the periphery of the trestle legs. The combination of these pressure variations gives rise to bending stresses in the support trestle legs. Furthermore, the cumulative, in-plane type compressive loads due to the shear stresses acting downward along the trestle legs can contribute to the buckling loads, thus exacerbating the lateral bending.

Referring to Fig. 9.13, it is assumed that the stockpile is filled to its maximum height so that the trestle leg structures are fully immersed in the coal with the highest loads occurring on the side of the legs nearest the tripper position during filling operation and on the side of the legs furthest from the reclaim openings during discharge.

9.7 Trestle Leg Load Analysis Model

The model for the trestle leg load analysis together with the load cross-sectional distribution is shown in Fig. 9.14. Figure 9.14a presents the model for the analysis of the active and passive stress fields on the upper side of the trestle leg. The dimension z_g defines the depth applicable to the relevant stress field. This dimension, in general, will be different for the active and passive cases. Figure 9.14b shows the normal stress distribution around the trestle leg from the upper value p_{nu} to the lower value p_{nl} on the rear side of the trestle leg.

Referring to Fig. 9.14a, the pressure P_{nc} is the peak pressure at depth z as determined for the active and passive cases, respectively.

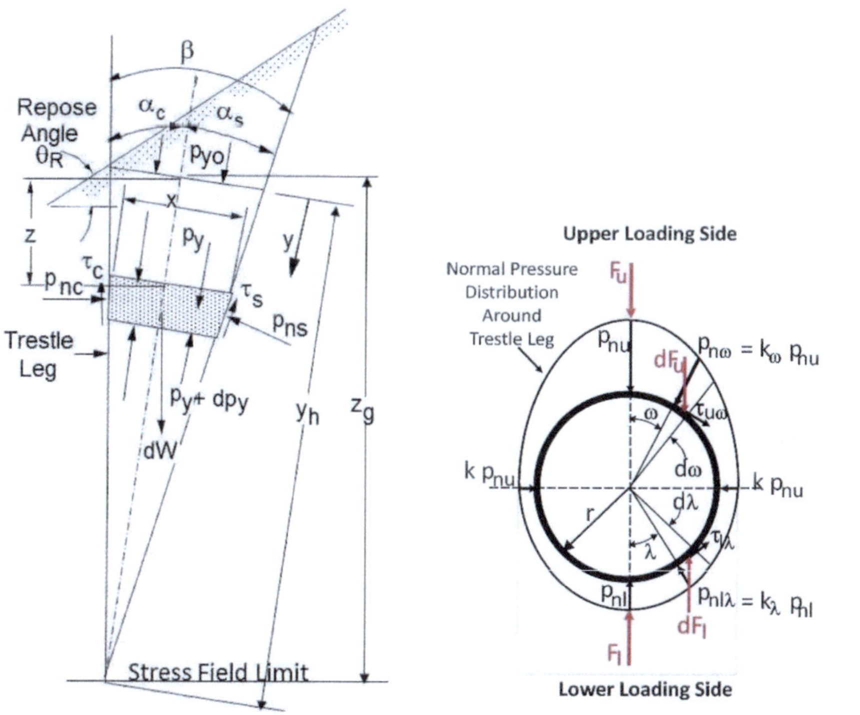

(a) Load analysis model **(b) Normal stress distribution around leg**

Fig. 9.14 Trestle leg load analysis model

9.7.1 Upper or Loading Half of Trestle Leg

At an arbitrary position defined by the angle ω, the normal pressure is defined by

$$P_{n\omega} = k_\omega P_{nu} \tag{9.26}$$

where k_ω is expressed by

$$k_\omega = 1 - (1 - k)\sin\omega \tag{9.27}$$

such that $0.35 \le k \le 1.0$. It is noted that for the lower bound value of $k = 0.35$, $k_\omega = 1$ for $\omega = 0$, and $k = 0.35$ when $\omega = \frac{\pi}{2}$. It is assumed there is some build-up on the trestle surface, so the friction coefficient μ_i defining the shear stress τ is $\mu_i = \sin\delta$.

The elemental force per unit length dF_u is given by

$$dF_u = r\,P_{un\omega}\cos\omega + \mu_i \sin\omega\, d\omega \tag{9.28}$$

The total force F_u per unit length is given by

$$F_u = 2P_n r \int_0^{\pi/2} [1 - (1 - k)\sin\omega][\cos\omega + \mu_i \sin\delta]\, d\omega \tag{9.29}$$

which becomes

$$F_u = P_{nu} D_e \tag{9.30}$$

where D_e is the effective trestle leg diameter and is given by

$$D_e = 2\,C\,r \tag{9.31}$$

and

$$C = \left[1 - \frac{(1 - k)}{2}\right] + \mu_i\left[1 - \frac{(1 - k)\pi}{4}\right] \tag{9.32}$$

where r = outside radius of trestle leg

The recommended lower bound value is $k = 0.35$, which, for $\delta = 50^\circ$, gives $C = 1.05$ and $D_e = 1.05D$, where $D = 2r$ = actual trestle leg outside diameter. The value $k = 1.0$ corresponds to the pressure P_{nu} being constant around the upper loading half of the trestle leg. For this maximum or upper bound case, $C = 1.7$, for which $D_e = 1.7D$.

9.7.2 Backfilling Half of Trestle Leg

This is determined in a similar way to the above. In this case, when $\lambda = 0$, $k_\omega = k_n$ as defined by Eq. (9.18).

9.7.3 Effective Trestle Leg Diameter for Load Calculations

It is noted that AS3774 [8] recommends that the load calculations for buried structural elements be based on the projected area corresponding to $D_e = 2D$. The Standard seems to be referring more to horizontal or lateral support tie beams in hoppers and silos rather than vertical or steeply inclined support legs buried in stockpiles. It then becomes a value judgement as to whether the factor of 2 should be applied in the case of the trestle leg loads. It needs to be noted that, over a period, progressive adhesion and build up may occur around the trestle legs, thus increasing the effective diameter. This is more likely in the case of bulk materials exhibiting a high level of cohesion. If this is a distinct likelihood, then the conservative approach is to select $D_e = 2D$. On the other hand, the analysis presented in Sect. 9.7.1 above shows that the lower and upper bound values to be $C = 1.05$ and $C = 1.7$, respectively. Weighing up the various factors, the recommended lower limit is $C = 1.2$.

9.7.4 Trestle Leg Compressive Loads

The vertical shear stresses acting along the column faces give rise to compressive loads which increase down the column. At an arbitrary depth location, z, corresponding to coordinate y, the vertical shear stress on the column face is given by

$$\tau = K\,P_y \tag{9.33}$$

where $K = z = \mu_i\,K_{ca}$ for the active case and $K = K_{cp}$ for the passive case.

At depth z, the cumulative compressive force per metre of trestle leg circumference is given by

$$F_C = \int_0^z \tau_z\,dz\ (\mathrm{kN/m}) \tag{9.34}$$

Noting that the vertical pressure P_y is given by Eq. (9.3), then substituting for P_y in (9.33), for τ_z in Eq. (9.34) then integrating gives the following expression for the compressive force per unit circumference

$$F_C = \frac{C_1}{2}\left[z_h^2 - (z_h - z)^2\right] + \frac{C_2}{j+1}\left[z_h^{j+1} - (z_h - z)^{j+1}\right] + F_{Z0} \tag{9.35}$$

where

$$C_1 = \frac{\mu\, K_c\, y}{j - 1} \tag{9.36}$$

and

$$C_2 = \frac{\mu\, K_c}{z_h^j}\left[p_{zo} - \frac{\gamma\, z_h}{j - 1}\right] \tag{9.37}$$

$\mu_i = \tan \phi_w$ which allows for some build-up of bulk solid on the trestle leg surfaces.

F_{Z0} = surcharge load on each leg due to the weight of the conveyor, its load and the support structure.

While the normal stress and corresponding shear stress will vary around the leg circumference as illustrated in Fig. 9.14, it is recommended that the conservative approach for the compressive loads be adopted by assuming the normal and shear stresses are the maximum values uniformly distributed around the periphery of the leg.

9.8 Trestle Leg Design Example

A particular case of the load-out coal stockpile and reclaim system of Fig. 9.13 is now considered. The relevant details are as follows:

Stockpile height fully loaded $H_s = 32$ m, trestle leg Height $H_T = 34$ m, leg slope angle $\varepsilon = 10^\circ$, leg spacing $B_T = 1.2$ m, leg diameter $D = 1.2$ m, active stress zone $z_{ga} = 3$ m, passive stress zone $z_{gp} = 22.4$ m, reclaim hopper offset $X = 7$ m, angle of repose $\theta_R = 38^\circ$, bulk density of coal $\rho = 1.0\ \mathrm{t/m^3}$, trestle friction angle $\phi = 30^\circ$, static internal friction $\phi_t = 40^\circ$, effective internal friction angle $\delta = 50^\circ$.

Relevant results are presented in Figs. 9.15, 9.16, 9.17 and 9.18. Figure 9.15 shows the normal pressures for the active and passive states for the upper loading face, and rear and backfilling face. The corresponding net normal pressures acting on the loading faces of the trestle legs are shown in Fig. 9.16. These pressures are used to calculate the corresponding loads in accordance with Sects. 9.6 and 9.7. Figure 9.17 shows the circumferential normal pressure distributions around the trestle legs at critical levels for the passive case at the depth 8.3 m and for the active case at the depth of 27 m. For the active case, the parameter $j = 0.2$ is assumed. Figure 9.18 shows the compressive loads per metre of circumference acting on the trestle legs.

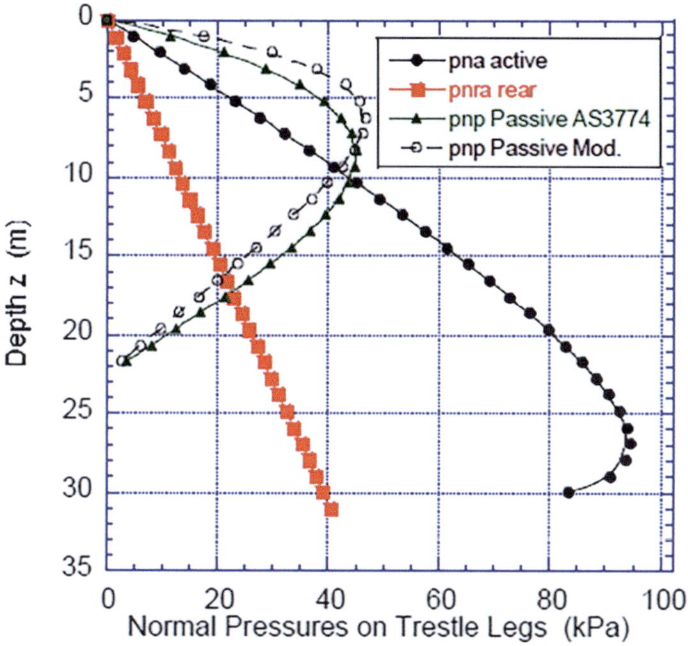

Fig. 9.15 Normal pressure on trestle leg

9.9 Concluding Remarks

The loading conditions experienced in conveyor support trestles buried in coal stockpiles are both complex and variable. The loads are a combination of lateral loads due to filling and reclaim, and in-plane compressive loads due to the draw-down forces. The magnitude and distribution of the loads over the trestle leg faces depend very much on the properties of the coal and the way the stockpile is formed and reclaimed. The loads result from an active stress state during filling and from a passive stress state during reclaim. The degree of lateral support due to the backfilling is quite uncertain and should be neglected if there is any doubt as to the possible level of support that is provided.

It is important that the compressive forces in the buried trestle legs due to the cumulated shear or traction forces acting downward around the leg surfaces be taken into consideration. These forces contribute to the compressive buckling loads in the legs by adding to the compressive structural loads that exist in supporting the load-out conveyor.

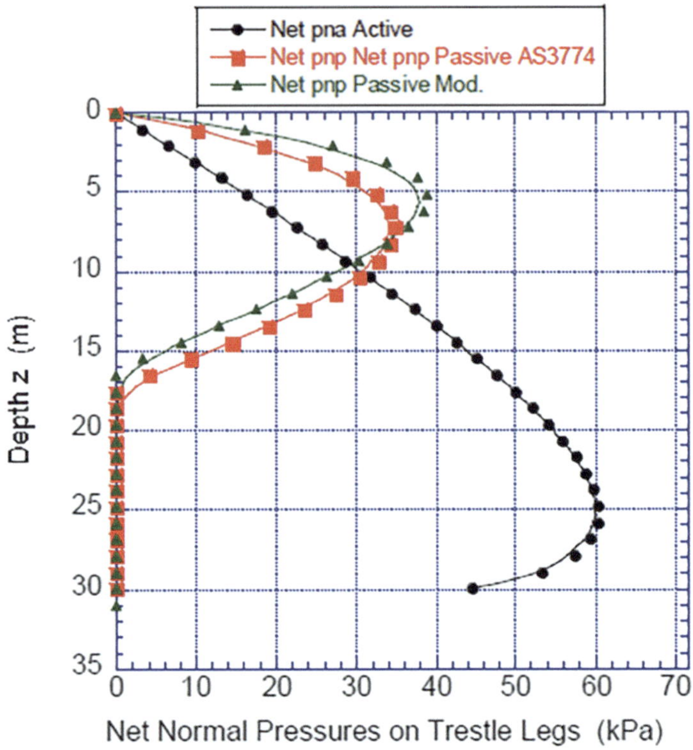

Fig. 9.16 Net normal pressure on trestle legs

Due to the uncertainty of the loading conditions, it is important that a conservative approach be adopted in the design. It is also important to note that any loads due to mobile equipment working on the stockpile, as well as dynamic effects, slumping and possible avalanching due to high increases in moisture are not included in the loads determined herein.

This chapter is part of ongoing research into the subject of loads on buried structures. Follow up research is currently being undertaken through the integration of DEM and FEA analysis of the buried structures and trestle leg loads. The results of this follow-up work can be correlated with the predictions of the continuum mechanics approach described herein with a view to establishing a reliable design methodology for predicting the loads in buried structures.

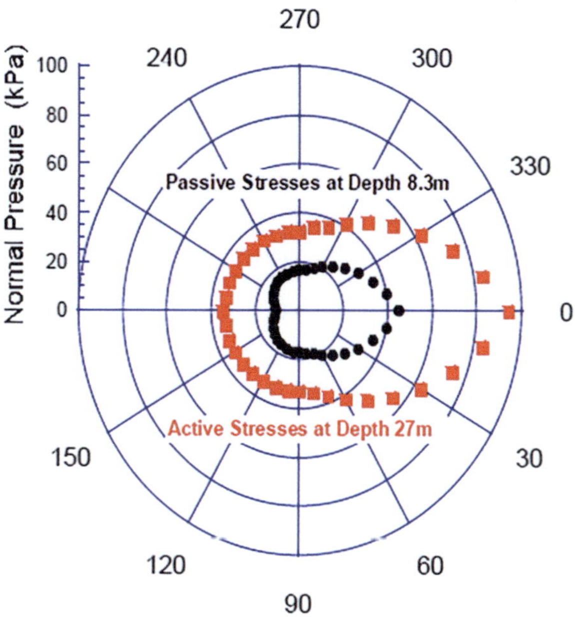

Fig. 9.17 Circumferential pressure distribution

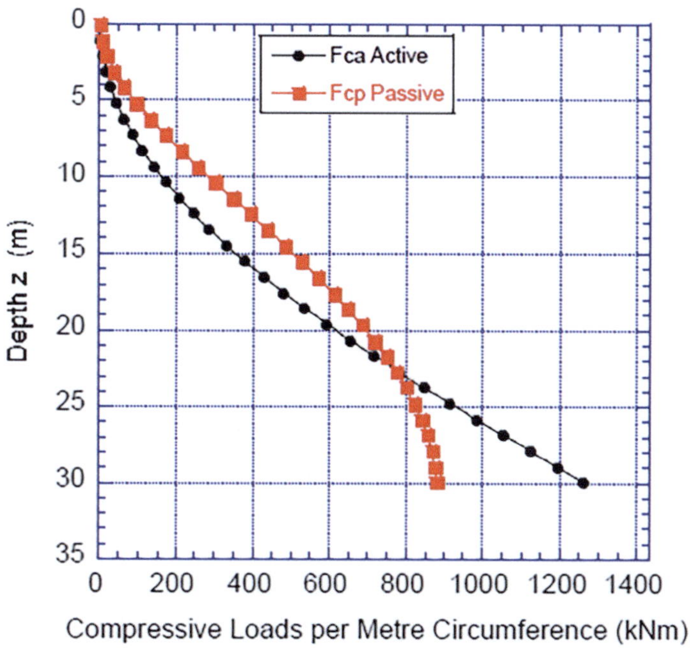

Fig. 9.18 Compressive loads per metre circumference

References

1. Roberts, A.W., *Loads on Support Structural Elements Buried in Stockpiles.* Particle & Particle Systems Characterization, 2007. **24**(4–5): p. 352–359.
2. Katterfeld, A. and A.W. Roberts, *Loads on buried structural elements during stockpile filling and reclaim.* Bulk Solids & Powder: Sci. Technol., 2009. **4**: p. 15–22.
3. He, Y., *Investigation of the mechanics of funnel flow in relation to draw-down and loads on buried structures in stockpiles.* 2025, Open Research Newcastle.
4. He, Y., A. Roberts, and J.-D. Prigge, *Pressures on the support columns buried in iron ore stockpiles.* Powder Technology, 2014. **258**: p. 358–369.
5. Roberts, A.W.D., T.J.; Chen B.; Munzenberger, P. *Load Analysis of Trestle Conveyor Support Structures Buried in Stockpiles.* in *The 12th International Conference on Bulk Materials Storage, Handling and Transportation (ICBMH 2016).* 2016. Darwin N.T.: Engineers Australia.
6. Roberts, A.W., *Characterisation for Hopper and Stockpile Design*, in *Characterisation of Bulk Solids.* 2005. p. 85–131.
7. Roberts, A.W., S.J. Wiche, and T. Krull, *Review of funnel-flow theory in relation to draw-down and live capacity of gravity reclaim stockpiles.* Bulk Solids Powder Sci. Technol., 2008. **3**: p. 97–107.
8. AS3774–1990, A.S., *Loads on Bulk Solids Containers.* Standards Association of Australia, 1990.

Chapter 10
Feeding of Bulk Solids—Part 1—Review of Design Objectives and Feeder Selection

10.1 Introduction

In bulk handling operations, feeders are employed extensively to control the flow of bulk solids from storage bins and stockpiles. While the scale of operations varies widely from large feeders in mining operations to small feeders in powder processing, the underlying principles of feeder design remain the same. To achieve efficient feeding, it is important to understand the interactive roles of the bin and feeder. In this respect the bin, or more particularly the hopper, together with the selected feeder, must be considered as one integral unit and designed accordingly. The role of discharge chutes used in the feeding process should also be fully appreciated. The combination of hoppers, feeders and chutes in the feeding process may be summarised as follows:

- **Gravity flow hopper**: The hopper geometry and initial wall friction characteristics in conjunction with the flow properties of the bulk solid establishes the type of discharge flow pattern and maximum potential rate of discharge.
- **Feeder**: The feeder controls and meters the flow of bulk solids from the hopper to meet the specified throughput. It is important that the interfacing of the feeder is such that the hopper outlet is fully active.
- **Chute**: Often a chute is used in conjunction with a feeder to transfer the bulk solid onto a conveyor. While primarily a flow directing device, the chute, when properly designed, can control the velocity of material. On discharge, this feature is of particular importance when feeding bulk solids onto conveyor belts where uniform deposition of material with minimum belt wear, spillage and power losses are essential.

The design and performance of feeders have been the subject of considerable research along with the strong support and feedback through industry consulting projects and experience. With this background, the essential aspects of feeder performance and design are the basis of material presented herein.

A. W. Roberts, *The Theory and Practice of Bulk Materials Handling*,
https://doi.org/10.1007/978-3-032-19702-3_10

10.2 Handling Plant Design—Basic Concepts

As has been well established from the original work of Jenike [1, 2], there are two basic modes of flow in bins namely, mass-flow and funnel-flow as illustrated in Fig. 10.1. The fundamental aspects of bin design based on these flow modes have been discussed in previous Chapters of this book with Chap. 4 focussing on mass-flow bin design and Chap. 5 on funnel-flow bin and stockpile design. In view of the importance to feeder design where the hopper and feeder interface and associated hopper geometry and feeder selection, the modes of flow are of particular significance. By way of background, the following summary is presented.

10.2.1 Modes of Flow in Bins of Symmetrical Geometry

The two principal modes of flow based on symmetrical geometry are reproduced in Fig. 10.1.

In mass-flow the bulk solid is in motion at every point within the bin whenever material is drawn from the outlet. The opening dimension B is determined from the bulk solid flow properties to meet the required discharge flow rate. There is flow of bulk solid along the walls of the cylinder (the upper parallel section of the bin) and the hopper (the lower tapered section of the bin). Mass-flow guarantees complete discharge of the bin contents at predictable flow rates. It is a 'first-in, first-out' flow pattern; when properly designed, a mass-flow bin can re-mix the bulk solid during discharge should the solid become segregated upon filling of the bin. Mass-flow requires steep, smooth hopper surfaces and no abrupt transitions or in-flowing

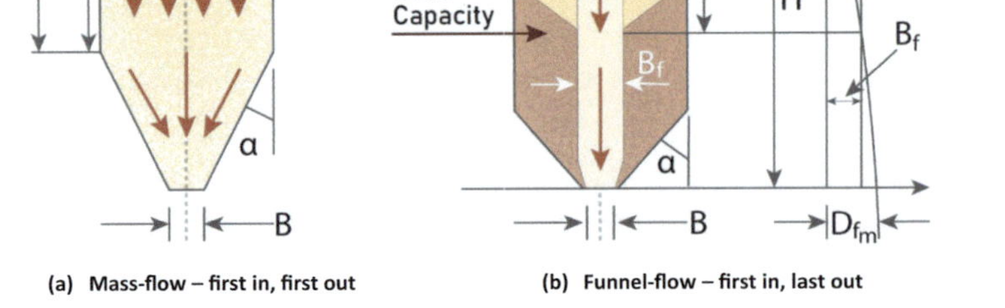

Fig. 10.1 Modes of flow

valleys. Mass-flow is particularly essential in the case of fine powders where it is most important to prevent air entrainment and flooding during discharge.

Mass-flow bins are classified according to the hopper shape and associated flow pattern. The two main hopper types are conical hoppers which operate with axi-symmetric flow and wedged-shaped or chisel-shaped hoppers in which plane-flow occurs. In plane-flow bins, the hopper half-angle "α" will usually be, on average, approximately 8 °–10 ° larger than the corresponding values for axi-symmetric bins with conical hoppers. Therefore, they offer larger storage capacity for the same head room than the axi-symmetric bin, but this advantage is somewhat offset by the long-slotted opening which can give rise to feeding problems. Pyramid shaped hoppers, while simple to manufacture, are undesirable in view of build-up of material that is likely to occur in the sharp corners or in-flowing valleys. This may be overcome by fitting triangular-shaped gusset plates in the valleys.

Funnel-flow occurs when the hopper does not meet the requirements for mass flow. That is the walls are not sufficiently steeply sloped, and/or the walls of the hopper are not smooth enough. In this case, the bulk solid sloughs off the top surface and falls through the vertical flow channel that forms above the opening. It is classified as a 'first-in last-out', that is, the bulk solid that enters the bin during filling is always the last to discharge. Flow is generally erratic and gives rise to segregation problems. Flow will continue until the level of the bulk solid in the bin drops an amount H_D equal to the draw-down. At this level, the bulk strength of the contained material is sufficient to sustain a stable rathole of diameter D_f as illustrated in Fig. 10.1b. Once the level defined by H_D is reached, there is no further flow and the material below this level represents 'dead' storage. This is a major disadvantage of funnel-flow.

For complete discharge, the bin opening needs to be at least equal to the critical rathole dimension determined at the bottom of the bin corresponding to the bulk strength at this level. However, for many cohesive bulk solids and for the normal consolidation heads occurring in practice, ratholes measuring several metres are often determined. This makes funnel-flow impracticable. Funnel-flow has the advantage of providing wear protection of the bin walls, since the material flows against stationary material. However, it is a 'first-in last-out' flow pattern which is unsatisfactory for bulk solids that degrade with time. It is also unsatisfactory for fine bulk solids of low permeability. Such materials may aerate during discharge through the flow channel, and this can give rise to flooding problems or uncontrolled discharge.

10.2.2 Expanded-Flow

The disadvantages of funnel-flow are overcome using expanded-flow, as illustrated in Fig. 10.2. This combines the wall protection of funnel-flow with the reliable discharge of mass-flow. Expanded flow is ideal where large tonnages of bulk solid are to be stored. For complete discharge, the dimension at the transition of the funnel-flow and mass-flow sections must be at least equal to the critical rathole dimension at that

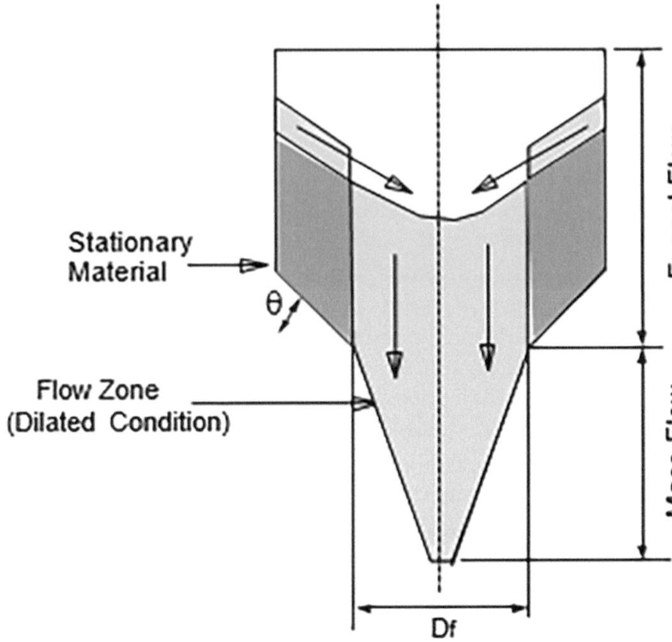

Fig. 10.2 Expanded flow

level. Expanded-flow bins are particularly suitable for storing large quantities of bulk solids while maintaining acceptable head heights.

The concept of expanded flow may be used to advantage in the case of bins or bunkers with multiple outlets. Expanded flow is also employed to advantage in the case of gravity reclaim stockpiles as illustrated in Fig. 10.3. Of particular importance is the feeder load control provided by the mass-flow hopper which is due to the *passive* stress field generated in the hopper rendering the load and corresponding torque and drive power to be independent of the high surge pressure of the stored bulk solids in the stockpile.

As a general comment, symmetric bin shapes provide the best performance. Asymmetric shapes often lead to segregation problems particularly with free-flowing material having a wide range of particle sizes.

10.2.3 Hopper and Feeder Interface

As has been stated, it is most important that the outlet of the mass-flow hopper be fully active. This is to preserve the necessary axi-symmetric stress field required for conical hoppers or plane symmetric stress field for plane-flow hoppers. Since the role of the selected feeder normally involves a uni-directional feeding action, the stress

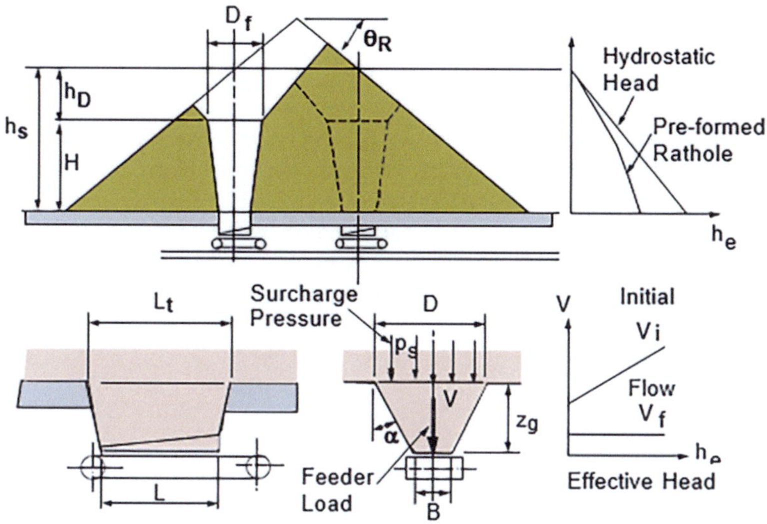

Fig. 10.3 Expanded flow principle in stockpile reclaim

field in the interface zone will undergo a re-distribution. This needs to be considered in the design and selection of the feeder to meet the necessary interface objectives along with the required discharge feed rate.

For these objectives to be achieved, it is necessary for the feeder capacity in the feed zone to increase from the rear to the front discharge end of the hopper. Ideally the draw-down of bulk solids in the hopper should be uniform, that is, the feeder capacity feed rate should increase incrementally in the direction of the feeding process. In effect, the feeder may be described as an *increasing variable capacity conveyor*.

10.3 Feeders for Bulk Handling

Feeders for controlling the flow of bulk solids onto conveyor belts must meet a certain criterion:

- Deliver the range of flow rates required.
- Handle the range of particle or lump size and flow properties expected.
- Deliver a stable flow rate for a given equipment setting.
- Permit the flow rate to be varied easily over the required range without affecting the performance of the bin or hopper from which it is feeding.
- Feed material onto the belt in the correct direction at the correct speed with the correct loading characteristic and under conditions which will produce minimum impact, wear, and product degradation. Often a feed chute is used in conjunction with the feeder to achieve these objectives.

- Fit into the available space.

It is important that the flow pattern be such that the whole outlet of the feed hopper is fully active. This is of fundamental importance in the case of mass-flow hoppers. When feeding along slotted outlets in wedge-shaped hoppers the maintenance of a fully active outlet requires the capacity of the feeder to increase in the direction of feed. To achieve this condition, special attention needs to be given to the design of the outlet as vertical skirts and control gates can often negate the effect of a tapered outlet. Gates should only be used as flow trimming devices and *not* as flow rate controllers. Low-rate control must be achieved by varying the speed of the feeder.

Noting the foregoing comments, a selection of various types of feeders used to feed bulk solids is presented.

10.4 Vibratory Feeders

10.4.1 General Description

Vibratory feeders are used extensively in controlling the discharge of bulk materials from bins and stockpiles and as commonly occurs, directing these materials onto belt conveyors. They are suitable for a broad range of bulk solids, being able to accommodate a range of particle sizes, as well as being particularly suitable for abrasive materials. However, they are generally not suited to fine powders (e.g. <100 microns) where flooding can be a problem. Also, *sticky* cohesive materials may lead to build-up on the pan leading to a reduction in flow rate.

A typical industrial application of a vibratory feeder is illustrated in Fig. 10.4. In this case the feeder is handling coal. As may be observed, the feeder is supported by steel cables attached to the feed bin.

Bulk solids are conveyed along the pan of the feeder because of the vibrating motion imparted to the particles. The pan of the feeder is driven in an approximate sinusoidal fashion at some angle to the trough.

The conveying velocity and throughput depend on the feeder drive frequency, amplitude or stroke, drive angle and trough inclination, coefficient of friction between the bulk solid and the pan as well as the bulk solid parameters such as bulk density, particle density and general flow properties. Figure 10.5 shows a general hopper/feeder interaction diagram with flow vectors indicated.

10.4.2 Hopper/Feeder Configurations

There are several aspects to note when designing feed hoppers for use with vibrating feeders. The effectiveness of the feeder, as with all feeders, depends largely on the hopper which must be capable of delivering material to the feeder in an uninterrupted

Fig. 10.4 Vibratory feeder conveying coal

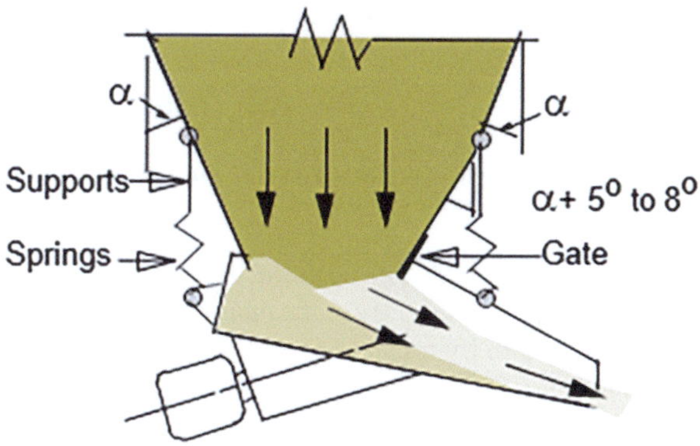

Fig. 10.5 Hopper and vibratory feeder interaction

way. For a symmetrical hopper there is a tendency for the feeder to draw material preferentially from the front of the hopper. Uniform draw can be achieved by making the hopper outlet asymmetrical with the back wall at the correct hopper half-angle α and the front wall at an angle of $\alpha + (5° - 8°)$. Alternatively, a symmetrical hopper may be made to feed approximately uniformly by using a rougher lining material on the front face and retaining a smoother surface for the rear face and side walls.

10.4.3 Types of Vibrating Feeders

In general vibrating feeders are classified as *brute force* or *tuned* depending on the way the driving force imparts motion to the pan. The two types of feeders are

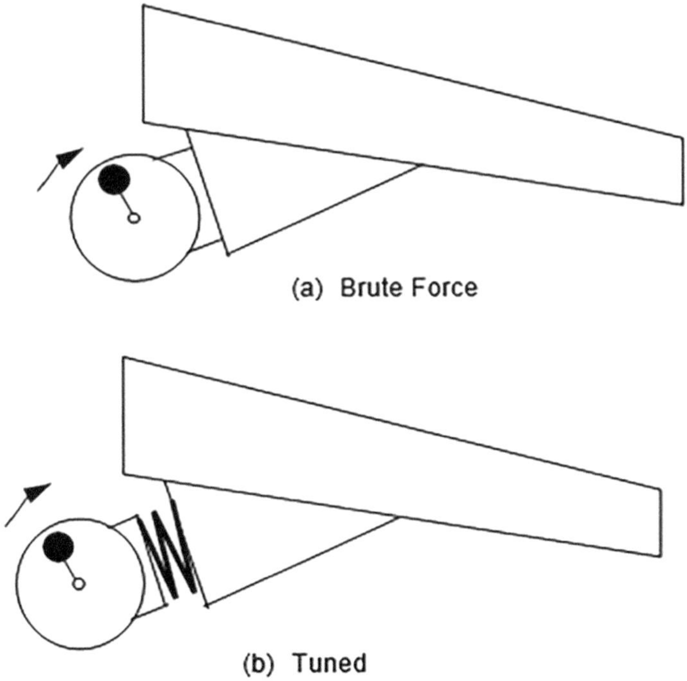

Fig. 10.6 *Brute Force* and *Tuned* type vibratory feeders

illustrated in Fig. 10.6. As the name implies *brute* type feeders involve the application of the driving force directly to the pan, while in the *tuned* feeder the exciter is attached to the feeder pan by means of springs.

Brute force feeders have the following characteristics:

- Lower initial cost but higher operating costs.
- Greater forces to be accommodated in the design.
- Impact loads on the pan are transmitted to bearings on which out-of-balance masses rotate.
- Delivery rates are dependent on the feeder load due to bulk solids.
- Generally confined to applications requiring only one feed rate.

On the other hand, *tuned* vibrating feeders are more sophisticated in their operation in as much as the driving force is transmitted to the pan via connecting springs. In this way they act essentially as a two-mass vibrating system and employ the principle of force magnification to impart motion to the pan. The primary driving force is provided by either an electro-magnet or by a rotating out-of-balance mass system.

Following the work of Rademacher [3], some general comments may be made. Normally the trough mass is designed to be 2.5–3 times as large as the exciter mass. Figure 10.7 shows typical magnification curves for the tuned feeder for two damping ratios ζ_1 and ζ_2. Normally the feeders operate below the resonance frequency with

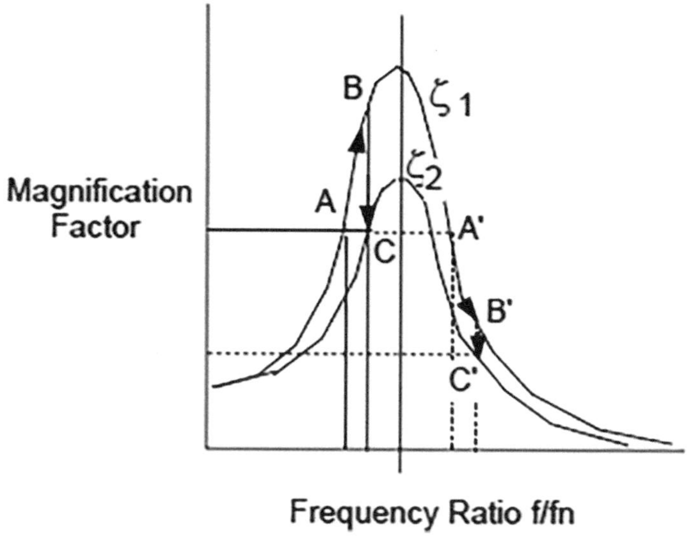

Fig. 10.7 Typical magnification curves for a *tuned* two mass system ($\zeta_1 < \zeta_2$)

$f/f_n = 0.9$ where f = driving frequency and f_n = natural frequency of the system. It is often claimed that the *tuned* feeder maintains its feed rate when the head load varies. The validity of this claim may be examined by reference to Fig. 10.7. The increased head load effectively increases the pan mass lowering the natural frequency f_n and increasing the frequency ratio f/f_n. This corresponds to a shift from A to B in the diagram. At the same time, the damping increases due to the increased load resistance causing a shift from B to C. Thus, the claim that the feeder maintains its performance regardless of the head load, basically, is not true. However, in many instances the negative effect of the increase in damping approximately compensates for the change in magnification factor. This means that points A and C in Fig. 10.7 are approximately of the same magnification factor so that the trough stroke is kept approximately constant.

A similar analysis in the case of the over critical operation with $f/f_n > 1$ indicates that increases in f/f_n and ζ leads to a smaller magnification factor corresponding to a smaller stroke. For this reason, sub-critical operation of this type of feeder must be provided.

10.5 Belt Feeders

Belt feeders, as illustrated in Fig. 10.8, are used to provide a controlled volumetric flow of bulk solids from storage bins and bunkers. They generally consist of a flat belt supported by closely spaced idlers and driven by end pulleys. While normally

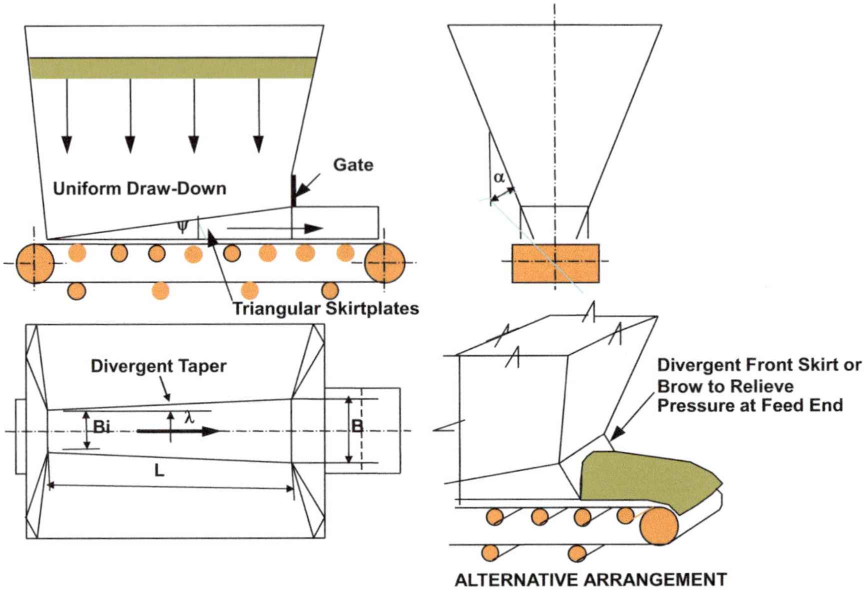

Fig. 10.8 Belt and apron feeder

belt feeders are installed horizontally, on some occasions a feeder may be designed to operate at a low inclination angle. Some features of belt feeders include:

- Suitable for withdrawal of material along slotted hopper outlets when correctly designed.
- Can sustain high impact loads from large particles.
- Flat belt surface can be cleaned quite readily allowing the feeding of cohesive materials.
- Suitable for abrasive bulk solids.
- Capable of providing a low initial cost feeder which is dependent on operation and suitable for automatic control.

The hopper and feeder interface geometry for long slots is critical if uniform draw is to be obtained. For this reason, a tapered opening is required, as illustrated in Fig. 10.8. The triangular skirtplates in the hopper bottom provide an effective way to achieve the required divergence angle λ. An important feature of these diverging skirts is the relief they provide to skirtplate drag. Figure 10.9 shows an actual hopper/feeder arrangement with triangular skirt plates.

The gate at the front of the feeder acts as a flow trimming device rather than a flow rate controller. The height of the gate is adjusted to give the required release angle and to achieve uniform draw along the slot. Once the gate is correctly adjusted, it should be fixed in position; the flow rate is then controlled by varying the speed of the feeder. This requirement places some limitations on belt feeders when very

Fig. 10.9 Industrial application of triangular skirt plates

low flow rates are required, especially if the bulk solid is at all cohesive or contains large lumps. The use of a diverging front skirt or brow as illustrated in Fig. 10.10 has the advantage of relieving the pressure at the feed end during discharge and forward flow.

Care is needed with the design of the hopper/feeder arrangement when handling fine powders to ensure that problems of flooding are avoided. If the bulk material tends to stick to the belt, spillage may be a problem with belt feeders. Therefore, if

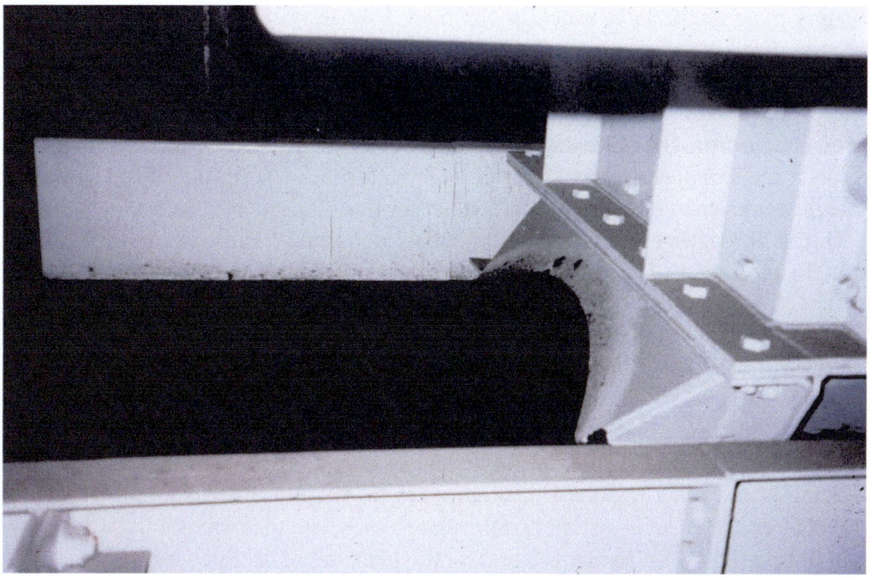

Fig. 10.10 Application of divergent brow in belt feeder handling coal

sufficient headroom is available, it is desirable to mount the feeder above the belt conveyor onto which it is feeding material so that any material falling from the return side of the belt will automatically fall onto the conveyor belt.

10.6 Apron Feeders and Reciprocating, Shear-Gate Plate Feeders

10.6.1 Apron Feeders

Apron feeders are a version of belt feeders and are useful for feeding large tonnages of bulk solids, being particularly relevant to heavy abrasive ore type bulk solids and materials requiring feeding at elevated temperatures. They are also able to sustain extreme impact loading. The remarks concerning the need for uniform drawdown and gate settings applicable to belt feeders are also applicable to apron feeders.

10.6.2 Reciprocating Shear-Gate Plate Feeders

A schematic arrangement of the reciprocating, shear gate type plate feeder is shown in Fig. 10.11. In some ways it resembles a belt feeder, as the basic hopper design requirements and associated feeder load determinations for a belt feeder and reciprocating plate feeder have much in common. Like the belt feeder, the objectives of optimising the interfacing of the hopper and feeder are of primary concern, together with the hopper and feeder interfacing and uniform draw-down. Obviously, the main difference between the reciprocating shear gate plate feeder and a belt feeder is that the discharge from the belt feeder is continuous whereas the reciprocating motion of the plate feeder causes intermittent discharge. A brief description of the shear gate plate feeder as shown in Fig. 10.11 is now given.

The reciprocating motion of the plate is commonly activated by a hydraulic ram. On the out stroke, bulk solids from the hopper are carried out as illustrated in Fig. 10.11. No feed takes place. On the return strokes the plate is retracted far enough for feed to take place with bulk solids discharged as a series of small heaps as shown. Thus, the feeding action is intermittent rather than continuous which is a disadvantage.

For the efficient operation of the reciprocating plate feeder, it is found desirable for the plate to be declined by the angle "θ" as shown. The reason for this is that the bulk solids and plate friction is more critical for the return stroke than for the out stroke. Hence the drag resistance of the plate along with the compression of the bulk solid during the return stroke is eased by the plate being declined by the angle "θ".

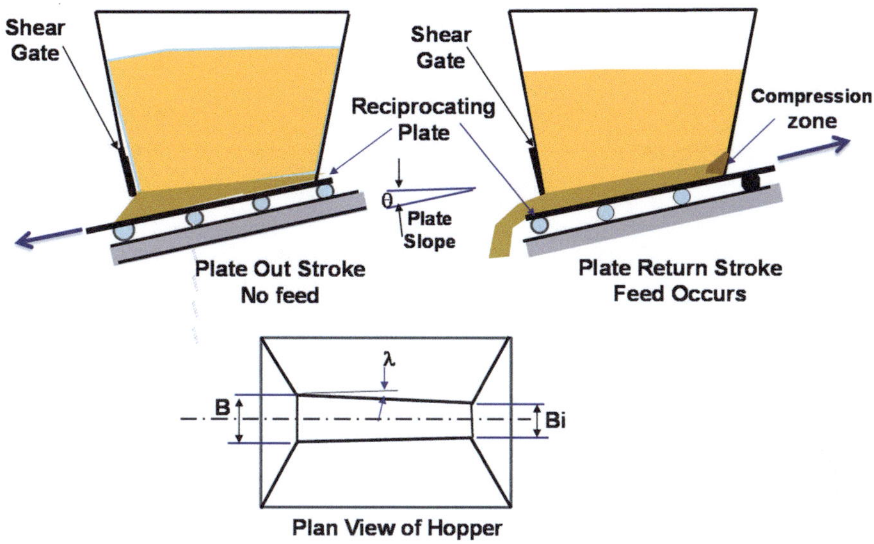

Fig. 10.11 Schematic Diagram of a Reciprocating Plate Feeder

For the reasons associated with the above comments, reciprocating plate feeders are not widely used. The mechanics of plate feeders has been published by Roberts [4].

10.7 Plough Feeders

Rotary plough feeders are generally used in long reclaim tunnels under stockpiles where they travel along the tunnel as shown in Fig. 10.12. In other cases, they are fixed under large storage bins. In the case of the stockpile slot reclaim, it is necessary for the diagonal dimension of the slot to be at least equal to the critical rathole diameter D_f of the bulk solid to prevent ratholes from forming under the high storage pressures [5, 6]. In this way the gravity reclaim efficiency is maximised. The tie beams between slots should be steeply capped. Furthermore, the slot width B must be large enough to prevent arching.

The basic concept of the travelling plough feeder is to allow bulk solids to flow by gravity onto a stationary shelf and then remove the solids from the shelf either with a linear drag plough or a travelling rotary plough. It is important that high penetration of the plough is achieved and that there is a small vertical section behind the plough to prevent material build- up on the sloping back wall.

Colijn and Vitunac [7] have reviewed the application of plough feeders and recommend the following limiting values for plough feeders:

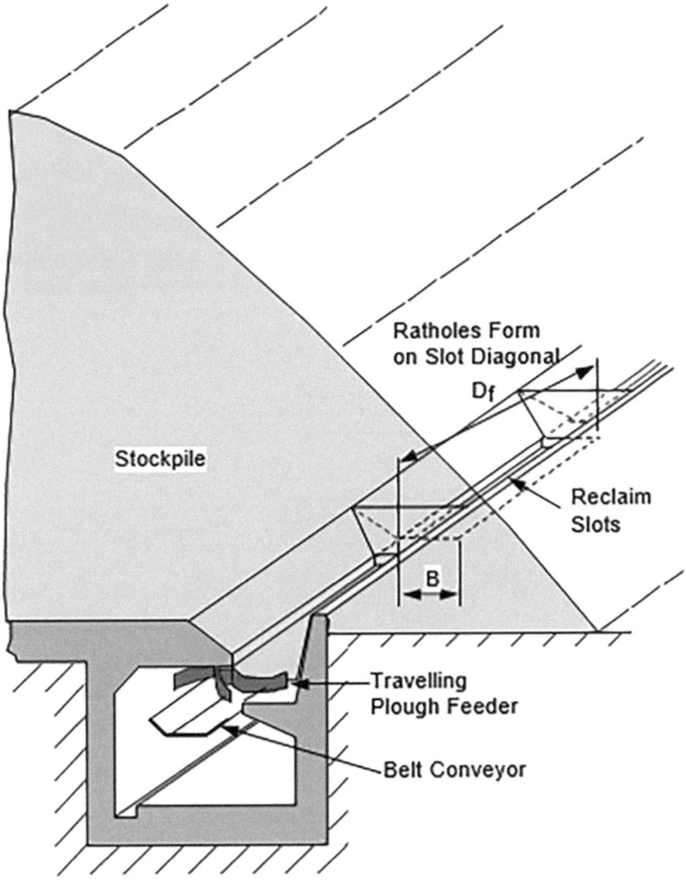

Fig. 10.12 Travelling rotary plough feeder

(a) Rotary Plough

- Tip speed <20 m/s
- Tip diameter range 1.8–4 m
- Carriage speed 0.01–0.08 m/s (does not contribute significantly to capacity).

(b) Linear Drag Plough

- Traversing speed 0.13–0.76 m/s
- Cut width 1.4–2.3 m.

Research into plough feeder performance has also been conducted by Roberts et al. [8].

10.8 Rotary Table Feeders

The rotary table feeder can be considered as an inverse of the plough feeder. It consists of a power-driven circular plate rotating directly below the bin opening, combined with an adjustable feed collar which determines the volume of bulk material to be delivered. A typical rotary feeder arrangement is shown in Fig. 10.13. The aim is to permit equal quantities of bulk material to flow from the complete bin outlet and spread out evenly over the table as it revolves. The material is then ploughed off in a steady stream into a discharge chute.

This feeder is suitable for handling cohesive materials which require large hopper outlets, at flow rates between 2 and 125 tonnes per hour. Feed rates, to some extent, are dependent on the degree to which the material will spread out over the table. This is influenced by the angle of repose of the material which varies with moisture content, size distribution and consolidation. These variations prevent high feed accuracy from being obtained.

Rotary table feeders, as shown in Fig. 10.13, are suitable for bin outlets with diameters up to 2.5 m; the table diameter is usually 50–60% larger than the hopper outlet diameter. With some materials a significant dead region can build up at the centre of the table. This can sometimes be prevented from becoming excessive by

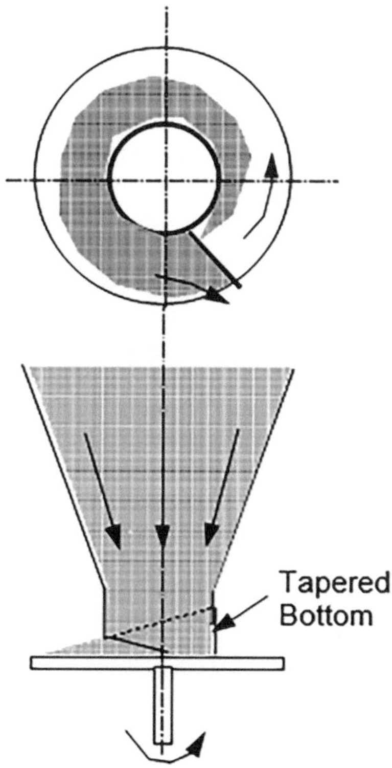

Fig. 10.13 Rotary table feeder

incorporating a scraping bar across the surface of the plate, severely curtailing or preventing removal of the bulk material.

10.9 Screw Feeders

Screw feeders are widely used for bulk solids of low or zero cohesion such as fine and granular materials which must be dispensed under controlled conditions at low flow rates. However, as with belt feeders, design difficulties arise when the requirement is to feed along a slotted hopper outlet. An equal pitch, constant diameter screw tends to draw material from the back of the hopper. To counteract this, several arrangements are advocated for providing an increasing screw capacity in the direction of feed such as:

- Stepped pitch
- Variable pitch
- Variable pitch and diameter
- Variable shaft diameter

Not all of these produce an ideal arrangement. For instance, Fig. 10.14a shows a screw feeder in which the screw and shaft diameters are each constant, while the pitch progressively increases from the rear to the front. This is not a satisfactory arrangement, mainly because the volumetric efficiency of the feeder decreases with the expanding pitch in the direction of feed. The feeder will draw preferentially from the rear as shown. To overcome this problem, the screw requires a tapered shaft in addition to the expanding pitch as illustrated in Fig. 10.14b.

Pitch variation is generally limited to a range between 0.3 diameters minimum to 1.5 diameters maximum. This limits the length to diameter ratio for a screw feeder to approximately six, making them unsuitable for long slots. The section of the screw leading from the hopper to the feeder outlet is fundamental in determining the quantity of material discharged per revolution of the screw. At the point when the screw leaves the hopper, it is essential for control purposes to cover the screw, normally by a *choke* section having the same radial clearance as the trough. This choke section should extend for at least one pitch to prevent material cascading over the flights.

As a screw feeder relies on friction to transport material it has a very low efficiency in terms of the energy requirements. Furthermore, the volumetric efficiency is impaired somewhat due to the rotary motion imparted to the bulk material during the feeding operation. Entrainment patterns in screw feeders have been examined by Bates [9], while Roberts has established a theory to predict draw-down patterns in screw feeder and hopper combinations [10].

Since screw feeders are generally fully enclosed, relatively good dust control is achieved. However, due to the high frictional losses abrasive type bulk solids can effectively reduce the life of the feeder due to abrasive wear. Fine powders that tend to flood are difficult to control in a screw feeder in flooding situations.

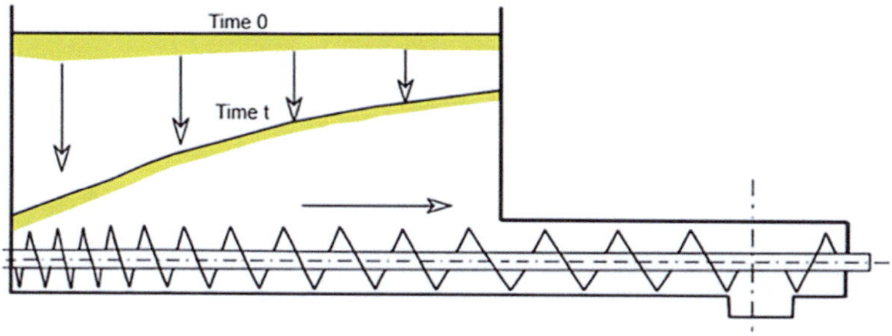

(a) Screw feeder with constant screw diameter, constant shaft diameter and expanding pitch. Feed occurs preferentially from rear of hopper.

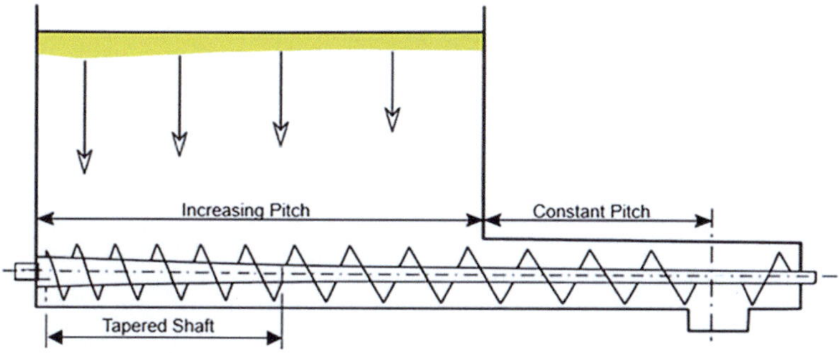

(b) Screw feeder with constant screw diameter, tapered shaft diameter and expanding pitch. Results in uniform draw-down in hopper.

Fig. 10.14 Examples of screw feeders **a** Less than optimum **b** Optimum

10.10 Tube Feeders

The tube type feeder, a Swedish invention marketed under the name *Tubefeeder*, consists of a rotating cylindrical tube or drum with a series of reclaim scoops and openings arranged in a helical pattern around the drum surface. Inside the tube is an independently driven, rotating screw conveyor to feed the reclaimed bulk solid to the required discharge point. The elements of the *Tubefeeder* are shown in Fig. 10.15. A typical application of a T*ubefeeder* is shown in Fig. 10.16. In this case the feeder is handling salt.

Tubefeeders usually operate in either one of the following two modes:

(i) For the reclaim of bulk materials from the bottom of flat bottom cylindrical bins or silos. In this case the rotating feeder moves in a circular orbital path, with the feed discharge point located at the centre of the silo bottom. This is illustrated in Figs. 10.15 and 10.16.

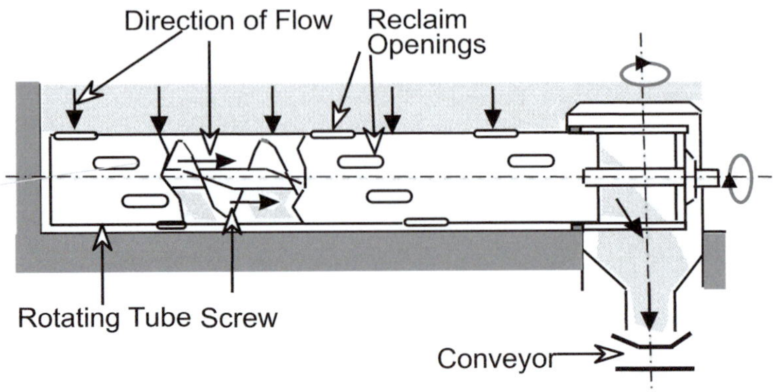

Fig. 10.15 Tube feeder

Fig. 10.16 Tube feeder handling raw salt

(ii) For the reclaim of bulk materials from long stockpiles. In this case the rotating feeder moves in translational motion forward and backward along the floor of the stockpile, discharging the reclaimed bulk material onto a conveyor located along one edge or side of the stockpile.

In terms of volumetric performance, the size, spacing and geometrical configuration of the reclaim openings in the tube needs to be based on the properties of the bulk material as well as on the need to achieve a uniform draw-down pattern in the bin or silo. Uniform draw- down is necessary to avoid segregation problems. Both the drive torque and tube strength requirements depend on the surcharge head as well as the normal and shear stresses developed around the tube surface.

10.11 Rotary Feeders

Rotary feeders, as illustrated in Fig. 10.17, also known as drum, vane, star and valve feeders, are generally used for the volumetric feeding of fine bulk solids which have reasonably good flowability. The drum, which is cylindrical in form, comprises a square to rectangular feed intake opening, and has enclosed ends to prevent the bulk material from flowing out. It discharges by rotation. This feeder is only suitable for materials with good flowability which are not prone to aeration. Similar considerations apply to the rotary valve feeder, which might be considered as an extremely short apron feeder. The rotary valve feeder is completely enclosed and aims at preventing powders or fine-grained materials from flooding. The star feeder, which is essentially a wide rotary valve type feeder, is used to obtain uniform withdrawal along a slotted opening.

These feeders are not suitable for abrasive bulk materials as clearances cannot be maintained, and the feeders tend to lose control especially when handling aerated

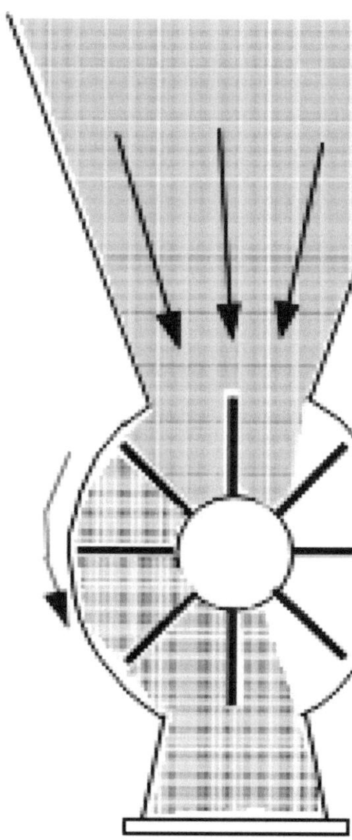

Fig. 10.17 Rotary feeder

powders. Cohesive powders will tend to clog the rotor pockets and reduce feeder capacity.

10.12 Role of Feeder in Controlling Segregation and Mixing

Segregation and mixing of bulk materials in feeding operations is influenced by the flow patterns generated in mass-flow hopper and feeder combinations because of a *feed-back* effect. This is demonstrated in the following two examples:

- Application of a flat belt feeder in conjunction with a mass-flow hopper (Fig. 10.18).
- Influence of rotational speed of a rotary valve feeder on mass-flow hopper flow pattern (Fig. 10.19).

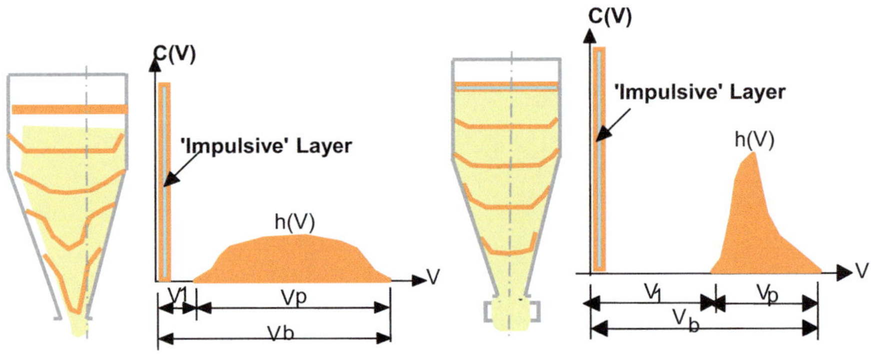

(a) Free discharge - good mixing - segregation (b) Belt feeder – non mixing – anti-segregation

Fig. 10.18 Hopper discharge and belt feeder discharge

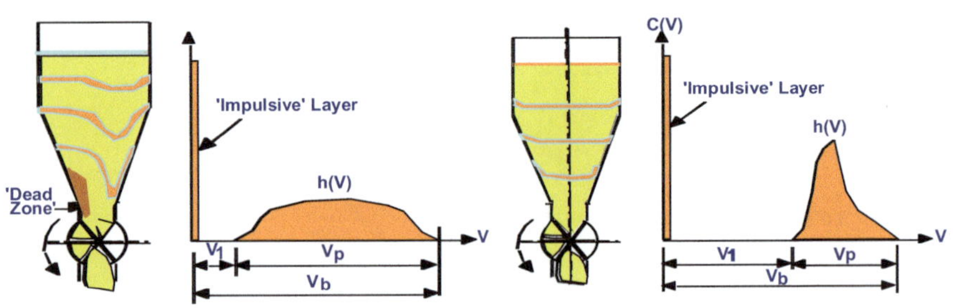

(a) Slow valve speed – good mixing – segregation (b) Faster valve speed – non mixing – anti-segregation

Fig. 10.19 Influence of rotational speed of a rotary valve feeder

The model upon which these two Figures are based follows from Systems theory based on the research of Roberts [11, 12]. To identify the flow pattern, a thin layer of marked particles, referred to as the *Impulse Layer*, is placed on the top surface of bulk materials contained in the bin. As the bulk materials commence discharging from the bin, the change in the flow pattern is depicted by the discharging marked particles.

The following variables and terminology are adopted:

C(V) = Particle volume concentration
V_1 = Volume of bulk particles discharging before marked particles appear
V_p = Volume of marked particles appearing
V_b = total volume of bulk particles in bin and hopper, full to empty
h(V) = System marked particles *Weighting Function*

10.12.1 Free Hopper Discharge and Belt Feeder Controlled Discharge

(a) Free Discharge from Hopper

In this case the particles in the central region of the hopper discharge faster than those from the outer regions adjacent to the hopper walls. Only a small volume V_1 appears before the first of the marked particles are observed at the hopper outlet. The mixture of the original *impulsive* layer with the hopper contents, labelled V_p, continues until the total hopper content discharges and the hopper is then empty. Hence, in the case depicted in Fig. 10.18a, the free discharge indicates *good mixing*. By contrast it could also be interpreted as a potential source of segregation if the hopper mass is well mixed to begin with. However, this latter comment is less significant than the greater benefit of "good mixing".

(b) Employment of Belt Feeder to Control Discharge

In this case, the feed-back from the flat feeder belt is to re-shape the flow pattern as illustrated by the series of more horizontal and wider spaced flow lines in Fig. 10.18b. The velocity profile is more uniform across the hopper at each level, as shown. A greater volume V_1 discharges from the hopper before the marked particles first appear than occurs in case (a). The net result is that the belt feeder assists in preventing segregation.

10.12.2 Influence of Rotational Speed of a Rotary Valve Feeder

This case is concerned with a hopper and rotary valve feeder, as commonly used in feeding powders into pneumatic conveying systems, with the objective of preventing air leakage. The system on which this example is based is depicted in Fig. 10.19, which shows a rotary valve with six blades linked to a mass-flow hopper. Of particular interest is the effect of the rotational speed in preventing good bulk powder mixing while avoiding segregation problems. As the results illustrate, the slower valve speed of the case depicted in Fig. 10.19a provides good mixing but may result in some segregation problems. On the other hand, the faster rotational speed of the case shown in Fig. 10.19b does not produce good mixing but, at the same time, helps to avoid segregation problems.

10.13 Feeder Selection

There are many different types of feeders, some of which have been reviewed in the preceding sections. The selection of a feeder for a particular situation may not be straightforward, particularly if more than one satisfactory solution appears possible. Although the type and size of feeder for a given application is primarily dictated by the characteristics of the bulk material to be handled and the required feed rate, Reisner [13] provides useful guidance for initial selection, as summarised in Table 10.1. In this table, the letter *E* indicates the feeder is an *excellent* choice for the application, *G* indicates that the feeder is *good,* or acceptable, and *F* means that the feeder is *fair*, and that special consideration may be required to handle the type of material indicated.

Feeders are also classified according to Table 10.2 (redrawn from Reisner [13]). By way of general observation, it is emphasised that both the interface region (geometry between hopper and feeder) and the material properties dominate both the feed rate and uniformity of feed. There are well established procedures for selecting the correct hopper and feeder geometry and for determining feeder loads and power requirements. Nonetheless, good feeder design relies on sound judgment and experience.

10.14 The Interfacing Problem—A Case Study

The fundamental difference between a conveyor and a feeder needs to be noted. A conveyor has constant capacity per unit length, whereas a feeder has variable or increasing capacity per unit length. For this reason, a conveyor should never be used as a feeder. Yet sometimes this happens, often with dire consequences, as illustrated in the case study example that follows.

Table 10.1 Feeder selection criteria (redrawn from Reisner [13])

Feeder type	Maximum material size					
	Very coarse +300 mm	Coarse – 200 mm	Medium – 50 mm	Fine – 13 mm	Very fine – 150 μ	Concentrate −75 μ
Belt		F	G	E	E	G
Apron	E	E	G	F		
Screw			F	G	E	E
Vibrating (Brute force)	E	E	G	F		
Vibrating (Tuned)	E	E	E	E	F	F
Rotary plow		F	E	E		
Rotary table				G	E	G

The coal storage and feed system of a particular thermal power station comprises conical hopper type boiler bunkers, with the lower end of each bunker being configured with twin outlets through a *trouser leg* arrangement. Two chain feeder conveyors are employed, one for each outlet. Coal is fed simultaneously by the feeders but in opposite directions, the coal is being directed to each end of a tube mill. Figure 10.20 shows, schematically, the arrangement for each feeder conveyor. Because of the constant capacity of the conveyor, filling takes place at the rear end, and no more coal can enter the conveyor, resulting in a rathole being formed at the rear end of the hopper.

The intention is for coal to feed in the front half of the hopper to the left of the horizontal baffle plate, however, the cohesive strength of the coal at critical moisture levels causes coal to block the spaces between the chain bars of the feeder. This is exacerbated by the limited opening size between the chain bars as indicated in Fig. 10.21. Coal is then dragged around the end of the feeder casing.

The net effect is for two ratholes to form, as illustrated in Fig. 10.22. These ratholes were typically about 1 m in diameter, greatly reducing the live capacity of the bunkers and necessitating remedial action through the installation of a flow promotion system. The more serious aspect of the problem was the non-uniform pressure distribution around the bunker walls leading to severe denting and buckling problems (Fig. 10.23).

A solution to the problem is to promote an increase in capacity in the direction of feed by fitting level bars, raising the front height of the casing and extending the horizontal baffle plate the full length of the hopper opening. The modifications are shown in Fig. 10.24. The walls of the hopper have also been made steeper to promote mass-flow.

Table 10.2 Feeder classification (redrawn from Reisner [13])

Material	Feeder type						
	Belt feeder	Apron feeder	Screw feeder	Vibratory feeder	Rotary flow feeder	Rotary table feeder	Rotary valve feeder
Aluminium oxide			✓				✓
Ammonium sulphate	✓			✓	✓		
Ashes (Dry)	✓		✓	✓		✓	✓
Ashes (Moist)	✓						
Barite	✓	✓		✓			
Blast furnace slag	✓			✓		✓	
Cement			✓				✓
Cement clinker	✓			✓		✓	
Chips			✓				
Coal (Normal)	✓	✓	✓	✓	✓	✓	
Coal slime	✓	✓	✓	✓	✓		
Coal dust							✓
Coke	✓	✓		✓	✓		
Dolomite	✓	✓		✓	✓	✓	
Fireclay		✓	✓			✓	✓
Flue dust			✓				✓
Grain			✓	✓	✓		
Gravel	✓			✓	✓	✓	
Iron sulphate	✓			✓	✓		
Limestone	✓	✓		✓	✓	✓	
Magnesite	✓	✓		✓	✓	✓	
Ore (Heavy)		✓		✓	✓		
Ore (Light)		✓		✓	✓		
Sand (Dry)	✓			✓	✓	✓	
Sawdust	✓		✓				
Soot			✓				✓
Titanium dioxide			✓				✓
Zinc oxide			✓				✓

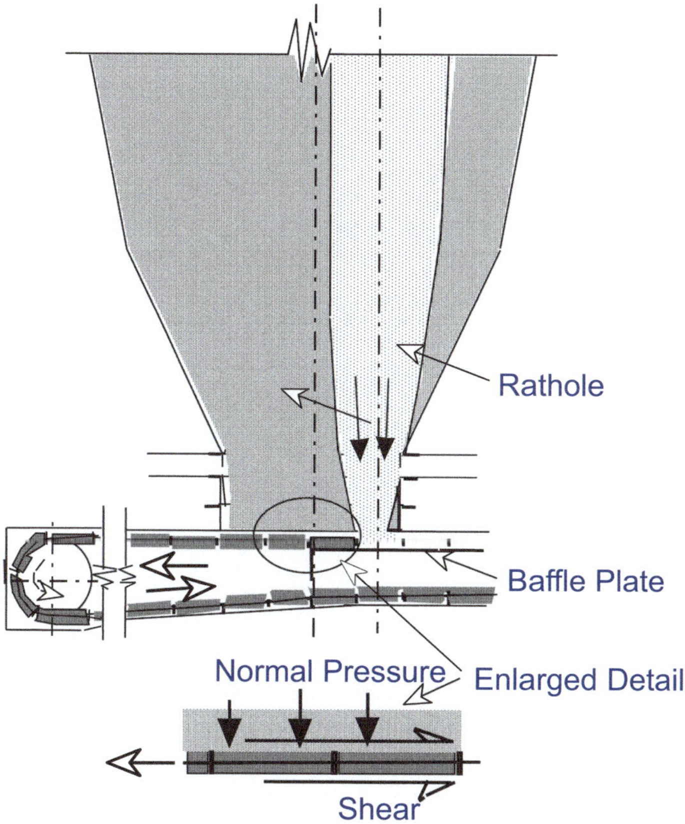

Fig. 10.20 Coal bunker with chain conveyor as feeder

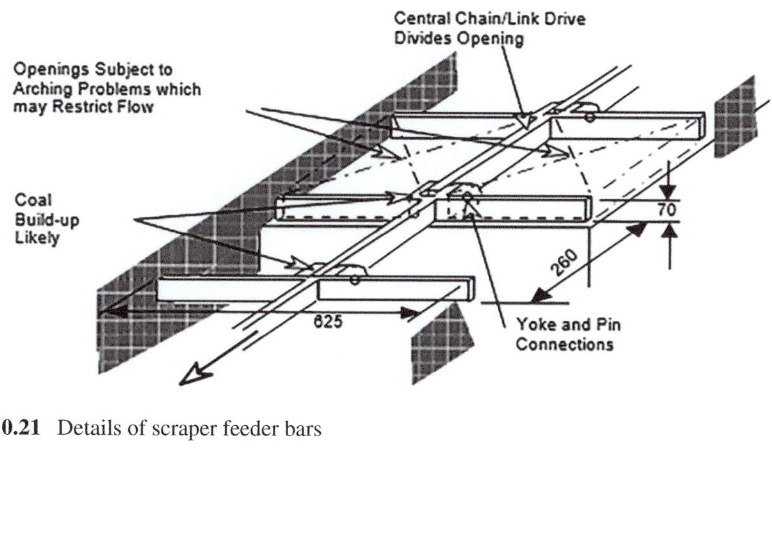

Fig. 10.21 Details of scraper feeder bars

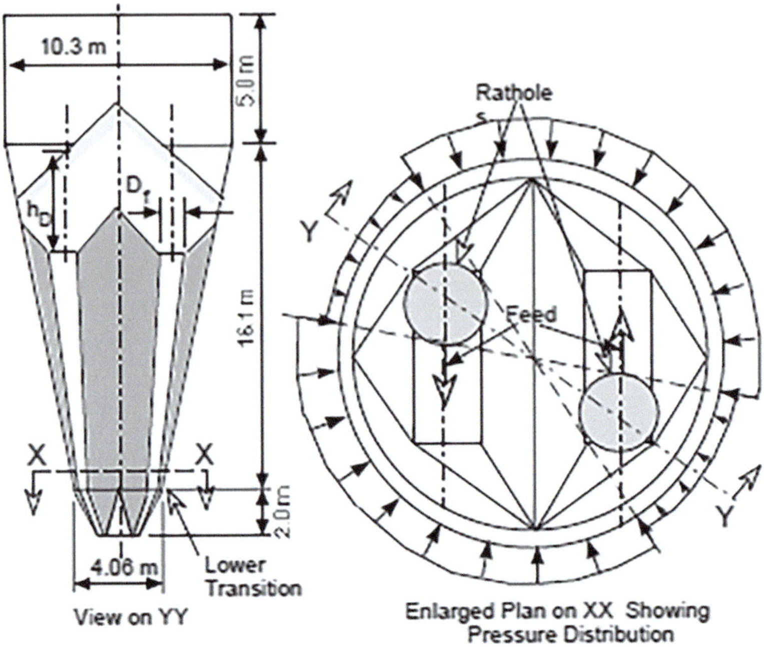

Fig. 10.22 Boiler bunkers and non-uniform pressures due to ratholes

Fig. 10.23 Creasing and buckling of bin

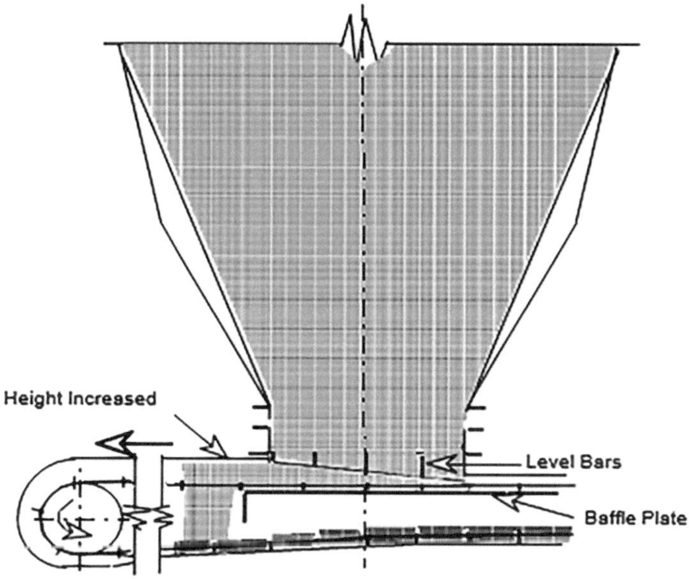

Fig. 10.24 Modified feeder and hopper

References

1. Jenike, A.W., *Gravity Flow of Bulk Solids.* Utah Agric. Exp. St. Bull, 1961. **108**.
2. Jenike, A.W., *Storage and Flow of Solids.* Utah Agric. Exp. St. Bull, 1964. **123**.
3. Rademacher, F.J.C., *Reclaim Power and Geometry of Bin Interfaces in Belt and Apron Feeders - An Analytical Approach.* International Journal of Bulk Solids Handling, 1982. **2**(2).
4. Roberts, A.W. and M. Ooms, *Performance Characteristics of Reciprocating Plate Feeders*, in *Third International Conference on Bulk Materials, Storage, Handling and Transportation: Preprints of Papers.* 1989, Institution of Engineers, Australia: Barton, ACT.
5. Arnold, P.C., A.G. McLean, and A.W. Roberts, *Bulk Solids: Storage, Flow and Handling.* 2nd ed. 1979: The University of Newcastle Research Associates (TUNRA).
6. Roberts, A.W., *Basic principles of bulk solids storage, flow and handling.* 1993: Institute for Bulk Materials Handling Research.
7. Colijn, H.V., E., *Application of Plow Feeders*, in *Paper 79-WA/MH - 1, presented at Annual Winter Meeting, ASME.* 1979: New York.
8. Roberts, A.H., A.; Wiche, S., *Performance Characteristics of Plough Feeders*, in *Proc. Intl. Powders and Bulk Solids Conf.* 1990: Chicago, USA.
9. Bates, L., *Entrainment Patterns of Screw Hopper Dischargers.* Journal of Engineering for Industry, 1969. **91**: p. 295.
10. Roberts, A.W., *Predicting the Volumetric and Torque Characteristics of Screw Feeders.* Bulk Solids Handling, 1996. **16**(2): p. 233–244.
11. Roberts, A.W., *Bin and Feeder Design Criteria for Anti-Segregation and Blending*, in *Proc. Conference on Cost Effective Bulk Materials Handling, Mech 91 Congress, Instn. of Engrs Aust.* 1991: Sydney.

12. Roberts, A.W., *Design of Bins and Feeders for Anti-Segregation and Blending*, in *Proc. Bulk 2000, Instn, of Mech. Engrs.* 1991: London.
13. Reisner, W. and M.v. Eisenhart Rothe, *Bins and bunkers for handling bulk materials: practical design and techniques*. Series on rock and soil mechanics; v. 1no. 1. 1971, Cleveland: Trans Tech Publications.

Chapter 11
Feeding of Bulk Solids—Part 2 Performance Analyses, Loads & Design Objectives

11.1 Introduction

By way of overview, the previous Chapter (Chap. 10) presented the essential aspects of feeder design along with a selection of feeders commonly employed in the handling of bulk solids. It is important that the feeders be chosen to suit the particular bulk solid and to provide the range of feed rates required. It is also important that feeders be used in conjunction with mass-flow hoppers to ensure both reliable flow and good control over the feeder loads and drive powers. The overriding requirement for the interfaced connection of the feeder with the mass-flow hopper is the need for the hopper outlet to be fully active. The consequences of major structural damage to a coal bin through the bad choice of feeder and associated restricted interfacing and outflow was demonstrated in the case study example presented in Sect. 10.14 of Chap. 10.

This current Chapter follows on with the detailed analyses of feeder design and performance. This includes hopper and feeder interfacing for controlled draw-down of bulk solids in the hopper along with the *active* and *passive* stress states linked to the operational modes and associated load determinations. This information is applied to the feeder design and drive torques and power determinations.

The hopper and feeder combination employed in industrial operations almost invariably involve a mass-flow, plane-flow hopper with the long-slotted opening, interfaced with the chosen feeder. This requires the feeder capacity to progressively increase in the direction of feed. In the case of a screw feeder, for example, this is achieved by combinations of increasing screw pitch and decreasing shaft diameter, as illustrate in Fig. 10.14 of Chap. 10.

Undoubtedly belt and apron feeders play a major role in bulk solids handling operations, particularly in applications requiring high tonnage feed rates such as in the mining and mineral processing industries. Their role is commonly linked to the loading of belt conveyors from large storage bins and stockpiles for long overland transportation and associated in-plant operations. A description and overview of

A. W. Roberts, *The Theory and Practice of Bulk Materials Handling*,
https://doi.org/10.1007/978-3-032-19702-3_11

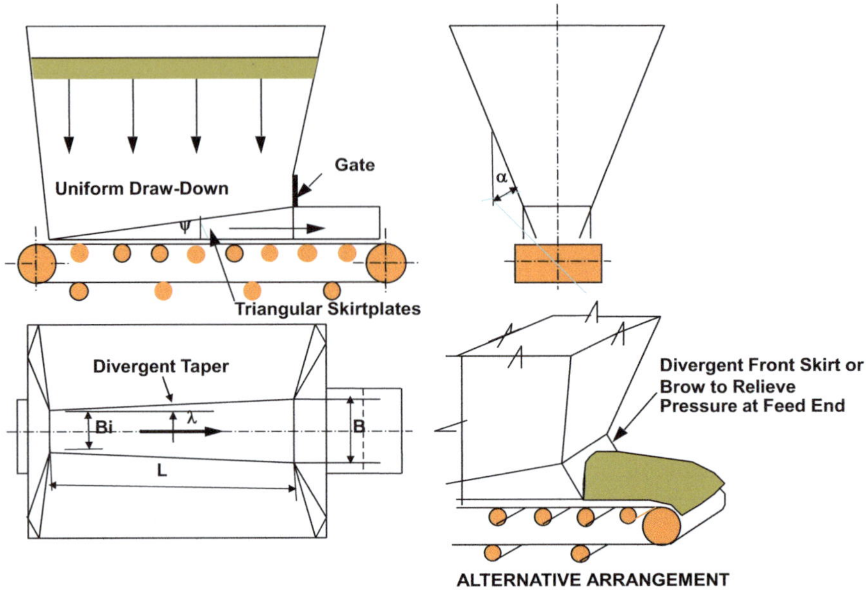

Fig. 11.1 Belt and Apron Feeder

belt/apron feeders is presented in Chap. 10 with the belt feeder model depicted in Fig. 10.8. In view of the focus of this current Chapter on belt and apron feeders, for convenience, this belt feeder model is reproduced as Fig. 11.1.

References [1–8] provide the supporting material for this Chapter.

11.2 Belt and Apron Feeders

11.2.1 Brief Description

Referring to Fig. 11.1, the mass-flow hopper is designed to operate under plane flow, that is the cross-sectional geometry requires the slope of the hopper walls to be based on the relevant hopper half-angle "α" and corresponding outlet dimension "B" based on the bulk solid flow properties and the specified feeder tonnage throughput. The hopper outlet with the feeder is defined as the *shear zone*.

The vertical triangular skirtplates in the hopper bottom section are an effective way to achieve the required diverging taper defined by the angle "λ". These skirtplates have the advantage of not only assisting the required progressive feed rate along the outlet, but also in reducing the skirtplate drag resistance. In addition, their intersection with the hopper walls defines the inclined slope of the required shear surface, as defined by the release angle "ψ".

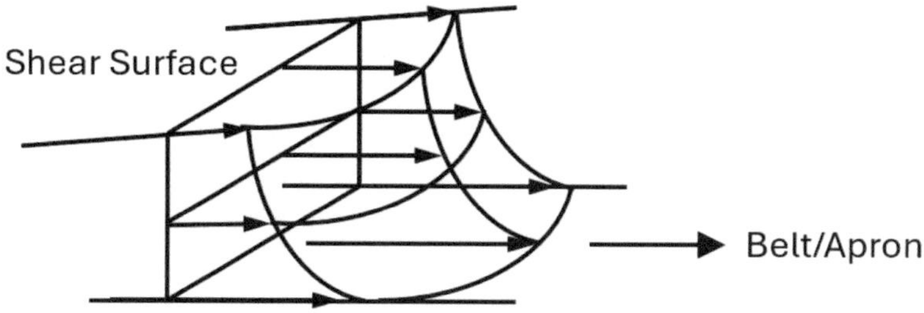

Fig. 11.2 Velocity profile in shear zone

The height of the gate at the feeder outlet is adjusted to give the required release angle to achieve the desired draw-down along the slot. It needs to be noted that the gate is a flow trimming device and not a flow rate controller. An alternative is to employ a diverging brow, as shown in Fig. 11.1, which has the advantage of relieving the pressure at the feed end.

11.2.2 Shear Zone

The complexity of the shear zone of belt feeders has been highlighted in a comprehensive study performed by Schulze and Schwedes [5]. They showed that the shear zone may be divided into three regions, the lengths of the regions being predicted based on the *Coulomb principle of minimal safety*. This assumes that the rupture surface in a consolidated bulk solid will develop in such a way that the bearing capacity of the solid is minimised.

The velocity profile at a particular location in the shear zone is illustrated in Fig. 11.2. The profile is 3-dimensional in form, that is, the longitudinal position along the feeder, and the corresponding height and width of the bulk solid at that location. This characteristic profile would vary in shape from position to position along the feed zone rendering complexities in the determination of the bulk solid draw-down profile in the hopper.

11.2.3 Belt/Apron Feeder Model

Based on the model in Fig. 11.1 and an associated pilot-scale hopper and belt feeder handling cohesive bulk solids, Roberts [8] demonstrated that the Fig. 11.3 model, with acceptable assumptions, can be applied under uniform hopper drawdown, as illustrated.

In the extended skirtplate zone, the velocity profile is substantially constant with the bulk solid moving at an average velocity equal to the belt velocity. Since the

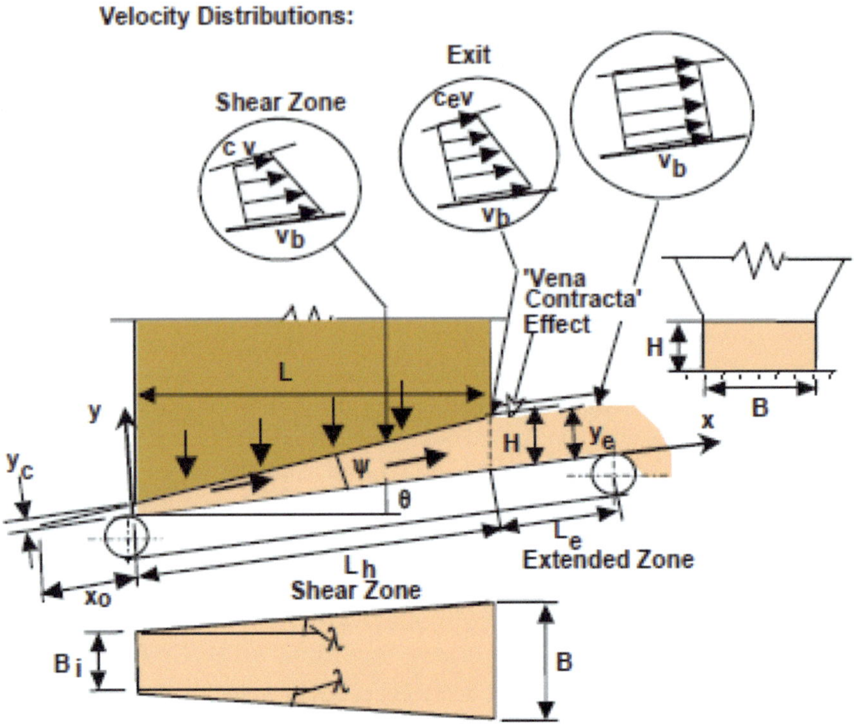

Fig. 11.3 Belt/Apron velocity profiles and throughput

average bulk solid velocity at the exit end of the hopper skirtplate zone is less than the average velocity in the extended skirtplate zone, there will be a *vena contracta* effect with the bed depth y_e less than the bed depth H at the exit end of the feeder.

Referring to Fig. 11.3, the mass throughput of the feeder will vary along the feed zone. At any location x, the throughput $Q(x)$ is given by:

$$Q(x) = \rho\, A(x)\, v_b\, \eta_v(x) \qquad (11.1)$$

where

$A(x) =$ cross sectional area
$v_b =$ velocity of the belt or apron
$\eta_v(x) =$ volumetric efficiency
$\rho =$ bulk density in feed zone (assumed constant)

$$A(x) = (B_i + 2\,x \tan \lambda)(y_c + x \tan \psi) \qquad (11.2)$$

The volumetric efficiency $\eta_v(x)$, which relates the actual throughput to the maximum theoretical throughput based on the bulk solid moving forward with the belt or apron without slip, is given by:

$$\eta_V(x) = \frac{\rho\, A(x)\, v_f(x)}{\rho\, A(x)\, v_b} = \frac{v_f(x)}{v_b} \tag{11.3}$$

where

$v_f(x) =$ average feed velocity at location x, given by

$$v_f(x) = (1 + C)\frac{v_b}{2} \tag{11.4}$$

$C =$ velocity distribution coefficient at location x

Equation (11.4) assumes there is no slip at the belt or apron surface. It has been shown [6–8] that the throughput from Eqs. (11.1) to (11.3) is given by the cubic Eq. (11.5):

$$Q(x) = \rho\, v_b\left[-a_2\, a_4\, x^3 + (a_2\, a_3 - a_1\, a_4)x^2 + (a_1\, a_3 - a_0\, a_4)x + a_0\, a_3\right] \tag{11.5}$$

$$a_o = y_c\, B_i \tag{11.6}$$

$$a_1 = 2y_c \tan\lambda + B_i \tan\psi \tag{11.7}$$

$$a_2 = 2\tan\lambda \tan\psi \tag{11.8}$$

$$a_3 = 1 - \frac{(1 - C_e)x_0}{2(L_h + x_0)} \tag{11.9}$$

$$a_4 = \frac{(1 - C_e)}{2(L_h + x_0)} \tag{11.10}$$

$$x_0 = \frac{y_c}{\tan\psi} \tag{11.11}$$

The parameters in Eqs. (11.6) through (11.11) are:

y_c = clearance at rear of feeder
B_i = width of opening at rear of feeder
λ = divergence angle
ψ = release angle
x_o = dimension defined in Fig. 11.3

L_h = length of hopper shear zone
C_e = velocity distribution factor at x = L_h

11.2.4 Feeder Throughput

At the discharge or feed end of the hopper the throughput is given by

$$Q = \rho\, B\, H\, v_b\, \eta_V(L) \tag{11.12}$$

where

$\eta_V(L)$ = volumetric efficiency at exit

Also:

$$Q = \rho_e\, B\, y_e\, v_b \tag{11.13}$$

where

ρ_e = bulk density in extended zone

It is noted that $\rho_e < \rho$ since the consolidation pressures are lower in the extended zone. Hence Eq. (11.14) applies.

$$y_e = H\left(\frac{1 + C_e}{2}\right)\left(\frac{\rho}{\rho_e}\right) \tag{11.14}$$

It is desirable that the ratio of the gate height H to the width of opening be such that $\frac{H}{B} < 1.0$.

Preferably $\frac{H}{B} < 0.75$ to ensure satisfactory flow in the extended skirtplate zone.

11.3 Optimum Hopper Geometry

11.3.1 Conditions for Uniform Draw-Down

Draw-down in the hopper is related to the feed in the feed zone by the continuity of the mass flow, as illustrated in Fig. 11.4. The condition for uniform draw-down, which represents the optimum performance, is such that:

$$Q'(x) = \frac{dQ(x)}{dx} = constant \tag{11.15}$$

That is, the gradient of the throughput along the feed zone is constant.

Fig. 11.4 Continuity of feed

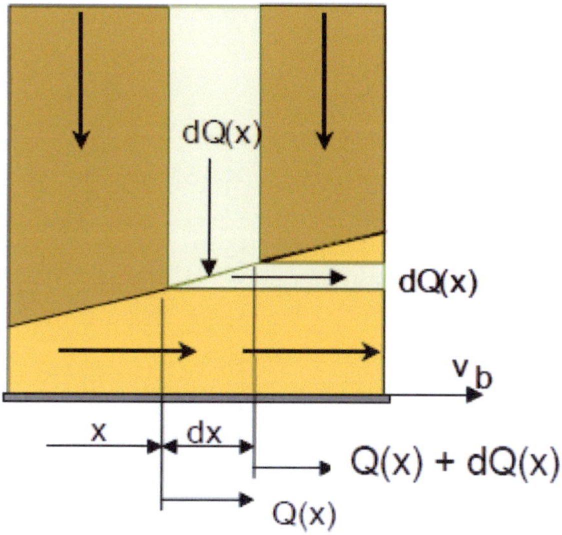

11.3.2 Optimum Divergence Angle

Often the requirement of Eq. (11.15) is impossible to achieve. In the case of a belt or apron feeder, for example, Eq. (11.5) for $Q(x)$ is cubic in form and $Q\prime(x)$ is quadratic, which means that Eq. (11.15) cannot be satisfied. To overcome this problem, an optimum performance may be achieved by setting:

$$Q''(x) = \frac{dQ'(x)}{dx} = 0\, at\, x = \frac{L_h}{2} \tag{11.16}$$

This is illustrated by the surface profile shown in Fig. 11.5.

Based on the foregoing analysis, it has been shown [6, 7] that the optimum divergence angle λ is given by:

Fig. 11.5 Condition for optimum draw-down

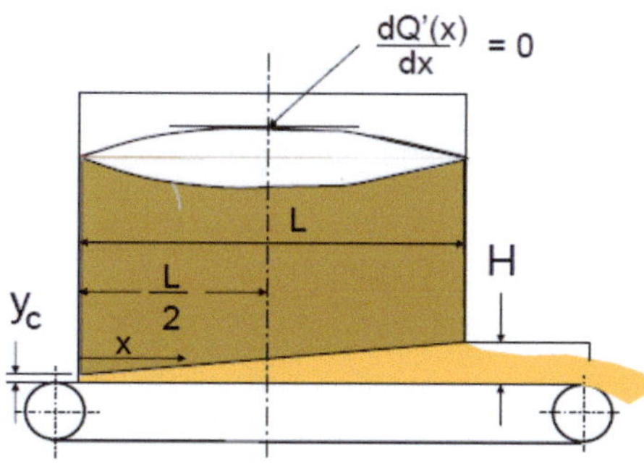

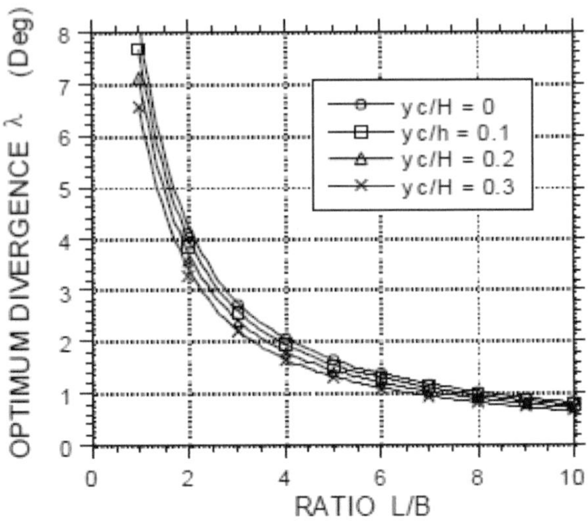

Fig. 11.6 Optimum divergence angle versus $\frac{L_h}{B}$ ratio for a range of clearance ratios with $C_e = 0.5$ and $\eta_V = 0.75$

$$\tan\lambda = \frac{B}{2L_h}\left[\frac{1}{\frac{2}{1-\frac{y_c}{H}}\left[\frac{1}{1-C_e} - \frac{y_c}{H}\right] - 0.5}\right] \tag{11.17}$$

The influence of the feeder $\frac{L_h}{B}$ ratio on the optimum values of λ for a range of clearance ratios is illustrated in Fig. 11.6. The optimum divergence angle λ for uniform draw-down is shown to decrease with an increase in the $\frac{L_h}{B}$ ratio, the rate of decrease being quite rapid at first but lessening as the $\frac{L_h}{B}$ ratio increases.

11.3.3 Use of Transverse Inserts

In the case of feeders employing long opening slots, that is the aspect ratio $\frac{L_h}{B} > 5$, the use of transverse inserts, as illustrated in Fig. 11.7, can assist in promoting uniform draw-down of bulk solids from the hopper along the length of the feeder as well as eliminating belt wear. With reference to the latter, the inserts assist in establishing the required release angle along the feeder. The inserts also assist in reducing the loads on the feeder. The inserts may incorporate diverging brows as illustrated to smooth the flow at the transitions.

11.3.4 Belt Feeder Example

As an example, the case of a belt or apron feeder with $\frac{L_h}{B} = 5$ is considered. For convenience, the throughput Q(x) and gradient of the throughput Q'(x) are expressed

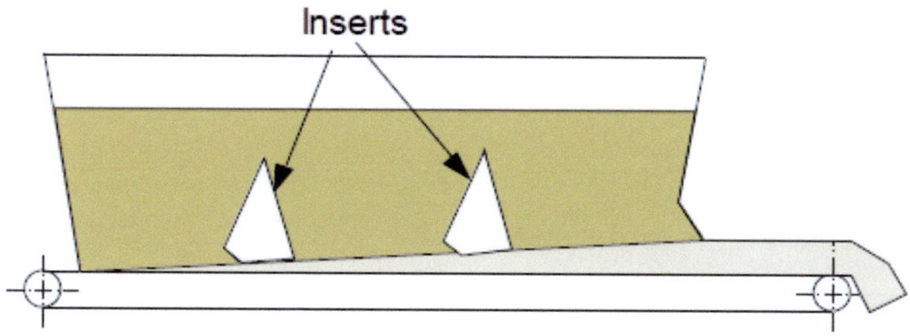

Fig. 11.7 Use of transverse inserts in long feeder

in normalised form as follows:

$$N_Q(x) = \frac{Q(x)}{\rho v_b B H} \text{ and } N_{Q'}(x) = \frac{dN_Q(x)}{dx} \tag{11.18}$$

Figure 11.8 shows the volumetric efficiency η_V (x), throughput parameter $N_Q(x)$ and gradient $N_{Q'}(x)$ for the case of $\frac{y_c}{H} = 0.1$ and $C_e = 0.5$. The full lines for $N_Q(x)$ and $N_{Q'}(x)$ correspond to the optimum divergence angle $\lambda = 1.54°$ and, as shown, the gradient $N_{Q'}(x)$ is virtually constant indicating uniform draw-down in the hopper. The volumetric efficiency decreases from the rear to the front of the feeder as is expected.

For comparison purposes, the performance of a feeder having the same feed rate as the optimum feeder but with a larger divergence angle of 3° is also presented. The

Fig. 11.8 Throughput Characteristics of Belt Feeder with $C_e = 0.5$ and $\frac{L_h}{B} = 5$ Case 1: Optimum $\lambda = 1.54°$ Case 2: $\lambda = 3°$

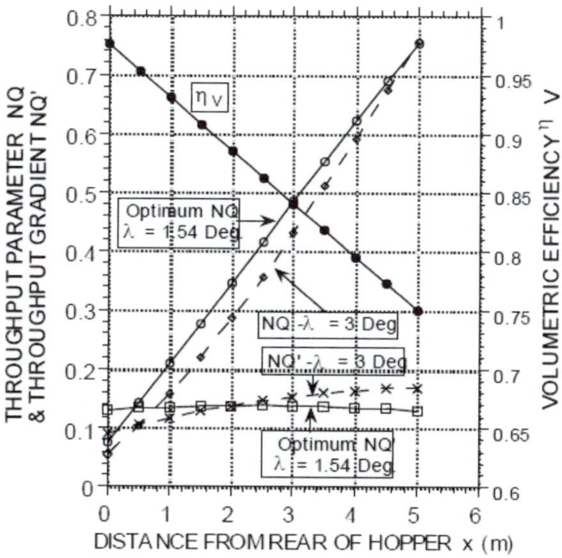

relevant graphs are shown by dotted lines. In this case, the gradient $N_{Q'}(x)$ for this case increases toward the feed discharge end which indicates that the hopper will draw down preferentially from the front.

11.4 Stress Fields and Loads

11.4.1 Stress Fields

The determination of feeder loads and drive powers requires knowledge of the stress fields generated in the hopper during the initial filling condition and during discharge. The relationship between the vertical pressure p_v generated in a mass-flow bin during both filling and flow and the corresponding feeder load V is illustrated in Fig. 11.9. Under filling conditions, a peaked or *active* stress field is generated throughout the entire bin as illustrated in Fig. 11.9a. Once flow is initiated, an arched or *passive* stress field is generated in the hopper and a much greater proportion of the bin surcharge load on the hopper is supported by the upper part of the hopper walls. Consequently, the load acting on the feeder is substantially reduced as shown in Fig. 11.9b.

It is quite common for the load acting on the feeder under flow conditions to be in the order of 20% of the initial load. The arched stress field is quite stable and is maintained even if the flow is stopped. This means that once flow is initiated and then the feeder is stopped while the bin is still full, the arched stress field is retained and the load on the feeder remains at the reduced value. The subject of feeder loads and performance is discussed in some detail in references [1–8].

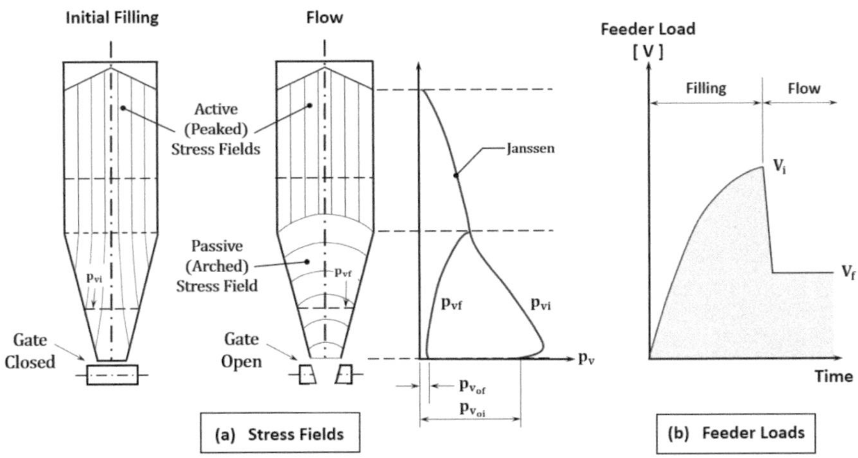

Fig. 11.9 Vertical pressure and load variations on a feeder

11.4.2 Loads at Hopper Outlet and Feeder Interface

The design equations used to determine the feeder loads in this Chapter are based, essentially, on the widely accepted horizontal slice model for bin and hopper load analyses such as what has been presented Chaps. 6 and 8 and shown in Fig. 11.10. It is important to note that the horizontal slice model and associated bin load equations were an integral part of Jenike's modified theory on bin loads [9].

11.4.2.1 Hopper Vertical Pressures Equilibrium Model

Considering the equilibrium of the model presented in Fig. 11.10, the following differential equation is derived:

$$\frac{dP_{vh}}{dz_h} + \frac{j\,P_{vh}}{(h_h - z_h)} = \gamma \tag{11.19}$$

The solution to this equation yields the vertical pressure P_{Vh}

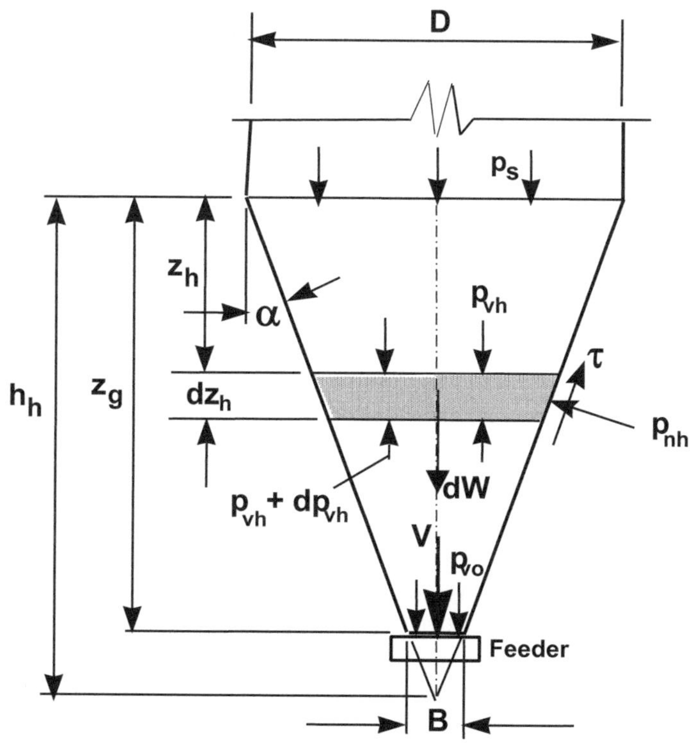

Fig. 11.10 Mass-flow hopper pressure and load model

$$P_{Vh} = \frac{\gamma(h_h - z_h)}{(j-1)} + \left[P_s - \frac{\gamma h_h}{(j-1)}\right]\left[\frac{(h_h - z_h)}{h_h}\right]^j \quad (11.20)$$

$$j = (m+1)\left[k_h\left(1 + \frac{\tan\phi_w}{\tan\alpha}\right) - 1\right] \quad (11.21)$$

where

m =	hopper flow symmetry factor
m = 0	for plane flow or wedged shaped hopper
m = 1	for axi-symmetric or conical hopper
$\phi_w =$	wall friction angle
$\alpha =$	hopper half-angle at outlet
$P_s =$	*surcharge* pressure at switch level
$\gamma = \rho g =$	bulk specific weight
$k_h = \frac{P_{nh}}{P_{vh}} =$	pressure ratio
$P_{nh} =$	normal pressure acting on hopper wall
$P_{vh} =$	corresponding average vertical pressure acting on slice at depth z_h

The loads acting on the feeder and corresponding power requirements vary according to the stress condition in the stored bulk mass. The general expression for the load V is

$$V = P_{v0} A_0 \quad (11.22)$$

where

$P_{v0} =$	Vertical pressure on feeder surface at hopper outlet
$A_0 =$	Area of hopper outlet

The vertical pressure acting at the hopper outlet is determined from Eq. (11.23) as follows

$$P_{VO} = \frac{\gamma B}{2(j-1)\tan\alpha} + \left[P_s - \frac{\gamma D}{2(j-1)\tan\alpha}\right]\left[\frac{B}{D}\right]^j \quad (11.23)$$

11.4.2.2 Significance of the Parameter "J"

The parameter j as defined by Eq. (11.21) is defined in AS3774 [10] as the *hopper exponent*. Its introduction in the equilibrium Eq. (11.19) is to link the bulk solid flow properties, notably the hopper wall friction angle ϕ_w with the hopper geometry and stress fields defined by the pressure ratio $k_h = \frac{P_{nh}}{P_{vh}}$. Perhaps j could be appropriately defined as the hopper *compliance parameter*.

The wall pressure ratio k_h is derived from Eq. (11.19) as:

$$k_h = \left(\frac{j}{m+1} + 1\right)\left[\frac{\tan \propto}{\tan \propto + \tan \phi_w}\right] \tag{11.24}$$

11.4.2.3 Vertical Pressures and Corresponding Non-Dimensional Surcharge Factors

While the loads on the feeder are usually determined from the vertical pressures, as described in Eqs. (11.22), it is often found useful to express the pressures and loads in non-dimensional form as illustrated by Arnold et al. [1] as follows:

$$V = q\gamma\, L^{(1-m)} B^{(2+m)} \tag{11.25}$$

where

$q =$	non-dimensional surcharge factor
$\gamma = \rho g =$	bulk specific gravity
$\rho =$	bulk density
$L =$	length of slotted opening
B =	width of slot or diameter of circular opening
$m =$	hopper symmetry factor
$m = 0$	for plane-flow hopper
$m = 1$	for conical hopper

It follows from Eqs. (11.22) and (11.30) that

$$q = \left(\frac{\pi}{4}\right)^m \frac{P_{v0}}{\gamma B} \tag{11.26}$$

From Eqs. (11.23) and (11.30), a general expression for the non-dimensional surcharge pressure may be obtained. That is,

$$q = \left(\frac{\pi}{4}\right)^m \frac{1}{2(j-1)\tan\alpha}\left\{1 + \left[\frac{2P_s(j-1)\tan\alpha}{\gamma} - D\right]\frac{B^{j-1}}{D^j}\right\} \tag{11.27}$$

Two cases are of importance, the initial filling condition and the flow condition, are now discussed in the following Sections.

11.5 Feeder Loads—Initial Filling Conditions

11.5.1 Introductory Remarks

This applies when the feed hopper is initially empty and then filled while the feeder is not operating. Research has shown that the initial filling loads can vary substantially according to such factors as:

- Rate of filling and height of drop of solids, as this may produce impact effects.
- Uniformity of filling over the length and breadth of the feed bin; asymmetric loading will produce a non-uniform pressure distribution along the feeder.
- Clearance between the hopper bottom and feeder surface.
- Degree of compressibility of bulk solid.
- Rigidity of feeder surface.

For the initial filling condition, the stress field in the hopper is active, that is, the major principal stress is almost vertical at any location. The determination of the initial surcharge factor q_i can be made by using an appropriate value of j in Eq. (11.27). The following cases are considered:

(a) For a totally incompressible bulk solid and a rigid feeder with minimum clearance, the upper bound value of q_i may be approached. The upper bound value corresponds to $j = 0$ for which the vertical pressure in the hopper is *hydrostatic*. In this case the ratio of normal pressure to vertical pressure is given by:

$$k_{hi} = \frac{\tan\alpha}{\tan\alpha + \tan\phi_w} \tag{11.28}$$

With $j = 0$, the upper bound value of q_i is obtained from Eq. (11.27), which becomes

$$q_i = \left(\frac{\pi}{4}\right)^m \left\{ \frac{1}{2\tan\alpha} \left[\frac{D}{B} + \frac{2P_s \tan\alpha}{\gamma B} - 1 \right] \right\} \tag{11.29}$$

It is worth noting that, from a practical point of view, the case of $j = 0$ is more relevant to the structural design for strength of the hopper rather than feeder loads.

(b) For feeder design the following three cases, as discussed in Chap. 8 and Australian Standard AS3774 [10] are relevant:

   i. For very incompressible bulk solids and stiff feeder $j = 0.1$
   ii. For moderately compressible bulk solids and flexibly supported feeders, $j = 0.45$
   iii. For very compressible bulk solids and a flexibly supported feeder, $j = 0.9$

(c) Recommended Value of q_i

While the value of q_i may be determined using an appropriate value of j in Eq. (11.27), from a practical point of view, it has been established that a satisfactory prediction of q_i may be obtained from Eq. (11.34).

$$q_i = \left(\frac{\pi}{4}\right)^m \left\{ \frac{1}{2\tan\alpha} \left[\frac{D}{B} + \frac{2P_s \tan\alpha}{\gamma D} - 1 \right] \right\} \tag{11.30}$$

The vertical load V_i is given by

$$V_i = q_i \rho g L^{(1-m)} B^{(2+m)} \tag{11.31}$$

(d) Limitations on Selection of j Value

As may be observed by reference to the solution of the hopper initial flow pressure, as given in Eq. (11.20), a *singularity* occurs when $j = 1.0$. It is for this reason that, for practical reasons, the limit for the fully developed *active* initial filling case, the maximum value of $j = 0.9$ is chosen. Values of j exceeding 0.9 and approaching 1.0 indicate there is region where the stress field is readjusting from active to passive, that is from peaked to arched as flow commences, or even load settlement as discussed in Chap. 8 which deals with Gate Loads.

Based on the studies of feeder loads, in the absence of test information on compressibility of the bulk solid and feeder stiffness, values of j in the range 0.3 to 0.45 may be selected.

11.5.2 Surcharge Load—Mass and Expanded-Flow Bins

The computation of the initial vertical load acting on a feeder requires knowledge of the surcharge pressure P_s acting at the transition of the feed hopper. It is to be noted the flow load acting on a feeder is independent of the surcharge head. The determination of the initial surcharge pressure P_s depends on the type of storage system employed.

Referring to Fig. 11.11, the surcharge pressure P_s is given by the Janssen equation:

$$P_s = \frac{\gamma R}{K_j \tan\phi}\left[1 - e^{-K_j \tan\phi \frac{H}{R}}\right] + \gamma h_s e^{-K_j \tan\phi \frac{H}{R}} \tag{11.32}$$

where

$R =$	Characteristic' or effective radius defined as:
$R = \frac{D}{2(1+m_c)}$	for mass-flow bin
$R = \frac{D_c}{2(1+m_c)}$	for expanded-flow bin
$m_c = 0$	for long rectangular cylinder

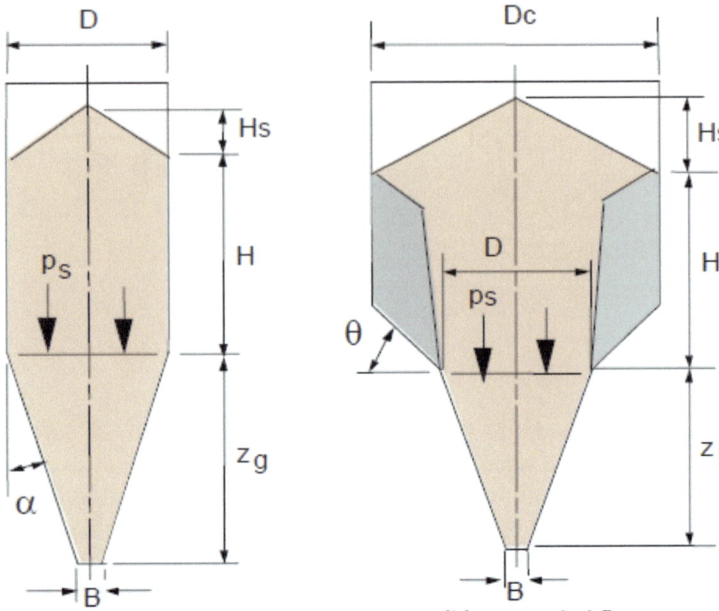

Fig. 11.11 Mass-flow and expanded-flow bins

$m_c = 1$	for square or circular cylinder
$H =$	height of bulk solid in contact with cylinder walls
$K_j =$	$\frac{P_n}{P_v}$ for cylinder. Normally, $K_j = 0.4$
$\phi =$	wall friction angle for cylinder

It is noted that, in the case of the expanded-flow bin, if the flow channel is pre-formed, then the dimension D may replace D_c in Eq. (11.30).

The effective surcharge head for the heap on top of the cylinder is given by

$$h_s = \frac{H_s}{m_s + 2} \tag{11.33}$$

where

$H_s =$	surcharge head
$m_s = 1$	for conical surcharge
$m_s = 0$	for triangular surcharge

11.5.3 Surcharge Load—Gravity Reclaim Stockpiles—Initial Filling Condition

The use of mass-flow reclaim hoppers and feeders under stockpiles is illustrated in Fig. 11.12. The initial load V_i on the reclaim feeder is dependent on the effective surcharge head, while the flow load V_f is independent of the head as illustrated.

The determination of surcharge head and pressure in the case of stockpiles is somewhat uncertain owing to the significant variations that can occur in the consolidation conditions existing within the stored bulk solid. The state of consolidation of the bulk solid is influenced by such factors as.

- Stockpile management and loading history
- Loading and unloading cycle times and length of undisturbed storage time
- Variations in moisture content
- Degree of segregation
- Variations in the quality of bulk solid over long periods of time
- Compaction effects of heavy mobile equipment that may operate on the surface of the stockpile

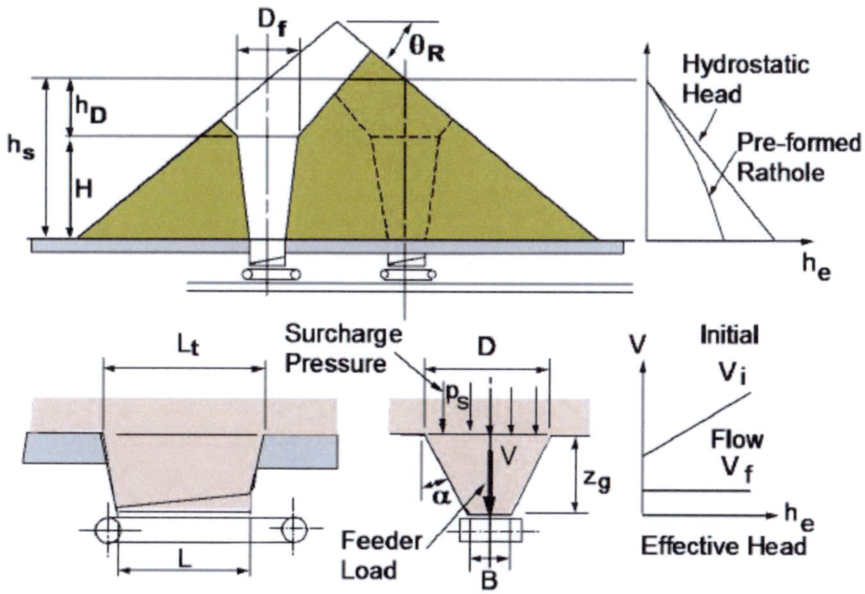

Fig. 11.12 Gravity-reclaimed stockpile

11.5.3.1 Case 1—Uniformly Consolidated Stockpile—Highly Incompressible Bulk Solid

The effective head is equal to the actual head, as per Eq. (11.34). This is the most conservative solution and would rarely occur in practice.

$$P_s = \gamma h_s \tag{11.34}$$

A less conservative solution may be applied through the use of the Rankine pressure or head, see Eq. (11.35)

$$P_s = \gamma h_s \cos \phi_R \tag{11.35}$$

where

$\phi_R =$ angle of repose.

11.5.3.2 Case 2—Pre-Formed Rathole or Flow Channel

During the initial filling process, there will be some deformation of the bulk solid in the flow channel relative to the stationary material adjacent to the flow channel at the hopper interface, therefore the surcharge pressure will be significantly reduced. Furthermore, during subsequent filling and emptying the rathole that is formed acts as a pseudo bin and serves to reduce the surcharge pressure. This is illustrated in Fig. 11.13 where the effective rathole diameter D_e is approximately equal to the diagonal of the rectangular transition.

In such cases, the effective head may be estimated using the Janssen equation following the procedures described in Chaps. 4 and 5 for an expanded flow bin. In this case the cylinder diameter is the actual rathole diameter D_f, and the wall friction angle is estimated on the assumption that the maximum shear stress occurs during flow. On this basis, ϕ is given by:

$$\phi = \tan^{-1}(\sin \delta) \tag{11.36}$$

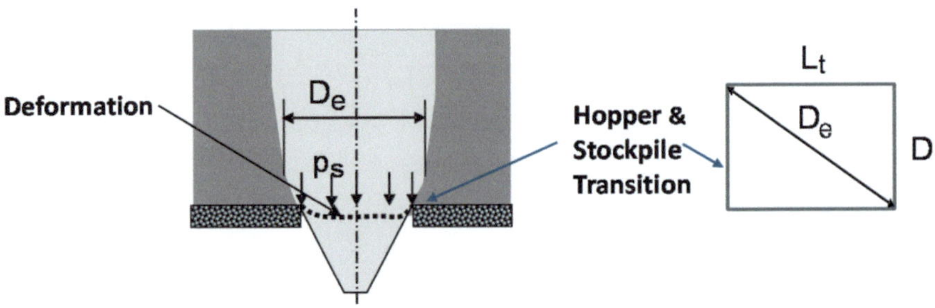

Fig. 11.13 Preformed Rathole and Stockpile and Hopper Transition

where

$\delta =$ effective angle of internal friction

In many cases, the $\frac{H}{R}$ ratio of the ratholes is such that the asymptotic value of the Janssen pressure may be applied. That is:

$$P_s = \frac{\gamma R}{K_j \tan \phi} \tag{11.37}$$

In this case, R is the effective or characteristic radius of the rathole or flow channel (defined in Chap. 6. In general, except for long slotted hopper outlets, the transition can be assumed to be approximately circular. That is $R \cong D_e/4$.

11.6 Feeder Loads—Flow Condition

Once flow has been initiated, an arched or passive stress field is set up in the hopper. Even if the feeder is started and then stopped, as illustrated in Fig. 11.9, the arched stress field in the hopper is preserved. In this case, the hopper is able to provide greater wall support, resulting in less load on the feeder, and the corresponding drive power being significantly reduced.

Like the analysis of the initial feeder loads, the horizontal slice method, depicted in Fig. 11.10, is also relevant to the analysis of the Flow Loads acting on the feeder. However, in this case the interaction of the horizontal slice to describe the loads developed in the *arched* or *passive* stress field is taken into account. By way of background, the relevant analysis and associated stress equations is discussed in Sects. 6.5.2, 6.5.3 and 6.5.4 of *Chapter 6Bin Wall Loads.*

11.6.1 Flow Load Equations

The stress field in the hopper and its feed zone Interface are illustrated in Fig. 11.14. As shown, the arched or passive stress field is established in the hopper while for the feed zone, which includes the weight of bulk solid, the stress field switches to an *active* or *peaked* state which is developed between the hopper outlet and the feeder surface.

The stresses acting at the wall of the hopper outlet are shown in Fig. 11.14. They comprise the normal pressure P_{nfo} and corresponding vertical pressure P_{vfo} (shown in red) along with the major consolidation stress σ_1. The major consolidation stress σ_1 acts tangentially to the circular arch, while its magnitude remains constant around the arch. The major consolidation stress σ_1 acts at the angle η with respect to the normal pressure at the wall as shown. The angle η is defined by the following equation:

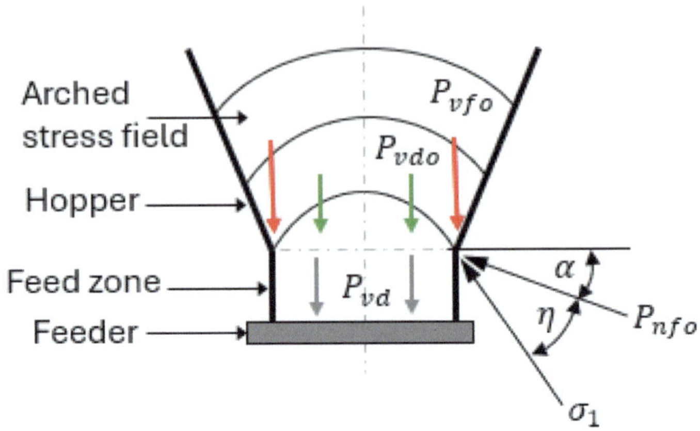

Fig. 11.14 Stress Fields at Hopper and Feed Zone

$$\eta = \frac{1}{2}\left[\phi_w + sin^{-1}\left(\frac{sin\,\phi_w}{sin\,\delta}\right)\right] \tag{11.38}$$

where

ϕ_w = bulk solids friction angle at the hopper wall
δ = effective angle of internal friction.

Equations (11.23) to (11.27) also apply in this case, with the appropriate variables to define the arched or passive stress field. That is the parameter $j = j_f$ and $k_h = k_{hfd}$.

Based on Fig. 11.14, two cases are considered in the determination of the feeder loads:

- The *Wall Boundary Value* which, by way of example, is adopted in the Australian Standard AS3774 [10]
- The *Modified Value* introduced by Roberts [11] and included in Chap. 6 dealing with Bin Loads

11.6.1.1 Wall Boundary Value

The pressure ratio k_{hfo} and corresponding $j = j_f$ value are given, respectively in the following equations:

$$k_{hfo} = \frac{P_{nfo}}{P_{vfo}} = \frac{1 + \sin\delta\cos 2\eta}{1 - \sin\delta\cos 2(\eta + \alpha)} \tag{11.39}$$

$$j_f = (m+1)\left\{k_{hfo}\left(1 + \frac{\tan\phi_w}{\tan\alpha}\right) - 1\right\} \tag{11.40}$$

where

$\alpha =$ hopper half-angle
$m =$ symmetry factor
m = 1 for axial symmetry
m = 0 for plane flow

11.6.1.2 Modified Value

The foregoing method assumes that the major consolidation stress does not vary across each respective horizontal slice. However, this is not the case, the vertical stress P_{vfo} is only applicable at the wall. As has been demonstrated by Roberts [11], the vertical stress decreases over each slice from the wall to the centre of the hopper. For practical reasons, it is assumed that the mean value of the vertical stress across the slice would provide an acceptable solution. The *modified* version, which is recommended, is given by

$$k_{hfd} = \frac{P_{nfo}}{P_{vfd}} = \frac{2(1 + \sin\delta\cos(2\eta))}{2 - \sin\delta(1 + \cos 2(\eta + \alpha))} \tag{11.41}$$

The pressure P_{vdo} at the outlet is given by

$$P_{vdo} = \left\{\frac{\gamma B}{2(j_f - 1)\tan\alpha} + \left[P_s - \frac{\gamma D}{2(j_f - 1)\tan\alpha}\right]\left[\frac{B}{D}\right]^{j_f}\right\} \tag{11.42}$$

where

$$j_f = (m + 1)\left\{k_{hfd}\left(1 + \frac{\tan\phi_w}{\tan\alpha}\right) - 1\right\} \tag{11.43}$$

The vertical force acting at the hopper outlet is

$$V_f = p_{vdo} A_0 \tag{11.44}$$

$$A_0 = \left(\frac{\pi}{4}\right)^m B^{(m+1)} L^{(1-m)} \tag{11.45}$$

where

$A_0 =$ area of hopper outlet
$B =$ width or diameter of outlet
$L =$ length of outlet of plane-flow or wedged shaped hopper
$j_f =$ parameter j for flow case

Other variables as previously defined
Alternatively, the non-dimensional surcharge factor q_f is

$$q_f = \left(\frac{\pi}{4}\right)^m \frac{P_{vo_d}}{\gamma B} \tag{11.46}$$

$$q_f = \left(\frac{\pi}{4}\right)^m \frac{1}{2(j_f - 1)\tan\alpha}\left\{1 + \left[\frac{2P_s(j_f - 1)\tan\alpha}{\gamma} - D\right]\frac{B^{j_f-1}}{D^{j_f}}\right\} \tag{11.47}$$

$$V_f = q_f \gamma L^{(1-m)} B^{(2+m)} \tag{11.48}$$

11.6.1.3 Design Load on Feeder

The vertical design pressure on the feeder, denoted by P_{vd} in Fig. 11.14 is, somewhat empirically, assumed to be based on the mean of the major and minor principal stresses at the hopper and feeder interface. The pressure design load factor k_{des} is introduced:

$$k_{des} = (m + 1)(1 + \sin\delta) \tag{11.49}$$

where

m = symmetry factor
δ = effective angle of internal friction.

Hence

$$p_{vd} = k_{des}\, p_{vdo} \tag{11.50}$$

11.6.2 Geometrical Aspect Ratio of Mass-Flow Hopper

Referring to Fig. 11.15, the geometrical aspects of the mass-flow hopper are required to fulfill two important functions:

- Ensure that the upper switch stress zone is sufficient for the surcharge load to be supported by the upper hopper walls of the switch stress zone.
- The linear converging radial stress zone is sufficient to ensure that the mass-flow during discharge of bulk solids is fully developed.

11.7 Belt and Apron Drive Resistances

The general layout of a belt or apron feeder is shown in Fig. 11.16. The components of the drive resistance are:

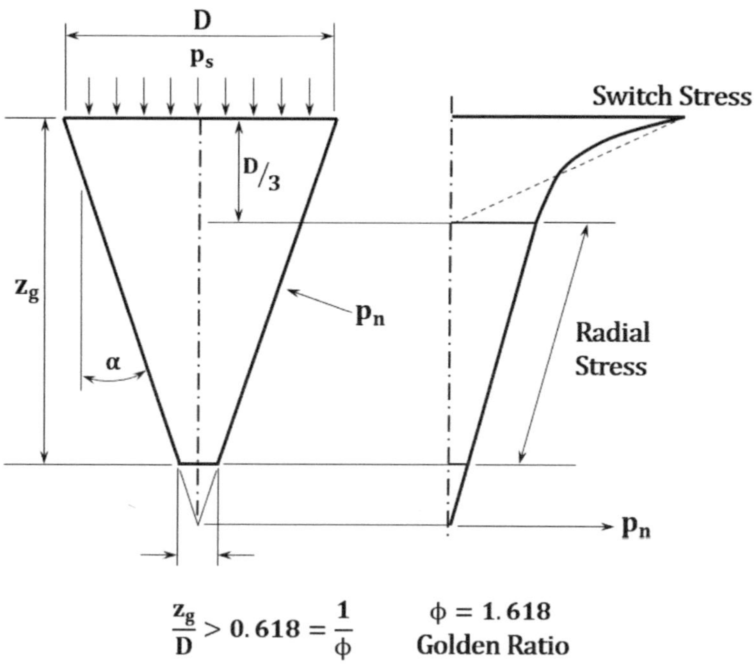

Fig. 11.15 Geometrical Aspects of Mass-Flow Hopper

i. shear resistance of bulk solid
ii. skirtplate friction in the hopper zone and in the extended zone beyond the hopper
iii. belt or apron support idler friction
iv. elevation of the bulk solid

Aspects concerned with the hopper/feeder interface are:

- The force to shear the bulk solid
- The bulk solid and belt/apron friction to prevent slip.

11.7.1 Force to Shear Bulk Solid

The forces acting in the feed zone are illustrated in Fig. 11.16. The vertical Details of the analysis of these various resistances are given in Refs [2–4]. Two particular pressure distributions on the shear plane are shown diagrammatically and will change from the initial filling case to the flow case. Under operating conditions, the resistance F parallel to the feeder surface is given by

$$F = \mu_E V \tag{11.51}$$

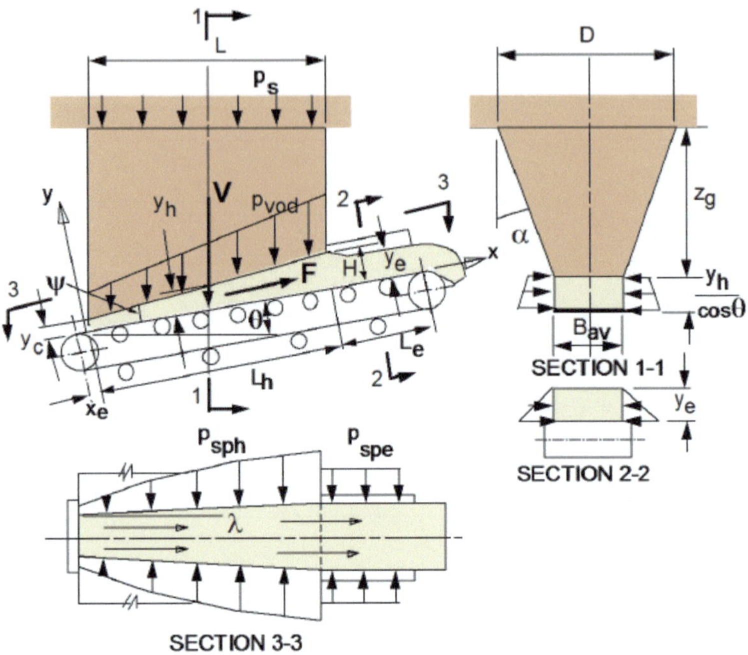

Fig. 11.16 Hopper geometry for feeder load determination

where

$\mu_E =$ equivalent friction coefficient
$V =$ vertical force on shear surface

Starting or breakaway conditions are more difficult to predict and depend on such factors as the hopper and feeder interface geometry, skirtplate geometry, feeder stiffness, the compressibility of the bulk solid and whether any load control is applied. In the absence of any of the foregoing information, a reasonable estimate of the breakaway force F_i is

$$F_i = \mu_E \left[\frac{V_i + V_f}{2} \right] \tag{11.52}$$

An expression for μ_E based on the geometry of the feed zone ([6–8]) is,

$$\mu_E = \frac{\mu_s \cos \psi - \sin \psi}{\cos(\theta + \psi) + \mu_s \sin(\theta + \psi)} \tag{11.53}$$

where

$\theta =$ feeder slope

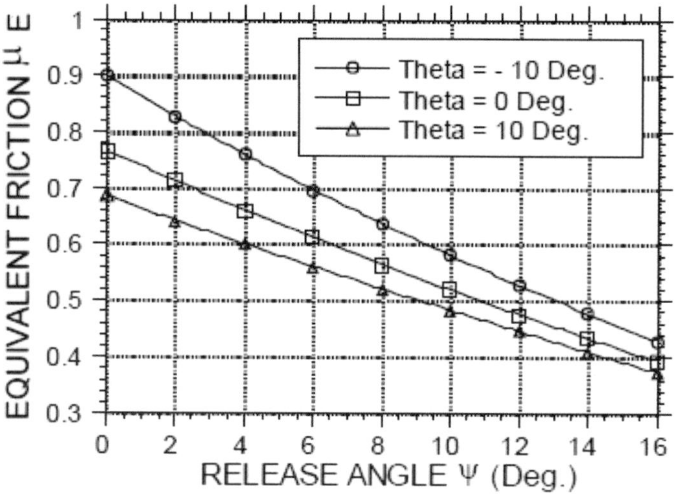

Fig. 11.17 Equivalent friction for belt and apron feeder for $\delta = 50°$

$\psi =$ release angle
$\mu_s =$ coefficient of internal friction on shear plane.

Assuming that the maximum shear stress corresponds to the failure condition, then:

$$\mu_s = \sin\delta \tag{11.54}$$

where

$\delta =$ effective angle of internal friction.

By way of example, a set of design curves for μ_E based on Eq. (11.53) is shown in Fig. 11.17. As indicated, μ_E is sensitive to both the feeder slope angle θ and the release angle ψ, decreasing with increase in both these angles.

11.7.2 Skirtplate Resistance

Assuming steady flow, the skirtplate resistance is determined for the hopper and extended sections (see Fig. 11.16) as follows:

i. Hopper Section

$$F_{sph} = \mu_{sph} K_v (2V + W_h) \frac{y_h}{B_{av}} \cos\theta \tag{11.55}$$

$$W_h = \rho g B_{av} L_h y_h \tag{11.56}$$

ii. Extended Section (beyond hopper)

$$F_{spe} = \mu_{sp} K_v W_e \frac{y_e}{B} \cos\theta \tag{11.57}$$

$$W_e = \rho\, g\, B\, L_e y_e \tag{11.58}$$

where

V = Feeder load
ρ = Bulk density
y_h = Average height of material against skirtplates for hopper section
y_e = Average height of material against skirtplates for extended section
K_v = Ratio of lateral to vertical pressure at skirtplates
g = Acceleration due to gravity = 9.81 (m/s^2)
θ = Slope angle
B_{av} = Average width between skirtplates
μ_{sph} = Equivalent skirtplate friction coefficient
μ_{sp} = Friction coefficient for skirtplates
L_h = length of skirtplates form hopper section
L_e = Length of skirtplates for extended section
W_h = Weight of material in skirtplate zone of hopper
W_e = Weight of material in extended skirtplate zone

It should be noted that, in the hopper zone, the skirtplates are diverging. Hence the frictional resistance, and hence the normal pressure on the skirtplates, will be less than in the case of parallel skirts. Referring to Fig. 11.16, μ_{sph} may be estimated from Eq. (11.59).

$$\mu_{sph} = \frac{\mu_{sp} - \tan\lambda}{1 + \mu_{sp} \tan\lambda} \tag{11.59}$$

where

λ = half-divergence angle of skirtplates

The pressure ratio K_v is such that $0.4 \le K_v \le 0.6$. The lower limit may be approached for the static case and the upper limit for steady flow. In the case of low feed velocities, as in the case of apron feeders, the value of K_v for flow may be in the middle range.

11.7.3 Load Slope Resistance

$$F_{slope} = (W_h + W_e)\sin\theta \tag{11.60}$$

11.7.4 Belt or Apron Load Resistance

i. Hopper Section

$$F_{bh} = (V + W_h)\mu_b\cos\theta \tag{11.61}$$

ii. Extended Section

$$F_{be} = W_e\mu_b\cos\theta \tag{11.62}$$

where

$\mu_b =$ idler friction

11.7.5 Empty Belt or Apron Resistance

$$F_b = w_b L_B \mu_b \tag{11.63}$$

where

$w_b =$ belt or apron weight per unit length
$L_B =$ total length of belt $\geq 2(L_h + L_e + x_e) + 1.5\,[m]$

11.7.6 Force to Accelerate Material onto Belt or Apron

$$F_A = Qv_b \tag{11.64}$$

where

$Q =$ Mass flow rate (given by Eq. (11.13)
$v_b =$ Belt or Apron speed.

Usually the force F_A is negligible.

It should be noted that, for good performance, belt and apron speeds should be kept low. Generally $v_b \leq 0.5$ m/s.

11.7.7 Drive Powers

The power is computed from Eq. (11.65)

$$P = \left(\sum Resistances\right)\frac{v_b}{\eta} \tag{11.65}$$

where η = efficiency and v_b = average belt or apron speed. For start-up, v_b may be approximated as half the actual speed. For the flow condition, v_b will be the actual belt or apron speed during running.

11.8 Condition for Non-Slip

The condition for non-slip between the belt and bulk solid under steady motion can be determined as follows:

$$\mu_{bs} = \tan\phi_b \geq \frac{F + F_{sp} + F_{slope} + F_a}{V\cos(\phi_s - \psi) + \beta W_T \cos\theta} \tag{11.66}$$

where

$\mu_{bs} =$	friction coefficient for bulk solid in contact with the belt or apron
$W_T = W_h + W_e =$	total weight of bulk solid in the skirplate zones
$F = \mu_E = V$	force to shear material at hopper outlet (normally F_f for flow is used)
$F_{sp} = F_{sph} + F_{spe} =$	total skirtplate resistance
$F_a =$	force to accelerate the bulk solid. For normal feeder speeds, $F_a \approx 0$
$V =$	feeder load acting on shear surface (normally the flow load is relevant)
$\beta =$	$\cos(\phi_s - \theta - \psi)$

Neglecting F_a, alternative expressions for μ_{bs} and $\tan\phi_b$ are:

$$\mu_{bs} = \tan\phi_b \geq \frac{V\sin(\phi_s - \psi) + \beta(W_T\sin\theta + F_{sp})}{V\cos(\phi_s - \psi) + \beta W_T\cos\theta} \tag{11.67}$$

or

$$\mu_{bs} = \tan\phi_b \geq \frac{\sin(\phi_s - \psi) + \beta A}{\cos(\phi_s - \psi) + \beta(C_v + C_{ve})\cos\theta} \tag{11.68}$$

where

$$A = (C_v + C_{ve})\sin\theta + K_v\cos\theta\left[0.5\frac{H}{B}\mu_{sph}(2 + C_v) + \mu_{sp}\frac{y_e}{B}C_{ve}\right] \tag{11.69}$$

Noting from Fig. 11.3, that for a small clearance y_c, $\frac{y_h}{B} \approx 0.5\frac{H}{B}$.

A more detailed analysis is given in Ref [8]. As an example, Fig. 11.18 illustrates the minimum belt or apron friction angle as a function of release angle to prevent slip for the case when.

$$\frac{L_h}{B} = 5;\ \frac{y_c}{B} \approx 0;\ \delta = 50°;\ \mu_s = \sin\delta = 0.76;\ C_v = C_{ve} = 0.05\frac{H}{B}$$

The volumetric efficiency is such that $\eta_v(L_b) = \frac{1+C_e}{2}$ and $C_e = 1 - 0.5\frac{H}{B}$. The graphs have been plotted for the feeder slope angles, $-10°$, $0°$, and $10°$. As indicated, the minimum belt friction angle $\phi_b = \tan^{-1}\mu_b$ are shown to be sensitive to both changes in feeder slope and release angles.

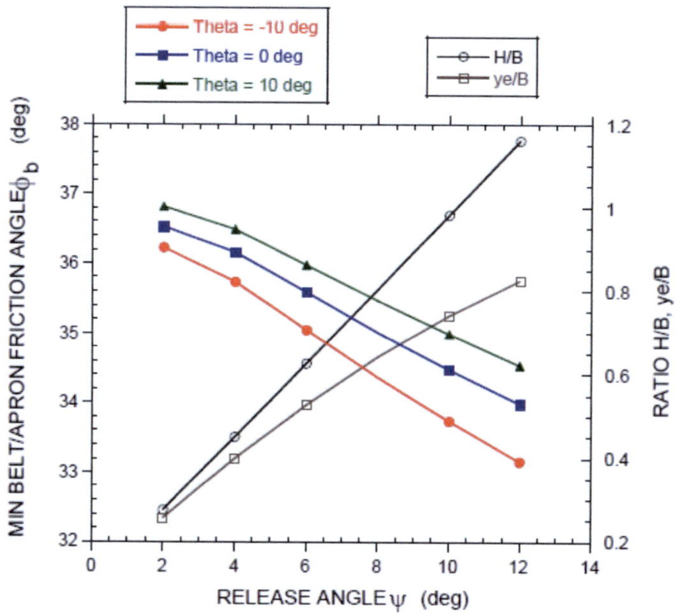

Fig. 11.18 Minimum belt/apron friction angle to prevent slip $\frac{L}{B} = 5$; $\frac{y_c}{B} = 0.1$; $\delta = 50°$; $\mu_s = \sin\delta = 0.76$; $Optimum\lambda = 1.54°$; $C_e = 0.5$

11.9 Controlling Feeder Loads

The loads on feeders and the torque during start-up may be controlled by ensuring that an arched or passive stress field fully or partially exists in the hopper just prior to starting. This may be achieved by such procedures as:

- Cushioning in the hopper, that is leaving a quantity of material in the hopper as buffer storage. This preserves the arched stress field from the previous discharge as illustrated in Fig. 11.19.
- Starting the feeder under the empty hopper before filling commences.
- Using transverse, triangular-shaped inserts.
- Raising the feeder up against the hopper bottom during filling and then lowering the feeder to the operating condition prior to starting. In this way an arched stress field may be partially established.

The choice of mounting arrangement for a feeder can assist in generating a preliminary arched stress field near the outlet sufficient to moderate both the initial feeder load and starting power. In some cases, belt feeders are mounted on helical springs, as illustrated in Fig. 11.20, where the initial compression of the springs during filling

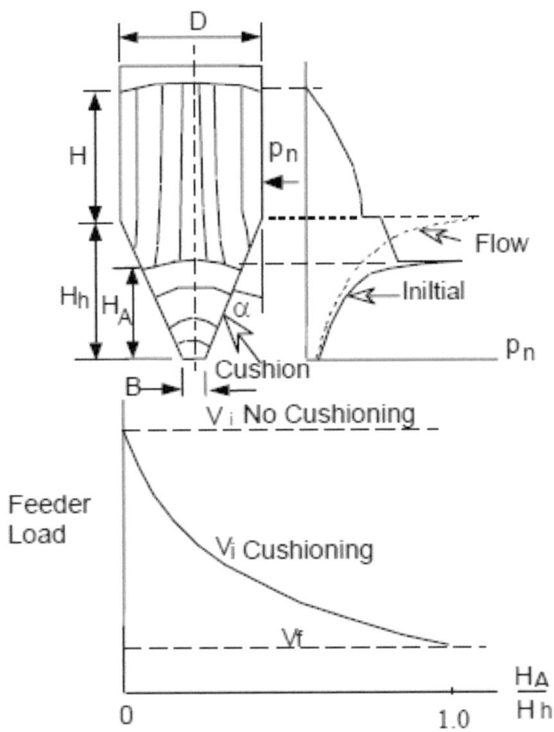

Fig. 11.19 Application of load cushioning to control feeder loads

Fig. 11.20 Helical springs supporting a large belt feeder handling coal

of the bin can assist in generating an arched pressure field near the outlet and reduce the initial load.

An alternative arrangement is to incorporate a jacking system to lift the feeder up against the bottom of the hopper during filling. Before starting, the feeder is released to its operating position sufficient to cause some movement of the bulk solid in order to generate a cushion effect. The use of a slide gate or valve above the feeder is another way of limiting the initial load and power. The gate is closed during filling and opened after the feeder has been started.

For *emergency* purposes, the provision of jacking screws, as illustrated in Fig. 11.21, can be used to lower the feeder should a peaked stress field be established on filling and there is insufficient power to start the feeder. Lowering the feeder can induce, either fully or partially, an arched stress field and allow the feeder to be started. This precaution is useful for feeders installed under stockpiles where surcharge pressures as high as 1000 kPa may be experienced.

11.10 Case Studies

To illustrate the design procedures presented in this Chapter two design examples are illustrated:

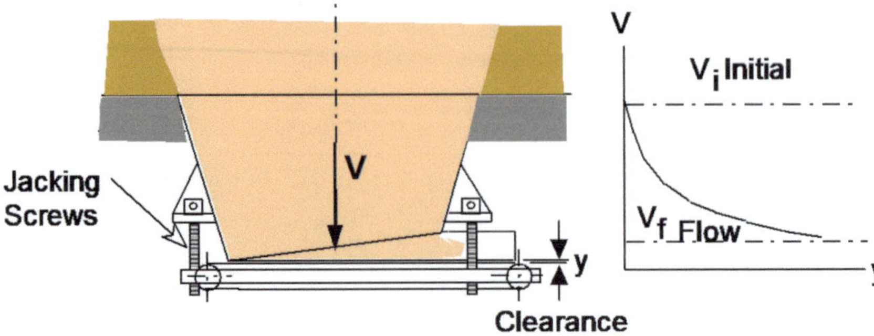

Fig. 11.21 Use of jacking screws to lower the feeder

11.10.1 Case Study 1—Pilot Scale Hopper and Belt Feeder Combination

Figure 11.22a shows a pilot scale laboratory test rig consisting of a plane-flow, wedged-shaped hopper and belt feeder combination. The hopper is constructed of Perspex, and with the aid of adjusting screws, provision is made for the geometry to be varied for the chosen test run. The feeder assembly is supported by cables incorporating load cells for recording the feeder loads. The drive system for the feeder belt is a torque mounted hydraulic motor which enabled the drive torque to be measured. The feeder assembly was restrained with provision for the horizontal feeder drive load to be recorded.

The test set-up for this case study is summarised as follows:

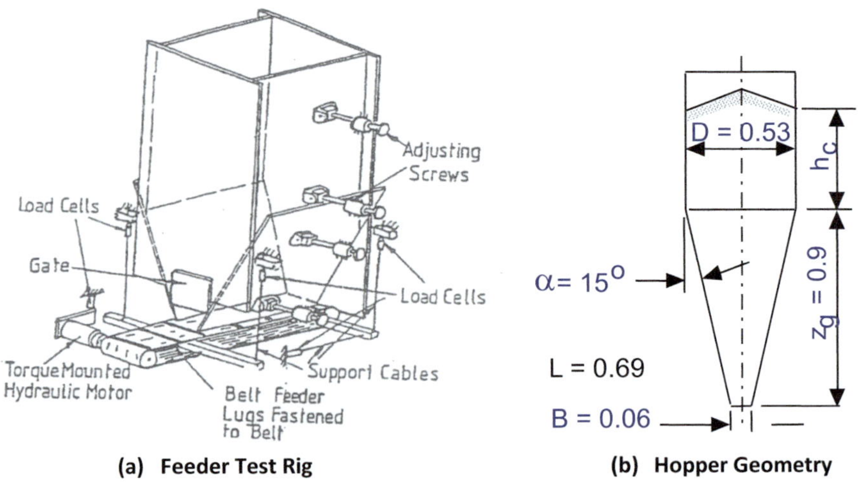

Fig. 11.22 Pilot scale hopper and belt feeder test rig

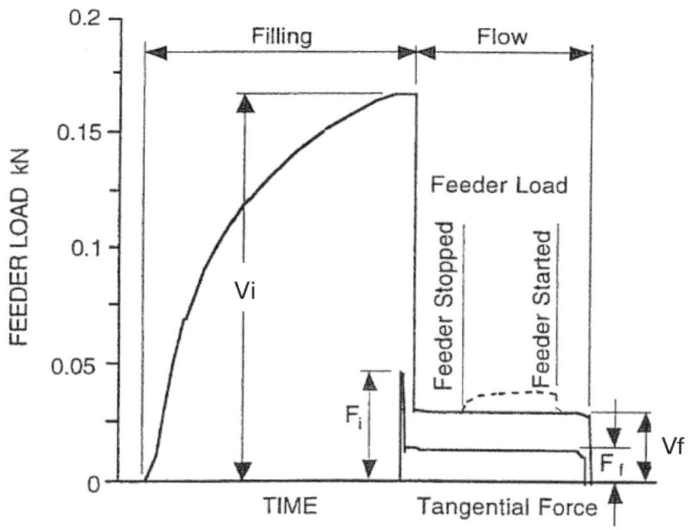

Fig. 11.23 Example test results

- Hopper geometry: As shown in Fig. 11.22b, fill height $h_c = 0.5$ m
- Bulk Solid: Plastic pellets; $\delta = 42°$; $\varnothing_w = 20°$; $\rho = 0.485$ t/m^3.

The measured loads for the initial and flow loads corresponded closely to the loads predicted as follows:

$$V_i = 0.17\,kN \quad V_f = 0.035\,kN \quad F_i = 0.05\,kN \quad F_f = 0.02\,kN$$

Once flow has commenced and the passive stress state established, the load on the feeder did not change except for a small readjustment of the stress field between the stopping of the feeder and then restarting as shown (Fig 11.23). When the feeder was started the tangential driving forced showed an initial spiked increase breakaway force F_i. The driving force then quickly reduced to the running force F_f.

11.10.2 Case Study 2—Bauxite Stockpile Reclaim Belt Feeder

A second case study is concerned with the design of a of belt/apron feeder for controlling the discharge of Bauxite from a stockpile as illustrated in Fig. 11.24. The height of the stockpile above the reclaim feeder is HT = 25 m.

The design of the feeder focusing on feeder loads and drive power is summarised in Table 11.1.

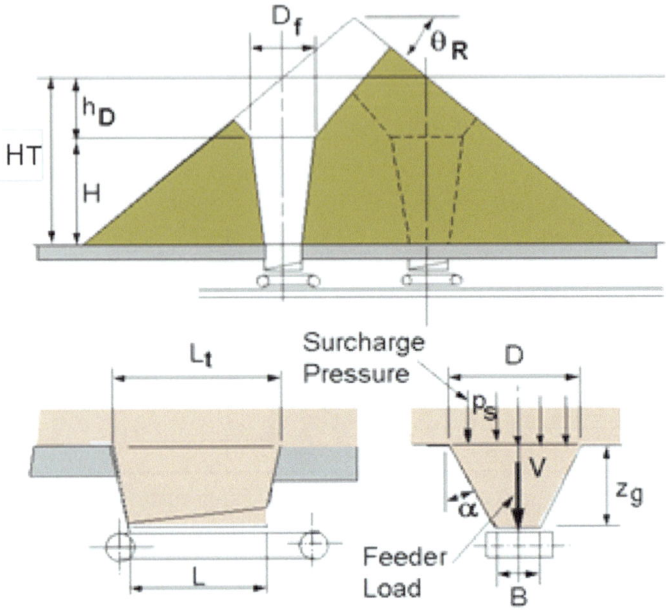

Fig. 11.24 Bauxite stockpile and belt/apron feeder reclaim

11.11 Concluding Remarks

The concepts of feeder design in relation to the loading of bulk solids onto belt and apron feeders have been presented. The importance of interfacing the mass-flow hopper with the feeder has been emphasised. The aim of achieving uniform draw-down in the hopper has been demonstrated.

The hopper stress fields during the initial filling of the hopper have been shown to correspond to the active or peaked stress state, while for the flow case with the feeder operating, the stress field switches to the arched or passive state, where the loads on the feeder are significantly reduced due to the load support provided by the hopper walls.

The horizontal slice model has been mainly adopted in the load equilibrium studies, which requires the integration of active and passive stress field analyses into the Slice model. The definition of the pressure ratios, the so-called "k" parameters and the associated "j" compliance factors have been discussed in some detail.

The application to feeder design has been illustrated by two case study examples. The design methodology presented herein has been successfully utilised in a number of industrial feeder design applications.

Table 11.1 Belt and Apron Feeder Load and Power Calculations—Gravity Reclaim Stockpile.

BIN DIMENSIONS AND PARAMETERS STOCKPILE SURCHARGE PRESSURE ON HOPPER		
Width of Hopper at Transition (m)	Dc	5.5
Height above Feeder (m)	HT	25
Length of Hopper at Transition (m)	Lt	6.25
Effective Rathole Diameter (m)	DR	8.742
Janssen Pressure Ratio	kj	0.4
Effective Head (m)	hs	7.132
Gravitational Acceleration (m/s^2)	g	9.810

HOPPER		
Symmetry Factor	m	0
Transition Diameter or Width (m)	D	5.5
Hopper Half Angle (deg)	Al	25
Height of Hopper (m)	zg	4.557
Hopper Opening - Rear (m)	Br	1.08
Hopper Opening - Front (m)	Be	1.42
Hopper Opening Dimension - Average (	B	1.25
Height to Hopper Apex (m)	hh	5.897
Velocity Parameter	Kvu	0.8
Constant	Pi	3.142

FEEDER DETAILS		
Feeder Speed (m/s)	vb	0.3
Slope Angle of Feeder (deg)	Th	0
Length of Feeder (m)	L	8.5
Skirtplate Length Hopper Zone (m)	Lh	6.25
Skirtplate Length Extended Zone (m) L	Le	1.5
Skirtplate Contact Height - Rear (m)	y	0.1
Skirtplate Contact Height - Front (m)	yf	0.8
Skirtplate Contact Height - Extended (m	ye	0.64
Belt or Apron Width (m)	Bb	1.85
Mass of Belt or Apron (kg/m)	Mb	400
Skirtplate Friction Angle (deg)	Phisp	30
Belt or Apron Idler Friction Coefficient	mub	0.05
Drive Efficiency Efd	Efd	90
Pressure Ratio for Skirtplate Loads	Kv	0.4
Release Angle (deg)	Psir	6.391
Skirtplate Divergence Angle (deg)	Lambd	1.557
Throughput (t/h)	Qm	1472.3

BULK SOLIDS PROPERTIES		
Bulk Density - Hopper Zone (t/m^3)	Rho	1.7
Bulk Density - Ext.Skirtplate Zone (t/m^3	Rhoe	1.5
Effective Angle of Internal Friction (deg)	Del	50
Wall Friction Angle for Cylinder (deg)	Phc	0
Wall Friction Angle for Hopper (deg)	Phh	30
Skirtplate Friction Angle (deg)	Phsp	30
Skirtplate Friction Coefficient	usp	0.577

CALCULATED PARAMETERS		
Bulk Specific Weight (kN/m^3) Ga	Ga	16.677
Conversion Degrees to Radians Qx	Qx	0.017
Principal Stress Angle (deg)	Eta	35.373
Flow Pressure Design Parameter Kfm	Kfm	1.269
Weight of Material in Skirtplate Zone:		
Hopper Section (kN)	Wh	58.630
Extended Section (kN)	We	22.734
Effective Skirtplate Friction Coefficient	usph	0.542

HOPPER		
Wall Friction Coefficient	mh	0.577
Initial Surcharge Pressure (kPa)	ps	118.940
Initial Non-Dimensional Surg. Factor	qi	4.942
Initial Pressure at Outlet (kPa)	pvoi	103.030
Pressure Ratio - Flow - AS3774	khf	0.900
Pressure Ratio - Flow - Modified	khfm	1.541
Exponent "j" AS3774	j	1.015
Exponent "j" Modified	jm	2.449
Height above Apex	hh	5.897
Flow Non-Dimensional Surg. Factor	qf	0.805
Flow Pressure at Outlet (kPa)	pvof	16.781
Design Flow Pressure (kPa)	pvofd	21.206

FEEDER LOADS		
Initial Vertical Feeder Load - Max.	Vi	804.925
Initial Vertical Feeder Load - Avge. (kN)	Viav	485.648
Flow Vertical Feeder Load (kN)	Vf	166.372

RESISTANCES		
Equivalent Friction mue	mue	0.602
Force to Shear Bulk Solid - Initial (kN)	Fsi	292.536
Flow (kN)	Fsf	100.216
Skirtplate Resistances		
Hopper Zone - Initial (kN)	Fsphi	130.142
Hopper Zone - Flow (kN)	Fsphf	30.5273
Extended Zone (kN)	Fspe	2.36629
Belt/Apron Resistances		
Hopper Zone - Initial (kN)	Fbrhi	43.1777
Hopper Zone - Flow (kN)	Fbrhf	11.2501
Extended Zone (kN)	Fbe	1.1367
Empty Belt (kN)	Fb	3.924
Slope Resistance	Fsl	0
Acceleration Resistance	Fa	0.12269

TOTAL RESISTANCES		
Initial (kN)	FTi	473.406
Flow (kN)	FTf	149.543
DRIVE POWERS		
Initial (kW)	PTi	157.802
Flow (kW)	PTf	49.8477

References

1. Arnold, P.C., A.G. McLean, and A.W. Roberts, *Bulk Solids: Storage, Flow and Handling*. 2nd ed. 1979: The University of Newcastle Research Associates (TUNRA).
2. Rademacher, F.J.C., *Reclaim Power and Geometry of Bin Interfaces in Belt and Apron Feeders - An Analytical Approach.* International Journal of Bulk Solids Handling, 1982. **2**(2).

3. Roberts, A.W., M. Ooms, and K.S. Manjunath, *Feeder Loads and Power Requirements in the Controlled Gravity Flow of Bulk Solids from Mass Flow Bins.* Mechanical Engineering Transactions, The Instn. of Engrs. Australia, 1984. **ME9**(1): p. 49–61.
4. Manjunath, K.S. and A.W. Roberts, *Wall Pressure-Feeder Load Interactions in Mass-Flow Hopper/Feeder Combinations -Part 1* Bulk Solids Handling, 1986. **6**(4): p. 769–775.
5. Schulze, D.S., J., *Bulk Solids Flow in the Hopper/Feeder Interface*, in *Proc. Symposium on Reliable Flow of Particulate Solids (RELPOWFLO II).* 1993: Oslo, Norway.
6. Roberts, A.W., *Interfacing Feeders with Mass-Flow Hoppers for Optimal Performance*, in *Proc. Intl. Conf. on Bulk Materials Storage, Handling and Transportation.* 1998, The Instn. of Engrs Australia: Wollongong. p. 459–468.
7. Roberts, A.W., *Feeders and Transfers - Recent Developments*, in *", Proc. Bulkex '99, Australian Society for Bulk Solids.* 1999, The Instn. of Engrs, Australia and the Centre for Bulk Solids and Particulate Technologies: Sydney.
8. Roberts, A.W., *Feeding of Bulk Solids - Design Considerations, Loads and Power*. 1998, Centre for Bulk Solids and Particulate Technologies, The University of Newcastle.
9. Jenike, A.W., *Load Assumptions and Distributions in Silo Design*, in *Conference on Construction of Concrete Silos.* 1977: Oslo, Norway.
10. AS3774–1990, A.S., *Loads on Bulk Solids Containers.* Standards Association of Australia, 1990.
11. Roberts, A.W., *Concepts of feeder design and performance in relation to loading bulk solids onto conveyor belts*, in *IMHC BELTCON 9* 1997: Republic of South Africa.

Chapter 12
Open Front Inclined Apron Feeders—Dynamic Simulation and Design

12.1 Introduction

Open front inclined apron feeders are used to feed run-of-mine (R.O.M.) ore into primary crushers. The feeders are usually quite large as illustrated in Fig. 12.1a which shows a feeder handling R.O.M. bauxite ore being loaded from a dump truck. Typically, R.O.M. ores have a very wide lump/particle size range from rocks of one to two metres average diameter to fine particles less than 100 μm. It is because of the large lump or rock size that conventional shear-gate type feeders, as used for handling processed ores and materials of controlled top particle size, cannot be used for R.O.M. ore. As illustrated in Fig. 12.1b, the controlled feeding action is achieved by the inclined feed zone in which the inclination angle is normally in the range of 20° to 26°.

Relevant research into mass-flow hopper design and feeder loads is included in Refs. [1–8]. Specifically of relevance to this Chapter, is the research into the design of inclined, open front apron feeders and associated dynamic simulation procedures as introduced by Roberts [9, 10] and subsequently extended by Donohue and Roberts [11].

12.2 Open Front Apron Feeder—Feed Zones

Based on the industrial feeder of Fig. 12.1, the feeder model, shown in Fig. 12.2 comprises the following four feed zones: Hopper Zone, Transition Zone, Shear Zone and Skirtplate Zone.

A. W. Roberts, *The Theory and Practice of Bulk Materials Handling*,
https://doi.org/10.1007/978-3-032-19702-3_12

(a) Open Front Inclined Apron Feeder Handling R.O.M Bauxite.

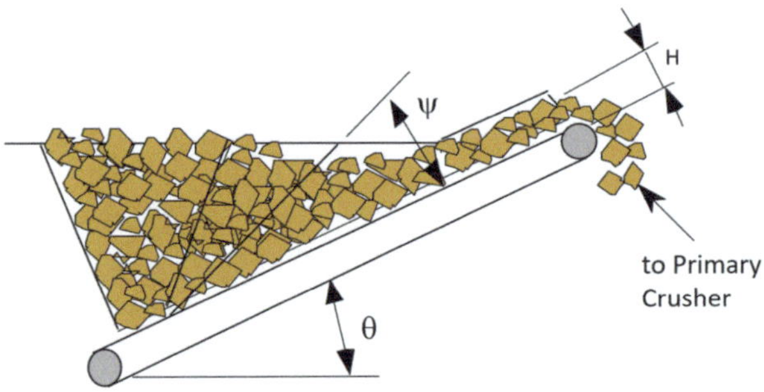

(b) Feeder Schematic Arrangement

Fig. 12.1 Open front inclined apron feeder handling R.O.M ore

12.3 Performance Equations

The use of continuum theory, coupled with assumptions for the mode of flow, can be used to predict the different components of the shear force acting to determine the torque and power required for the feeder drive. These components are presented below.

12.3.1 Throughput

The throughput may be measured in terms of the flow through the skirtplate zone. Referring to Fig. 12.2 the throughput, Q, is given by

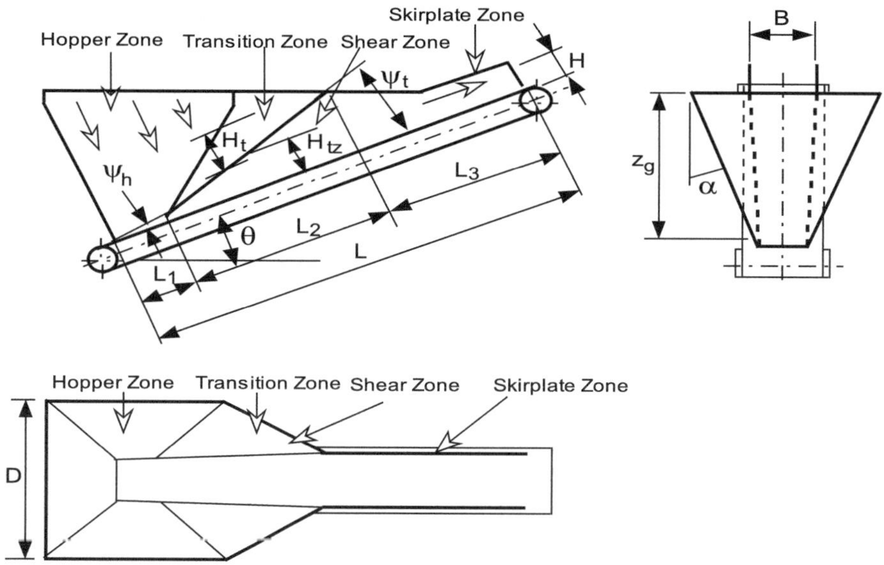

Fig. 12.2 Relevant details of dump hopper and apron feeder

$$Q = \rho BHv \tag{12.1}$$

where ρ = bulk density and v = belt velocity.

12.3.2 Feeder Load and Power

The feeder loads and power may be determined in accordance with references [2, 3].

12.3.3 Hopper Zone Resistance

Assuming a flow stress field exists in the mass-flow hopper, the vertical load on the feeder is given by Eq. (12.2).

$$F_{vf} = q_f \rho g L_1 B^2 \tag{12.2}$$

where q_f = non-dimensional surcharge factor, g = acceleration due to gravity and L_1 = length of hopper. For the model hopper, q_f is determined in accordance with [2, 3]. The shear force, F_{sh}, can then be calculated via Eq. (12.3) with the coefficient of friction, μ_{Eh}, calculated via Eq. (12.4).

$$F_{sh} = \mu_{Eh} F_{vf} \tag{12.3}$$

$$\mu_{Eh} = \frac{\mu_s \cos \psi_h - \sin \psi_h}{\cos(\theta + \psi_h) + \mu_s \sin(\theta + \psi_h)} \tag{12.4}$$

Assuming that the yield condition on the shear plane is governed by the maximum shear stress, the friction coefficient for internal shear is given by $\mu_s = \sin(\delta)$, where δ is the effective angle of internal friction of the bulk material.

12.3.4 Transition Zone

In the transition zone, the converging section of the hopper resists forward movement against the hopper wall causing a stationary boundary layer. There is likely to be some forward movement of a central wedge-shaped block of width B due to internal shear along vertical faces. If this motion occurs, then the resistance is estimated using the following analysis for the average vertical load (F_{tv}) and the shear load (F_{ts}).

$$F_{tv} = \rho g B L_2 \tag{12.5}$$

$$F_{ts} = \mu_{tz} K F_{tv} cos\theta \tag{12.6}$$

where K = ratio of lateral to vertical pressure = 0.4 (based on Janssen), H_t = average height of material above the shear plane in the transition zone and $\mu_{tz} = \sin(\delta)$ for internal shear.

12.3.5 Shear Zone

In the transition zone, the vertical load on the shear plane is F_{tv} as per Eq. (12.5). The force parallel to the feeder surface, F_{ss}, to shear the bulk solid is given by

$$F_{ss} = \mu_{Et} F_{tv} \tag{12.7}$$

where μ_{Et} is given by Eq. (12.4) with ψ_h replaced by ψ_t.

12.3.6 Skirtplate Zone

Here the resistance is due to the skirtplates for the two sections, the shear zone and the extended zone beyond the hopper transition section.

12.3.7 Shear Zone Section

In the shear zone section, the weight of the bulk solid in the skirtplate section below the shear plane, F_{spv}, is given by Eq. (12.8), and the skirtplate shear resistance (F_{sp1}) is given by Eq. (12.9).

$$F_{spv} = \rho g B L_2 H_{tz} \tag{12.8}$$

$$F_{sp1} = \mu_{sp} K_v (2F_{tv} + F_{spv}) \frac{H_{tz}}{B} cos\theta \tag{12.9}$$

where H_{tz} = average height of material in transition zone below shear plane, K_v = ratio of lateral to vertical pressure, F_{tv} is defined by Eq. (12.5) and μ_{sp} is the skirtplate friction. Normally $K_v = 0.4$ but may be as high as 0.7 if the bulk solid is in a fairly loose state. Given the likely state of the bulk solid in the skirtplate zone, a value of $K_v = 0.7$ is generally assumed in this case.

12.3.8 Extended Zone Section

In the extended zone section, the vertical load due to weight of the bulk material, F_{spev}, is given by Eq. (12.10), and the skirtplate shear resistance (F_{sp2}) is given by Eq. (12.11).

$$F_{spev} = \rho g L_3 H B \tag{12.10}$$

$$F_{sp2} = \mu_{sp} K_v F_{spev} \frac{H}{B} cos\theta \tag{12.11}$$

12.3.9 Load Slope Resistance

A load slope resistance, F_{sl}, is also added to the total force due to the bulk solid in contact with the vertical walls of the feeder, shown in Eq. (12.12).

$$F_{sl} = \frac{Q}{v} g L_s \sin\theta \tag{12.12}$$

$$L_s = \frac{L_1 + L_2}{2} + L_3 \tag{12.13}$$

12.3.10 Resistance Due to the Apron

The frictional drag caused by the motion of the resolved weight of the apron sliding over the idlers is considered. The total resistance can be expressed as:

$$F_{bs} = \mu_b F_{fb} \cos\theta \tag{12.14}$$

where μ_b is the idler coefficient of friction and F_{fb} is the resolved weight of the apron feeder.

12.3.11 Total Resistance

The total resistance is then the summation of the individual resistances:

$$F_T = F_{sh} + F_{ts} + F_{ss} + F_{sp1} + F_{sp2} + F_{sl} + F_{bs} \tag{12.15}$$

12.3.12 Torque

$$T = F_T R \tag{12.16}$$

where R is the radius of the drive pulley.

12.3.13 Power

The power required can then be calculated using F_T, the apron feeder speed v and the drive efficiency (η).

$$P = \frac{F_T v}{\eta} \tag{12.17}$$

12.4 Scale Model Test Rig

To verify the predicted performance determinations, tests were performed on a pilot scale model dump hopper and feeder test rig as shown in Fig. 12.3. The dump hopper has a mass-flow half-angle of 22.5°. The opening width between skirt plates is $B =$ 0.145 m and the length of the feeder is 2.1 m. The scale of the test rig relative to the full-size prototype feeder of the case study is one 1:14. The test rig incorporates the following details.

- Feeder is supported on load cells to permit forces in both vertical and horizontal planes to be measured.
- The whole hopper/feeder assembly is supported on load cells so that the throughput can be measured by the loss in weight.
- Feeder is driven by a variable speed hydraulic motor which is suitably instrumented for torque measurements
- Rotational speed of the motor is continuously monitored by means of a transducer
- Dump hoppers constructed of clear Perspex to allow the flow pattern to be observed
- The test setup includes a receiving conveyor bucket elevator an off-loading storage bin and feeder which allowed the dump hopper to be completely emptied during each test run and then filled again for each new run

The bulk material used in the test was bauxite, which had been conditioned in accordance with the scaling laws described in Sect. 12.4. The tests were performed at the inclination angles of 18°, 20°, 22.5°, 24° and 26° and for the feeder speeds of 0.055 m/s, 0.085 m/s and 0.127 m/s. During each test, continuous records of feeder speed, feeder loads, drive torque and throughput were taken.

Fig. 12.3 Scale model, inclined, open front feeder test rig

12.5 Dynamic Similarity Performance Analysis

The application of dimensional analysis and dynamic similarity linked to the model and prototype are now outlined.

12.5.1 System Variables

The following variables are relevant to the investigation.

B = width of hopper/feeder at outlet (m)

ff = critical flow factor

g = gravitational acceleration

Q = throughput (kg/sec)

T = torque (Nm)

α = hopper half-angle (deg)

σ = normal stress (kPa)

$\overline{\sigma}_1$ = stress in arch at hopper outlet (kPa)

ω = angular velocity of feeder drive (rad/s)

FF = flow function of bulk solid

ff_a = actual flow factor

L = length of feeder (m)

P = power (W)

v = feeder speed (m/s)

ρ = bulk density (kg/m^3)

σ_c = unconfined yield strength (kPa)

θ = feeder slope angle (deg)

ψ = release angle (deg)

12.5.2 Non-dimensional Parameters

(a) ***Speed parameter N_v***

$$N_v = \frac{v}{\sqrt{Bg}} \tag{12.18}$$

It is noted that N_v is the Froude Number. The corresponding speed for model tests is given by

$$v_m = v_p\sqrt{\frac{B_m}{B_p}} \tag{12.19}$$

where the subscripts "*m*" and "*p*" refer to the "model" and "prototype" respectively.

(b) **Throughput Parameter N_Q**

$$N_Q = \frac{Q}{\rho g^{0.5} B^{2.5}} \tag{12.20}$$

(c) ***Torque Parameter N_T***

$$N_T = \frac{T}{\rho g B^4} \tag{12.21}$$

(d) ***Power Parameter N_P***

$$N_P = \frac{P}{\rho g^{1.5} B^{3.5}} \tag{12.22}$$

(e) ***Major consolidation stress parameter $N_{\sigma 1}$***

$$N_{\sigma 1} = \frac{\sigma_1}{\rho g B} \tag{12.23}$$

$$\therefore \sigma_{1m} = \sigma_{1p}\left(\frac{\rho_m}{\rho_p}\right)\left(\frac{B_m}{B_p}\right) \tag{12.24}$$

(f) ***Arching stress parameter $N_{\overline{\sigma}_1}$***

During feeding, the arching stress in relation to the unconfined yield is relevant

$$N_{\overline{\sigma}_1} = \frac{\overline{\sigma}_1 H(\alpha)}{\rho g B\left(\frac{ff}{ff_a}\right)} \tag{12.25}$$

where $H(\alpha)$ = arch thickness parameter is defined in the mass-flow hopper design discussed in Chapter 4 of this book, and α = hopper half-angle.

Equation (12.25) assumes that, for geometric similarity, the hopper half angle α is the same for the model and prototype. Hence the function $H(\alpha)$ will also be the same for both.

$$\overline{\sigma}_{1m} = \overline{\sigma}_{1p}\left(\frac{B_m}{B_p}\right)\left(\frac{\rho_m}{\rho_p}\right)\frac{\left(\frac{ff}{ff_a}\right)_m}{\left(\frac{ff}{ff_a}\right)_p} \tag{12.26}$$

12.5.3 *Conditioning of Bulk Solid for Model Tests*

Experience has shown that very good results in model tests are obtained by using the same bulk solid as in the full-scale prototype. It is noted that the major consolidation stress of the bulk solid used in the model tests, compared with that of the prototype, is reduced by the scale ratio. This greatly facilitates the model tests. This makes it

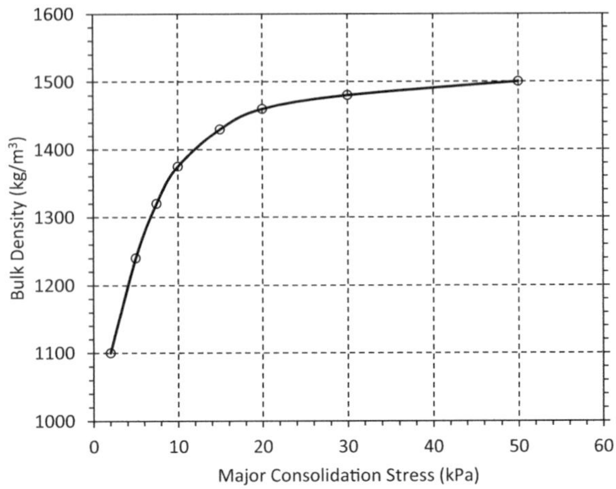

Fig. 12.4 Bulk density for bauxite

quite feasible to achieve similar arching conditions in the model as in the prototype. It is necessary to reduce the maximum lump size in the bulk solid of the model tests in accordance with the scale ratio and to adjust the moisture content of the bulk solid to provide corresponding arching stress conditions in accordance with Eq. (12.6). This procedure was followed in the case of the investigation of the inclined, open front apron feeder.

The variation in bulk density with major consolidation stress σ_1 needs to be taken into account. In the case of bauxite, as used in the case study under consideration, the bulk density plot is shown in Fig. 12.4. For the model tests, the bulk density corresponding to the consolidation pressures relevant to the model is taken to be $\rho_m = 1250$ kg/m^3, while for the hopper section of the full-size prototype feeder, the bulk density is assumed to be the maximum value of 1500 kg/m^3.

12.6 Non-Dimensional Parameters

Using the model feeder test results, the relevant non-dimensional parameters defined in Sect. 12.4 are plotted in Figs. 12.5, 12.6 and 12.7. These graphs show, respectively, the non-dimensional throughput, torque N_T and power N_P as a function of the non-dimensional speed N_V for the elevation angles 18°, 20°, 22°, 24° and 26°.

In the model feeder, the resistance F_{bs} due to the belt sliding on the pressure plate is disproportionably high due to the high friction coefficient $\mu_b = 0.577$. In an actual feeder, the belt or apron runs on rollers for which the friction coefficient is quite low; usually $\mu_b \le 0.05$. For this reason, the torque and power results from the model have been adjusted accordingly in the determination of N_T and N_P in Figs. 12.6 and 12.7, respectively.

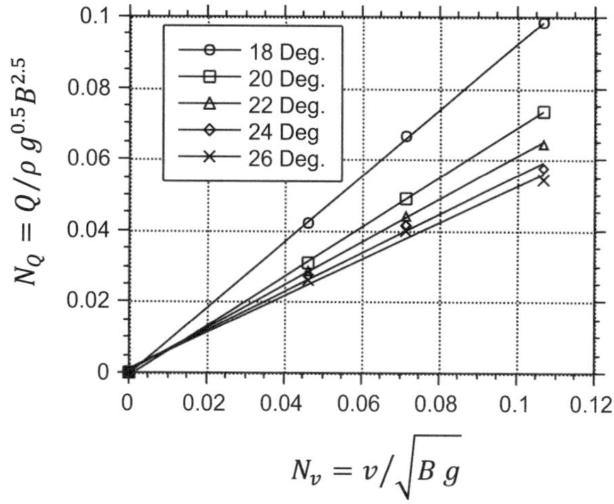

Fig. 12.5 Non-dimensional N_Q versus N_P for bauxite

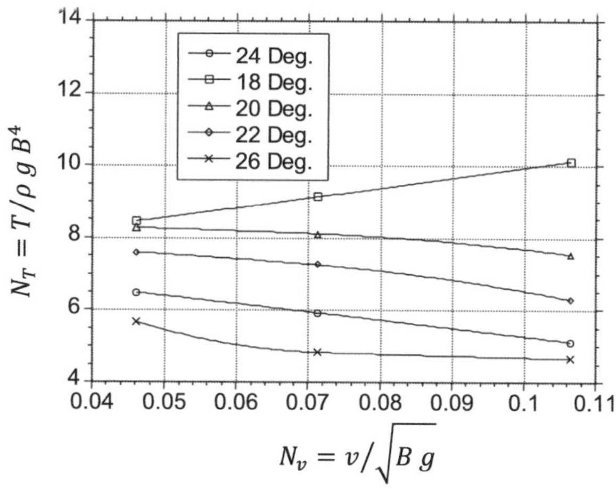

Fig. 12.6 Non-dimensional N_T versus N_V for bauxite

It is noted that both N_T and N_P decrease with increase in feeder angle. Even though the slope resistance increases with an increase in feeder angle, there is a reduction in shear resistance due to the benefit of the increase in combined slope angle and release angle. There is also a reduction in throughput with an increase in feeder angle.

12.7 Full Size Industrial Feeder—Predicted Performance

The full scale feeder of the case study is to operate at a speed of 0.30 m/s and angle of 24° and is to feed bauxite of bulk density 1500 kg/m^3. The feeder is geometrically similar to the model feeder and is such that $B = 2.0$ m. The non-dimensional speed

Fig. 12.7 Non-dimensional N_P versus N_V for bauxite

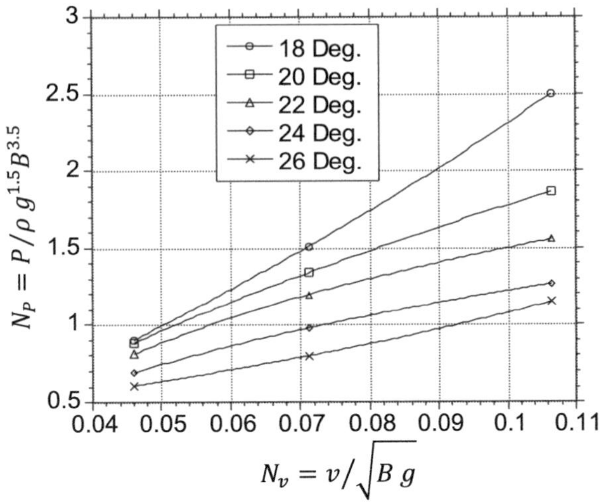

is $N_V = 0.067$. From Eq. (12.1), the corresponding speed of the model feeder is 0.08 m/s. From Fig. 12.5, $N_Q = 0.037$. Using Eq. (12.20), the predicted throughput is

$$Q = N_Q \rho g^{0.5} B^{2.5} = 983 \frac{kg}{s} = 3540 \frac{t}{h}$$

From Fig. 12.7, $N_P = 0.9$. Using Eq. (12.22) the predicted power is

$$P = N_p \rho g^{1.5} B^{3.5} = 469\text{kW}$$

As a check, the procedures outlined in Sect. 12.5 have also been used to calculate the power. The relevant hopper/feeder geometry (Fig. 12.2) and bauxite properties are as follows:

$B = 2.0$ m; $\alpha = 22.5°$; $\delta = 50°$; $D = 8.4$ m; $\rho = 1.5$ t/m^3; $\theta = 24°$; $\phi = 25°$ for hopper and skirtplates; $L_1 = 4.55$ m; $L_2 = 7.35$ m; $L_3 = 14.7$ m; $v = 0.3$ m/s; $Q = 3440$ t/h; $K = K_V = 0.4$; $H_t = 3.64$ m; $H_{tz} = 2.52$ m; $z_g = 7.7$ m; $\psi_h = 1°$; $\psi_t = 20°$; plane-flow assumed for hopper.

$$\begin{aligned} F_T &= F_{sb} + F_{ts} + F_{ss} + F_{sp1} + F_{sp2} + F_{sl} + F_{bs} \\ &= 179.2 + 220.4 + 237.7 + 455.4 + 44.0 + 240.3 + 77.7 = 1455\,\text{kN} \\ P &= 1455 \times 0.3 = 436\,\text{kW} \end{aligned}$$

This is reasonably close to the above value of P = 469 kW. The difference is mainly due to the variation in effective angle of internal friction from $\delta = 60°$ for the model to $\delta = 50°$ for the prototype.

It should be noted that the power requirements for an inclined open-front feeder are significantly influenced by the hopper geometry and skirtplate friction. For instance,

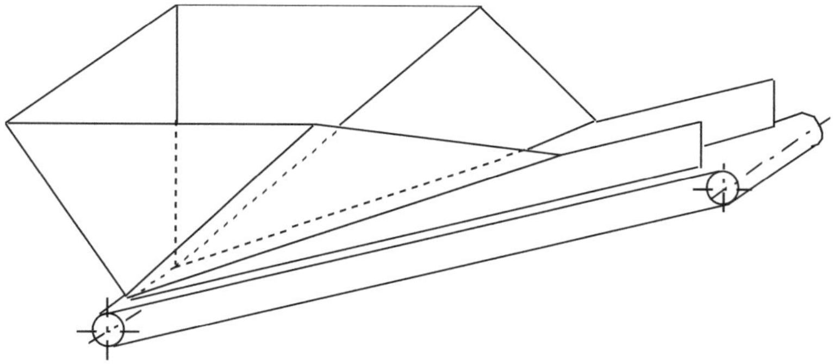

Fig. 12.8 Alternative dump hopper geometry

referring to Fig. 12.2, if the hopper of the full scale feeder is closer to axisymmetric than plane-flow, that is, $\frac{L_1}{B} \approx 1.0$, the release angle ψ_h is reasonably large, say $\psi_h = 8^\circ$, and skirtplate friction angle $\phi = 20^\circ$, then the feeder power is reduced to 351 kW, a reduction of 20%. One design of dump hopper for and inclined open-front apron feeder which has been proposed is shown in Fig. 12.8. In this case $L_1 = 0$, which has the advantage of reducing the loads and power.

12.8 Concluding Remarks

The application of pilot scale model testing has been illustrated by reference to a case study example involving an inclined, open front apron feeder for Run-of-Mine (R.O.M.) bauxite. Tests performed on a 1:14 scale model feeder were conducted over a range of inclination angles and speeds and measurements of feeder throughput, torque and power were made. The results were shown to correlate very well with those derived theoretically. The application of dimensional analysis and dynamic similarity have been shown to be an effective way for predicting the performance of large feeding equipment.

References

1. Arnold, P.C., A.G. McLean, and A.W. Roberts, *Bulk Solids: Storage, Flow and Handling*. 2nd ed. 1979: The University of Newcastle Research Associates (TUNRA).
2. Rademacher, F.J.C., *Reclaim Power and Geometry of Bin Interfaces in Belt and Apron Feeders - An Analytical Approach.* International Journal of Bulk Solids Handling, 1982. **2**(2).
3. Roberts, A.W., M. Ooms, and K.S. Manjunath, *Feeder Loads and Power Requirements in the Controlled Gravity Flow of Bulk Solids from Mass Flow Bins.* Mechanical Engineering Transactions, The Instn. of Engrs. Australia, 1984. **ME9**(1): p. 49–61.
4. Manjunath, K.S. and A.W. Roberts, *Wall Pressure-Feeder Load Interactions in Mass-Flow Hopper/Feeder Combinations -Part 1* Bulk Solids Handling, 1986. **6**(4): p. 769–775.

5. Schulze, D.S., J., *Bulk Solids Flow in the Hopper/Feeder Interface*, in *Proc. Symposium on Reliable Flow of Particulate Solids (RELPOWFLO II)*. 1993: Oslo, Norway.
6. Roberts, A.W., *Interfacing Feeders with Mass-Flow Hoppers for Optimal Performance*, in *Proc. Intl. Conf. on Bulk Materials Storage, Handling and Transportation*. 1998, The Instn. of Engrs Australia: Wollongong. p. 459–468.
7. Roberts, A.W., *Feeders and Transfers - Recent Developments*, in *", Proc. Bulkex '99, Australian Society for Bulk Solids*. 1999, The Instn. of Engrs, Australia and the Centre for Bulk Solids and Particulate Technologies: Sydney.
8. Roberts, A.W., *Feeding of Bulk Solids - Design Considerations, Loads and Power*. 1998, Centre for Bulk Solids and Particulate Technologies, The University of Newcastle.
9. Roberts, A.W., *Predicting the Performance of Open Front Inclined Apron Feeders*, in *Proc. International Conference "From Powders to Bulk"*. 2000: London. p. 443–452.
10. Roberts, A.W., *A Dynamic Modelling of Bulk Solids Storage, Discharge and Handling Systems*. Bulk Solids Handling, 2000. **20**(2): p. 183-189.
11. Donohue, T. and A.W. Roberts, *Dynamic analysis and dem simulation of R.O.M. ore inclined apron feeders*. ICBMH 2013 - 11th International Conference on Bulk Materials Storage, Handling and Transportation, 2013.

Chapter 13
Chute Design Considerations for Feeding and Transfer

13.1 Introduction

Chutes are key components in industries involved in bulk solids handling. While such industries are widely variable in scale and production characteristics, the overriding objective is to achieve efficient movement of bulk solids both in-plant, and in many cases, overland transportation. Such is typically the case in the mining and mineral processing industries where belt conveyors play a major role, as illustrated in Fig. 13.1a, b.

Figure 13.1a shows the feeding of bulk solids via a slow-moving belt or apron feeder onto a much faster moving belt conveyor. Typically, the feeder speed will be in the order of 0.3 m/s and the belt speed 5 m/s or higher. In other cases, transfer chutes are employed to direct the flow of bulk material from one conveyor to another often via a three-dimensional path as illustrated by a 90° transfer in Fig. 13.1b. The purpose of the dribble chute is to collect and transfer the cohesive carry-back material from the belt cleaning system.

Most industrial chute applications, such as those described above, involve rapid or accelerated flow conditions. Such conditions are characterised by *shallow bed* stream flow in which the thickness or depth of the flowing stream of bulk material remains less than the width of the chute. Under these conditions, it is possible to describe the stream flow by means of a lumped parameter model which considers the frictional drag around the chute boundaries as well as making allowance for inter-particle friction.

The importance of correct chute design to ensure efficient transfer of bulk solids without spillage and blockages and with minimum chute and belt wear cannot be too strongly emphasised. This objective is accentuated with the trend towards higher conveying speed.

Like all areas of Bulk Solids Handling, the current approach to chute design is based primarily on numerical simulation notably Discrete Element Modelling (DEM). However, it is important that the fundamental, *classical mechanics* approach

A. W. Roberts, *The Theory and Practice of Bulk Materials Handling*,
https://doi.org/10.1007/978-3-032-19702-3_13

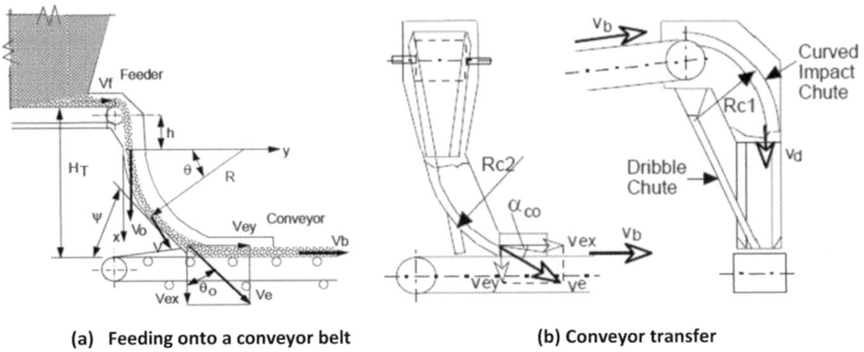

Fig. 13.1 Bulk solid feeding and transfer

not be overlooked. This is the basis of the material presented in this Chapter. By way of background, a selection of relevant research publications are cited [1–28].

13.2 Boundary Friction, Cohesion and Adhesion

13.2.1 Boundary or Wall Yield Locus

For chute design, wall or boundary surface friction has a major influence. It has been shown that friction depends on the interaction between the relevant properties of the bulk solid and lining surface, with external factors including loading conditions and environmental parameters such as temperature and moisture having a significant influence.

The determination of wall or boundary friction is usually performed using the Jenike direct shear test as illustrated in Fig. 13.2. The cell diameter is 95 mm. The shear force S is measured under varying normal force V and the wall or boundary yield locus, S versus V, or more usually shear stress τ versus normal stress σ is plotted.

The Jenike test was originally established for hopper design for which the normal stresses or pressures are always compressive. In the case of chute design, the pressures are normally lower than in hoppers, and often tensile, particularly where adhesion occurs due to the cohesive nature of the bulk solid. The Jenike test of Fig. 13.2a does not allow low compressive pressures to be applied since there is always the weight of the bulk solid in the shear cell, the shear ring and lid which forms part of the normal load. To overcome this shortcoming, the inverted shear tester shown, schematically, in Fig. 13.2b was developed at the University of Newcastle making it possible to measure the shear stress under low compressive and even tensile stresses. The inverted shear tester has been manufactured with a cell diameter of 300 mm to allow more representative size distributions of bulk solids to be tested.

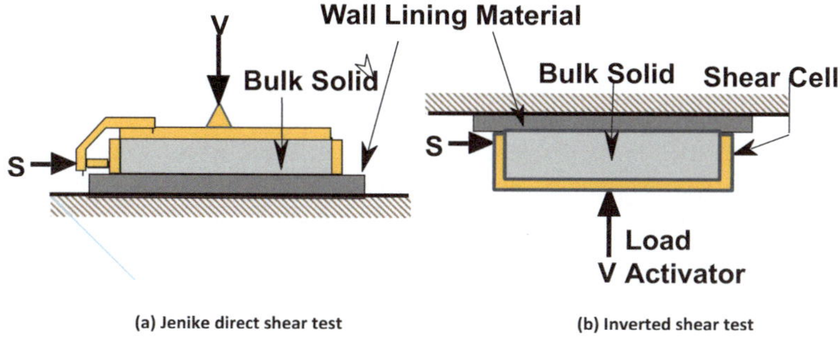

Fig. 13.2 Boundary or wall friction measurement

The τ versus σ boundary or wall yield loci (WYL) for most cohesive bulk solids and lining materials is illustrated in Fig. 13.3. For the compressive stress zone, the locus tends to be slightly convex upward in shape tending towards a sloping linear graph as the consolidation load increases. The wall yield locus intersects the shear stress axis at $\tau = \tau_o$ corresponding to the normal stress $\sigma = 0$ as the normal load $V = 0$. The stress τ_o represents the cohesion of the bulk solid. As the bulk solid consolidation stress decreases below zero ($\sigma = 0$) it switches to the tensile zone with a corresponding further reduction in the shear stress which becomes zero when $\sigma = \sigma_o$, the adhesive stress, as depicted in Fig. 13.3. The wall or boundary friction angle ϕ_w is defined by

$$\phi_w = \tan^{-1}\left[\frac{\tau_w}{\sigma_w}\right] \tag{13.1}$$

where

τ_w = shear stress at the wall
σ_w = pressures acting normal to the wall

Fig. 13.3 Wall or boundary friction and adhesion characteristics

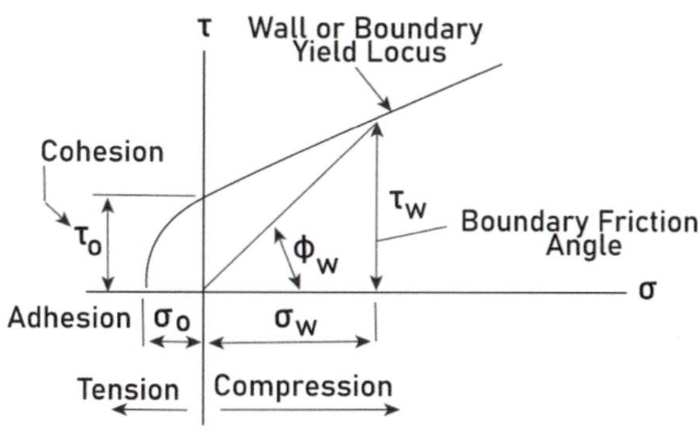

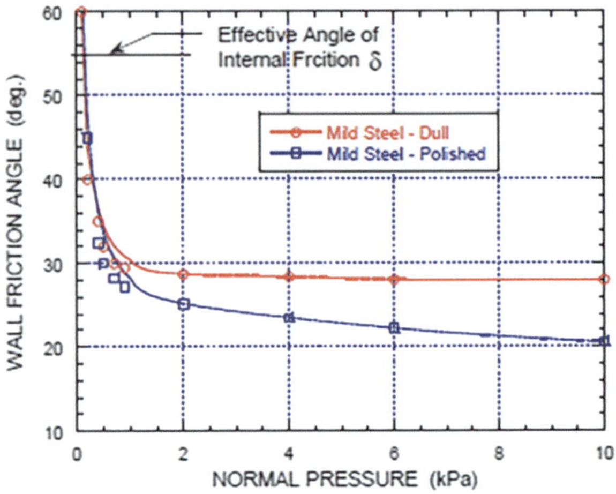

Fig. 13.4 Friction angles for a particular coal on Mild Steel surfaces

The Wall Friction Angle ϕ_w will then decrease with an increase in normal pressure. This is illustrated in Fig. 13.4 which shows the wall friction angles for a representative cohesive coal in contact with dull and polished mild steel surfaces. It is to be noted that the wall friction angle cannot be larger than the effective angle of internal friction δ which is an upper bound limit for ϕ_w. Thus, for very low normal pressures where the friction angle can be quite large, the bulk solid will fail by internal shear rather than by boundary shear, leaving a layer of material on the surface.

The characteristics of adhesion and cohesion are quite complex. A study of this subject would require a detailed understanding of the physics and chemistry of bulk solid and surface contact. It is known, for example, that cohesion and adhesion generally increase as the wall surface becomes smoother relative to the mean particle size of the adjacent bulk solid. Also, adhesion and cohesion generally increase as moisture content of the bulk solid increases, particularly in the case of very smooth surfaces. No doubt, in such cases, surface tension has a significant influence. Cohesion and adhesion can cause serious flow blockage problems when corrosive bonding occurs, such as when moist coal is in contact with carbon steel surfaces. The bonding action can occur after relatively short contact times. Impurities such as clay can also seriously aggravate the behaviour due to adhesion and cohesion.

13.2.2 Adhesion Related Problems

So that build-up and hence blockages can be avoided, it is necessary for the body forces generated in the bulk mass to be sufficient to overcome the forces due to adhesion and shear. Figure 13.5 illustrates the types of build-up that can occur.

The body forces are normally those due to the weight component of the bulk solid but may also include inertia forces in dynamic systems such as in the case

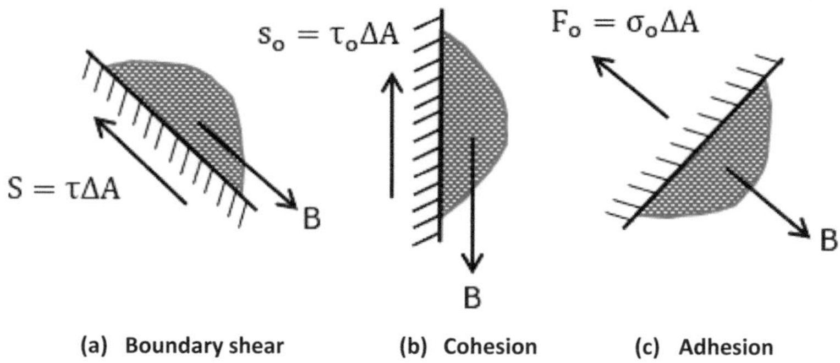

Fig. 13.5 Build-up on surfaces, S = Shear Force; B = Body Force; F_O=Adhesive Force

of belt conveyor discharge or, in other cases, when vibrations are applied as a flow promotion aid.

13.2.3 Mechanisms of Failure

When the body forces are sufficient to cause failure and hence flow, the mode of failure will depend on the relative strength versus shear conditions existing at the boundary surface and internally within the bulk solid [29].

(a) Failure Envelopes—General Case

In this case the shear stress versus normal stress failure envelope for a cohesive bulk solid is always greater than the failure envelope at the boundary. This is illustrated in Fig. 13.6. For such cases, it is expected that failure will occur at the boundary surface rather than internally within the bulk solid.

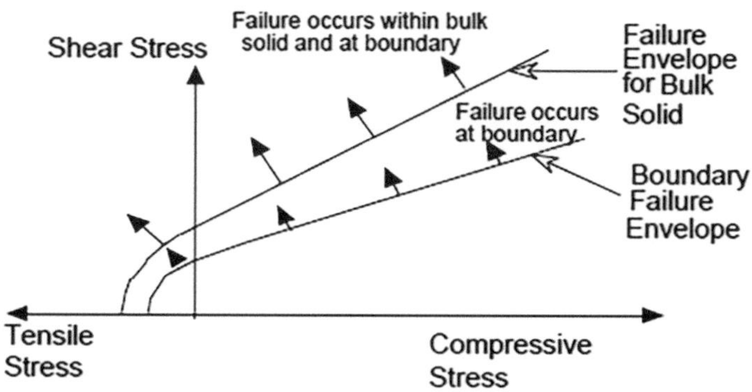

Fig. 13.6 Failure envelopes—General case

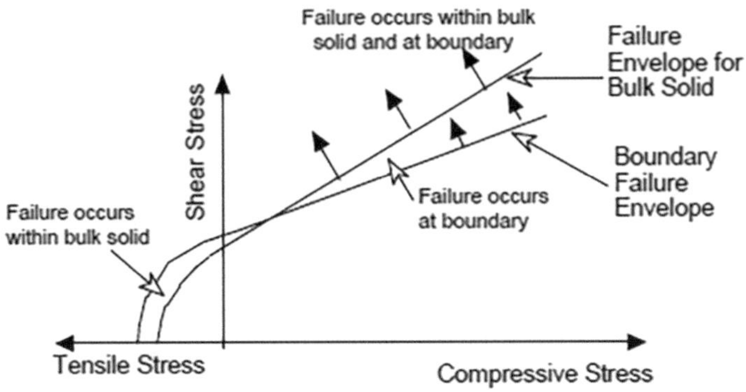

Fig. 13.7 Failure envelopes—Special case

(b) Failure Envelopes—Special Case

In cases of high moisture content cohesive, it is possible for the failure envelope of the bulk solid at lower consolidation stresses or pressures to give lower internal strength than the corresponding strength conditions at the boundary. This is depicted in Fig. 13.7. The body forces may then cause failure by internal shear leaving a layer of bulk solid adhering to the chute surface. This layer may build up progressively over a period. This characteristic has some similarity with the mass-flow and funnel flow limits in hoppers.

Often such problems arise in cases where the bulk solid is transported on belt conveyors leading to segregation with the fines and moisture migrating to the belt surface as the belt moves across the idlers. The segregation condition may then be transferred to chute surfaces. Other cases occur when the very cohesive carry-back material from conveyor belts is transferred to chute surfaces.

(c) Failure Envelope—Free Flowing Bulk Solids.

For free flowing, dry bulk solids with no cohesion, the boundary surface failure envelope is higher than the bulk solid failure envelope. In this case, adhesion of the bulk solid to a chute surface will not occur. Figure 13.8 illustrates this condition.

The foregoing cases indicate that for failure and hence flow to occur, the shear stress versus normal stress state within the bulk solid near the boundary must lie above the failure envelope.

13.2.4 Example Depicting Influence of Adhesion

Consider the case of a cohesive coal of bulk density $\rho = 1\ \text{t/m}^3$ which has a measured adhesive stress of $\sigma_0 = 1$ kPa for contact with mild steel, a typical value. The coal is attached to the underside of the mild steel surface as illustrated in Fig. 13.9.

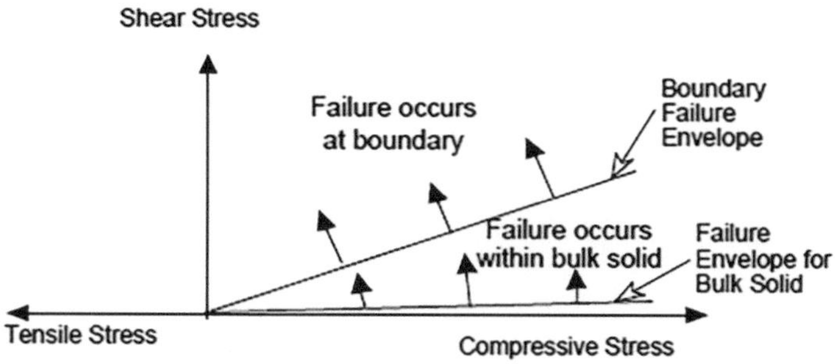

Fig. 13.8 Failure envelopes—Free flowing bulk solids

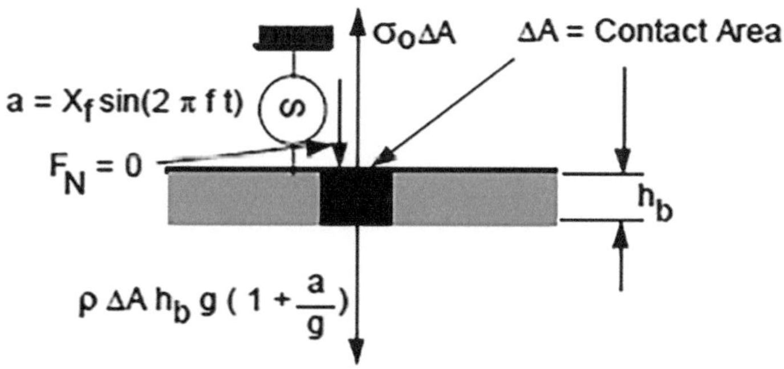

Fig. 13.9 Adhesion problem

A vibrator is proposed as a means of removing the coal. Assuming a failure condition as depicted by Fig. 13.9, the stable build up, denoted by the h_b is given by:

$$h_b = \frac{\sigma_o}{\rho g\left(1 + \frac{a}{g}\right)} \tag{13.2}$$

where

$a = (2\pi\ f)^2\ X_f =$ amplitude of applied acceleration

Without vibration (that is $a = 0$), $h_b = 0.1$ m $= 100$ mm, a substantial amount. By way of example, to reduce h_b to say 10 mm, an acceleration $a = 9.2$ g is required. There are many combinations of frequency and amplitude to achieve this. For instance, a frequency of $f = 151$ Hz and amplitude of $X = 0.1$ mm would suffice. This example indicates the difficulty of overcoming adhesion problems.

A practical case of adhesion is displayed in Fig. 13.10 which shows the influence of prolonged build-up in a coal chute. Apart from the major adhesion problem, the coal has *hardened* to such an extent to make ongoing contact with the belt causing scoring and serious wear. This case study reinforces the importance of correct chute design.

Fig. 13.10 Sustained build-up in coal chute

13.3 Free Fall of Bulk Solids During Loading

A typical example of the free fall of a bulk solid is illustrated in Fig. 13.1a which depicts the application of a gravity feed chute to direct the discharge from a belt or apron feeder to a conveyor belt. The bulk solid is assumed to fall vertically through a height h before contacting the curved section of the feed chute. Since, normally, the belt or apron speed $v_f \leq 0.5\,m/s$, the velocity of impact v_i with the curved section of the feed chute will be, essentially, in the vertical direction.

As a comment the alternative to the use of an accelerating chute is to employ a short accelerating conveyor. These are high maintenance devices and still require head room. Feed chutes may be regarded as the better proposition.

For the free fall section, the velocity v_i may be estimated from

$$v_i = \sqrt{{v_{fo}}^2 + 2\,g\,h} \tag{13.3}$$

Equation (13.3) neglects air resistance, which in the case of a chute, is likely to be small. If air resistance is taken into account, the relationship between height of drop and velocity v_i is

$$h = \frac{v_\infty^2}{g} \log_e\left[\frac{1 - \frac{v_{fo}}{v_\infty}}{1 - \frac{v_i}{v_\infty}}\right] - \left(\frac{v_i - v_0}{g}\right) v_\infty \tag{13.4}$$

where

$v_\infty =$ terminal velocity
$v_{fo} =$ vertical component of velocity of bulk solid discharging from feeder

$v_i =$ velocity corresponding to drop height h at point of impact with chute

13.4 Chute Flow Model

Studies performed by Roberts [1] showed that, under thin stream accelerated flow through chutes, approximately 82% of the energy losses are attributed to the bulk material sliding along the chute bottom, about 9% of the losses being due to sliding against the side walls, with the remaining 9% due to inter granular friction. Under these conditions the motion can be described by a lumped parameter model as show in Fig. 13.11. The basic theory is now reviewed.

13.4.1 Equivalent Friction for Chutes of Constant Width

The drag force F_D due to Coulomb friction is expressed by

$$F_D = \mu_e F_N \tag{13.5}$$

F_N is the normal force and μ_e is the equivalent friction factor which takes into account the friction coefficient between the bulk solid and the chute surface, the stream cross-section and the internal shear of the bulk solid. For a chute of rectangular cross-section, μ_e is approximated by

$$\mu_e = \mu\left[1 + K_v \frac{H}{B}\right] \tag{13.6}$$

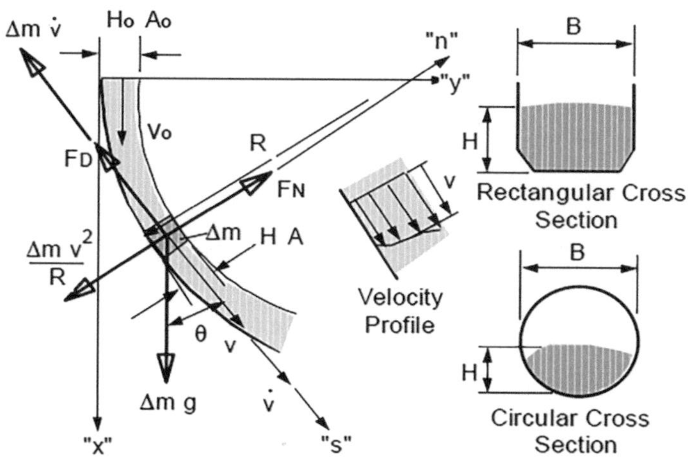

Fig. 13.11 Chute flow model

where

$\mu =$ actual friction coefficient for bulk solid in contact with chute surface
$K_v =$ pressure ratio. Normally $K_v = 0.4$ to 0.6
$H =$ depth of flowing stream at a particular location
$B =$ width of chute

The equivalent friction for chutes of circular cross-section has been analysed by Roberts et al. [15, 23] It has been shown that μ_e can be expressed by the power series $\mu_e = \mu f\left(\frac{H}{B}\right)$ as

$$\mu_e = \mu\left[1 + a_1\frac{H}{B} + a_2\left(\frac{H}{B}\right)^2 + \ldots\right] \tag{13.7}$$

where $a_1 = 0.4$, $a_2 = 0.2057$.

For shallow bed conditions, the higher order terms may be neglected so that the remaining linear relationship becomes identical to Eq. (13.6).

The conditions applicable to curved chutes of circular cross-section are illustrated in Fig. 13.12. For the acceleration zone commencing at the vertical feed point, it is possible for the chute to be more than 50% full, but certainly less than completely full, that is, H /B less than 100%. The acceleration zone commences at the initial feed point; the velocity then increases followed by a corresponding decrease in the stream thickness. The aim is to establish rapid flow in the lower half of the chute, that is H/B < 50%. Once the acceleration switches to deceleration or retardation, the stream thickness starts to build-up with the potential for a blockage to occur. Generally, the aim is to maintain rapid, streamlined flow with H/B < 50%.

13.4.2 Equivalent Friction for Chutes of Varying Width

In some cases, it is necessary to converge the flow, as illustrated for chutes of rectangular cross-section in Fig. 13.13. In this case, the equivalent friction is given by

$$\mu_e = \mu\left[1 + K_v\left(\frac{H}{B_o - 2\,s\,\tan\lambda}\right)\left(1 + \frac{\tan\lambda}{\mu}\right)\right] \tag{13.8}$$

13.4.3 Continuity of Flow

For continuity of flow,

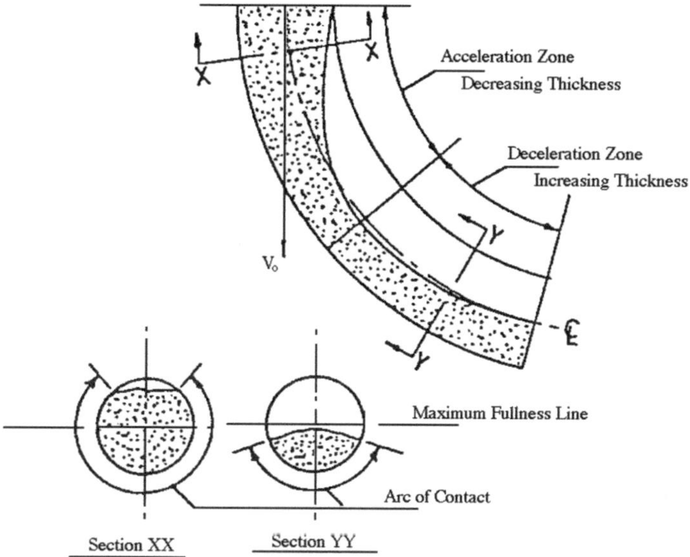

Fig. 13.12 Flow through curved chute of tubular profile

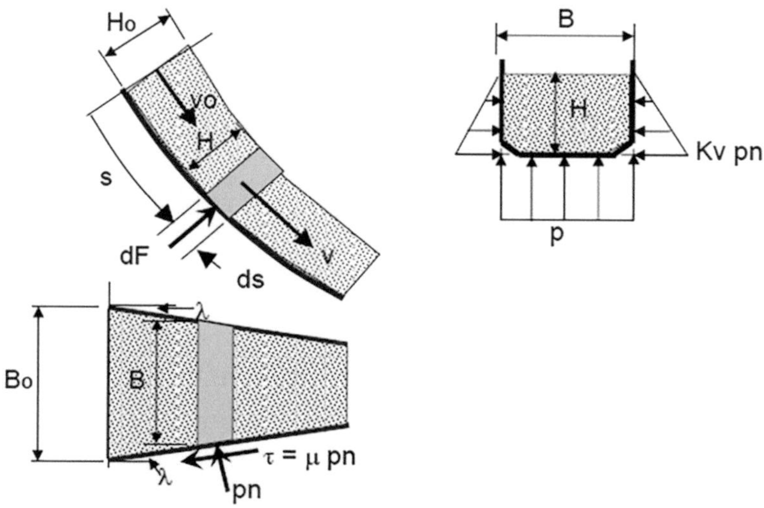

Fig. 13.13 Chute of varying width

$$\rho A\ v = \text{Constant} \tag{13.9}$$

where

ρ = bulk density

A = cross-sectional area of flowing stream
v = velocity of the flowing stream

13.5 Equations of Motion

Referring to the chute model shown in Fig. 13.11 and analysing the dynamic equilibrium conditions, the following differential equations for the stream velocity may be derived:

Moving Coordinates, Tangential "s" and "n" Components

$$\frac{dv}{ds} + \mu_e \frac{V}{R} - \frac{g}{v}(\cos\theta - \mu_e \sin\theta) = 0 \tag{13.10}$$

Cartesian Coordinates, "x" and "y" Components

$$\frac{d\dot{x}}{dx} + \frac{\dot{x}y''(x)\left[y'(x) + \mu_e\right] - g\left[1 - \mu_e y'(x)\right]}{1 + y'(x)^2} = 0 \tag{13.11}$$

$$\dot{y} = y'(x)\dot{x} \tag{13.12}$$

$$v = \sqrt{\dot{x}^2 + \dot{y}^2} \tag{13.13}$$

The radius of curvature of the path at any arbitrary location is

$$R = \frac{\left[1 + y'(x)^2\right]^{1.5}}{y''(x)} \tag{13.14}$$

13.5.1 Complete Solution Involving Variation of μ_e

Solutions of the equations of motion need to take account of the variation of the equivalent friction along the chute. Assuming the bulk density remains approximately constant, the combination of Eqs. (13.6) and (13.7) allows the equivalent friction to be expressed as a function of stream velocity. That is, for a chute of constant width,

$$\mu_e = \mu\left(1 + \frac{C_1}{v}\right) \tag{13.15}$$

$$C_1 = K_v \frac{v_o H_o}{B} \tag{13.16}$$

where

$v_o =$ initial velocity
$H_o =$ initial stream thickness

For the converging chute of Fig. 13.13;

$$\mu_e = \mu\left[1 + \frac{C_2}{v\,B^2}\right] \tag{13.17}$$

$$C_2 = K_v B_o H_o v_o\left(1 + \frac{\tan\lambda}{\mu}\right) \tag{13.18}$$

$$B = B_o - 2\,s\tan\lambda \tag{13.19}$$

where

$B_o =$ initial chute width

μ_e given, as appropriate, by Eqs. (13.15) or (13.17) and substituted into Eqs. (13.10) or (13.11) which may then be solved numerically.

13.6 Chutes of Constant Curvature

To illustrate the solution of the flow equations, the case of chutes of constant curvature and rectangular cross-section is considered. The more exact solution involves the variation of μ_e to be considered. However, an approximate solution will often give satisfactory results.

If the curved section of the chute is of constant radius R and μ_e is assumed constant at an average value for the stream, it may be shown that the solution of Eq. (13.10) leads to the equation below for the velocity at any location θ.

$$v = \sqrt{\frac{2\,g\,R}{4\,\mu_e^2 + 1}\left[\left(1 - 2\,\mu_e^2\right)\sin\theta + 3\,\mu_e\cos\theta\right] + K\,e^{-2\mu_e\theta}} \tag{13.20}$$

For $v = v_o$ at $\theta = \theta_o$,

$$K = \left\{v_o^2 - \frac{2\,g\,R}{4\,\mu_e^2 + 1}\left[\left(1 - 2\mu_e^2\right)\sin\theta_o + 3\,\mu_e\cos\theta_o\right]\right\}e^{2\mu_e\theta_o} \tag{13.21}$$

Special case, when $\theta_o = 0$ and $v = v_o$

$$K = v_o^2 - \frac{6\mu_e\,g\,R}{1 + 4\,\mu_e^2} \tag{13.22}$$

Equation (13.20) becomes:

$$v = \sqrt{\frac{2\,g\,R}{4\,\mu_e^2+1}\left[\left(1-2\,\mu_e^2\right)\sin\theta + 3\,\mu_e\cos\theta\right] + e^{-2\mu_e\theta}\left[v_o^2 - \frac{6\mu_e\,R\,g}{4\,\mu_e^2+1}\right]} \tag{13.23}$$

13.6.1 Chutes of Constant Width—Complete and Approximate Solutions

The complete and approximate solutions are compared in Fig. 13.14 for the case of a chute of radius R = 2.0 m for the initial velocity $v_o = 4$ m/s. The friction angle for the bulk solid in contact with the chute surface is $\phi_s = 25°$. Two initial bed depths are considered, $\frac{H_o}{B} = 0.2$ and $\frac{H_o}{B} = 0.6$.

For the two bed depths, Fig. 13.14a shows the variation of equivalent friction, while Fig. 13.14b compares the exact and approximate solutions for the velocity distributions. As shown, there is little variation in μ_e for the smaller H_o/B, which explains why the complete and approximate solutions for the velocity profiles are virtually identical. Even for the thicker stream for which $\frac{H_o}{B} = 0.6$, there is close agreement between the exact and approximated velocity distributions. The cut-off angle shown in Fig. 13.14b is the angle at which the chute profile should be terminated for it to be self-cleaning.

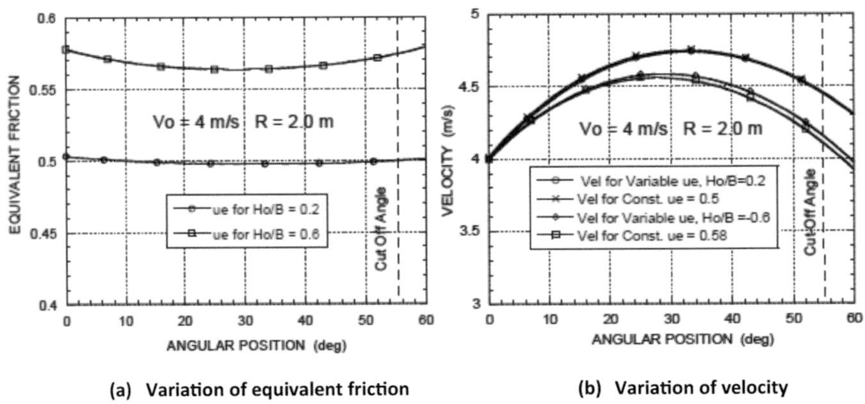

Fig. 13.14 Flow through chute of constant radius. v_o= 4 m/s; R = 2.0 m; μ = tan 25° = 0.466

13.6.2 Chutes of Constant Radius and Converging Width

The case of a chute profile of the type shown in Fig. 13.13 is now examined. As an example, the chute radius is R = 3.0 m, the initial chute width is $B_o = 1.0$ m, the initial velocity is $v_o = 4$ m/s at $\theta_o = 0°$ and the friction angle is $\phi_s = 25°$. Three profiles are compared, $\lambda = 0°$, that is, constant width, $\lambda = 2.5°$ and $\lambda = 5°$.

Figure 13.15a shows the stream velocities, while Fig. 13.15b shows the variation in chute width H and stream thickness or depth as a ratio of chute width, that is, H/B. The angles $\lambda = 0°$ to 2.5° give acceptable performance. However, the convergence angle of $\lambda = 5°$ causes the stream thickness to increase significantly beyond the angular position of $\theta = 40°$. For $\lambda = 5°$ the cut-off angle for the converging section should be not greater than 40°.

13.7 Inverted Curved Chute Sections

Inverted curved chute sections are often employed in the impact zone of transfer chutes as illustrated in Fig. 13.16b. The method outlined above for curved chutes may be readily adapted to inverted curved chute sections as depicted by the model of Fig. 13.16.

Noting that $F_D = \mu_e F_N$, it may be shown that the differential equation in moving coordinates is given by

$$-\frac{dv}{d\theta} + \mu_e v = \frac{g R}{v}(\cos\theta + \mu_e \sin\theta) \tag{13.24}$$

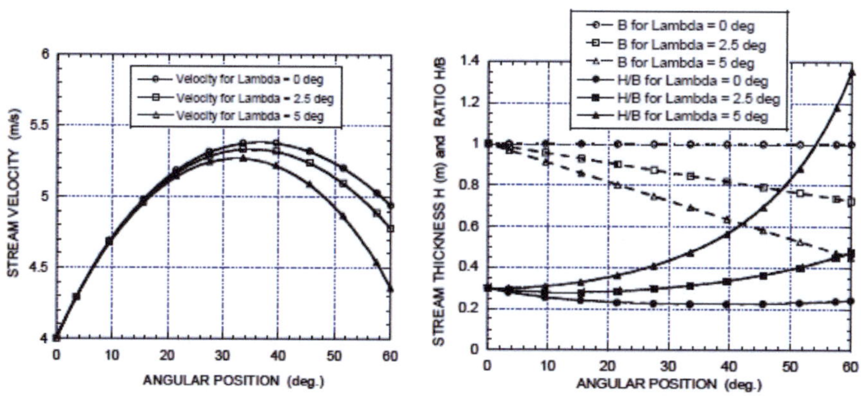

(a) Variation of stream velocity **(b) Chute width and ratio H/B**

Fig. 13.15 Flow through converging chute of constant radius. v_o= 4 m/s; R = 3.0m; μ = tan 25° = 0.466

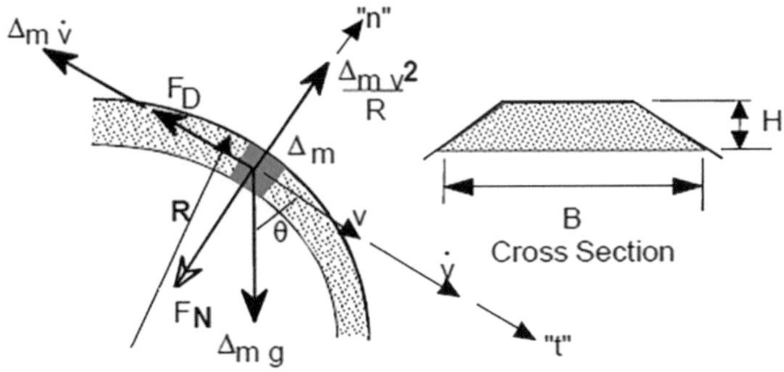

Fig. 13.16 Inverted curved chute model

where

$\mu_e =$ equivalent friction factor as previously defined.

For a constant radius and assuming μ_e is constant at an average value for the stream, the solution of Eq. (13.24) is

$$v = \sqrt{\frac{2\,g\,R}{4\mu_{e^2+1}}[\sin\theta(2\mu_{e^2} - 1) + 3\,\mu_e\cos\theta] + K\,e^{2\mu_e\theta}} \tag{13.25}$$

For $v = v_o$ at $\theta = \theta_o$:

$$K = \left\{v_o^{\,2} - \frac{2\,g\,R}{1 + 4\,\mu_e^{\,2}}\left[3\mu_e\cos\theta_o + \left(2\,\mu_e^{\,2} - 1\right)\sin\theta_o\right]\right\}e^{-2\mu_e\theta_o} \tag{13.26}$$

Special Case, $v = v_o$ at $\theta_o = \frac{\pi}{2}$:

$$K = \left\{v_o^{\,2} - \frac{2\,g\,R}{1 + 4\,\mu_e^{\,2}}\left[2\,\mu_e^{\,2} - 1\right]\right\}e^{-\mu_e\pi} \tag{13.27}$$

and

$$v = \left\{\begin{array}{l}\dfrac{2gR}{4\,\mu_e^{\,2} + 1}\left[\sin\theta\left(2\,\mu_e^{\,2} - 1\right) + 3\,\mu_e\cos\theta\right] \\ +e^{-\mu_e(\pi-2\theta)}\left[v_o^{\,2} - \dfrac{2Rg\left(2\,\mu_e^{\,2} + 1\right)}{4\,\mu_e^{\,2} + 1}\right]\end{array}\right\}^{0.5} \tag{13.28}$$

Equations (13.25) to (13.28) apply during positive contact, that is, when

$$\frac{v^2}{Rg} \geq \sin\theta \tag{13.29}$$

13.8 Wear in Chutes

Chute wear is a combination of abrasive and impact wear. Abrasive wear may be analysed by considering the mechanics of chute flow as will be now described.

13.8.1 Abrasive Wear Factor for Chutes

In cases where the bulk solid moves as a continuous stream under rapid flow conditions, the abrasive or rubbing wear is a function of the normal pressure and rubbing velocity. Consider the general case of a curved chute of rectangular cross-section as illustrated by the chute flow model of Fig. 13.11. An abrasive factor W_c for wear of the chute bottom surface is given by:

$$W_c = \frac{Q_m K_c \tan \phi_s}{B} N_{WR} \tag{13.30}$$

W_c has units of kg/s^3. N_{WR} is the relative contact acceleration wear parameter and is given by

$$N_{WR} = \frac{v^2}{R} + g \sin\theta \tag{13.31}$$

where

$Q_m =$ mass flow rate (kg/s)
$K_c = \frac{v_s}{v}$
$v =$ average stream velocity at section considered (m/s)
$v_s =$ rubbing velocity for bulk solid on chute bottom surface (m/s)
$\phi_s =$ friction angle for bulk solid on chute surface (deg)
$B =$ chute width (m)
$R =$ radius of curvature (m)
$\theta =$ chute slope angle measured from the vertical (deg)

The factor $K_c < 1.0$. For rapid thin stream flow $K_c \approx 0.8$. As the stream thickness increases K_c will reduce.

13.8.2 Special Case—Straight Inclined Chute

Equations (13.30) and (13.31) apply to any chute profile for a straight inclined chute, $R = \infty$. In this case, Eq. (13.30) becomes:

$$W_c = \frac{Q_m K_c \tan \phi_s g \sin \theta}{B} \tag{13.32}$$

For a constant flow rate and assuming that K_c is approximately constant, then, theoretically, the wear along a straight inclined chute is constant and independent of the velocity variation.

13.8.3 Abrasive Wear of Chute Side Walls

Equation (13.32) applies to the chute bottom surface. For the side walls, the wear will be much less, varying from zero at the stream surface to a maximum at the chute bottom. Assuming the side wall pressure increases linearly from zero at the stream surface to a maximum value at the bottom, then the average wear on the side walls can be estimated from

$$W_{csw} = \frac{W_c K_v}{2K_c} \tag{13.33}$$

where K_c and K_v are as previously defined. If, for example, $K_c = 0.8$ and $K_v = 0.4$, then the average side wall wear is 25% of the chute bottom surface wear.

13.8.4 Impact Flow Effects, Wear and "Granular Jumps" in Chutes

(a) Impact Wear

Impact wear may occur at points of entry of points of sudden change in direction as depicted in Fig. 13.17 and described by Eq. (13.34).

Fig. 13.17 Impact effects

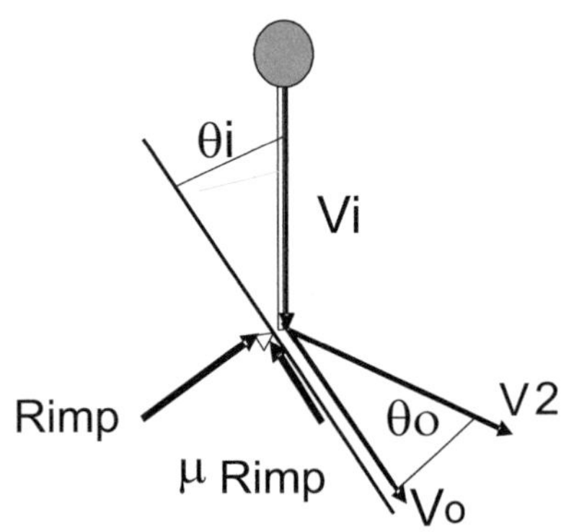

$$\frac{v_o}{v_i} = \cos\theta_i - (1 - e)\mu \sin\theta_o \qquad (13.34)$$

where

$v_i =$ Initial velocity
$v_o =$ rebound velocity
$\mu =$ Friction coefficient
$e =$ Coefficient of restitution
Rimp = Impulse force

For ductile materials, the greatest wear is caused when impingement angles are low, that is in the order of 15° to 30°. For hard brittle materials, the greatest impact damage occurs at steep impingement angles, that is, angles in the vicinity of 90°.

(a) The "Granular Jump" Phenomenon

The *Granular Jump* effect, shown in Fig. 13.18, typically occurs in the case of free-flowing bulk materials. When a flow disturbance occurs, such as a result of a localised imperfection in the chute surface, causing a sudden reduction in the stream velocity and a corresponding increase instream the stream thickness. Often the disturbances are short-lived and the streamline flow is re-established. In cases of fully enclosed chutes, the sudden increase in stream thickness should not lead to contact with the top surface of the chute giving rise to choked flow and complete blockage. By way of explanation, the term *Granular Jump* is adopted in view of its similarity with the term *Hydraulic Jump* as occurs in water stream flow.

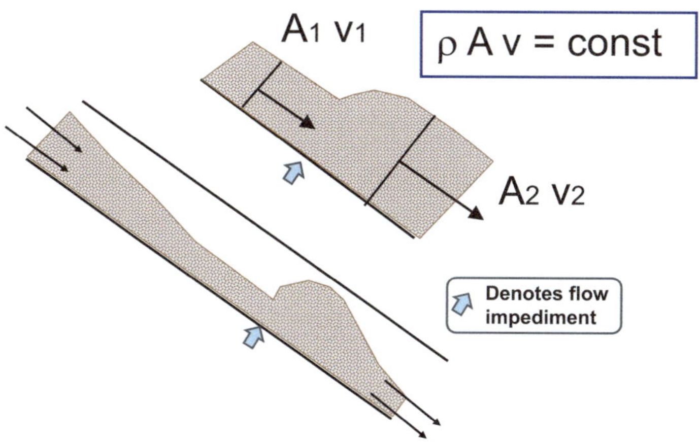

Fig. 13.18 *Granular Jump* effect

13.9 Feeding onto Belt Conveyor

A common application of gravity flow chutes is in the loading of bulk solids onto belt conveyors. The application is illustrated in Fig. 13.1a which shows a gravity feed chute used in conjunction with a belt or apron feeder. The normal operating speeds of such feeders is quite low with $v_f \leq 0.3\,\mathrm{m/s}$. Hence the chute must not only direct the bulk solid onto the belt without spillage, but it must also allow the bulk solid to be accelerated so that at the point of discharge onto the belt, the horizontal component, v_{ex}, of the discharge velocity matches, as close as possible, the belt speed.

The bulk solid falls vertically through a height 'h' before making contact with the curved section of the feed chute. In view of the very low speed of the feeder, the velocity, v_o, of first contact with the curved section of the feed chute will be, essentially, in the vertical direction.

(a) Free Fall of Bulk Solid

For the free fall section, the velocity v_o at the point of contact with the curved chute may be estimated in accordance the procedures described in Sect. 13.3, Eqs. (13.3) and (13.4)

(b) Chute Profile and Belt Wear at Feed Point

It is essential to select the best geometry to meet the performance requirements within the practical constraints of the installation. An overriding consideration is needed to minimise chute wear as well as belt wear at the feed point. This aspect of chute design is depicted in Fig. 13.19. It is also necessary to ensure that the chute slope angle ψ at the belt conveyor feed point is sufficiently large for the chute to be self-cleaning during start-up following each shutdown with bulk solid still retained in the chute. The latter objective is governed by the condition:

$$\psi > \tan^{-1}(\mu_e) + 5^\circ \tag{13.35}$$

Since the discharge slope angle of the chute $\theta_o = (90^\circ - \psi)$, of necessity, there will be an appreciable vertical component v_{ex} of the exit velocity. This component, when combined with the bulk density, gives rise to the impact pressure, $\rho {v_{ex}}^2$, on the belt. When linked with the relative rubbing velocity $(v_b - v_{ey})$, the relative wear parameter W_a is obtained, where the units are $\frac{kPa}{ms}$.

$$W_a = \mu_b\, \rho\, {v_{ex}}^2 (v_b - v_{ey}) \tag{13.36}$$

where

$\mu_b =$ friction coefficient between bulk solid and belt surface
$\rho =$ bulk density
$v_b =$ belt velocity

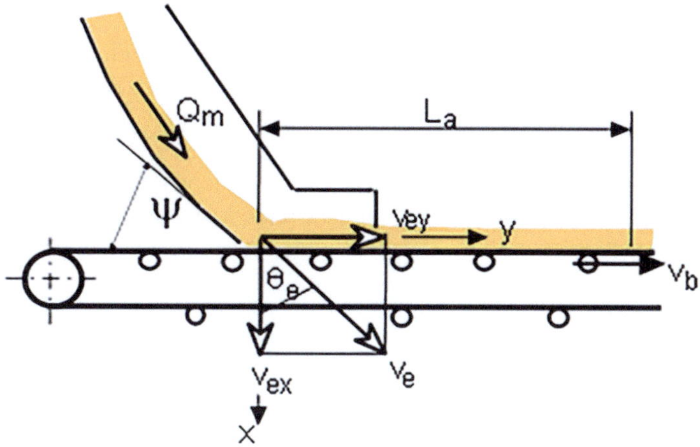

Fig. 13.19 Chute and belt feed point

v_{ey} = bulk solid velocity component parallel to belt at chute discharge

(c) **Acceleration Length**

The acceleration length L_a over which slip occurs is given by

$$L_a = \frac{{v_b}^2 - v_{ey}^2}{2\,g\,\mu_b} \qquad (13.37)$$

13.10 Chute Design Examples

These examples involve a coal feed system based on Fig. 13.1a. of Sect. 13.1. Coal is being fed at the rate of 1000 t/h from a belt feeder moving at 0.3 m/s. The overall feed height $H_T = 5$ m, the bulk density $\rho = 0.8\ t/m^3$ and the receiving belt conveyor speed $v_b = 6$ m/s. The friction angle for the coal on the chute surface is $\phi_s = 25°$ and the coefficient of friction for the coal in contact with the conveyor belt surface $\mu_b = 0.6$. The chute cut-off angle $\psi = 35°$ for which $\theta_o = 55°$.

13.10.1 Vertical Drop and Constant Curvature Chutes

Various combinations of drop heights, h, and corresponding radii of curvature, R, to meet the geometrical constraints of the system are examined. These arrangements are illustrated in Fig. 13.20a. For each of these combinations, the initial velocity

v_o for the point of contact with the chute profile following the free fall have been determined and the velocity distribution around the constant radius chute have been computed together with the belt wear at the feed point. The results of this exercise are presented in Fig. 13.20b and c. All chutes shown in Fig. 13.20a terminate with the same slope angle and same height end point.

It is interesting to observe from Fig. 13.20b that the velocities corresponding to $\theta = 45°$ are the same for all chutes. Despite the wide variation in the initial velocities at the first contact point with the curved chute sections, the velocities at discharge

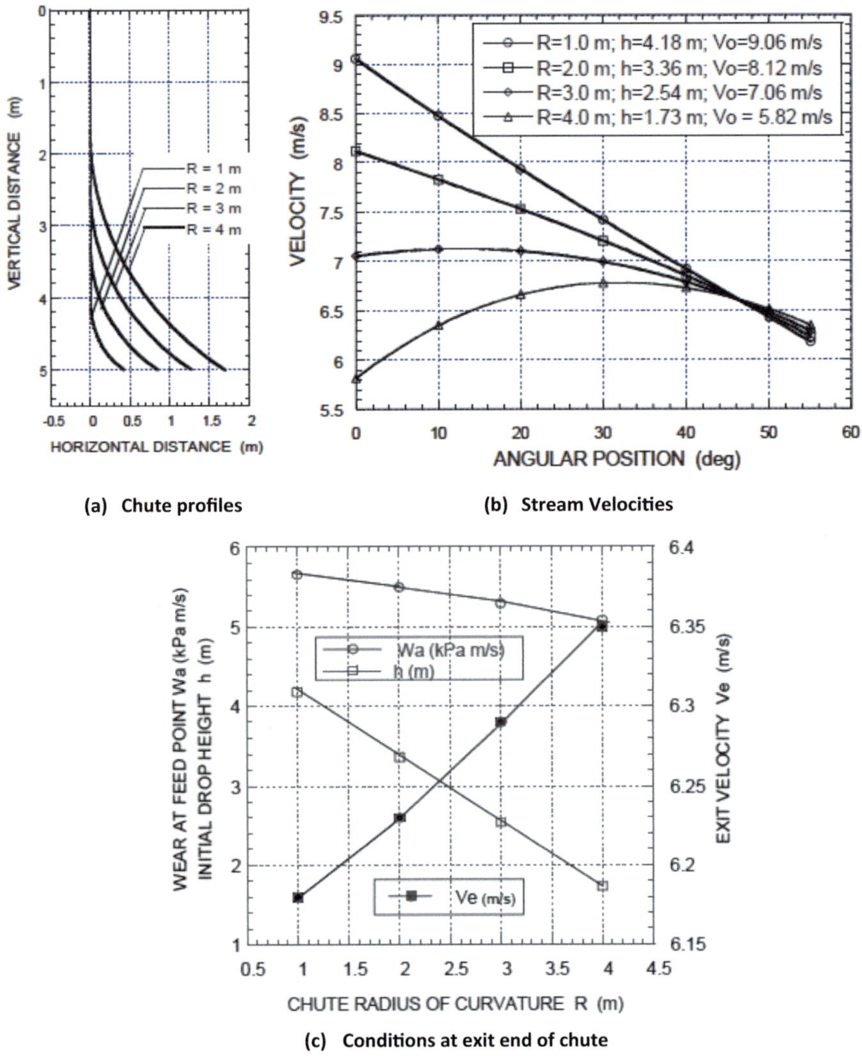

(a) Chute profiles **(b) Stream Velocities**

(c) Conditions at exit end of chute

Fig. 13.20 Feed through drop height h followed by chute of constant radius R

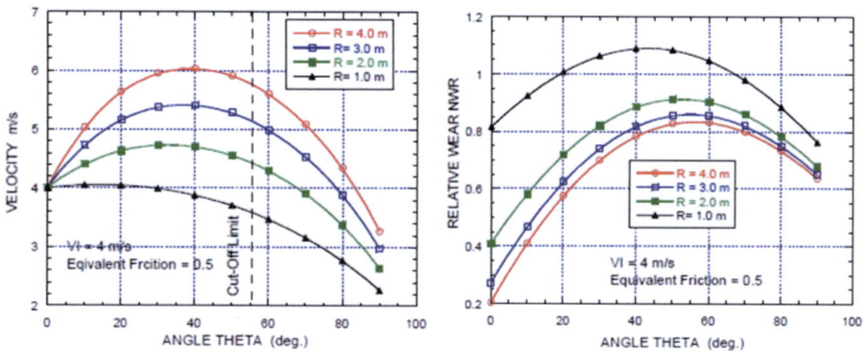

(a) Velocity variation (b) Abrasive wear

Fig. 13.21 Velocities and wear in chutes of constant curvature. Q = 30 t/h; v_o= 0.2 m/s; $\rho = 1$ t/m^3; b = 0.5m; μ_E= 0.6; ϕ= 30°

are almost the same varying from 6.18 m/s for the drop of h = 4.18 m with the R = 1.0 m radius chute to v_e = 6.35 m/s for the drop h = 1.73 m with the R = 4.0 m radius chute. This is shown in Fig. 13.20c which presents the discharge velocities, v_e, belt wear W_a and initial drop height, h, as functions of chute radius R. It is evident that the smaller the chute radius, the more energy the chute absorbs, and this implies greater wear of the chute surface. The best performance is obtained with a smaller drop height and larger chute radius.

13.10.2 *Influence of Chute Radius on Solids Flow Velocity and Chute Wear*

Following on from the previous case, the relationship between the curved chute radius and corresponding abrasive wear are examined. The results are presented in Fig. 13.21 as functions of the slope angle q which is measured with respect to the vertical 'x' axis, Fig. 13.21a depicts the sliding velocities while Fig. 13.21b presents the relative abrasive wear expressed by the parameter Wa.

13.10.3 *Constant Curvature and Parabolic Chute Profiles*

The foregoing examples highlight the sensitivity of interaction between chute radius and stream velocity. For rapid *fast* flow, the adoption of a curved chute radius has a detrimental effectual effect if the radius is too small. This is due to the v^2/R drag wear influence, that is, the abrasive wear is inversely proportional to the radius. This highlights the importance of varying the chute geometry and associated radius of curvature. A parabolic chute profile has some advantages as illustrate in the following example:

The example discussed in the previous section is again considered in this case, a comparison is made between a chute of constant curvature and a chute of parabolic geometry. The coal is allowed to drop a distance of h = 1.73 m before coming into contact with each chute, the velocity of initial chute contact being 5.82 m/s. The constant curvature chute has a radius of R = 4.0 m. The total drop height is $H_T =$ 5.0 m and the cut off angle $\theta_o = 55°$. The equation of the parabolic chute is

$$y = C\,x^2 \text{ where } C = 0.218 \tag{13.38}$$

This value of C satisfies the cut-off condition of y'(x) = tan(55°) corresponding to x = 3.27 m which is based on $H_T = 5.0$ m. Substituting $\dot{y}'(x) = 2\,C\,x$, $y''(x) = 2C$, and solving the relative flow equations to obtain $\dot{x}$ and $\dot{y}$ as functions of x. The velocity $v = \sqrt{\dot{x}^{2+}\dot{y}^2}$ is then obtained.

The results are presented in Fig. 13.22. Figure 13.22a compares the chute profiles, while Fig. 13.22b presents the velocity profiles. The constant curvature chute shows that the velocity increases to a maximum and then decreases as the coal is subjected to retardation. On the other hand, the parabolic chute shows the velocity increasing steadily approaching a maximum value asymptotic value towards the cut-off point. This indicated that the coal is being accelerated over the entire chute. The velocity distribution for the parabolic chute is more favourable than for the constant radius chute. This is because of the increasing radius of curvature of the parabolic chute as given by

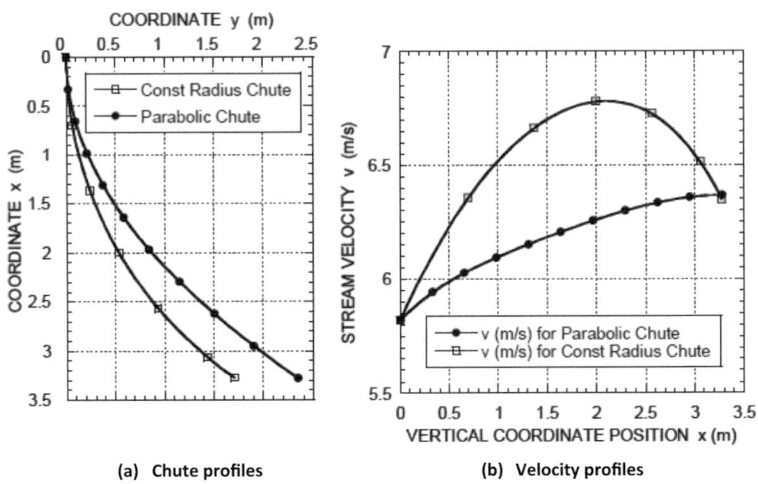

Fig. 13.22 Comparison of constant radius and parabolic chutes

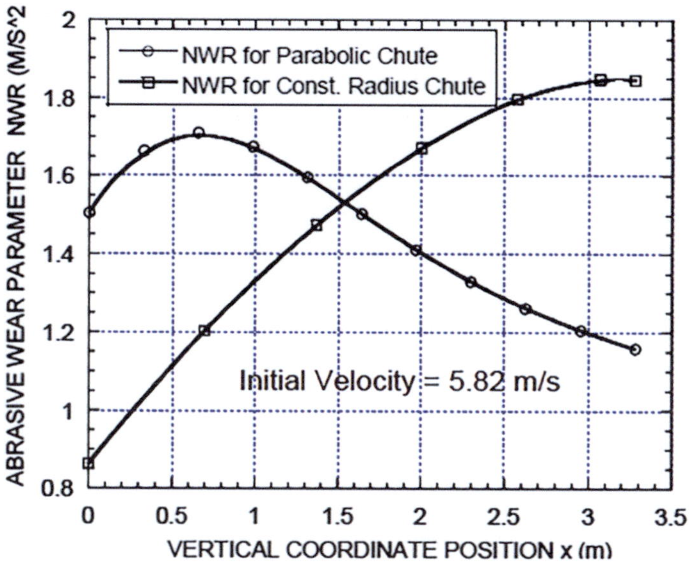

Fig. 13.23 Non-dimensional wear profiles for constant radius and parabolic chutes

$$R = \frac{1}{2C}\left[1 + (2\,C\,x)^2\right]^{1.5} \tag{13.39}$$

The energy loss has a more favourable distribution for the parabolic chute profile as is the chute wear. The abrasive wear parameters N_{WR} determined in accordance with Eq. (13.31) are plotted in Fig. 13.23. The exit velocity for the parabolic chute is $v_e = 6.37$ m/s which is almost identical with $v_e = 6.35$ m/s for the constant curvature chute.

13.10.4 Discussion

The foregoing examples may be examined in terms of the exit velocity v_e determined from the energy and frictional work relationship. For the system of Fig. 13.11, v_e is given by:

$$v_e = \left\{\left[v_{fo}{}^2 + 2\,g\,H_T\right] - \left[2\,g\,\mu_e y_e\right] - \left[2\,\mu_e \int_o^{x_e} \dot{x}^2\, y''dx\right]\right\}^{0.5} \tag{13.40}$$

where

$v_{fo} =$ initial vertical velocity of bulk solid discharging from feeder
$y_e =$ horizontal coordinate of curved chute at exit point

The first term in square brackets of Eq. (13.40) depends on v_{fo} and H_T, which are the same for all chutes considered. The second term in square brackets depends on the horizontal exit coordinate y_e which, in turn, depends on the chute profile and cut-off angle θ_e. The latter is the same for all chutes considered. The last term containing the integral depends on the vertical component of the chute velocity, $\dot{x}$, and chute curvature term y". Despite the wide variation of the chute profiles in the examples considered, there was little variation in the final exit velocity v_e.

13.11 Wear Measurement

The abrasive wear of chute lining and conveyor belt samples may be determined using the wear test apparatus illustrated in Fig. 13.24. This test rig was developed by Roberts [30]. The rig has been replaced by the circular rotating rig of Fig. 3.35 of Chap. 3.

As illustrated, the rig incorporates a surge bin to contain the bulk material, which feeds onto a belt conveyor. The belt delivers a continuous supply of the bulk material at a required velocity to the sample of material to be tested, which is held in position by a retaining bracket secured to load cells that monitor the shear load. The bulk material is drawn under the sample to a depth of several millimetres by the wedge action of the inclined belt. The required normal load is applied by weights on top of the sample holding bracket. The bulk material is cycled back to the surge bin via a bucket elevator and chute. The apparatus is left to run for extended periods interrupted at intervals to allow measurement of the test sample's weight and surface roughness if required. The measured weight loss is then converted to loss in thickness using the relationship given in Eq. (13.41).

$$\text{Thickness Loss} = \frac{M.10^3}{A\rho}\,\text{mm} \tag{13.41}$$

where

M = Mass Loss (g)

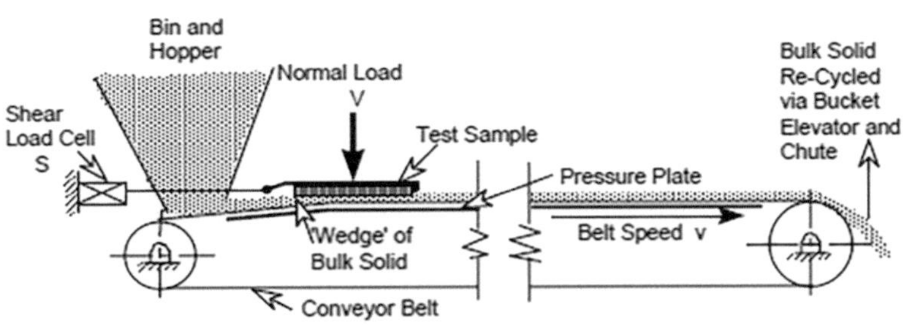

Fig. 13.24 Wear test apparatus [31]

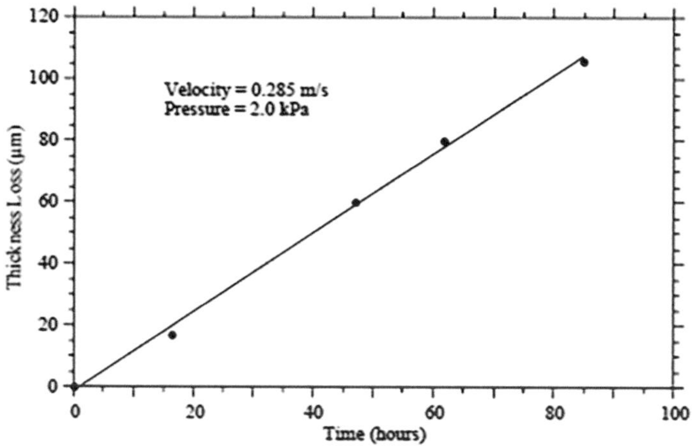

Fig. 13.25 Wear test results

A = Contact Surface Area (m^2)
ρ = Test Sample Density (kg/m^3)

Tests have been conducted on samples of solid woven PVC Conveyor Belt using black coal as the abrading agent. A typical test result for a normal pressure of 2 kPa and a velocity of 0.285 m/s is given in Fig. 13.25. The graph indicates a wear rate of approximately 1.3 μm/hour. This information may be used to estimate the wear expected to take place due to loading of coal on this type of conveyor belt.

13.12 Concluding Remarks

The objectives of this Chapter have been to present an overview of the basic fundamentals of chute design to meet specified design requirements. While the applications are many and varied, the general aim is to achieve accelerated, *thin stream* flow such as in the feeding of bulk solids from a slow moving belt or apron feeder to a fast moving belt conveyor as depicted in Fig. 13.1.

The importance of the bulk solids flow properties along with the associated minimisation of chute wear and adhesion problems is emphasised. The advantage of the accelerated, streamlined flow of the bulk solids is to ensure that the fine particles are encapsulated, without segregation or dispersion, in the fast-moving stream, a necessary objective for dust control.

As a final comment related to the current application of computer simulation programs, notably Discrete Element Modelling (DEM) which is now widely accepted for chute design, the overriding importance of the *classical* theory as presented in this Chapter cannot be too strongly emphasised. This point is further stressed in Chap. 14 which follows on with the design of transfer chutes.

References

1. Roberts, A.W., *The Dynamics of Granular Material Flow through Curved Chutes.* Mech. and Chem. Eng. Transactions, The Instn. of Engrs. Aust., 1967. **MC3**(2).
2. Roberts, A., *An Investigation of the Gravity Flow of Noncohesive Granular Materials Through Discharge Chutes.* Transactions of A.S.M.E, Journal of Engineering for Industry, 1969. **91**: p. 373.
3. Charlton, W., C. Chiarella, and A.W. Roberts, *Gravity flow of granular materials in chutes: Optimising flow properties.* Journal of Agricultural Engineering Research, 1975. **20**(1): p. 39–45.
4. Charlton, W.H. and A.W. Roberts, *Chute profile for maximum exit velocity in gravity flow of granular material.* Journal of Agricultural Engineering Research, 1970. **15**(3): p. 292–294.
5. Charlton, W.H. and A.W. Roberts, *Gravity Flow of Granular Materials: Analysis of Particle Transit Time. Paper No. 72, MH-33, A.S.M.E.*, in *presented at 2nd Symposium of Storage and Flow of Solids.* 1972: Chicago, U.S.A.
6. Chiarella, C. and W. Charlton, *Chute profile for minimum transit time in gravity flow of granular material.* Journal of Agricultural Engineering Research, 1972. **17**(2): p. 149–153.
7. Chiarella, C., W. Charlton, and A.W. Roberts, *Optimum chute profiles in gravity flow of granular materials: a discrete segment solution method.* 1975.
8. Chiarella, C., W.H. Charlton, and A.W. Roberts, *Gravity Flow of Granular Materials: Chute Profiles for Minimum Transit Time*, in *Solids and Slurry Flow of and Handling in Chemical Process Industries, AIChE, 77th National Meeting.* 1974: Pittsburg, U.S.A.
9. Rozenthals, J., *Chute Design*, in *First International Chute Design Conference*, T.B.R. Institute, Editor. 1991: Johannesburg, South Africa.
10. Rozenthals, J. and R.P. Stahura, *Chute Problems - Causes and Solutions*, in *Bionic Research Institute - Chute Design Conference.* 1992: Johannesburg, South Africa.
11. Lonie, K.W., *The Design of Conveyor Transfer Chutes*, in *Third International Conference on Bulk Materials, Storage, Handling and Transportation: Preprints of Papers.* 1989, Institution of Engineers, Australia: Barton, ACT.
12. Marcus, R.D., W.J. Baller, and P. Barthel, *The Design and Operation of the Weber Chute.* Bulk Solids Handling, 1996. **16**(3): p. 405–409.
13. Roberts, A.W., *Bulk Solids Handling - Recent Developments and Future Directions.* Bulk Solids Handling, 1991. **11**(1): p. 17–35.
14. Roberts, A.W., M. Ooms, and S.J. Wiche, *Concepts of Boundary Friction, Adhesion and Wear in Bulk Solids Handling Operations.* Intl. Jnl. of Bulk Solids Handling, 1988. **10**(2).
15. Roberts, A.W., O.J. Scott, and R.D. Parbery, *Gravity Flow of Bulk Solids through Transfer Chutes of Variable Profile and Cross-Sectional Geometry*, in *Proc of Powder Technology Conference.* 1984, Hemisphere Publ Corp: Washington, DC. p. pp 241–248.
16. Roberts, A.W. and W.H. Charlton, *Application of Pseudo-Random Test Signals and Cross Correlation to the Identification of Bulk Handling Plant Dynamic Characteristics.* Journal of Engineering for Industry, 1973. **95**(1): p. 31–36.
17. Roberts, A.W., C. Chiarella, and W.H. Charlton, *Optimisation and Identification of Flow of Bulk Granular Solids*, in *Proceedings IFAC Symposium on Automatic Control in Mining, Mineral and Metal Processing.* 1973, Inst. of Engrs., Aust.: Sydney, Aust.
18. Roberts, A.W. and P.C. Arnold, *Discharge-Chute Design for Free-Flowing Granular Materials.* Transactions of the ASAE, 1971. **14**(2): p. 304–0308.
19. Roberts, A.W. and O.J. Scott, *Flow of Bulk Solids through Transfer Chutes of Variable Geometry and Profile*, in *Proceedings of Powder Europa 80 Conference.* 1980: Wiesbaden, West Germany.
20. Roberts, A.W. and G.J. Montagner, *Identification of Transient Flow Characteristics of Granular Solids in a Hopper Discharge Chute System*, in *Paper presented at Symposium on Solids and Slurry Flow and Handling in Chemical Process Industries, AIChE, 77th National Meeting.* 1974: Pittsburg, Pa., U.S.A.
21. Roberts, A.W. and G.J. Montagner, *Flow in a Hopper Discharge Chute System.* Chem. Eng. Prog., 1975. **71**(2).

22. Roberts, A. and O. Scott, *Flow of bulk solids through transfer chutes of variable geometry and profile.* Bulk Solids Handling, 1981. **1**(4): p. 715–727.
23. Roberts, A.W., O.J. Scott, and R.D. Parbery, *Gravity Flow of Bulk Solids through Transfer Chutes of Variable Profile and Cross-Sectional Geometry*, in *Proceedings of International Symposium on Powder Technology*. 1981: Kyoto, Japan.
24. Roberts, A.W., *Basic principles of bulk solids storage, flow and handling.* 1993, Newcastle, Australia: Institute for Bulk Materials Handling Research.
25. Roberts, A.W. and S.J. Wiche, *Interrelation Between Feed Chute Geometry and Conveyor Belt Wear.* Bulk Solids Handling, 1999. **19**(1).
26. Savage, S.B., *Gravity flow of cohesionless granular materials in chutes and channels.* Journal of Fluid Mechanics, 1979. **92**(1): p. 53–96.
27. Savage, S.B., *The Mechanics of Rapid Granular Flows.* Advances in Applied Mechanics, 1984. **24**: p. 289–366.
28. Roberts, A.W.H., J.W., *Economic Analysis in the Optimal Design of Conveyors.* 1980, Australia: TUNRA Bulk Solids Handling Research Associates, The University of Newcastle, New South Wales.
29. Scott, O.J., *Conveyor Transfer Chute Design, in Modern Concepts in Belt Conveying and Handling Bulk Solids.* The Institute for Bulk Materials Handling Research, University of Newcastle, 1992: p. 11.11–11.13.
30. Roberts, A.W. and S.J. Wiche, *Prediction of lining wear life of bins and chutes in bulk solids handling operations.* Tribology International, 1993. **26**(5): p. 345–351.
31. Wiche, S., S. Keys, and A. Roberts, *Abrasion wear tester for bulk solids handling applications.* Wear, 2005. **258**: p. 251–257.

Chapter 14
Belt Conveyor Transfers

14.1 Introduction

Following on from the previous chapter, which covered the fundamentals of chute design, this chapter focuses on the analysis and design of transfer chutes as applied to belt conveyor systems. These systems involve the delivery of bulk solids from one conveyor to another.

In many cases, the transfer includes two receiving conveyors, as illustrated in Fig. 14.1, which depicts the conveying of Bauxite. In this example, bauxite is conveyed at 7.3 m/s over a distance of 9.3 km, before being transferred through a 90° transfer chute to two receiving belt conveyors, each measuring 5.6 km in length and operating at 6.2 m/s.

Figure 14.2 compares the discharge trajectories of cohesive iron ore and dry, free-flowing potash. The bulk solid will have the tendency to spread laterally as the belt troughing angle decreases through the transition. The amount of spreading is more pronounced for free-flowing bulk solids than for cohesive bulk solids and is more pronounced at lower belt speeds.

There will also be a variation in the adhesive stress across the depth of the stream. In most cases, however, the segregation that occurs during the conveying process will result in the moisture and fines migrating to the belt surface to form a thin boundary layer at the surface. This layer will exhibit higher adhesive stresses than will occur for the remainder of the discharging bulk solids stream. The magnitude of the adhesive stresses at the interface of the boundary layer and the belt surface will determine the extent of the carry-back on the belt as illustrated for the iron ore in Fig. 14.2.

The design of the transfer chute assembly involves two sequential chute functions, the first being to receive the bulk solid from the delivering belt conveyor, and the second, to direct the bulk solids to the receiving belt conveyor. The transfer chute assembly is often referred to, colloquially, as the "Hood and Spoon" due to their respective inverted and normal profiles.

A. W. Roberts, *The Theory and Practice of Bulk Materials Handling*,
https://doi.org/10.1007/978-3-032-19702-3_14

Fig. 14.1 Belt conveying of Bauxite

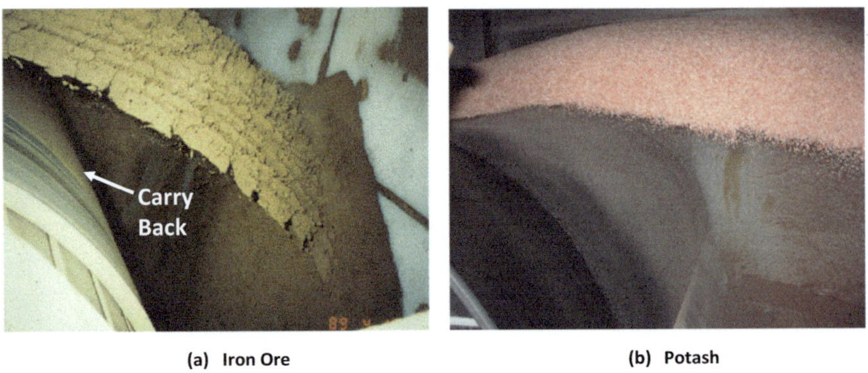

Fig. 14.2 Discharge trajectories of cohesive iron ore and free flowing dry potash

14.2 Conveyor Discharge and Chute "Hood" Geometry

The relevant elements of the conveyor discharge are shown in Fig. 14.3. It depicts the following actions:

- transfer of bulk solid over the distance L_t from the last set of troughed idlers to the first point of contact with the discharge drum, and
- formation of the discharge trajectory of the bulk solids as it leaves the belt and contacts the curved chute or hood as illustrated.

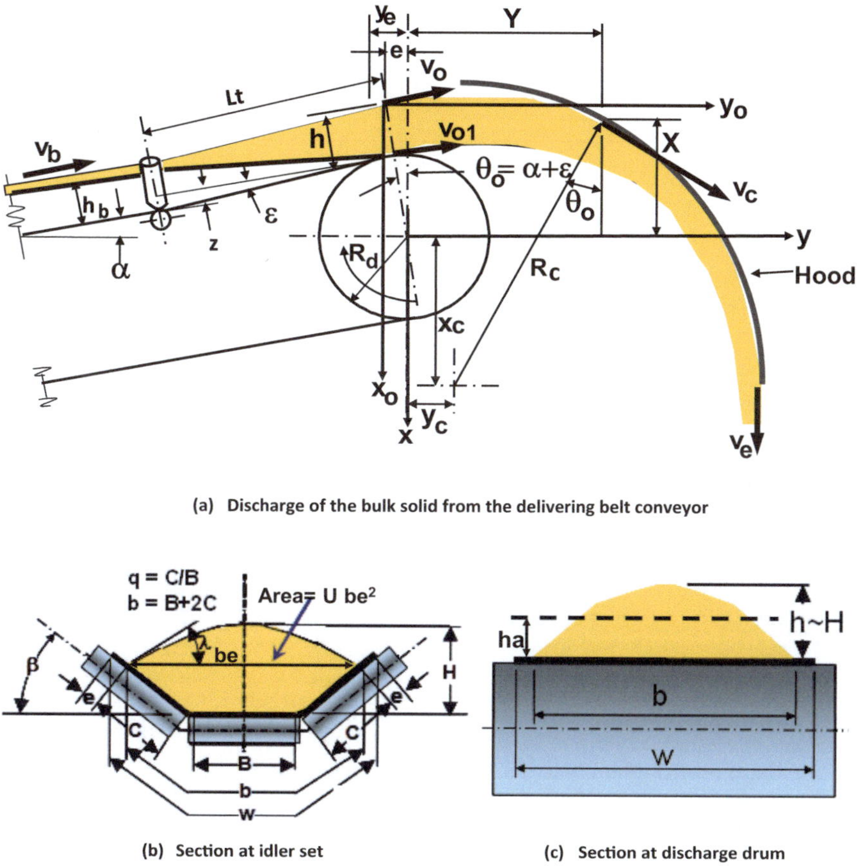

Fig. 14.3 Conveyor belt transition geometry and load profiles

14.2.1 Bulk Solid Discharge Geometry and Load Profiles

Figure 14.3a shows the longitudinal transitional profile of the bulk solid on the belt as the belt traverses over the distance L_t during the unloading operation. Figure 14.3b shows the load profile for the troughed section of the belt at the commencement of discharge and Fig. 14.3c depicts the belt as it contacts the drive drum. A three-roll idler set is assumed, with the contact profile width denoted by "b" and the bulk solid profile illustrated accordingly. As the belt makes contact with the drum, it transitions to a flat configuration, resulting in a change to the cross-sectional profile of the bulk material, as shown in the accompanying illustration. Based on the work of Roberts [1], the surface of each profile is assumed to be of parabolic geometry. For the flat belt, the contact width is represented by "be" which may be approximated by the idler contact width "b", that is, be $\cong$ b.

Discharge of the bulk solid from the delivering belt conveyor.

14.2.2 Load Profiles of Bulk Solid on Belt

The conveyor throughput is given by the following equation

$$Q_m = \rho A v_b \tag{14.1}$$

where

$\rho =$ bulk density under conveying conditions
$v_b =$ belt velocity
$A =$ Cross-sectional area bulk solid on belt
$\varepsilon =$ belt slope angle as belt transforms from troughed to flat profile
$\theta =$ conveyor slope angle at drive end.

Roberts [1] assumed that the load surface profile is parabolic in form and, for convenience, the cross-sectional area can be represented with the inclusion of the non-dimensional shape factor "U" as follows.

$$A = Ub_e^2 \tag{14.2}$$

where

$U =$ non-dimensional cross-sectional area factor
$b_e =$ effective belt width.

For a flat, non-troughed belt, such as at the drive drum

$$U = \frac{\tan \lambda}{6} \tag{14.3}$$

For a three-roll idler set, as depicted in Fig. 14.3b, assuming a parabolic surface profile

$$U = \frac{1}{(1+2q)^2}\left\{q \sin \beta + \frac{q^2}{2} \sin 2\beta + \frac{\tan \lambda}{6}\left[1 + 4q \cos \beta + 2q^2(1 + \cos 2\beta)\right]\right\} \tag{14.4}$$

where

$U =$ non-dimensional cross-sectional area factor
$q =$ $\frac{C}{B}$
$\beta =$ troughing angle
$\lambda =$ Surcharge angle
$b = B + 2C =$ belt contact width
$b_e = B + 2C \cos \beta =$ effective trough contact width
B = centre idler contact
C = wing idler contact.

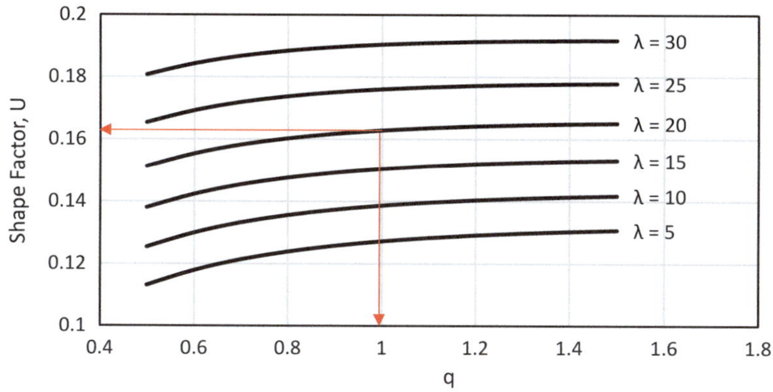

Fig. 14.4 Example of cross sectional area on the trough belt as a function of non-dimensional q = C/B, and surcharge angle, λ, with $\beta = 35°$ b = 1m

Figure 14.4 illustrates, the U versus $q - C/D$ values for a range of surcharge angles λ based on the troughing angle $\beta = 35°$ and belt contact width of $b = 1$ m.

Design Example

The case of a troughed belt conveyor conveying Bauxite is considered. The conveying parameters are given as:

- Conveyor throughput $Q_m = 2500$ t/h
- belt speed $v_b = 5$ m/s
- bulk density $\rho = 1200$ kg/m^3.

Assumed parameters include

- Belt width b = 1.0 m
- Profile ratio q = C/B = 1.0
- Trough angle $\beta = 35°$.

The effective troughed contact width is calculated as $b_e = 0.877$ m.

Based on these parameters, and for a surcharge angle $\lambda = 15°$, the corresponding value of the cross-sectional area factor is: U = 0.1514.

To account for slight variations in the bulk solids properties during conveying, the values of U = 0.162 and surcharge angle $\lambda = 20°$ are selected. The values are depicted in Fig. 14.4 for q = 1.0.

14.2.3 Cross Sectional Profile at Drive Drum

The profile shape at the drive drum is difficult to determine precisely. It depends on the conveyor speed, cohesive strength of the bulk solid and the troughing configuration.

The mean height h_a may be assumed to be:

$$h_a = U\frac{b_e^2}{b} \tag{14.5}$$

For the above example, $h_a = 0.124$ m.

14.3 Conveyor Discharge and Curved Impact Hood Plate Design

The discharge of bulk solids from a belt conveyor is illustrated in Fig. 14.3a. Relevant details of the discharge model are reproduced in Fig. 14.5. The following discussion is focused on the design and positioning of the curved impact plate, or "hood" as it is commonly called.

14.3.1 Angle at Which Discharge Commences

To evaluate the conditions governing the discharge of bulk material from a conveyor, the model shown in Fig. 14.6 is considered. In general, slip may occur before lift-off takes place. Therefore, the model includes the acceleration $\dot{v}$ and the associated inertial force $\Delta m\dot{v}$, where v represents the relative velocity of the bulk material.

However, under typical operating conditions, slip is unlikely to be significant and may therefore be neglected in the analysis.

For an arbitrary radius r, the condition for discharge will commence when the normal force N becomes zero.

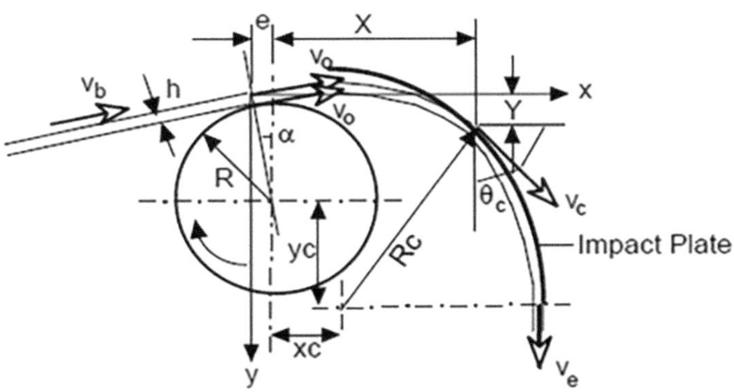

Fig. 14.5 Conveyor discharge and "hood" design

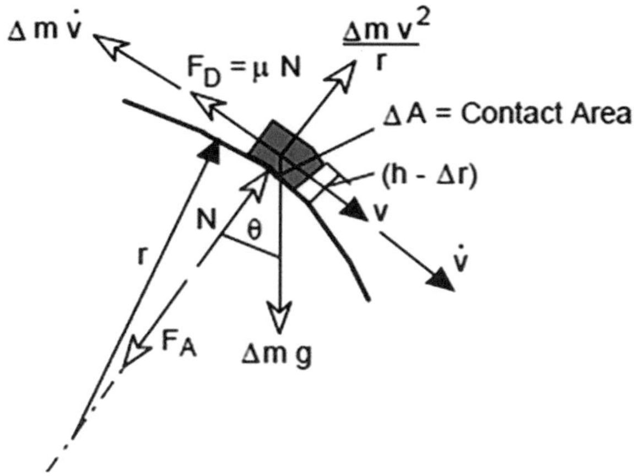

Fig. 14.6 Conveyor discharge model

$$\frac{v^2}{r} = gcos\theta + \frac{F_A}{\Delta m} \tag{14.6}$$

where

$\Delta m = \rho \Delta A(h - \Delta r) =$	mass of element
$F_A = \sigma_o \Delta A =$	adhesive force
$\sigma_o =$	adhesive stress
$\rho =$	bulk density.

14.3.2 *Minimum Belt Speed for Discharge at First Point of Drum Contact*

In most cases, the speed of the conveyor is such that discharge will commence as soon as the belt contacts the discharge drum. In this case $\theta = -\alpha$, where α is the slope of the belt at the contact point with the drum. The critical case will be for the belt surface, that is, when $\Delta r = 0$. The minimum belt speed for discharge at the first point of drum contact is given by Eq. 14.7.

$$v_b = \sqrt{Rg\left(\cos\alpha + \frac{\sigma_0}{\rho g h}\right)} \tag{14.7}$$

The graph in Fig. 14.7 applies to the case when $\alpha = 0^\circ$ and $\sigma_o = 1\,\text{kPa}$. The need for higher belt speeds to achieve lift-off as the layer thickness decreases is highlighted. This indicates the difficulty of removing the thin layer of cohesive bulk solid. This becomes the carry-back that is required to be removed by belt cleaners.

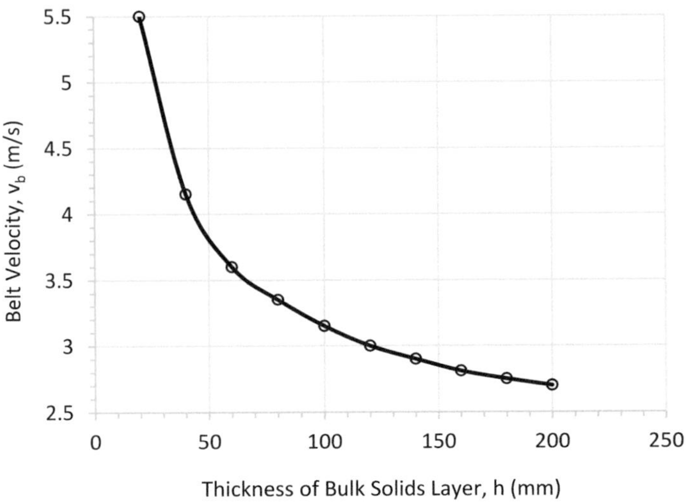

Fig. 14.7 Belt speed for discharge at first point of drum contact. R = 0.5m, $\rho = 1\,\text{t/m}^3 = 1$, $\alpha = 0^\circ$, $\sigma_o = 1$ kPa

14.3.3 Discharge Trajectories

In most cases the influence of air drag is negligible. Hence the equation of motion, defined by the "x" and "y" in Fig. 14.5, is simplified as follows.

$$y = x\tan\theta + \frac{1}{2}g\frac{x^2}{v^2\cos^2\theta} \tag{14.8}$$

Referring to Fig. 14.5, the bounds for the trajectories may be determined for the two radii $R + h$ and R. The radius of curvature of the discharge trajectory of the bulk solids during *free flight* is given by Eq. 14.9.

$$R_c = \frac{\left[1 + \left(\frac{gx}{vb^2\cos^2\theta}\right)^2\right]^{1.5}}{\frac{g}{v_b^2\cos^2\theta}} \tag{14.9}$$

For contact to be made with a curved impact plate of constant radius, the radius of curvature of the trajectory at the point of contact must be such that the condition given by Eq. 14.10 is satisfied.

$$R_c \geq R \tag{14.10}$$

where

$R_c =$ Chute radius

The curved impact plate or *hood* is positioned so that the chute radius matches the radius of curvature at the point of contact. This is illustrated in Fig. 14.5. For the hood design, a constant radius is assumed, where the centre of curvature coordinates xc and yc are defined with respect to the centre of the drive drum.

14.3.4 *Conceptual* Hood and Spoon *Transfer Chute Design Example*

The selected example from Sect. 4.1 involves the transfer of Bauxite from a delivering belt conveyor travelling at a velocity of 5 m/s. The focus of this example is on the design and positioning of the curved impact plate or *hood*, in accordance with the transfer chute model of Fig. 14.5.

From a practical point of view there are a number of constraints that will impinge on the chute design, notably those that are linked to available space and the required transfer geometry. It is quite common for the transfer of bulk solids from the deliv ering conveyor to the receiving conveyor to require a three-dimensional transfer. In this chosen example, a 90° transfer is presented.

(a) **Hood Design**

The design and positioning of the curved impact *hood* chute, along with the chosen variables are illustrated Fig. 14.9. Relevant details of the design and selection process are summarised below:

- Discharge trajectory commences at first point of contact of the belt with the drum.
- For the hood, a constant curvature is adopted. Positioning of the hood is defined with respect drum centre
- The trajectory is assumed to make contact with the hood at a distance of X = 1.0 m at which the required chute radius is Rc = 2.8 m. The coordinates of the centre point of the chute radius are Xc = 0.238 m and Yc = 1.96 m.
- At the point of contact, the slope of the hood and the trajectory are the same, θc = 74.2^{O} and the bulk solid velocity Vc = 5.12 m/s.

The velocity of the bulk solid at exit is Ve = 6.75 m/s (Fig. 14.8).

(B) **Spoon Design**

Figure 14.9 shows the schematic arrangement of the 90° *hood and spoon* conveyor transfer relevant to this design example. The procedures upon which the design of the *spoon* chute are based, relate to the theory and applications of chute design presented in Chap. 13.

The key parameters from the detailed analysis of the spoon are summarised as follows:

- Selected Spoon Radius Rc2 = 3.0 m

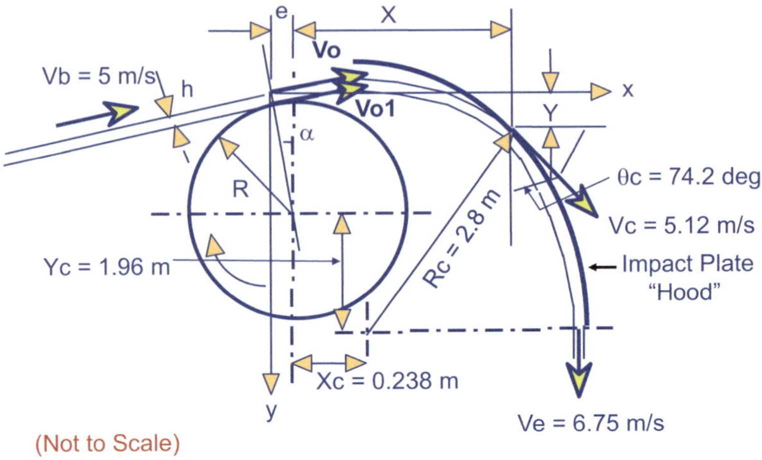

Fig. 14.8 Design and positioning of impact *hood* plate

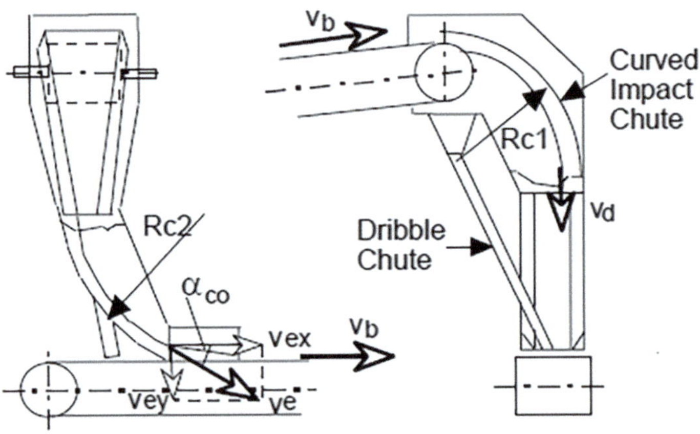

Fig. 14.9 Transfer chute example

- Velocity at intake from hood Vd = 6.75 m/s
- Velocity at discharge from spoon Ve = 5.57 m/s
- Cut-off angle θ = 55°
- Velocity components Vex = 4.57 m/s and Vey = 3.12 m/s
- Belt wear at feed point Wa = 2.26 kPa·m/s
- Receiving belt speed Vb = 5 m/s.

(C) **Comments**

The performance results presented in this example serve as a conceptual guide for transfer chute design. Given the complexities of hood and spoon geometrical combinations, the use of simulation techniques, particularly Discrete Element Modelling (DEM), plays a critical role in achieving effective designs. Nevertheless, it is essential to maintain a strong connection between classical mechanics principles and computer-based simulations to ensure robust and reliable outcomes.

14.4 Three-Dimensional Transfer from Wide to Narrow Belts

The need to transfer bulk materials from a wide, slow speed belt feeder or conveyor to a faster moving receiving belt conveyor requires particular attention to ensure that blockage problems are avoided. To illustrate this objective, the following case study example is presented. It is concerned with the problem of transferring Potash from a wide, slow moving belt conveyor feeder onto a fast, inclined belt conveyor as illustrated, schematically, in Fig. 14.10. The study demonstrates the importance of DEM.

In view of the difficulties of achieving the necessary transfer within the limited headroom exacerbated by the conveyor inclination, it was considered that a single chute transfer would not be feasible. This led to the adoption of the two-chute arrangement illustrated in Fig. 14.10. For obvious reasons, the cut-off height h_{cl} of Chute-2 needs to be sufficient to permit the efficient overlaying of the discharge from Chute-2 with that of Chute-1. The DEM simulation shown in Fig. 14.11 was carried out to examine the feasibility of this arrangement.

The DEM study clearly demonstrates a complete flow blockage occurring at 7 seconds of simulated time when a single spoon configuration was used. In contrast, the twin spoon chute arrangement maintained continuous operation without blockages. The benefits of DEM simulation are further enhanced through subsequent fine-tuning of the model to mitigate or eliminate localised flow impediments identified during analysis. However, the role of classical mechanics in optimising the governing parameters of transfer chute design remains fundamental and should not be overlooked.

14.5 Concluding Remarks

The focus of this Chapter is on the *classical* fundamentals of *hood and spoon* transfer chute design. Such chutes play a major role in industrial belt conveying applications. Together with the basic theory of chute analysis presented in Chap. 13, the overriding objective has been to reinforce the importance of chute flow theory and associated

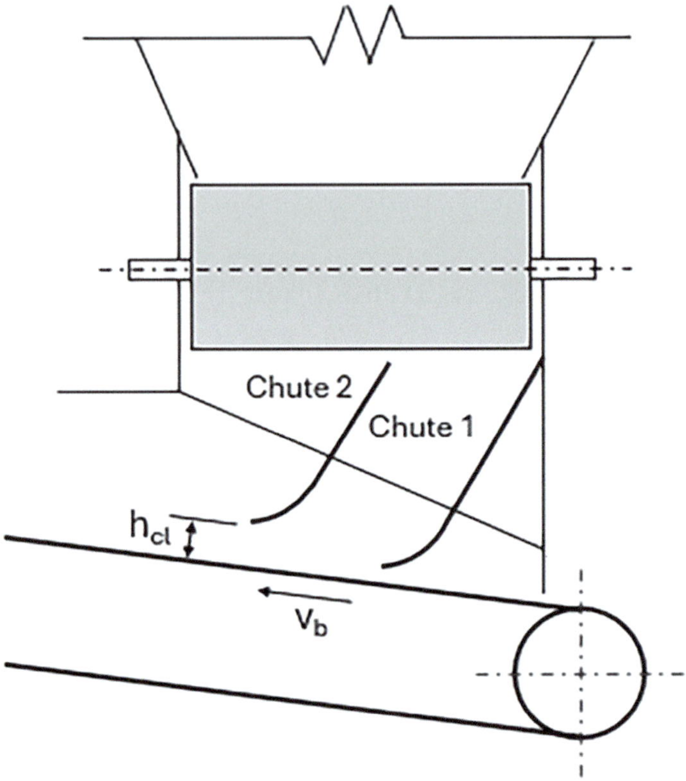

Fig. 14.10 Schematic arrangement of 3-D potash belt transfer

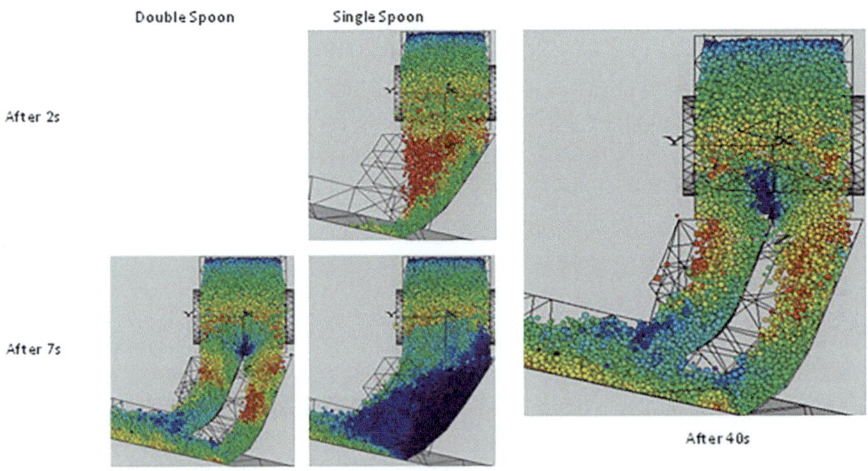

Fig. 14.11 Results of DEM of simulation of the potash 3-D belt conveyor transfer

analysis that underlines the widely varying applications and performance reliability of chutes in industry.

It is not surprising that the current chute design applications depend to a great extent on computer simulation, notably DEM, along with other programs including Finite Element Analysis (FEA) for stress analysis and Computational Fluid Dynamics (CFD) for chute design involving dust control. In all such applications, the importance of linking the *classical* fundamentals of chute design with the numerical computer-based simulations is strongly stressed.

As a final comment, the presented examples of chute design underpin the philosophy and widely varying industrial applications of bulk solids handling and conveying upon which the contents of this book is based.

Reference

1. Roberts, A.W. Hayes, J.W., *Economic Analysis in the Optimal Design of Conveyors*. 1980, Australia: TUNRA Bulk Solids Handling Research Associates, The University of Newcastle, New South Wales.

MIX
Papier aus verantwortungsvollen Quellen
Paper from responsible sources
FSC® C105338

If you have any concerns about our products,
you can contact us on
ProductSafety@springernature.com

In case Publisher is established outside the EU,
the EU authorized representative is:
Springer Nature Customer Service Center GmbH
Europaplatz 3, 69115 Heidelberg, Germany

Printed by Libri Plureos GmbH
in Hamburg, Germany